AF341607

White Biotechnology for Sustainable Chemistry

RSC Green Chemistry

Editor-in-Chief:
Professor James Clark, *Department of Chemistry, University of York, UK*

Series Editors:
Professor George A. Kraus, *Department of Chemistry, Iowa State University, Ames, Iowa, USA*
Professor Andrzej Stankiewicz, *Delft University of Technology, The Netherlands*
Professor Peter Siedl, *Federal University of Rio de Janeiro, Brazil*

How to obtain future titles on publication:
A standing order plan is available for this series. A standing order will bring
delivery of each new volume immediately on publication.

For further information please contact:
Book Sales Department, Royal Society of Chemistry, Thomas Graham
House, Science Park, Milton Road, Cambridge, CB4 0WF, UK
Telephone: +44 (0)1223 420066, Fax: +44 (0)1223 420247
Email: booksales@rsc.org
Visit our website at www.rsc.org/books

White Biotechnology for Sustainable Chemistry

Edited by

Maria Alice Z. Coelho
Universidade Federal do Rio de Janeiro, Rio de Janeiro, Brazil
Email: alice@eq.ufrj.br

Bernardo D. Ribeiro
Universidade Federal do Rio de Janeiro, Rio de Janeiro, Brazil
Email: bernardo@eq.ufrj.br

RSC Green Chemistry No. 45

Print ISBN: 978-1-84973-816-3
PDF eISBN: 978-1-78262-408-0
ISSN: 1757-7039

A catalogue record for this book is available from the British Library

Published by The Royal Society of Chemistry,
Thomas Graham House, Science Park, Milton Road,
Cambridge CB4 0WF, UK

Registered Charity Number 207890

For further information see our web site at www.rsc.org

Printed in the United Kingdom by CPI Group (UK) Ltd, Croydon, CR0 4YY, UK

Preface

White biotechnology can be regarded as applied biocatalysis with enzymes and microorganisms, aiming at industrial production from bulk and fine chemicals to food and animal feed additives. In turn, biocatalysis has many attractive features in the context of sustainable chemistry: mild reaction conditions (at physiological pH and temperature), and environmentally compatible catalysts and solvents (often water) combined with high activities and chemo-, regio- and stereoselectivity in multifunctional molecules. This affords processes which are shorter, generate less waste and are, therefore, both environmentally and economically more attractive than conventional routes. The main contribution of this book will be the use of white biotechnology (enzymes, microorganisms and plant tissues) within the green chemistry concept for: waste minimization, the use of alternative solvents (supercritical fluids, pressurized gases, ionic liquids and micellar systems) and energy sources (microwaves and ultrasound), besides providing more sustainable approaches for the production of fine and bulk chemicals (aromas, polymers, pharmaceuticals and enzymes), such as the use of renewable resources or agroindustrial residues, and biocatalyst recycling.

This text was driven by considering the concepts involved in both the subjects white biotechnology and sustainable chemistry, so that it could be possible to combine the knowledge obtained in each chapter herein presented. In addition, a contribution from the industrial point of view is also presented to demonstrate the feasibility of bioproduction systems. This last aspect can be considered unique!

RSC Green Chemistry No. 45
White Biotechnology for Sustainable Chemistry
Edited by Maria Alice Z. Coelho and Bernardo D. Ribeiro
© The Royal Society of Chemistry 2016
Published by the Royal Society of Chemistry, www.rsc.org

We would like to thank all of the authors who agreed to participate in this book project, giving the readers a broad spectrum on the state of the art, since this book combines people from different parts of the world, as well as providing a glance at Brazilian reality.

Maria Alice Z. Coelho and Bernardo D. Ribeiro

Contents

RSC Green Chemistry No. 45
White Biotechnology for Sustainable Chemistry
Edited by Maria Alice Z. Coelho and Bernardo D. Ribeiro
© The Royal Society of Chemistry 2016
Published by the Royal Society of Chemistry, www.rsc.org

Principles of Green Chemistry and White Biotechnology

BERNARDO DIAS RIBEIRO[a], MARIA ALICE Z. COELHO[a], AND ALINE MACHADO DE CASTRO*[b]

[a]Biochemical Engineering Department, School of Chemistry, Federal University of Rio de Janeiro, Brazil; [b]Biotechnology Division, Research and Development Center, PETROBRAS, Brazil
*E-mail: alinebio@petrobras.com.br

1.1 Green Chemistry: Could Chemistry be Greener?

Since the Second World War, world industrialization has been accelerated without caring about its effects on the environment, and peoples' safety and health. This has led to increased global warming, depletion of the ozone protective layer which protects against harmful UV radiation, contamination of land and waterways due to the release of toxic chemicals by industry, and the reduction of nonrenewable resources such as petroleum. Nevertheless, there is a growing awareness amongst end-users of the risks that chemicals are often associated with, and of the need to dissociate themselves from any chemical in their supply chain that is recognized as being hazardous.[1,2]

In the 1990s, the idea of developing new or improving existing chemical products and processes to make them less hazardous to human health and the environment had already been contemplated. Initially, in 1991, the Office of Pollution Prevention and Toxics (OPPT) of the United States launched a research grant program named "Alternative Synthetic Pathways for Pollution

RSC Green Chemistry No. 45
White Biotechnology for Sustainable Chemistry
Edited by Maria Alice Z. Coelho and Bernardo D. Ribeiro
© The Royal Society of Chemistry 2016
Published by the Royal Society of Chemistry, www.rsc.org

Prevention". In 1993, the program was expanded to include other topics, such as greener solvents and safer chemicals, and was renamed "Green Chemistry".[3]

Nowadays, green chemistry has as main objective the promotion of innovative chemical technologies that reduce or eliminate the use or generation of hazardous substances in the design, manufacture, and use of chemical products, meaning the use of more environmentally acceptable chemical processes and products.[1,4,5]

In 1998, Paul Anastas and Warner announced a set of 12 principles as a useful guide for designing environmentally benign products and processes or to evaluate already existing processes,[4] and in 2003, this promulgated another 12 principles on Green Engineering, which correlates Chemical Engineering with Green Chemistry, aiming to achieve sustainability (in the three dimensions: ecological, economic and social), maximize efficiency, minimize waste and increase profitability,[5,6] as shown in Table 1.1.

To achieve greener chemical processes, besides the more intensive use of renewable feedstocks, several technologies have been developed, some old and some new, which are becoming proven clean technologies, such as the use of alternative solvents (supercritical fluids, ionic liquids, fluorous liquids), non-thermal energetic sources (microwaves, ultrasounds, electrical fields, solar energy), environmentally-friendly separation processes such as membranes (ultrafiltration, nanofiltration and pervaporation), and biological catalysts, such as micro-organisms and enzymes, allowing the creation of more energy-efficient processes.[2,4,7]

1.2 White Biotechnology

Biotechnology is a very broad area which embraces five main sectors:

Blue Biotechnology – Also known as Marine and Fresh-water Biotechnology,[8] this sector includes bioprospecting in marine environments and the use of molecular biology and microbial ecology tools in marine organisms.[9]

Green Biotechnology – Is the biotechnology for agricultural applications. As input, plants are genetically modified to have resistance to insects or diseases, and as outputs, plants present improved agronomic behavior (yield, withstanding environmental stress) and can be used as green factories.[10]

Red Biotechnology – Is the area that focuses on humans and is used to develop alternative solutions to medical problems and issues from diagnosis to therapy.[11] Also named Pharmaceutical Biotechnology.[12]

White Biotechnology – Related to the use of living cells (yeasts, molds, bacteria, plants) and enzymes to synthesize products at industrial scale. Also known as Industrial Biotechnology.[13]

Yellow Biotechnology – Also known as Insect Biotechnology, this emerging field in applied entomology covers the use of insects in drug discovery, their study for plant defense, and the use of insects as a source of enzymes and cells for biotransformations and as a source of biosensors for online detection of compounds at industrial scale. Therefore, this area interacts with White and Green Biotechnology areas.[14]

Table 1.1 Comparative framework of principles of Green Chemistry and Green Engineering.

Green Chemistry	Green Engineering
1. **Prevention**: It is better to prevent waste than to treat or clean up waste after it has been created	1. **Inherent rather than circumstantial**: Designers need to strive to ensure that all materials and energy inputs and outputs are as inherently nonhazardous as possible
2. **Atom economy**: Synthetic methods should be designed to maximize the incorporation of all materials used in the process into the final product	2. **Prevention instead of treatment**: It is better to prevent waste than to treat or clean up waste after it is formed
3. **Less hazardous chemical syntheses**: Wherever practicable, synthetic methods should be designed to use and generate substances that present low or no toxicity to human health and the environment	3. **Design for separation**: Separation and purification operations should be designed to minimize energy consumption and materials use
4. **Designing safer chemicals**: Chemical products should be designed to effect their desired function while minimizing their toxicity	4. **Maximize efficiency**: Products, processes, and systems should be designed to maximize mass, energy, space, and time efficiency
5. **Safer solvents and auxiliaries**: The use of auxiliary substances (*e.g.*, solvents, separation agents, *etc.*) should be made unnecessary whenever possible and should be innocuous when used	5. **Output-pulled *versus* input-pushed**: Products, processes, and systems should be "output pulled" rather than "input pushed" through the use of energy and materials
6. **Design for energy efficiency**: Energy requirements of chemical processes should be recognized for their environmental and economic impacts and should be minimized. If possible, synthetic methods should be conducted at ambient temperature and pressure	6. **Conserve complexity**: Embedded entropy and complexity must be viewed as an investment when making design choices on recycle, reuse, or beneficial disposition
7. **Use of renewable feedstocks**: A raw material or feedstock should be renewable rather than depleting whenever technically and economically practicable	7. **Durability rather than immortality**: Targeted durability, not immortality, should be a design goal
8. **Reduce derivatives**: Unnecessary derivatization (use of blocking groups, protection/deprotection, temporary modification of physical/chemical processes) should be minimized or avoided if possible, because such steps require additional reagents and can generate waste	8. **Meet need, minimize excess**: Design for unnecessary capacity or capability (*e.g.*, "one size fits all") solutions should be considered a design flaw
9. **Catalysis**: Catalytic reagents (as selective as possible) are superior to stoichiometric reagents	9. **Minimize material diversity**: Material diversity in multicomponent products should be minimized to promote disassembly and value retention

(continued)

Table 1.1 (*continued*)

Green Chemistry	Green Engineering
10. **Design for degradation**: Chemical products should be designed so that at the end of their function they break down into innocuous degradation products and do not persist in the environment	10. **Integrate material and energy flows**: Design of products, processes, and systems must include integration and interconnectivity with available energy and materials flows
11. **Real-time analysis for pollution prevention**: Analytical methodologies need to be further developed to allow for real-time, in-process monitoring and control prior to the formation of hazardous substances	11. **Design for commercial "afterlife"**: Products, processes, and systems should be designed for performance in a commercial "afterlife"
12. **Inherently safer chemistry for accident prevention**: Substances and the form of a substance used in a chemical process should be chosen to minimize the potential for chemical accidents, including releases, explosions, and fires	12. **Renewable rather than depleting**: Material and energy inputs should be renewable rather than depleting

Enzymes are classified into 6 classes (as described below) and they receive a classification number, based on their class, subclass and the specific chemical groups participating in the reaction.[15]

1. **Oxidoreductases**: All enzymes catalyzing oxidoreduction reactions belong to this class. The substrate that is oxidized is regarded as a hydrogen donor.
2. **Transferases**: Transferases are enzymes which catalyze the transfer of a group, *e.g.* a methyl group or a glycosyl group, from one compound (generally regarded as a donor) to another compound (generally regarded as an acceptor).
3. **Hydrolases**: These enzymes catalyze the hydrolytic cleavage of C–O, C–N, C–C and some other bonds, including phosphoric anhydride bonds.
4. **Lyases**: Enzymes catalyzing the cleavage of C–C, C–O, C–N, and other bonds by elimination, leaving double bonds or rings, or conversely adding groups to double bonds.
5. **Isomerases**: These enzymes catalyze geometric or structural changes within one molecule.
6. **Ligases**: Enzymes that catalyze the linkage of two molecules, coupled with the hydrolysis of a diphosphate bond in ATP or a similar triphosphate.

White biotechnology is a continuously growing sector, with an average annual growth in the period 2007–2012 of 10.4%.[16] The industry embraces the large-scale production of molecules for several sectors, such as: fertilisers and gases, organic chemicals, polymers and fibers, agrochemicals, adhesives and sealants, paints and coatings, food additives, detergents, cosmetics, active pharma ingredients,[17] as well as the enzymes involved in the production of final molecules, such as in textiles processing,[18] beverages, foods, biofuels[19] and pulp and paper.[20]

The worldwide market for white biotechnology involved transactions on the order of €92 billion in 2010. In late 2011, it was estimated that sales would increase to around €228 billion in 2015 and to around €515 billion in 2020.[17] Specifically, in the field of enzyme catalysis, the global estimated market size of enzymes in 2010 was USD2.82 billion, with food and feed being the major end-user market (USD1.19 billion) and textiles the fastest growing end-user market (4.99%).[20] In 2010, carbohydrases (hydrolases acting on carbohydrates) were the fastest growing product segment (7.6%), and proteases alone accounted for 48% (USD1.35 billion) of the total enzyme market.[20] Additionally, lipases, a group of enzymes of paramount importance in green processes, have also shown growth in their market, which increased from USD235 million in 2001 to USD429 million in 2010,[20] mainly focused on the production of pharmaceuticals, foods and beverages and cleaning products.[21] The projected global market for lipases in 2015 is USD634 million.[20]

Enzyme-catalyzed reactions are indicated to be very promising to meet green chemistry criteria. In the context of the principles of green chemistry, catalysts as a whole provide not only a solution for the problem of waste, but additionally create more energy efficient and less raw material consuming processes. Biocatalysts, specifically, present some positive points: they can act as non-toxic catalysts; they generally operate with high selectivity, yielding high product purity; they operate under moderate reaction conditions at near ambient temperature, pressure and pH, thus resulting in reduced energy consumption; the reaction medium is commonly aqueous, which *per se* is considered non-toxic; biocatalysts have the potential to prevent high consumption of metals and organic solvents; as natural catalysts, enzymes can be considered as renewable catalysts.[22] It should be highlighted, however, that even for biocatalytic processes, each procedure must be evaluated for its environmental friendliness and economic feasibility.[23] Some important remarks on the use of biocatalysts in industrial processes are given in Table 1.2.

1.3 Concluding Remarks

With the above considerations, the interaction between green chemistry and white biotechnology will have a relevant role in the construction of a new industrial concept based on technologies (described herein in this book) that, in the near future, will become the basis of a new paradigm. Some examples of the development of sustainable production processes based on such principles can be seen nowadays all over the world. They can help to save energy and the environment.

Table 1.2　Important remarks about biocatalysis in green chemistry (adapted from ref. 23).

Remarks	Critical analysis
1. Use of water as green solvent	Some organic compounds present low solubility in water; downstream processing of aqueous solutions often demands extraction with organic solvents
2. Enzyme engineering	Promising field which, however, requires a long time for development from the idea to implementation at an industrial scale; new high throughput methods can accelerate development
3. Productivity	Considering that a minimum volumetric productivity of $0.1\ \mathrm{g\ L^{-1}\ h^{-1}}$ and a minimum final product concentration of $1\ \mathrm{g\ L^{-1}}$ is acceptable for implementation at an industrial scale, process optimization in terms of the increase of the substrate concentration and its feeding form and the stability of the biocatalyst is required
4. Low substrate concentration	Due to enzyme inhibition problems, low substrate concentrations are commonly adopted, resulting in oversized reactors and inefficient downstream processing
5. Potential as an alternative process	Although biocatalytic processes are often greener than chemical ones, for industry, ecological reasons are not the only subjects to be addressed for the replacement of an existing process. On the other hand, sometimes there are no chemical alternatives to a biotechnological pathway
6. Pharmaceutical development	The combination of chemical and biocatalytic steps is the most promising path for specific and functionalized products; for the obtainment of chiral molecules, if the separation of racemates is complex and not reliable, enantioselective biotransformations should be used
7. Price of the catalyst	The cost contribution of the biocatalyst is strongly related to the value of the products. They may vary from USD $0.05\ \mathrm{kg^{-1}_{product}}$ (bulk chemicals) to up to USD $10\ \mathrm{kg^{-1}_{product}}$ (pharma products)[24]
8. Downstream processing	Aqueous solutions, commonly used in biotransformations, require a significant amount of solvent for product isolation; strategies such as *in situ* product removal and engineering of solvent-tolerant enzymes could overcome this issue
9. Use of ionic liquids	Functional fluids often improve substrate solubility, but incur additional expense in downstream processing; more information about their toxicity is needed
10. Substrate spectrum for biocatalysis	Although specificity is claimed to be one advantage of enzymes over chemical catalysts, some biocatalysts, such as lipases, present substrate versatility and diverse catalytic function

Especially concerning to Brazil, it is generally recognized that the country has competitive advantages related to: the available area and favorable climate; the efficient production of biomass (sugar cane, eucalyptus, soy, *etc.*); the pioneering production of biofuels on a large scale; the productivity of agriculture which grew at twice the global average from 2001 to 2009, and it is the country with the highest biodiversity in the world, through the multiplicity of species and habitats.

Nevertheless, improvement in bioprocess efficiency needs considerable effort before bioprocesses can be considered a serious alternative to petrochemical industrial processes. Challenges related to the conversion of sugars contained in biomass into the required compounds as effectively as possible will lead to new biocatalyst characteristics, as well as novel operation strategies.

References

1. C. A. M. Afonso and J. G. Crespo, *Green Separation Processes*, Wiley-VCH Verlag, Weinheim, Germany, 2005, pp. 3–19.
2. M. Doble and A. K. Kruthiventi, *Green Chemistry and Engineering*, Academic Press, San Diego, USA, 2007, pp. 1–23.
3. http://www.epa.gov/greenchemistry/pubs/epa_gc.html, accessed 23 October 2012.
4. J. García-Serna, L. Pérez-Barrigón and M. J. Cocero, *Chem. Eng. J.*, 2007, **133**, 7.
5. J. A. Tao and R. Kazlauskas, *Biocatalysis for Green Chemistry and Chemical Process Development*, John Wiley & Sons, Hoboken, New Jersey, 2011, pp. 3–28.
6. P. T. Anastas and J. B. Zimmerman, *Environ. Sci. Technol.*, 2003, **37**(5), 94A.
7. J. Clark and D. Macquarrie, *Handbook of Green Chemistry and Technology*, Blackwell Science, Londres, 2002, pp. 1–26.
8. European Commission, http://ec.europa.eu/research/bioeconomy/biotechnology/research/marine_fresh/index_en.htm, accessed 26 January 2013.
9. Bigelow, http://www.bigelow.org/catt/bigelow-center-for-blue-biotechnology/, accessed 26 January 2013.
10. P. Oakley, 2005, http://www.basf.com/group/corporate/en/function/conversions:/publishdownload/content/investor-relations/calendar/images/050831/Presentation_Oakley_Biotechl.pdf, accessed 26 January 2013.
11. BIO NRW, http://www.bio.nrw.de/en/red, accessed 26 January 2013.
12. Linde, http://www.linde-engineering.com/en/process_plants/biotechnology_plants/pharmaceutical_biotechnology/index.html, accessed 26 January 2013.
13. G. Frazzetto, *EMBO Rep.*, 2003, **4**(9), 835.
14. A. Vilcinskas, *Yellow Biotechnology II – Insect Biotechnology in Plant Protection and Industry*, Springer, Dordrecht, 2013, pp. v–vi.
15. IUBMB, Last update November 2012, http://www.chem.qmul.ac.uk/iubmb/enzyme/, accessed 26 January 2013.
16. IBIS World, Oct 2012, http://www.ibisworld.com/industry/global/global-biotechnology.html?partnerid=prweb, accessed 26 January 2013.
17. G. Festel, C. Detzel and R. Maas, *J. Commer. Biotechnol.*, 2012, **18**(1), 11.

18. Europabio., 2003, www.europabio.org, accessed 26 January 2013.
19. B. Sarrouh, T. M. Santos, A. Miyoshi, R. Dias and V. Azevedo, *J. Bioprocess. Biotech.*, 2012, **S4**, 002.
20. Global Industry Analysts, Inc., 2011, available by purchase from www. strategyR.com.
21. Freedonia, 2010, available by purchase from http://www.freedoniagroup. com/DocumentDetails.aspx?DocumentId=509396.
22. R. Wohlgemuth, *Curr. Opin. Biotechnol.*, 2010, **21**, 713.
23. S. Wenda, S. Illner, A. Mell and U. Kragl, *Green Chem.*, 2011, **13**, 3007.
24. P. Tufvesson, J. Lima-Ramos, M. Nordblad and J. M. Woodley, *Org. Process Res. Dev.*, 2011, **15**, 266.

Sustainability, Green Chemistry and White Biotechnology

ROGER A. SHELDON*[a]

[a]Department of Biotechnology, Delft University of Technology, Netherlands
*E-mail: r.a.sheldon@tudelft.nl, roger@sheldon.nl

2.1 Introduction to Green Chemistry and Sustainability

The roots of industrial organic synthesis can be traced back to the preparation of the first synthetic dye, mauveine (aniline purple) by Perkin in 1856.[1] This serendipitous discovery (Perkin's goal was the synthesis of the anti-malarial drug, quinine) marked the advent of the synthetic dyestuffs industry based on coal tar, a waste product from steel manufacture. The modern pharmaceutical and allied fine chemical industries evolved as spin-offs of this industry. The target molecules were initially relatively simple, but in the ensuing decades they became increasingly complicated, as exemplified by the introduction of semi-synthetic beta-lactam antibiotics and steroid hormones in the 1940s and 1950s. To meet this and subsequent challenges, synthetic organic chemists have developed increasingly sophisticated methodologies. However, many of these time-honoured and widely applied synthetic methodologies were developed at a time when the toxic properties of

RSC Green Chemistry No. 45
White Biotechnology for Sustainable Chemistry
Edited by Maria Alice Z. Coelho and Bernardo D. Ribeiro
© The Royal Society of Chemistry 2016
Published by the Royal Society of Chemistry, www.rsc.org

many reagents and solvents were not known and waste minimisation and sustainability were not significant issues.

The publication of Rachel Carson's "Silent Spring" in 1962[2] and Barry Commoner's "The Closing Circle" in 1971[3] focused the attention of the general public on the problem of the negative side effects of the products of the chemical industry on our natural environment. This formed the basis for the environmental movement. Chemistry was perceived to be the source of the problem rather than as the means to a solution to it. However, the solution to the environmental problem is not a world without chemistry, but one with new and better chemistry that produces environmentally friendly products without the use or generation of hazardous and/or toxic substances. Another watershed was the publication in 1987 of the report "Our Common Future" by the World Commission on Environment and Development, otherwise known as the Brundtland report.[4] This report recognised that industrial and societal development were necessary to provide a growing global population with a satisfactory quality of life, but that such development must also be sustainable over time. In the following decades, the concept of sustainability became the focus of considerable attention both in industry and in society as a whole. There is even a sustainability index (http://www.djindexes.com/sustainability/) that ranks companies on the basis of their sustainability performance. For example, the 2012 supersector leader for Chemicals in the Dow Jones Sustainability Index is Akzo Nobel (Netherlands). Furthermore, 2005–2014 was declared by the United Nations[5] as the "Decade of Education for Sustainable Development".

Sustainable development is defined as development that meets the needs of the present generation without compromising the needs of future generations to meet their own needs. As Graedel has pointed out,[6] it is based on two central tenets: (i) using natural resources at rates that do not unacceptably deplete supplies over the long term and (ii) generating and dissipating residues at rates no higher than can be assimilated readily by the natural environment. Sustainability consists of three components: societal, ecological and economic, otherwise referred to as the three Ps: people, planet and profit. One important issue from the viewpoint of the chemical and allied industries is the sustainable use of chemical feedstocks. It is abundantly clear that a society based on non-renewable fossil resources – oil, coal and natural gas – is not sustainable over the longer term. However, it is worth pointing out that 97% of crude oil is processed to fuels and only *ca.* 3% serves as a feedstock for chemicals manufacture.

At roughly the same time that the concept of sustainable development was emerging, in the mid-1980s, there was mounting concern regarding the copious amount of waste being generated by many industrial chemical processes, particularly in the fine chemicals and pharmaceuticals industries. An illustrative example is provided by the manufacture of phloroglucinol, a reprographic chemical and pharmaceutical intermediate.[7] Up until the mid-1980s, it was produced mainly from 2,4,6-trinitrotoluene (TNT) by the process shown in Figure 2.1, a perfect example of vintage nineteenth century organic chemistry.

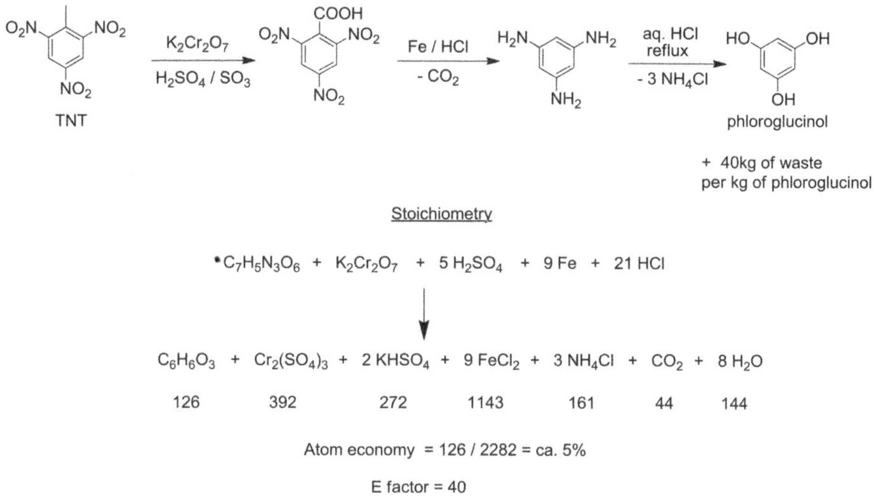

Figure 2.1 A production process for phloroglucinol.

The phloroglucinol product is obtained in >90% overall yield over the three reaction steps and, according to classical concepts of selectivity and reaction efficiency, would generally be considered to be a selective and efficient process. However, for every kilogram of phloroglucinol produced, *ca.* 40 kg of solid waste, containing $Cr_2(SO_4)_3$, NH_4Cl, $FeCl_2$ and $KHSO_4$ are formed. This process was eventually discontinued because of the prohibitive costs associated with the disposal of the chromium-containing waste. It is immediately clear, from an examination of the reaction stoichiometry, that this largely inorganic waste is a consequence of the use of inorganic reagents in stoichiometric amounts. The reaction stoichiometry predicts the formation of *ca.* 20 kg of waste per kilogram of phloroglucinol, assuming 100% chemical yield and exactly stoichiometric quantities of the various reagents. The observed formation of 40 kg of waste in practice is a direct consequence of the use of an excess of the oxidant and reductant and a large excess of sulfuric acid, which has to be subsequently neutralised with base, and an isolated yield of phloroglucinol of less than 100%.

It was clear from this and many other examples that a paradigm shift was needed from the traditional concepts of reaction efficiency and selectivity that focus largely on chemical yield to one that assigns value to the maximisation of raw materials utilisation, the elimination of waste and avoidance of the use of toxic and/or hazardous substances. By the same token, there was a pressing need for alternative, cleaner chemistry in order to minimise these waste streams. This led to the emergence of the concepts of waste minimisation, zero waste plants and green chemistry.[8] Green chemistry can be succinctly defined as:[7]

Green chemistry efficiently utilises (preferably renewable) raw materials, eliminates waste and avoids the use of toxic and/or hazardous reagents and solvents in the manufacture and application of chemical products.

Raw materials include the source of the energy used in the process as this leads to waste in the form of carbon dioxide emissions. The term "Green Chemistry" was introduced in the early 1990s by Anastas and colleagues[9] at the US Environmental Protection Agency (EPA). This does not mean that research on green chemistry did not exist before the early 1990s, merely that it did not have the name (it was generally referred to as clean chemistry). The guiding principle is *benign by design*[10] as embodied in the 12 Principles of Green Chemistry of Anastas and Warner[9] which can be paraphrased as:

The Twelve Principles of Green Chemistry

1. Waste prevention instead of remediation
2. Atom efficiency
3. Less hazardous chemicals
4. Safer products by design
5. Innocuous solvents and auxiliaries
6. Energy efficient by design
7. Preferably renewable raw materials
8. Shorter syntheses (avoid derivatisation)
9. Catalytic rather than stoichiometric reagents
10. Design products for degradation
11. Analytical methodologies for pollution prevention
12. Inherently safer processes

More recently, Poliakoff and coworkers proposed the mnemonic, PRODUCTIVELY[11] to capture the spirit of the twelve principles of green chemistry:

Condensed Principles of Green Chemistry

P – Prevent waste
R – Renewable materials
O – Omit derivatisation steps
D – Degradable chemical products
U – Use of safe synthetic methods
C – Catalytic reagents
T – Temperature, pressure ambient
I – In-process monitoring
V – Very few auxiliary substrates
E – E factor, maximise feed in product
L – Low toxicity of chemical products
Y – Yes, it is safe

Green chemistry eliminates waste at source, *i.e.* it is primary pollution prevention rather than end-of-pipe waste remediation, as is inherent to the first principle of green chemistry: prevention is better than cure. In the last two decades, the concept of green chemistry has been widely embraced in both industrial and academic circles. One could say that sustainability is our ultimate common goal, and green chemistry is a means to achieving it.

2.2   Green Chemistry Metrics

In order to know whether one process or product is greener than another one, we need meaningful metrics to measure greenness. As Lord Kelvin astutely observed: "to measure is to know". The most widely accepted measures of the environmental impact of chemical processes are, probably not coincidentally, the two most simple metrics: the *E factor*,[12,13,14,15,16] defined as the mass ratio of waste to desired product and the *atom economy*[17,18] or *atom utilisation*,[19] defined as the molecular weight of the desired product divided by the sum of the molecular weights of all substances produced in the stoichiometric equation, expressed as a percentage.

Armed with knowledge of the stoichiometric equation, one can predict, without performing any experiments, the theoretical amount of waste that will be formed. Atom economy (AE), introduced by Trost in 1991[17] has become the widely accepted terminology although *atom efficiency* (also abbreviated as AE) is often used. In Figure 2.2, the AE of the classical chlorohydrin route to propylene oxide is compared with that of oxidation with the green oxidant hydrogen peroxide.[20] It is interesting to note that the former process produces, on a weight basis, more calcium chloride than propylene oxide.

Atom economy is a theoretical number. It is based on the assumption that a chemical yield of 100% of the theoretical yield is obtained and that reactants are used in exactly stoichiometric amounts. Furthermore, it disregards substances, such as solvent and acids or bases used in work-up, which do not appear in the stoichiometric equation. The *E* factor, in contrast, is the actual amount of waste produced in the process, defined as everything but the desired product. It takes the chemical yield into account and includes all reagents, solvent losses, all process aids and, in principle, even the energy consumed. Originally,[12] water was excluded from the calculation of the *E* factor as it was thought that its inclusion would lead to exceptionally high *E* factors in many cases and make meaningful comparisons of processes difficult. There is a definite trend, however, especially in the pharmaceutical industry, towards the inclusion of water in the *E* factor. It is also worth noting that water usage can be a crucial issue in biomass conversion and in fermentation processes in general (see later).

The ideal *E* factor is zero, *i.e.* zero waste. A higher *E* factor means more waste and, consequently, greater negative environmental impact. Alternatively, one can view the *E* factor as: kilograms (of raw materials) in minus kilograms of the desired product, divided by kilograms of product out. It is easily calculated from a knowledge of the number of tons of raw materials purchased and the number of tons of product sold. This method of calculation automatically excludes the water used in the process, but not the water

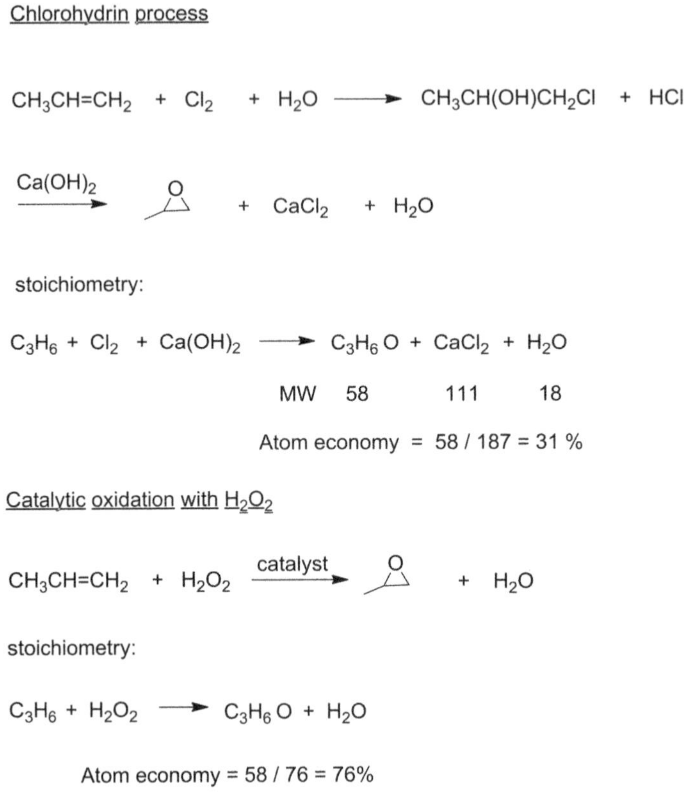

Chlorohydrin process

$$CH_3CH=CH_2 \ + \ Cl_2 \ + \ H_2O \ \longrightarrow \ CH_3CH(OH)CH_2Cl \ + \ HCl$$

$$\xrightarrow{Ca(OH)_2} \ \triangle\!\!\!\!O \ + \ CaCl_2 \ + \ H_2O$$

stoichiometry:

$$C_3H_6 + Cl_2 + Ca(OH)_2 \ \longrightarrow \ C_3H_6O + CaCl_2 + H_2O$$

MW   58   111   18

Atom economy = 58 / 187 = 31 %

Catalytic oxidation with H_2O_2

$$CH_3CH=CH_2 \ + \ H_2O_2 \ \xrightarrow{catalyst} \ \triangle\!\!\!\!O \ + \ H_2O$$

stoichiometry:

$$C_3H_6 + H_2O_2 \ \longrightarrow \ C_3H_6O + H_2O$$

Atom economy = 58 / 76 = 76%

Figure 2.2   Atom efficiencies of two processes for propylene oxide.

formed. The sheer magnitude of the waste problem in chemicals manufacture is readily apparent from a consideration of typical E factors in various segments of the chemical and allied industries (Table 2.1).

The substantial increase in E factor on moving from bulk chemicals to fine chemicals and pharmaceuticals is a direct consequence of the more widespread use of stoichiometric reagents in these industry segments. In bulk chemicals manufacture, in contrast, because of the enormous production volumes, the use of many stoichiometric reagents is economically prohibitive. The increase is also a consequence of the fact that pharmaceuticals, for example, are more complicated molecules involving production *via* multistep syntheses which can be expected to generate more waste. Consequently, waste generation can be reduced by developing processes that are more step economic, as advocated by Wender *et al.*[21] As shown in Table 2.1, the E factors of processes for the manufacture of pharmaceuticals are greater than 100 in many instances. We also note that the E factors for the production of therapeutic proteins (biopharmaceuticals) on a commercial scale are even higher (see later).[22]

The E factor has been widely embraced by the chemical industry and in particular by the pharmaceutical industry,[23] as a useful metric for assessing

Table 2.1 E factors in the chemical and allied industries.

Industry segment	Volume[a] (tons per annum)	E factor[b] (kg waste per kg product)
Bulk chemicals	104–106	<1–5
Fine chemicals	102–104	5– >50
Pharmaceuticals	10–103	25– >100
Therapeutic proteins	<1	400– >2000

[a]Range of typical production volumes for a single product in the industry segment.
[b]Not including process water.

E factor

$$E = \frac{\text{kg waste}}{\text{kg product}}$$

Atom efficiency (AE)

$$\text{AE (\%)} = \frac{\text{m.w of product} \times 100\%}{\Sigma \text{ m.w. of reactants}}$$

Mass intensity (MI)

$$\text{MI} = \frac{\text{total kg used in a process}}{\text{kg product}}$$

Reaction mass efficiency (RME)

$$\text{RME (\%)} = \frac{\text{kg product} \times 100\%}{\text{total kg of reactants}}$$

Mass Productivity (MP)

$$\text{MP} = \frac{\text{kg product}}{\text{total kg used in process}}$$

Carbon efficiency (CE)

$$\text{CE (\%)} = \frac{\text{kg carbon in product} \times 100\%}{\text{kg carbon in reactants}}$$

Effective mass yield (EMY)

$$\text{EMY (\%)} = \frac{\text{kg product} \times 100\%}{\text{kg of non-benign reagents}}$$

Figure 2.3   Green chemistry metrics.

the environmental impact of manufacturing processes.[24,25] Indeed, the Green Chemistry Institute Pharmaceutical Round Table has conducted an inventory of the waste generated in processes used by its members (see http:// www.epa.gov/greenchemistry/pubs/gcinstitute.html). The latter include several leading pharmaceutical companies (Eli Lilly, Glaxo Smith Kline, Pfizer, Merck, AstraZeneca, Schering Plough and Johnson & Johnson). The aim was to use this data to drive the greening of the pharmaceutical industry.

Other green metrics have been proposed[26,27,28,29] and they can be divided into two types: (i) metrics that are a refinement of the AE concept and (ii) metrics that are variations on the E factor (see Figure 2.3). Examples of the former are *reaction mass efficiency* (RME) and *carbon efficiency* (CE) introduced by Constable and coworkers[30] at Glaxo Smith Kline (GSK). RME is defined as the mass of product obtained divided by the total mass of reactants in the stoichiometric equation, expressed as a percentage. It is a refinement of AE that takes the chemical yield of the product and the actual quantities of reactants used into account. A disadvantage compared to AE is that experimental data is needed to calculate the RME which, therefore, cannot be used for rapid analysis of different processes prior to experimental work being performed. CE is similar to RME but takes only carbon into account, *i.e.* it is the mass of carbon in the product obtained divided by the total mass of carbon present in the reactants.

An example of the second type is *mass intensity* (MI),[31] defined as the total mass of materials used in a process divided by the mass of product obtained *i.e.* MI = E factor + 1. The same authors[31] also suggested the use of the so-called *mass productivity* which is the reciprocal of the MI. Hudlicky and coworkers[32] proposed an analogous metric: the *effective mass yield* (EMY), defined as the mass of the desired product divided by the total mass of non-benign reactants used in its preparation. The EMY does not include so-called environmentally benign compounds such as NaCl, acetic acid, *etc.* However, the definition of non-benign is difficult and arbitrary and it was concluded,[30] therefore, that EMY suffers from a lack of definitional clarity.

None of these alternative metrics offers any particular advantage over atom economy and the E factor for assessing how wasteful a process is. Atom economy and the E factor are complementary metrics. The former is a quick tool that can be used before conducting any experiments, and the latter is a measure of the total waste that is actually formed in practice. The ideal E factor is zero which is a better reflection of the ultimate goal of zero waste manufacturing plants than an ideal mass intensity of 1. The AE of the phloroglucinol process (see Figure 2.1) is *ca.* 5%, which would predict an E factor of *ca.* 20. In practice, the E factor is 40 because the overall yield is not 100%, a molar excess of the various reactants is used, and the sulfuric acid (which is used in large excess) has to be neutralised with base during down-stream processing. Indeed, the large amounts of waste generated in the processes for the manufacture of fine chemicals and pharmaceuticals, and even some bulk chemicals, consist primarily of inorganic salts, such as sodium chloride, sodium sulfate and ammonium sulfate, formed in the reaction or in subsequent neutralisation steps.

2.3   Environmental Impact and Sustainability Metrics

As noted above, sustainability encompasses the conservation of the natural resources of the planet and the minimisation of the effect of industrial activities, *e.g.* chemicals manufacture, on the health of its inhabitants and the natural environment. However, these are only two of the three pillars of sustainability. The third is economic viability. Green chemistry embodies essentially the same two components: (i) efficient utilisation of raw materials and elimination of waste, and (b) health, safety and environmental aspects of chemicals and their manufacturing processes, but without the economic component. However, the metrics discussed in the preceding section take only the mass of waste generated into account, whereas the environmental impact of this waste is also determined by its nature. One kg of sodium chloride is obviously not equivalent to one kg of a chromium(VI) compound, for example. Hence, the term 'environmental quotient', EQ was introduced[12] to take the nature of the waste into account. EQ is the product of the E factor and an unfriendliness multiplier, Q. The latter is dependent on various factors such as toxicity, ease of recycling, *etc.* Q for a particular substance can also be influenced by both the production volume and the location of the production facilities. For example, the generation of 100–1000 tons per annum of sodium chloride is unlikely to present a waste problem but 10 000 tons per annum, in contrast, may already present a disposal problem, thus warranting an increase in Q. Ironically, when very large quantities of sodium chloride are generated, the Q value could decrease again as recycling by electrolysis becomes viable, *e.g.* in propylene oxide manufacture *via* the chlorohydrin route (see earlier). Hence, the Q value of a particular waste will be determined by, *inter alia*, its ease of disposal or recycling. Generally speaking, organic waste is more easily remediated than inorganic waste, which can be important when considering the green metrics of biocatalytic processes (see later).

The magnitude of Q is obviously debatable and difficult to quantify, but one can conclude that a 'quantitative assessment' of the environmental impact of chemical processes is, in principle, possible. In the last decade, several groups have addressed the problem of quantifying Q. For example, Eissen and Metzger[33] developed the EATOS (Environmental Assessment Tool for Organic Synthesis) software in which metrics related to health hazards and persistence, bioaccumulation and ecotoxicity were used to determine the environmental index of the input (substrates, solvents, *etc.*) and output (product and waste). Similarly, Saling and coworkers at BASF[34,35,36] introduced eco-efficiency analyses which took both economic and environmental aspects into account, including energy, raw materials, emissions, toxicity, hazards and land use.

The basis for such an analysis is *Life Cycle Assessment (LCA)*[37,38] which is used to assess the environmental impact and sustainability of products and processes within defined domains, *e.g.* cradle-to-gate, cradle-to-grave and gate-to-gate, on the basis of quantifiable environmental impact indicators, such as energy usage, global warming, ozone depletion, acidification, eutrophication, smog formation, and ecotoxicity, in addition to waste generated. The outcome of an LCA resembles the EQ in that it constitutes an integration of the amount of waste with quantifiable environmental indicators based on the nature of the waste. Jessop and coworkers,[39] for example, used a combination of nine LCA environmental impact indicators – acidification, ozone depletion, smog formation, global warming, human toxicity by ingestion and inhalation, persistence, bioaccumulation, and abiotic resource depletion – in a gate-to-gate assessment of the greenness of alternative routes to a particular product.

2.4 The Role of Catalysis in Waste Minimisation

As noted above, the waste generated in the manufacture of fine chemicals and pharmaceuticals is primarily a consequence of the use of stoichiometric inorganic and organic reagents that are not (or are partially) incorporated into the product. Typical examples include oxidations with inorganic oxidants such as chromium(VI) salts, permanganate and manganese dioxide, and stoichiometric reductions with metals (Na, Mg, Zn, Fe) and metal hydride reagents ($LiAlH_4$, $NaBH_4$). A classic example is the phloroglucinol process discussed above, which combines an oxidation using stoichiometric amounts of chromium(VI) with a stoichiometric reduction using Fe/HCl. Similarly, a plethora of reactions, *e.g.* sulfonations, nitrations, halogenations, diazotisations and Friedel–Crafts acylations, employing stoichiometric amounts of mineral acids (H_2SO_4, HF, H_3PO_4) and Lewis acids ($AlCl_3$, $ZnCl_2$, BF_3) are major sources of waste. The solution is evident: substitution of antiquated stoichiometric methodologies with cleaner catalytic alternatives.[40,41,42] This is true elegance and efficiency in organic synthesis.[43] For example, catalytic hydrogenation, oxidation and carbonylation (Figure 2.4) are highly atom efficient, low-salt processes. The generation of copious amounts of inorganic salts can similarly be largely circumvented by replacing stoichiometric mineral acids, such as H_2SO_4, and Lewis acids and stoichiometric bases, such as NaOH and KOH, with recyclable solid acids and bases, preferably in

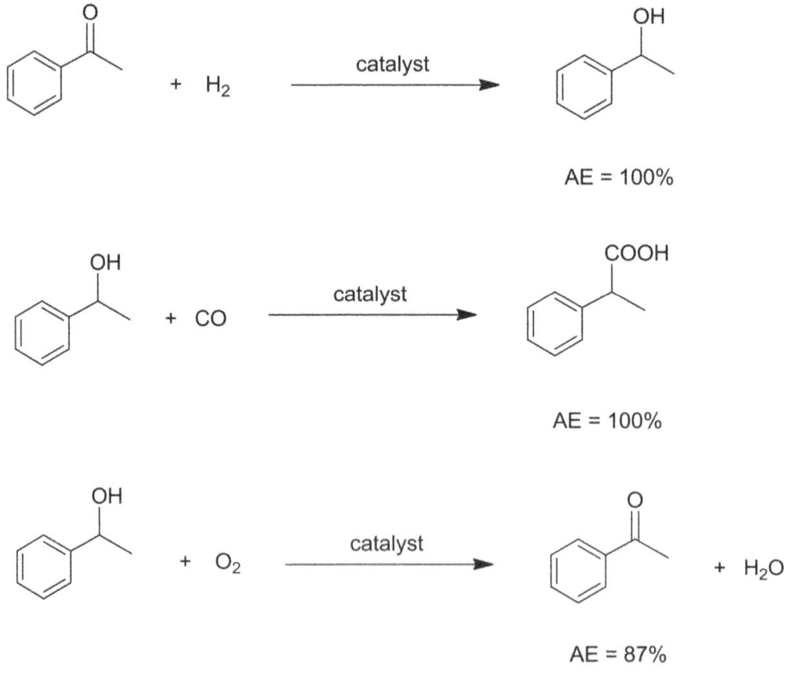

Figure 2.4 Atom efficient catalytic processes.

catalytic amounts.[44] Indeed, recyclable solid (heterogeneous) catalysts have an important role to play in waste minimisation of many basic reactions in industrial organic synthesis.[45]

The relatively large *E* factors in pharmaceuticals manufacture can also be ascribed to the multi-step syntheses that are widely used in this industry segment. This led Wender *et al.*[21] to advocate the development of more step economic syntheses. The ultimate in step economy is the development of catalytic cascade processes whereby several catalytic steps are integrated in one-pot procedures without the need for isolation of intermediates.[46] Such 'telescoping' of multi-step syntheses into catalytic cascades has several advantages – fewer unit operations, less solvent, lower reactor volume, shorter cycle times, higher volumetric and space time yields and less waste (lower *E* factor) – which afford substantial economic and environmental benefits. Furthermore, coupling of reactions can be used to drive equilibria towards the product, thus avoiding the need for excess reagents. On the other hand, there are problems to be overcome: catalysts are often incompatible with each other (*e.g.* an enzyme and a metal catalyst), rates and optimum conditions can be very different, and catalyst recovery and recycling can be complicated. Nature solves the problem of compatibility by compartmentalisation of enzymes in different parts of the cell, which suggests that the key to developing catalytic cascades is the immobilisation of the different catalysts.

2.5   Solvents and Multiphase Catalysis

Another major source of waste in chemicals manufacture is solvent loss, with solvents generally ending up in the atmosphere or in ground water. Indeed, solvent losses are a major contributor to the high *E* factors of pharmaceutical manufacturing processes.[37] Moreover, health and/or safety issues associated with many traditional organic solvents, such as chlorinated hydrocarbons, have led to their use being severely curtailed.

The FDA has issued guidelines for solvent use in the pharmaceutical industry (see http://www.fda.gov/cder/guidance/index.htm). Solvents are divided into four classes:

Class 1   solvents should not be used in the manufacture of drug substances because of their unacceptable toxicity or deleterious environmental effects. They include benzene and various chlorinated hydrocarbons.

Class 2   solvents should be used only sparingly in pharmaceutical processes because of their inherent toxicity. They include acetonitrile, dimethyl formamide, methanol and dichloromethane.

Class 3   solvents may be regarded as less toxic and of lower risk to human health. They include many lower alcohols, esters, ethers and ketones.

Class 4   solvents, for which no adequate data are available, include di-isopropyl ether, methyl tetrahydrofuran and isooctane.

Consequently, many pharmaceutical companies are focusing their attention on minimising solvent use and in replacing many traditional organic solvents, such as chlorinated and aromatic hydrocarbons, by more environmentally friendly alternatives such as lower alcohols, esters and some ethers such as methyl *tert* butyl ether (MTBE). Pfizer scientists[23,47] for example, have produced a solvent selection guide for medicinal chemists, dividing solvents into three categories: undesirable, usable and preferred, as shown in Table 2.2. Solvents derived from renewable feedstocks, such as ethanol, methyl tetrahydrofuran and ethyl lactate are becoming popular as reaction media.

In the original inventory of E factors of various processes, it was assumed, if details were not known, that solvents would be recycled by distillation and that this would involve a 10% loss. However, the organic chemist's penchant for using different solvents for the various steps in multi-step syntheses makes recycling difficult owing to cross contamination. The best solvent is no solvent, but if a solvent is needed, it should be safe to use and there should be provisions for its efficient removal from the product and reuse.

Since the major sources of waste in chemicals manufacture are stoichiometric reagents and solvent losses, the solution to the waste problem is evident: catalytic reactions in alternative reaction media[48] such as water,[49,50] supercritical CO_2,[51] fluorous biphasic[52] and ionic liquids,[53] alone or in liquid–liquid biphasic combinations.[54] We note that the use of water and supercritical carbon dioxide as reaction media fits well with the current drive towards the use of renewable raw materials, which are ultimately derived from carbon dioxide and water.

Table 2.2   Solvent selection guide for medicinal chemists.

Undesirable	Usable	Preferred
Pentane	Cyclohexane	Water
Hexane	Isooctane	Ethanol
Benzene	Methylcyclohexane	Methanol
Carbon tetrachloride	Toluene	1-Propanol
Dichloromethane	Xylenes	2-Propanol
Dichloroethane	Tetrahydrofuran	1-Butanol
Chloroform	Methyltetrahydrofuran	*tert*-Butanol
Diethyl ether	Methyl-*tert*-butyl ether	Acetone
Di-isopropyl ether	Dimethyl sulfoxide	2-Butanone
Dimethoxyethane	Acetic acid	Ethyl acetate
Dioxane	Acetonitrile	Isopropyl acetate
Pyridine	Ethylene glycol	Heptane
Dimethylformamide		
Dimethylacetamide		
N-Methylpyrollidone		

2.6 Green Chemistry and White Biotechnology

The term white biotechnology refers to the industrial application of biocatalysis, using whole cells or isolated enzymes, as opposed to green, red and blue biotechnologies which comprise agricultural, medical and marine applications, respectively. Biocatalysis has many attractive features in the context of green chemistry. The catalyst (an enzyme) is derived from renewable resources and is biodegradable and essentially non-hazardous, *i.e.* it fulfils the criteria of sustainability admirably well. Reactions are performed in an environmentally compatible solvent (water) under mild conditions (physiological pH and ambient temperature and pressure). Furthermore, reactions of multifunctional molecules proceed with high activities and chemo-, regio- and stereoselectivities and generally without the need for the functional group activation, protection and deprotection steps required in traditional organic syntheses. This affords processes which are more step economic and more efficient in energy and raw materials consumption, generate less waste and are, therefore, both environmentally and economically more attractive than conventional routes. Moreover, these reactions often afford products in higher purity than traditional chemical or chemo-catalytic processes, as a result of the high selectivities and milder reaction conditions. For example, they avoid the problem of contamination with traces of (noble) metals which is often a serious issue in pharmaceuticals manufacture. Finally, enzymatic processes (but not fermentations) can be conducted in standard multi-purpose batch reactors and, hence, do not require any extra investment, *e.g.* in high-pressure equipment.

In short, biocatalysis fits very well with the principles of green chemistry and sustainability. As Barry Commoner aptly remarked:[3] "In nature there is no such thing as waste, everything is recycled". Consequently, in the last two decades, biocatalysis has emerged as an important technology to meet the growing demand for green and sustainable chemicals manufacture,[55,56] particularly in the synthesis of pharmaceuticals,[57,58] flavours and fragrances,[59] cosmetic ingredients[60] and other fine chemicals.[61,62]

Thanks to advances in biotechnology and protein engineering techniques such as *in vitro* evolution,[63] it is now possible to produce most enzymes for commercially acceptable prices, and to manipulate them such that they exhibit the desired properties with regard to, *inter alia*, substrate specificity, activity, selectivity, stability and optimum pH.[64,65] This has made it possible to optimise an enzyme to fit a pre-defined optimum process, *i.e.*, genuine benign by design. Furthermore, the development of effective immobilisation techniques has paved the way for optimising the storage and operational stability and the recovery and recycling of enzymes.[66] Moreover, since most biocatalytic processes are performed under roughly the same conditions of (ambient) temperature and pressure, it is eminently feasible to integrate multiple steps into enzymatic cascade processes.[67] Co-immobilisation of two or more enzymes then affords multifunctional solid biocatalysts capable of catalysing such cascade processes.[68]

2.7   Green and Sustainability Metrics of White Biotechnology

Biocatalytic processes are performed using isolated enzymes or as whole cell biotransformations. Isolated enzymes have the advantage of not being contaminated with other enzymes present in the cell. The use of whole cells, on the other hand, is less expensive as it avoids the separation and purification of the enzyme. In the case of dead cells, the E factors of the two methods are essentially the same: the waste cell debris is separated before or after the biotransformation. In contrast, when growing microbial cells are used, *i.e.* in fermentation processes, substantial amounts of waste biomass can be generated. We note, however, that this waste is generally easy to dispose of, *e.g.* as animal feed, or it can, in principle, be used as a source of energy for the process. Many fermentation processes also involve the formation of copious amounts of inorganic salts that may even be a major contributor to the waste. To our knowledge, there are no reported E factors for fermentation processes. This would seem to be a gap which needs to be filled.

2.7.1   Fermentation processes

Mass balances, from which E factors can be calculated, of a few fermentation processes have been documented by Petrides *et al.*[69] For example, the E factor for the bulk fermentation product citric acid is 1.4, which compares well with the E factor range of <1–5 typical of bulk petrochemicals (see Table 2.1). Interestingly, *ca.* 75% of the waste is accounted for by an inorganic salt, calcium sulfate. During the process, calcium hydroxide is added to control the pH, affording calcium citrate which is reacted with sulfuric acid to produce citric acid and calcium sulfate. If water is included in the calculation, the E factor becomes 17.

Small volume fermentation processes for biopharmaceuticals can have extremely high E factors, even when compared with those observed in the production of small molecule drugs, and water usage also tends to be high. The fermentative production of recombinant human insulin,[69] for example, involves an E factor of *ca.* 6600. The most important contributors to the waste were relatively simple organic compounds and inorganic salts: (in kg per kg insulin) urea (1692), acetic acid (1346), formic acid (968), phosphoric acid (713), guanidine hydrochloride (445), glucose (432), sodium chloride (430), acetonitrile (424) and sodium hydroxide (140). If water is included, the E factor becomes an astronomical 50 000.

2.7.2   Enzymatic Production of an Atorvastatin Intermediate

In contrast, biotransformations involving the use of isolated enzymes tend to involve significantly higher substrate concentrations and combine a higher productivity with a lower water usage compared to fermentations. An illustrative example is the recently developed chemoenzymatic process (Figure 2.5)

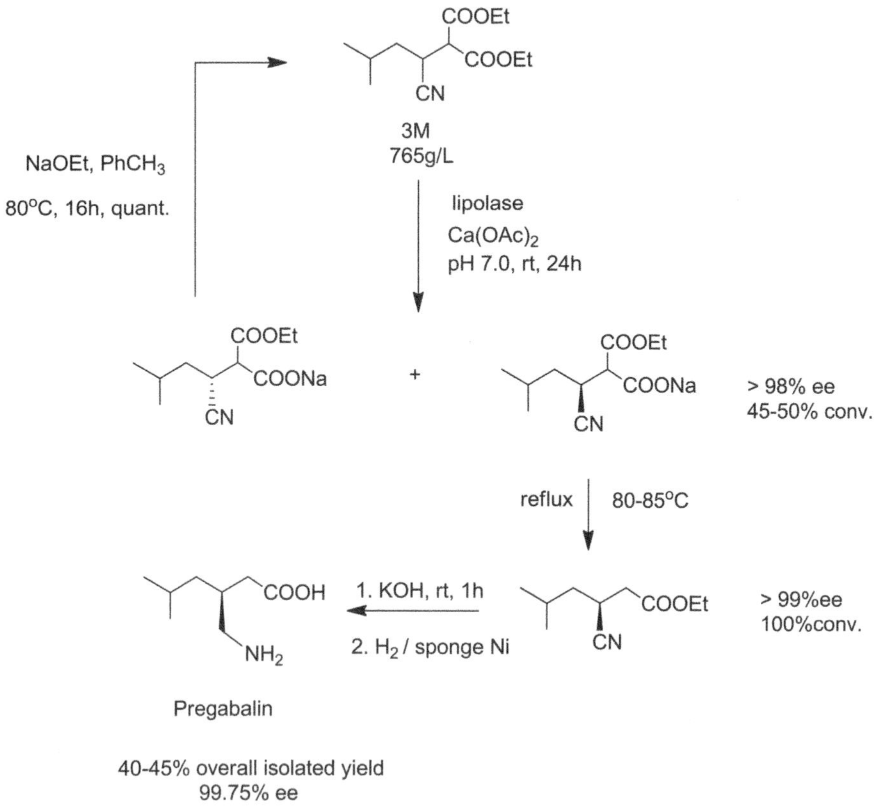

Figure 2.5 Chemoenzymatic process for pregabalin.

for the manufacture of pregabalin,[70] the active ingredient of the CNS drug Lyrica. The new route afforded a dramatic improvement in process efficiency by setting the stereocentre early in the synthesis (the golden rule of chiro-technology[71]) and enabling the facile racemisation and re-use of the wrong enantiomer. The key enzymatic step was conducted with an inexpensive, readily available laundry detergent, lipase, at a 3 M substrate concentration, that is at a quite staggering 765 g L^{-1}. Moreover, organic solvent usage was dramatically reduced in a largely aqueous process. Compared to the first generation manufacturing process, the new process afforded a higher yield and a five-fold reduction in the E factor from 86 to 17.

Another example of a biotransformation for the synthesis of a pharmaceutical intermediate is the green-by-design, three-enzyme process for the synthesis of a key intermediate (Figure 2.6) in the manufacture of atorvastatin, the active ingredient of the cholesterol-lowering drug Lipitor.[72,73] The process has been successfully commercialised on a multi-ton scale by Codexis. In the first step, ethyl-4-chloroacetoacetate undergoes highly enantioselective reduction catalysed by a ketoreductase (KRED). Cofactor regeneration was achieved with glucose and a NADP-dependent glucose dehydrogenase (GDH). The (*S*) ethyl-4-chloro-3-hydroxybutyrate product was obtained in 96% isolated yield and >99.5% e.e. In the second step, a halohydrin dehalogenase (HHDH) was employed to catalyse a nucleophilic substitution of chloride by cyanide using HCN at neutral pH and ambient temperature.

All previous manufacturing routes to the hydroxynitrile product involved, as the final step, a standard S_N2 substitution of a halide with a cyanide ion in alkaline solution at elevated temperature, resulting in extensive by-product formation owing to the base sensitivity of both substrate and product. Since the product is a high boiling point oil, a troublesome high-vacuum fractional distillation is required to recover the product in acceptable quality, resulting in further yield losses and more waste. Hence, the key to designing an economically and environmentally attractive process was to conduct the cyanation reaction under mild conditions at neutral pH. This was accomplished using the enzyme halohydrin dehalogenase (HHDH) to afford an elegant two-step, three enzyme process for the hydroxynitrile product.

Unfortunately, the wild-type KRED and GDH exhibited prohibitively low activities, and large enzyme loadings were required to obtain an economically viable reaction rate, resulting in troublesome emulsion formation in downstream processing. Thus, although the analytical yield was >99%, the recovered yield was only 85%. To enable a practical large-scale process, the enzyme loadings needed to be drastically reduced. This was achieved by *in vitro* evolution using the DNA shuffling technique[74] to improve the activity and stability of KRED and GDH, while maintaining the near perfect enantioselectivity exhibited by the wild-type KRED. The GDH activity was improved by a factor of 13 and the KRED activity by a factor of 7, while maintaining the enantioselectivity at >99.5%. With the improved enzymes, the reaction was complete in 8 h with a substrate loading of 160 g L^{-1}, using substantially reduced enzyme loadings and, consequently, with no emulsion problems. Phase separation required less than one minute and provided the chlorohydrin in >95% isolated yield of >99.9% e.e.

Similarly, the activity of the wild-type HHDH in the non-natural cyanation reaction was extremely low and the enzyme exhibited severe product inhibition and poor stability under the operating conditions. As a result of the

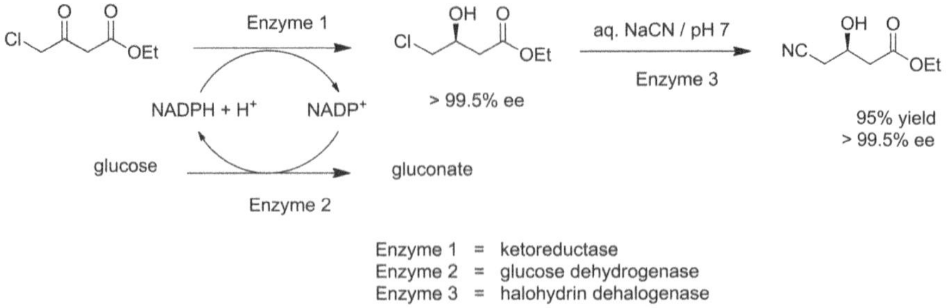

Figure 2.6   A two-step three-enzyme process for atorvastatin intermediate.

large enzyme loadings, downstream processing was challenging. However, after many iterative rounds of DNA shuffling, the inhibition was largely overcome, and the HHDH activity was increased more than 2500-fold compared to the wild-type enzyme.

The greenness of the process was assessed according to the twelve principles of green chemistry.

Principle 1 – waste prevention: The highly selective biocatalytic reactions afforded a substantial reduction in waste and, by avoiding by-product formation, the need for yield-sacrificing fractional distillation was circumvented. The butyl acetate and ethyl acetate solvents, used in the extraction of the product from the aqueous layer in the first and second step, respectively, were recycled with an efficiency of 85%. The E factor for the overall process is 5.8 if process water is excluded (2.3 for the reduction and 3.5 for the cyanation). If process water is included, the E factor for the whole process is 18 (6.6 for the reduction and 11.4 for the cyanation). The main contributors to the E factor (Table 2.3) are solvent (EtOAc and BuOAc) losses (51%), sodium gluconate (25%), NaCl and Na_2SO_4 (*ca.* 22% combined). The three enzymes and the NADP cofactor account for <1% of the waste. Furthermore, the main waste streams are aqueous and directly biodegradable.

Principle 2 – atom economy: The use of glucose as the reductant for cofactor regeneration is cost effective but the atom economy is only 45%. However, glucose is an inexpensive renewable raw material and the gluconate co-product is fully biodegradable.

Principle 3 – less hazardous chemical syntheses: The reduction reaction uses essentially non-toxic starting materials and avoids the use of potentially hazardous hydrogen and heavy metal catalysts, obviating concern about their removal from waste streams and/or contamination of the product. While cyanide must be used in the second step, as in all practical routes to the product, it is used more efficiently (higher yield) and under less harsh conditions compared to previous processes.

Principle 4 – design safer chemicals: This is not applicable as the hydroxynitrile product is the target molecule.

Table 2.3 E factors of the process for atorvastatin intermediate.

Waste	Quantity (kg per kg product)	% of E (excl. water)	% of E (incl. water)
Substrate losses (8%)	0.09	<2	<1
Triethanolamine	0.04	<1	<1
NaCl and Na_2SO_4	1.29	22	*ca.* 7
Na-gluconate	1.43	*ca.* 25	*ca.* 9
BuOAc (85% recycle)	0.46	*ca.* 8	*ca.* 0.3
EtOAc (85% recycle)	2.50	*ca.* 43	*ca.* 14
Enzymes	0.023	<1	<1
NADP	0.005	0.1	<0.1
Water	12.250	—	67
E factor		5.8 (Without H_2O)	18 (With H_2O)

Principle 5 – safer solvents and auxiliaries: Safe and environmentally accept-able ethyl acetate and butyl acetate are used, together with water, as cosolvent in the biocatalytic reduction reaction and extraction of the hydroxynitrile product; no auxiliaries are used.

Principle 6 and 9 – design for energy efficiency, and catalysis: The process con-stitutes very efficient biocatalysis with turnover numbers of >10^5 for KRED and GDH and >5×10^4 for HHDH. In contrast with previous processes which employ elevated temperature for the cyanation step and high pressure hydro-genation for the reduction step, both steps in the biocatalytic process are run at or close to ambient temperature and pressure and pH 7 and the very high energy demands of high vacuum distillation are dispensed with altogether, resulting in substantial energy savings.

Principles 7 and 10 – use of renewable feedstocks, and design for degra-dation: The enzyme catalysts and the glucose co-substrate are derived from renewable raw materials and are completely biodegradable. The by-products of the reaction are gluconate, NADP (the cofactor) and residual glucose, enzyme, and minerals and the waste water is directly suitable for biotreatment.

Principle 8 – reduce derivatisation: The process avoids derivatisation steps, *i.e.* it is step economic and involves fewer unit operations than previous pro-cesses, most notably by obviating the troublesome product distillation or bisulfite mediated separation of dehydrated by-products.

Principle 11 and 12 – real time analysis for pollution prevention, and inher-ently safer chemistry: The reactions are run in pH-stat mode at neutral pH by the computer-controlled addition of base. Gluconic acid generated in the first reaction is neutralised by aq. NaOH, and HCl generated in the second step is neutralised by feed-on-demand aq. NaCN, regenerating HCN ($pK_a \sim 9$) *in situ*. This minimises the overall concentration of HCN, affording an inher-ently safer process. The pH and the cumulative volume of the added base are recorded in real time.

In short, this process is an excellent example of a *benign by design* biocata-lytic process for the synthesis of an important pharmaceutical intermediate, the successful commercialisation of which has been enabled by the employ-ment of modern protein engineering to optimise enzyme performance.

2.7.3 Enzymatic Synthesis of Sitagliptin

Another relevant example is the recently reported[75] enzymatic synthe-sis of the antidiabetic sitagliptin, which replaced a rhodium-catalysed, high-pressure, asymmetric hydrogenation of an enamine. It involves an overall enantioselective reductive amination of a ketone using an (*R*)-trans-aminase-catalysed reaction with isopropylamine (Figure 2.7). At the outset no known enzyme showed any activity towards the ketone substrate. The starting point, therefore, was an (*R*)-selective transaminase which showed no activity towards the ketone substrate. First, *in silico* studies were used to identify what was needed to be able to fit the ketone into the binding

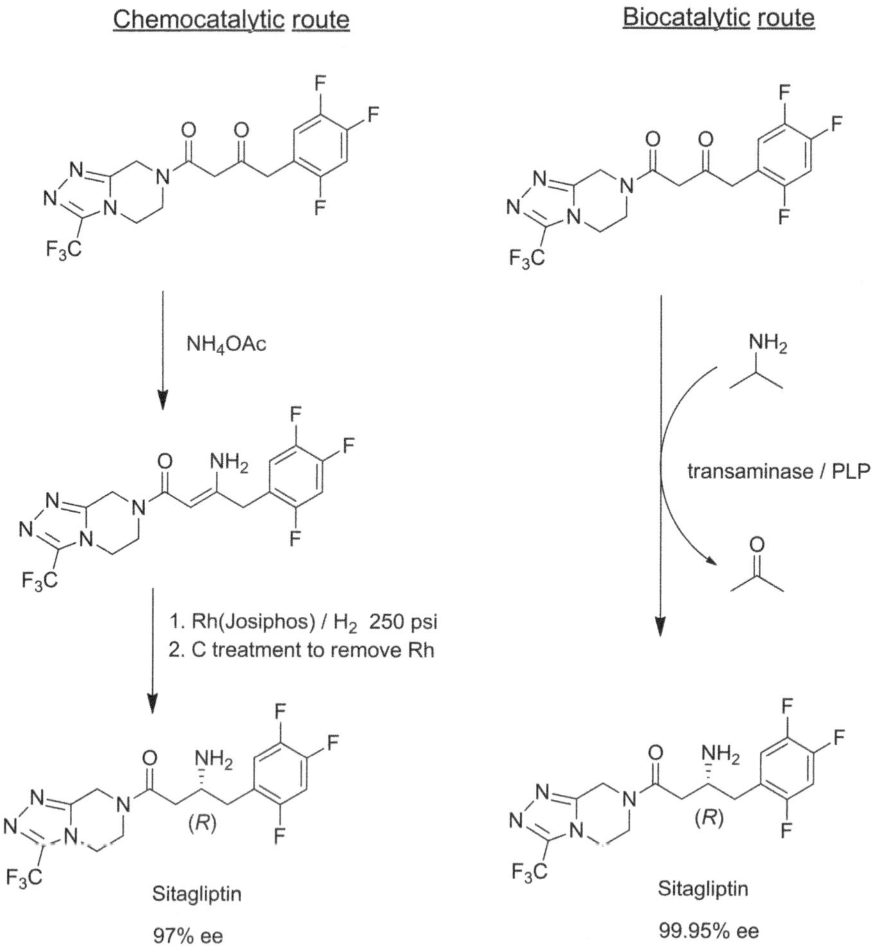

Figure 2.7 Two processes for sitagliptin.

pocket of the enzyme. The amino residues surrounding the binding pocket were then engineered to provide the extra space. When some, albeit, low activity was observed, the respective enzyme was further improved using *in vitro* evolution up to a commercially viable level. Under optimised conditions, 6 g L^{-1} of the best variant in 50% aq. DMSO converted 200 g L^{-1} of the ketone substrate to sitagliptin in >99.95% e.e. in 92% yield. Compared with the rhodium-catalysed asymmetric hydrogenation, the biocatalytic process afforded sitagliptin with a 10–13% increase in overall yield, a 53% increase in productivity (kg L^{-1} per day), a 19% reduction in total waste, the elimination of all heavy metals, and a reduction in the total manufacturing cost. Furthermore, the enzymatic reaction is run in multipurpose vessels, avoiding the need for specialised high-pressure hydrogenation equipment.

2.7.4   Enzymatic Production of Myristyl Myristate

White biotechnology is being widely applied in cosmetic ingredients where the label 'natural' can add considerable value to a product. The application of biocatalysis can be beneficial, even with relatively simple products such as emollients consisting of fatty acid esters. A pertinent example is the industrial production of the widely used emollient ester, myristyl myristate.[76] The classical process involves tin(ii) oxalate catalysed esterification at 240 °C for 4 h. The biocatalytic process employs an immobilized form of *Candida antarctica* lipase B, Novozyme 435, as the catalyst at 60 °C for 12 h. The two processes are compared in Figure 2.8.

The atom economy of the process is 96% and the E factor is <0.1 in both cases. Consequently, the environmental impacts of the two processes were compared in a cradle-to-gate environmental LCA based on five impact categories: (i) energy consumption (ii) global warming (iii) acidification (iv) nutrient enrichment and (v) smog formation. Substantial reductions in all categories were achieved (Table 2.4). Energy consumption was reduced by more than 60% and the formation of undesirable pollutants by up to 90%. Replacement of an environmentally unattractive tin catalyst by an enzyme and the considerably milder conditions were the major factors responsible for the significantly more eco-friendly profile of the biocatalytic process.

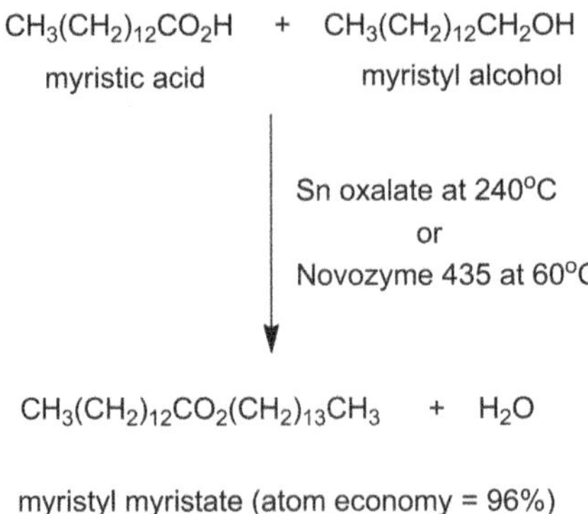

Figure 2.8   Chemo- *vs.* biocatalytic production of myristyl myristate.

Table 2.4   Key environmental indicators of chemo- *vs.* biocatalytic esterification.

Environmental parameter	Units	Chemocatalytic (Sn(ii) oxalate)	Biocatalytic NOV 435	Reduction (%)
Energy	GJ	22.5	8.63	62
Global warming	kg CO_2 eq.	1518	582	62
Acidification	kg SO_2 eq.	10.58	1.31	88
Eutrophication	kg PO_4 eq.	0.86	0.24	74
Smog formation	kg C_2H_4 eq.	0.49	0.12	76

Other added benefits of the biocatalytic *vs.* the chemocatalytic process are better product quality and process simplification. Product quality is much higher in the biocatalytic process as a result of the much milder reaction conditions, thereby obviating the need for purification steps and enabling simpler downstream processing.

2.8 White Biotechnology, Green Chemistry and the Utilisation of Waste Biomass

One of the great challenges of the 21st century is the implementation of the transition from an economy that is largely based on non-renewable fossil fuels as raw materials to a more sustainable bio-based economy based on renewable resources. Among various sustainable energy options (solar, wind, geothermal) only biomass, which encompasses agricultural food and feed crops, dedicated energy crops and trees, agriculture and forestry residues, aquatic plants, and animal and municipal waste, is a source of carbon-based fuels and chemicals. Hence, another important goal of green chemistry and sustainability is the substitution of fossil resources – oil, coal and natural gas – by biomass as the primary feedstock. Surely this is the ultimate green and sustainable organic synthesis: harnessing the energy of the sun to synthesise fuels and chemicals from carbon dioxide and water. Nature already does this, of course, but it takes too long; fossil resources laid down as biomass millions of years ago are being consumed at a much faster rate than they can be renewed.

The *utilisation of biomass for sustainable fuels and chemicals* has become a top priority on the international political agenda. The switch from non-renewable fossil fuels to renewable biomass as a feedstock for liquid fuels and commodity chemicals will afford an environmentally beneficial reduction in the carbon footprint of chemicals and liquid fuels. A key strategy for success is the development of integrated biorefineries that can coproduce a range of platform chemicals and materials in addition to biofuels, in order to generate the necessary added value.[77] An additional benefit will be that many existing products will be substituted by alternatives that are inherently safer and have a reduced environmental footprint, for example, biocompatible and biodegradable plastics.[78] First generation biofuels (bioethanol and biodiesel) and bio-based commodity chemicals such as lactic acid and 1,3-propane diol, and bioplastics are currently being produced from maize and edible oil seeds, such as rapeseed, as feedstocks. However, such first generation feedstocks compete, directly or indirectly, with food production, thereby affecting the price of food. This is obviously not a sustainable long term solution and the next generation of bio-based fuels and platform chemicals will utilise lignocellulosic biomass, inedible oilseed crops, as feedstocks in integrated biorefineries. This could involve the deliberate cultivation of fast-growing, non-edible crops, such as grasses, but it is probably even more attractive to utilise waste biomass generated in the production of edible crops, *e.g.* bagasse and corn stover, formed in the processing of sugar cane and maize, respectively.[79] Indeed, the drive to avoid waste and find new renewable resources for fuels and chemicals has focused attention on a new and promising approach: the use of food supply chain waste (FSCW) as a renewable feedstock for biorefineries.[80]

An illustrative example of the current drive towards bio-based platform chemicals and materials manufacture is the recent focus of attention on replacing the ubiquitous petrochemical-based polyethylene terephthalate (PET) bottles for, *inter alia*, water and soft drinks, by a bio-based alternative, the so-called "plant bottle", produced from renewable biomass. As shown in Figure 2.9, PET is currently produced from two basic petrochemicals derived from crude oil *via* naphtha: ethylene and *p*-xylene. The latter are converted to ethylene glycol and terephthalic acid, respectively, the raw materials for PET. The easiest way to produce a bio-based PET is to produce the ethylene glycol from bio-ethylene which, in turn, has been produced from bioethanol. This affords PET that is *ca.* 27% bio-based and is currently being used by, for example, Coca Cola. It is also possible to produce bio-*p*-xylene from biomass and the combination of this with bio-ethylene glycol, in principle, affords 100% bio-PET. One approach which is being developed by the company Gevo[81] is to convert bio-isobutanol, produced by fermentation, to *p*-xylene *via* existing chemocatalytic technologies.

As noted above, the switch from crude oil-based to biomass-based chemicals manufacture not only provides alternative, more sustainable routes to existing products, but it can also afford alternative products with an improved environmental footprint and/or better properties. That is the case with PEF. The potentially key platform chemical, hydroxymethyl furfural (HMF), is produced by acid-catalysed dehydration of hexoses, such as glucose and fructose.[82] Subsequent chemocatalytic[83] or biocatalytic[84] aerobic oxidation of HMF affords furan 2,5-dicarboxylic acid (FDCA). Reaction of the latter with ethylene glycol produces polyethylene furanoate (PEF), a potential substitute for PET. The company Avantium is building a platform technology based on so-called furanics and agreements are in place with Coca Cola and Danone to produce PEF bottles from FDCA derived from the chemical dehydration of hexoses in the presence of an alcohol and subsequent oxidation of the resulting HMF ether (see Figure 2.9).

Obviously, processes for biomass conversion should involve optimum utilisation of raw materials and minimisation of waste, *i.e.* they should have low *E* factors, by employing green catalytic methodologies. They should also employ environmentally friendly solvents, preferably water, alone or in combination with carbon dioxide. Ionic liquids are also of interest since they can, depending on their structure, dissolve large amounts of carbohydrates, including polysaccharides and can be used as solvents for primary conversion of lignocellulose.[85,86,87] If they are also derived from renewable raw materials, are biodegradable and have low ecotoxicity, all the better.

Since *E* factors and atom economy have been widely used for assessing the environmental footprint of chemical manufacturing processes, they would

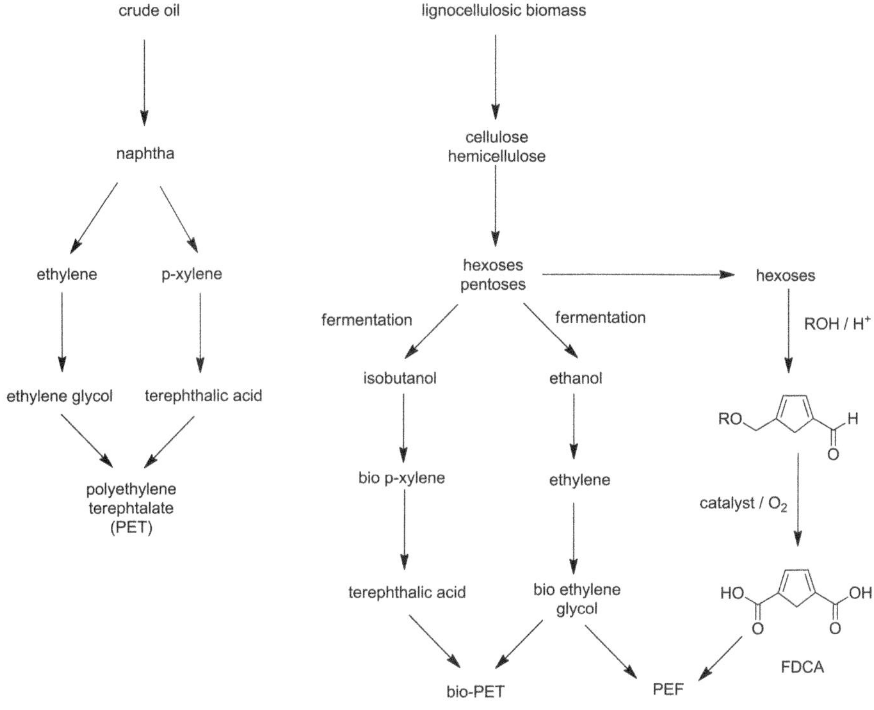

Figure 2.9　Bio-based route to polyethylene terephthalate (PET) and polyethylene furanoate (PEF).

appear to be a good starting point for evaluating processes for biomass utilisation. According to one report,[88] the E factor of cellulosic ethanol is 42. However, the major components of this waste are water (36.8 kg per kg ethanol) and carbon dioxide (4.1 kg per kg ethanol). If these are excluded, the E factor amounts to 1.1. However, the formation of 4 kg of molasses, 0.1 kg of furfural and 0.2 kg of acetic acid as by-products was not included in the E factor calculation. It was further noted that a cellulosic ethanol plant processing 10 000 tons of lignocellulose feedstock per day to produce 870 tons of ethanol a day would generate 32 million litres of wastewater daily. This water contains several organic by-products, the concentrations of which have to be decreased to the ppm level or below in order to enable reuse of the water. This means that a rather sophisticated industrial waste water treatment is mandatory. It is clear that devising meaningful metrics for biomass utilisation is fraught with many complicating factors,[89] many of which are a consequence of the enormous scale involved in biofuels production. It is not just a question of comparing the green and sustainability metrics of two processes. Sustainability metrics have to be devised for a meaningful comparison of a fossil fuel-based production with a biomass-based one. In addition to the material and energy efficiencies, and fixed and variable costs, indicators such as land and water use have to be taken into account.

2.9   Conclusions & Future Prospects

Hopefully, we have shown that catalytic processes play an important role in building the road to green and sustainable chemicals manufacture. Moreover, meaningful greenness and sustainability metrics are a must for monitoring progress. The concepts of atom economy and E factors constitute complementary and widely accepted metrics in this context. Together with life cycle assessment, they provide a sound basis for assessing the greenness and sustainability of different processes and products. All of the various sub-disciplines of catalysis—homogeneous, heterogeneous, organocatalysis and biocatalysis—have a role to play in this quest for sustainability. Biocatalysis, or white biotechnology, in particular, has many added benefits: mild reaction conditions employing a catalyst that is biocompatible, biodegradable and derived from renewable resources. Processes are step economic and highly selective, resulting in higher product quality and reduced waste generation, coupled with superior economics. Furthermore, catalysis in general, and white biotechnology in particular, will play a key role in the transition from a fossil fuel-based economy to a sustainable bio-based one based on renewable resources.

References

1. W. H. Perkin, *J. Chem. Soc.*, 1862, 232; W. H. Perkin, *Brit. Pat.*, 1856, 1984.
2. R. Carson, *Silent Spring*, Houghton Mifflin, New York, 1962.
3. B. Commoner, *The Closing Circle*, Bantam Books, New York, 1971.
4. C. G. Brundtland, Our Common Future, *The World Commission on Environmental Development*, Oxford University Press, Oxford, 1987.
5. http://www.desd.org; see also M. A. Fisher, *J. Chem. Educ.*, 2012, **89**, 179.
6. T. E. Graedel, in *Handbook of Green Chemistry and Technology*, ed. J. Clark and D. J. Macquarrie, Wiley, New York, 2002, pp. 56–61.
7. R. A. Sheldon, *C. R. Acad. Sci.*, 2000, **3**, 541.
8. R. A. Sheldon, *Chem. Soc. Rev.*, 2012, **41**, 1437–1451.
9. *Green Chemistry: Theory and Practice*, ed. P. Anastas and J. C. Warner, Oxford University Press, Oxford, 1998.
10. *Benign by Design: Alternative Synthetic Design for Pollution Prevention*, ed. P. T. Anastas and C. A. Farris, American Chemical Society, Washington DC, 1994, vol. 577, ACS Symposium Series.
11. S. L. Y. Tang, R. L. Smith and M. Poliakoff, *Green Chem.*, 2005, **7**, 761.
12. R. A. Sheldon, *Chem. Ind., London*, 1992, 903; see also R. A. Sheldon, *Chem. Ind., London*, 1997, 12.
13. R. A. Sheldon, *CHEMTECH*, 1994, 38–47.
14. R. A. Sheldon, *Pure Appl. Chem.*, 2000, **72**, 1233–1246.
15. R. A. Sheldon, *J. Chem. Technol. Biotechnol.*, 1997, **68**, 381.
16. R. A. Sheldon, *Green Chem.*, 2007, **9**, 1273.
17. B. M. Trost, *Science*, 1991, **254**, 1471.
18. B. M. Trost, *Angew. Chem., Int. Ed.*, 1995, **34**, 259–281.

19. See C. M. Caruana, *Chem. Eng. Prog.*, 1991, **87**(12), 11.
20. B. Notari, *Stud. Surf. Sci. Catal.*, 1998, **37**, 413.
21. P. A. Wender, M. P. Croatt and B. Witulski, *Tetrahedron*, 2006, **62**, 7505; P. A. Wender, S. T. Handy and D. L. Wright, *Chem. Ind., London*, 1997, 765.
22. S. V. Ho, in *Green Chemistry in the Pharmaceutical Industry*, ed. P. J. Dunn, A. S. Wells and M. T. Williams, Wiley-VCH, Weinheim, 2010, pp. 311–331.
23. K. Alfonsi, J. Collberg, P. J. Dunn, T. Fevig, S. Jennings, T. A. Johnson, H. P. Kleine, C. Knight, M. A. Nagy, D. A. Perry and M. Stefaniak, *Green Chem.*, 2008, **10**, 31.
24. S. K. Ritter, *Chem. Eng. News*, 2008, 59.
25. A. N. Thayer, *Chem. Eng. News*, 2007, 11.
26. F. G. CalvoFlores, *ChemSusChem.*, 2009, **2**, 905.
27. J. Auge, *Green Chem.*, 2008, **10**, 225.
28. J. Andraos, *Org. Process Res. Dev.*, 2005, **9**, 149.
29. *Green chemistry metrics: measuring and monitoring sustainable processes*, ed. A. Lapkin and D. J. C. Constable, Wiley, Oxford (UK), 2008.
30. D. J. C. Constable, A. D. Curzons and V. L. Cunningham, *Green Chem.*, 2002, **4**, 521.
31. A. D. Curzons, D. J. C. Constable, D. N. Mortimer and V. L. Cunningham, *Green Chem.*, 2001, **3**, 1; C. Jimenez-Gonzalez, C. S. Ponder, Q. B. Broxterman and J. B. Manley, *Org. Process Res. Dev.*, 2011, **15**, 912.
32. T. Hudlicky, D. A. Frey, L. Koroniak, C. D. Claeboe and L. E. Brammer, *Green Chem.*, 1999, **1**, 57.
33. M. Eissen and J. O. Metzger, Environmental performance metrics for daily use in synthetic chemistry, *Chem.–Eur. J.*, 2002, **8**, 3580.
34. R. Landsiedel and P. Saling, *Int. J. Life Cycle Assess.*, 2002, **5**, 261.
35. P. Saling, *Appl. Microbiol. Biotechnol.*, 2005, **68**, 1.
36. D. R. Shonnard, A. Kircherer and P. Saling, *Environ. Sci. Technol.*, 2003, **37**, 5340.
37. C. Jimenez-Gonzalez, A. D. Curzons, D. J. C. Constable and V. L. Cunningham, *Int. J. Life Cycle Assess.*, 2004, **9**, 114.
38. J. G. Moretz-Sohn Monteiro, O. de Queiroz Fernandes Araujo and J. Luiz de Medeiros, *Clean Technol. Environ. Policy*, 2009, **11**, 209–459.
39. S. M. Mercer, J. Andraos and P. G. Jessop, *J. Chem. Educ.*, 2012, **89**, 215.
40. C. J. Li and B. M. Trost, *Proc. Natl. Acad. Sci. U. S. A.*, 2008, **105**, 13197–13202.
41. R. A. Sheldon, I. W. C. E. Arends and U. Hanefeld, *Green Chemistry and Catalysis*, Wiley-VCH, Weinheim, 2007.
42. R. A. Sheldon, *Chem. Commun.*, 2008, 3352.
43. See also R. Noyori, *Chem. Commun.*, 2005, 1807.
44. *Fine Chemicals Through Heterogeneous Catalysis*, ed. R. A. Sheldon and H. van Bekkum, Wiley-VCH, Weinheim, 2001, ch. 3–7.
45. J. M. Thomas, J. C. Hernandez-Garrido and R. G. Bell, *Top. Catal.*, 2009, **52**, 1630; K. Kaneda and T. Mizugaki, *Energy Environ. Sci.*, 2009, **2**, 655–673.
46. A. Bruggink, R. Schoevaart and T. Kieboom, *Org. Process Res. Dev.*, 2003, **7**, 622.

47. P. J. Dunn, *Chem. Soc. Rev.*, 2012, **41**, 1452.

48. R. A. Sheldon, *Green Chem.*, 2005, **7**, 267.

49. M. O. Simon and C. J. Li, *Chem. Soc. Rev.*, 2012, **41**, 1415.

50. *Science of Synthesis, Water in Organic Synthesis*, ed. S. Kobayashi, Georg Thieme Verlag, Stuttgart, 2012.

51. D. J. Cole-Hamilton, *Adv. Synth. Catal.*, 2006, **348**, 1341; X. Han and M. Poliakoff, *Chem. Soc. Rev.*, 2012, **41**, 1428.

52. I. T. Horvath, *Acc. Chem. Res.*, 1998, **31**, 641.

53. R. A. Sheldon, *Chem. Commun.*, 2001, 2399; V. I. Parvulescu and C. Hardacre, *Chem. Rev.*, 2007, **107**, 2615–2665.

54. M. J. Muldoon, *Dalton Trans.*, 2010, **39**, 337.

55. J. Tao and R. J. Kazlaukas, *Biocatalysis for Green Chemistry and Chemical Process Development*, Wiley, Hoboken, New Jersey, 2011.

56. R. Wohlgemuth, *Curr. Opin. Biotechnol.*, 2010, **21**, 713.

57. D. Munoz Solano, P. Hoyos, M. J. Hernaiz, A. R. Alcantara and J. M. Sanchez-Montero, *Bioresour. Technol.*, 2012, **115**, 196.

58. G.-W. Zheng and J.-H. Xu, *Curr. Opin. Biotechnol.*, 2011, **22**, 784.

59. M. C. R. Franssen, L. Alessandrini and G. Terraneo, *Pure Appl. Chem.*, 2005, **77**, 273; E. J. Vandamme and W. Soetaert, *J. Chem. Technol. Biotechnol.*, 2002, **77**, 1323; S. Serra, C. Fuganti and E. Brenna, *Trends Biotechnol.*, 2005, **23**, 193.

60. T. Veit, *Eng. Life Sci.*, 2004, **4**, 508; V. Heinrichs and O. Thum, *Lipid Technol.*, 2005, **17**, 82.

61. C. M. Clouthierz and J. N. Pelletier, *Chem. Soc. Rev.*, 2012, **41**, 1585.

62. B. M. Nestl, B. A Nebel and B. Hauer, *Curr. Opin. Chem. Biol.*, 2011, **15**, 187.

63. F. A. Arnold, *Curr. Opin. Chem. Biol.*, 2009, **13**, 3; N. J. Turner, *Nat. Chem. Biol.*, 2009, **5**, 567; M. T. Reetz, *J. Org. Chem.*, 2009, **74**, 5767; S. Luetz, L. Giver and J. Lalonde, *Biotechnol. Bioeng.*, 2008, **101**, 647.

64. U. T. Bornscheuer, G. W. Huisman, R. J. Kazlauskas, S. Lutz, J. C. Moore and K. Robins, *Nature*, 2012, **485**, 185.

65. A. Illanes, A. Cauerhff, L. Wilson and G. R. Castro, *Bioresour. Technol.*, 2012, **115**, 48.

66. R. A. Sheldon, *Adv. Synth. Catal.*, 2007, **349**, 1289.

67. R. A. Sheldon, in *Multi-Step Enzyme Catalysis: Biotransformations and Chemoenzymatic Synthesis*, ed. E. Garcia-Junceda, Wiley-VCH, Weinheim, 2008, pp. 109–135.

68. For pertinent examples see: C. Mateo, A. Chmura, S. Rustler, F. van Rantwijk, A. Stolz and R. A. Sheldon, *Tetrahedron: Asymmetry*, 2006, **17**, 320–323; S. van Pelt, F. van Rantwijk and R. A. Sheldon, *Adv. Synth. Catal.*, 2009, **351**, 397–404.

69. B. Petrides, in *Bioseparations Science and Engineering*, ed. R. G. Harrison, P. W. Todd, S. R. Rudge and D. Petrides, Oxford University Press, 2003, ch. 11.

70. C. A. Martinez, S. Hu, Y. Dumond, J. Tao, P. Kellcher and L. Tully, *Org. Process Res. Dev.*, 2008, **12**, 392.

71. R. A. Sheldon, *Chirotechnology: the Industrial Synthesis of Optically Active Compounds*, Marcel Dekker, New York, 1993.
72. R. J. Fox, S. C. Davis, E. C. Mundorff, L. M. Newman, V. Gavrilovic, S. K. Ma, L. M. Chung, C. Ching, S. Tam, S. Muley, J. Grate, J. Gruber, J. C. Whitman, R. A. Sheldon and G. W. Huisman, *Nat. Biotechnol.*, 2007, **25**, 338.
73. S. K. Ma, J. Gruber, C. Davis, L. Newman, D. Gray, A. Wang, J. Grate, G. W. Huisman and R. A. Sheldon, *Green Chem.*, 2010, **12**, 81.
74. W. P. Stemmer, *Nature*, 1994, **370**, 389–391.
75. C. K. Savile, J. M. Janey, E. C. Mundorff, J. C. Moore, S. Tam, W. R. Jarvis, J. C. Colbeck, A. Krebber, F. J. Fleitz, J. Brands, P. N. Devine, G. W. Huisman and G. J. Hughes, *Science*, 2010, **329**, 305.
76. O. Thum and K. M. Oxenbøll, *SOFW J.*, 2008, **134**, 44.
77. P. Gallezot, *Chem. Soc. Rev.*, 2012, **41**, 1538; A. Corma, S. Iborra and A. Velty, *Chem. Rev.*, 2007, **107**, 2411; P. Gallezot, *Catal. Today*, 2007, **121**, 76.
78. E. De Jong, A. Higson, P. Walsh and M. Wellisch, *Biofuels, Bioprod. Biorefin.*, 2012, **6**, 606.
79. C. O. Tuck, E. Perez, I. T. Horvath, R. A. Sheldon and M. Poliakoff, *Science*, 2012, **337**, 695.
80. L. A. Pfaltzgraff, M. De Bruyn, E. C. Cooper, V. Budarin and J. H. Clark, *Green Chem.*, 2013, **15**, 307.
81. T. J. Taylor, D. J. Taylor, M. W. Peters and D. E. Henton, US Pat. Apl. US 20120190089 A1 20120726 to GEVO Inc., 2012.
82. R. J. van Putten, J. C. van der Waal, E. de Jong, C. B. Rasrendra, H. J. Heeres and J. G. de Vries, *Chem. Rev.*, 2013, **113**, 1499.
83. E. Taarning, I. S. Nielsen, K. Egeblad, R. Madsen and C. Christensen, *ChemSusChem*, 2008, **1**, 75; O. Casanova, S. Iborra and A. Corma, *ChemSusChem*, 2009, **2**, 1138.
84. N. Wierckx, F. Koopman, H. J. Ruijssenaars and J. H. de Winde, *Appl. Microbiol. Biotechnol.*, 2011, **92**, 1095.
85. M. E. Zakrzewska, E. Bogel-Lukasik and R. Bogel-Lukasik, *Energy Fuels*, 2010, **24**, 737.
86. H. Wang, G. Gurau and R. D. Rogers, *Chem. Soc. Rev.*, 2012, **41**, 1519.
87. K. Nakashima, K. Yamaguchi, N. Taniguchi, S. Arai, R. Yamada, S. Katahira, N. Ishida, H. Takahashi, C. Ogino and A. Kondo, *Green Chem.*, 2011, **13**, 2948.
88. R. Rinaldi and F. Schüth, *Energy Environ. Sci.*, 2009, **2**, 610.
89. R. A. Sheldon, *Catal. Today*, 2011, **167**, 3.

CHAPTER 3

Biocatalysis in Organic Media

A. ILLANES*[a]

[a]Pontificia Universidad Católica de Valparaíso, School of Biochemical
Engineering, Chile
*E-mail: aillanes@ucv.cl

3.1 Enzyme Structure and Function

Each one of the chemical reactions of cell metabolism in all living forms is
catalyzed by an enzyme. Enzymes are quite evolved forms of catalysts that
have been tailored to perform efficiently under the mild conditions required
for the integrity and functionality of biological systems. Enzymes are protein
molecules, ranging from one hundred to several hundreds of amino acid residues, linked by peptide bonds (a bond between the carbon atom in the carboxyl group of one amino acid and the nitrogen atom in the α-amino group
of a neighboring amino acid), and both their advantages and disadvantages
as catalysts stem from this nature. Their catalytic potential (activity) resides
in a very small portion of the protein structure comprising only a few amino
acid residues (usually three or four), called the active site or active center,
that is responsible for the mechanism of reaction. These catalytic amino acid
residues are usually far apart in the primary structure of the protein molecule, so that a special three dimensional configuration needs to be adopted
for the molecule to be active. This native active configuration is the result of
post-translational events, including both intramolecular interactions among
the side chains of the amino acid residues of the protein, and intermolecular interactions of those residues with the medium, so that the activity of

RSC Green Chemistry No. 45
White Biotechnology for Sustainable Chemistry
Edited by Maria Alice Z. Coelho and Bernardo D. Ribeiro
© The Royal Society of Chemistry 2016
Published by the Royal Society of Chemistry, www.rsc.org

the enzyme is very much conditioned by the properties of the medium surrounding it. Enzyme activity is conditioned by the maintenance of this complex native structure, which can be challenged by denaturing agents of all sorts (temperature, pH, chemicals, shear). As such, the active center is a very fragile structure, so that most of the protein molecule acts as a scaffold to confer the active site with the required structural stability. Even so, enzymes are, because of their complex protein nature, quite labile catalysts and also quite expensive molecules to build up. On the other hand, their molecular complexity allows them to be exquisitely tuned in terms of selectivity and activity under mild environmental conditions. Both the advantages and disadvantages derived from their protein nature are strong determinants of their potential use as process catalysts, as analyzed later.

In some instances, the protein moiety is self-sufficient to perform catalysis, but many enzymes are conjugated proteins (holoenzymes) where the protein structure is associated with other molecules that may or may not play a direct role in catalysis. The ones having an active role in catalysis are particularly relevant and can be either low molecular weight organic molecules that reversibly bind to the enzyme structure during catalysis (coenzymes), or simple inorganic compounds and ions that are tightly bound to the enzyme structure and are not dissociated from it (cofactors)[1]. The requirement for these non-protein catalytic effectors has a profound influence on the technological use of enzymes as catalysts, as analyzed later.

From a physiological standpoint, most enzymatic reactions are carried out in a cell environment which is preponderantly aqueous, so that it is to be assumed that the enzyme structure is conditioned to perform in water-rich surroundings. This is for the most part true, and it was a paradigm that enzymes were catalysts that performed only in aqueous medium, with their use restricted to it. This vision of enzymes began to change in the 1980s when several pieces of experimental evidence accumulated, sustaining that enzymes, under appropriate conditions, are able to express catalytic activity in predominantly organic media.[2] This is not to contradict the fact that the reported catalytic activities in organic media were frequently orders of magnitude lower than in aqueous medium.[3] However, non-aqueous biocatalysis has come a long way since then, and it is now possible, by judicious choice of the enzyme and the reaction conditions, to achieve catalytic activities of the same order of magnitude as in aqueous medium.[4]

3.2   Enzymes and Biocatalysts

Biocatalysts are catalysts of biological origin so, *sensu stricto*, biocatalysts are enzymes, because enzymes are the catalysts of all biochemical reactions. However, from a technological perspective, biocatalysts are sometimes referred to, *sensu lato*, as any biological entity catalyzing one or more reactions, so that whole cells in different physiological states are also considered biocatalysts, even though it is ultimately their enzymatic patrimony that is

Table 3.1 Strategies for improving the stability of enzyme biocatalysts.

	System	References
Screening	Biotic pools	6 and 7
	Metagenomic pools	8 and 9
Genetic improvement	Site-directed mutagenesis	10 and 11
	Directed evolution	12 and 13
Biocatalyst engineering	Chemical modification	14 and 15
	Immobilization	16 and 17
	Auto-aggregation	18 and 19
Medium engineering	Organic solvents	20 and 21
	Ionic liquids	22 and 23
	Supercritical fluids	24 and 25
	Semi-solid systems	26 and 27

responsible for catalysis.[5] Within the context of this chapter, biocatalysis refers to processes conducted by enzymes that are dissociated from the cell system where they were synthesized. Enzymes are the catalysts of metabolism, tailored to perform under physiological conditions; however, for technological purposes, enzymes are required to efficiently perform under conditions that are usually far from physiological. The challenge of biocatalysis is then to convert these physiological catalysts into process catalysts that, retaining the good properties of enzymes (high specificity and activity under mild conditions), can withstand the usually harsh process conditions. From a technological perspective, a biocatalyst is an enzyme that has been conditioned to work as a process catalyst. This may imply significant manipulation of the basal enzyme to conform to the biocatalyst. Poor stability under process conditions is certainly a major weak spot of enzymes as process catalysts, so most efforts have been devoted and strategies developed to tackle this problem, as summarized in Table 3.1.

All of the above strategies represent significant contributions for the production of robust biocatalysts which are well-endowed to perform efficiently under process conditions. Of them, enzyme immobilization stands out, representing a major breakthrough in biocatalysis that has widened its field of application considerably.[28]

3.3   Enzyme Catalysis in Aqueous Media

Enzymes have been used as process catalysts since the early years of the 20th century. Such enzymes were for the most part crude preparations extracted or recovered from plant and animal tissues and fluids. The development of submerged fermentation technology and industrial microbiology in the 1950s produced a shift toward microbial enzymes, clearly advantageous from a technological perspective.[29] This trend has been reinforced in recent decades because of the impressive advances in genetic and protein engineering that allow cloning and expressing improved genes from different cells and metagenomic pools in suitable microbial hosts.[30,31]

Most conventional uses of enzymes refer to hydrolases acting on aqueous media as catalysts for reactions of degradation of molecules (frequently polymers) to simpler structures. Even though the potential added value in these hydrolytic reactions is low, many applications based on the use of glycosidases, proteases and lipases are still of important industrial value,[32] and new fields of application are still emerging. Phytases for farm and pet animal nutrition,[33] improved α-amylases[34] and cellulases[35] for first and second generation biofuels, and optimized hydrolase cocktails for detergent formulation[36] are just a few of the actively growing fields of the conventional applications of hydrolases in aqueous media. Hydrolases are, in a certain way, ideal biocatalysts as they are, among all classes of enzymes, the best endowed for performing under the usually harsh conditions of industrial processes. In fact, hydrolases are robust proteins, many of them extracellular, not requiring coenzymes, all attributes that are well appreciated from a technological perspective. However, as mentioned before, these enzymes acting in aqueous media will only catalyze reactions of molecular degradation. Enzyme immobilization bloomed in the 1970s as a powerful tool for producing significant stabilization under process conditions, allowing the development of hydrolytic and non-hydrolytic processes which were otherwise not technologically feasible. This is well illustrated by the processes developed in the subsequent years for the production of 6-amino penicillanic acid from penicillin G or V as a precursor of semi-synthetic β-lactam antibiotics,[37] the production of high-fructose syrup using glucose (xylose) isomerase,[38] and the production of acrylamide from acrylonitrile with nitrile hydratase-containing immobilized cells.[39] However, these reactions represented no gain in molecular complexity so that the added value was still modest.

For decades, enzymes were considered as catalysts for water environments and the paradigm was that enzymes were only active in such media. This was a severe restriction, since many reactions of industrial interest need to be conducted in non-aqueous environments where the substrates and products of reactions are soluble enough. Reactions of organic synthesis are seldom carried out in aqueous media, and such reactions are the ones of most significant added value. On the other hand, enzymes whose metabolic role is related to synthesis are complex and labile intracellular, coenzyme-requiring proteins whose application as process catalysts has to confront many technological difficulties. In fact, enzymes have to be stabilized, usually by immobilization, and coenzymes co-immobilized or derivatized in the case of membrane reactors; beyond that, coenzymes have to be reconverted by the use of auxiliary reactions (usually enzymatic) to close the catalytic cycle. Many of those technological challenges have been, however, overcome by significant advances in both reactor and biocatalyst design.[40,41] Despite the progress in this field, there are still very few medium to large scale processes using these types of enzymes, with their use being restricted to small scale processes of significant added value.[42] The prospects for the use of enzymes as catalysts for reactions in organic synthesis was then confronted by several technological difficulties and, above all, by an industry rather reluctant to use catalysts of biological nature, which were reputedly inadequate for performing under the required conditions for synthesis, and which were tailored to act only on natural substrates. This situation began to change in the mid-1980s and this turning point was directly related to the use of enzyme catalysts in non-conventional (non-aqueous) media.

3.4 Enzyme Biocatalysis in Non-Aqueous (Non-Conventional) Media

In the early 1980s, significant information arose with respect to enzyme behavior in non-aqueous media. The pioneering work of Klibanov and co-workers on enzyme catalysis in organic solvents[43–45] opened up a vast field of research on enzyme behavior in all sorts of non-aqueous environments. Even though enzymes were significantly less active than in water, under appropriate conditions, their activity was significant and served as a starting point for exploring the potential of enzymatic catalysis in such media. Moreover, even though poorly active, enzymes were in some cases significantly more heat stable than in aqueous medium.[46] Enzyme catalysis in non-aqueous media is of paramount technological significance since in such media, reactions of hydrolysis can be reverted, hydrolases being able to catalyze reactions of bond formation instead of cleavage. In this way, robust, inexpensive and readily available enzymes became tools for organic synthesis. A necessary condition for their use in such processes is a low water activity reaction medium. In this respect, lipases are a distinctive class of enzyme because their metabolic function is usually performed at interfaces or within membranes, so they are particularly well-endowed to perform in non-aqueous or poorly aqueous media, and are by far the most used class of enzymes in organic synthesis.[47]

The reward is very attractive, since now traditional enzymes could be used in non-traditional processes or in organic synthesis: glycosidases in reactions of glycosidic bond formation for the synthesis of oligosaccharides and glycosides, proteases in the synthesis of peptides, acylases in peptide bond formation for the synthesis of bioactive molecules and lipases in reactions of esterification, transesterification and interesterification.

Most of the initial work on enzyme catalysis in non-conventional media was conducted in organic solvents and up to now, organic solvents represent the most important system for enzyme biocatalysis in reactions of organic synthesis[48,49] despite that, in principle, the use of organic solvents as reaction media for biocatalysis may seem somewhat contradictory, since one of the assets of enzyme biocatalysts is their environmental benignity. These aspects will be analyzed in more detail in the following section.

Several other non-conventional media have been proposed and studied to perform enzyme biocatalysis. They share in common the idea of increased enzyme stability and less offensive reaction media, and are briefly reviewed below.

3.4.1 Gases

The feasibility of conducting enzymatic reactions in gas phase was proven some thirty years ago, highlighting the potential benefit of high mass transfer rates, the removal of volatile inhibitors, increased enzyme stability and aseptic conditions.[50] Such systems are obviously limited to gases or highly volatile compounds, with the drawbacks of difficult control of water activity, and complex reactor design and operation.[51] Enzyme biocatalysis in the gas phase has been seldom considered as a technological option, beyond its potential in the specific removal of highly toxic pollutants in some industrial gas effluents.[52]

3.4.2 Supercritical Fluids

Sometimes termed as dense gases, supercritical fluids (SCFs) are those fluids at temperatures and pressures above their corresponding critical values. Under such conditions, properties lie somewhere in between those of gases and liquids: the densities of SCFs are similar to those of liquids and much higher than those of gases, while viscosities are somewhat higher than those of gases and lower than those of liquids, and diffusivities in SCFs are lower than in gases but much higher than in liquids. These characteristics make SCFs very promising reaction media for biocatalysis by allowing a solubility of substrates and products similar to that in liquid media but with mass transfer rates that are much higher and similar to those in gases. Recovery of catalyst and products is simple because the medium can be easily removed by pressure relief, although enzyme inactivation by sudden depressurization may occur, hampering catalyst recovery.[53] Having many advantages, the use of SCFs as reaction media for enzyme biocatalysis is restrained to those fluids whose critical pressure and temperature are compatible with enzyme activity. Some examples are chloro-trifluoro methane, ethane, ethylene, fluroform and sulphur hexafluoride. However, carbon dioxide, while not being the best suited SCF, is the most extensively used because of cost, safety and environmental considerations.[54] Its critical temperature is 31 °C (quite appropriate), but its critical pressure is high (73 bars), which means that highly pressurized reactors are required. Even so, enzyme catalysis in supercritical carbon dioxide medium is quite promising for the synthesis of fine chemicals in compliance with the green chemistry principles.[55] In general, high equipment and operation costs, and difficult control of key operational variables, have precluded the more extended use of SCF in biocatalysis, especially at a large scale of operation.[56]

3.4.3 Ionic Liquids

Enzyme catalysis in ionic liquids (ILs) has received considerable attention in the last decade as a greener alternative to the more popular organic solvents. ILs are polar solvents composed of a complex organic cation and a simple (often inorganic) anion.[57] With negligible vapor pressure, tunable physico-chemical properties[58,59] and being compliant with enzyme activity and stability,[60] ILs have been considered as the green alternative of the future for biocatalysis;[61] increased enantioselectivity has even been reported in asymmetric reactions.[62] However, biocatalysis in ILs is not yet a mature technology and several problems remain to be solved: difficulty in control of water activity, high viscosity, difficulty in purification, lack of a theoretical basis for enzyme behavior in such media and high cost.[61,63] Research in the field is intense, and there are good prospects for the increased impact of ILs as green reaction media for enzyme biocatalysis.[64]

3.4.4 Semisolid Systems

Enzymatic reactions in semi-solid and nearly solid media have been thoroughly investigated. These systems are characterized by the existence of a very small amount of liquid (water or organic) where the substrates become saturated and react with the enzyme to form the products that precipitate out.[65,66] It is indeed an appealing option because water activity can be quite low (very much as in other non-conventional media), hence favoring reactions of synthesis and depressing the competing reactions of hydrolysis, with the advantage that no noxious solvent is included and the enzyme surrounding is essentially an aqueous milieu; at the end of reaction, most of the reactor volume is product, which means an extremely high volumetric productivity.[67,68] The main drawback of semi-solid systems is related to mass transfer limitation and mixing, which becomes a critical issue when scaling-up.[69]

3.4.5 Reactions Conducted at Very High Substrate Concentration

Reactions conducted at very high substrate concentrations (up to saturation and beyond) may also depress water activity and represent a greener alternative to most non-aqueous systems. This strategy is also important when substrates are poorly soluble in non-aqueous solvents. This is the case of the production of semi-synthetic β-lactam antibiotics, where substantial increases in ampicillin and cephalexin yields were obtained when working under substrate supersaturation,[70,71] and the kinetically controlled synthesis of galacto-oligosaccharides using β-galactosidases.[72,73]

3.5 Enzyme Biocatalysis in Organic Solvents

Enzyme biocatalysis in organic solvents is the most studied system in non-aqueous enzymology and represents a most important one,[74,75] despite increasing environmental concern about it.[76,77] Substitution of water by an organic solvent offers many potential advantages for biocatalysis in organic synthesis: reversal of hydrolytic reactions using hydrolases, reactions which are thermodynamically unfavorable in water and reactions with water insoluble or poorly soluble substrates become possible, thermal stability may be significantly increased, product inhibition can be reduced, and substrate affinity and specificity and selectivity can be tuned.[78] However, the enzyme activity in most organic solvents is severely reduced with respect to water,[3] so that major effort has been devoted to both medium and biocatalyst engineering to improve the enzyme performance in such media.[31] Lipases represent a distinctive class of enzymes that are structurally conditioned for non-aqueous environments,[79] so it is usual that these enzymes have better performance in hydrophobic organic solvents than in water; in addition, lipases are highly regio- and enantioselective in such media.[80,81] These are salient properties making lipases the most important class of enzymes in non-aqueous biocatalysis.[82–84]

Based on the characteristics of the liquid phase, organic media for biocatalysis can be divided into homogeneous and heterogeneous, as presented in Table 3.2.

Homogeneous systems refer to mixtures of water and water-miscible solvents (cosolvents) in which the enzyme is either dissolved or suspended. Most cosolvents, when used at high concentrations, are detrimental to enzyme activity since they intrude in the aqueous layer surrounding the enzyme, stripping water off and interacting directly with the enzyme, distorting its native structure.[85] As a consequence, most enzymes are poorly active and unstable in such mixtures, so that immobilization or aggregation is in most cases mandatory.[86,87] There are some notable exceptions, like polyols[88] and glymes,[89] where many enzymes are active and stable, sometimes very much like in aqueous medium.

Microheterogeneous systems are those in which an organic liquid phase appears to be homogeneous to the naked eye, but is however microscopically heterogeneous, which may be because the enzyme is intrinsically insoluble in such a medium and very small enzyme aggregates surrounded by a water shell are suspended in the hydrophobic solvent, or because the enzyme is immobilized. The former, in which an enzyme powder (obtained by lyophilization or drying) is merely suspended in a hydrophobic solvent where

Table 3.2 Reaction systems for biocatalysis in inorganic media.

System	Liquid phase	Solid phase	
Homogeneous			
	Water-miscible solvent	None (soluble enzyme)	Immobilized enzyme
Heterogeneous			
Microheteroge-neous	Water-immiscible solvent	None (reverse micelles) (solubilized enzyme)	Solid enzyme Immobilized enzyme
Macroheteroge-neous	Water–water immiscible solvent	None (soluble enzyme)	Immobilized enzyme

the enzyme protein is completely insoluble, is the most studied and probably the most simple and useful non-aqueous system for biocatalysis, with the advantages of non-aqueous catalysis being fully exploited: high thermal stability,[90] easy biocatalyst and product recovery[91] and tunable enzyme specificity.[92] Immobilization is not strictly necessary since the enzyme is insoluble *per se* in the reaction medium; however, immobilization might be beneficial for increasing the surface of contact with the substrate, impeding direct contact of the solvent with the enzyme molecules, and providing additional stabilization.[93] Water is present in very small amounts in this system, and water activity becomes a key determinant in the enzyme performance.[4,94] A small amount of water should always be present to provide the enzyme with a minimal hydration level, allowing activity expression, but that amount may vary considerably from one enzyme to another, and in certain cases may be as low as less than one monolayer of water molecules surrounding the enzyme molecule.[95] It has been claimed that, in theory, enzymes may express activity in the complete absence of water if a competent conformation is attained and a stable transition state is achieved;[96] however, this has yet to be demonstrated. In these hydrophobic water-immiscible solvents, a minimal hydration is required for the enzyme to acquire the flexibility required for catalysis; however, too much water may be detrimental by distorting the acquired configuration; hence, a water activity will exist where the enzyme activity and stability are optimally balanced. The hydrophobicity of the solvent is a strong determinant of enzyme behavior in this system, and more hydrophobic solvents are preferred since the protecting water layer surrounding the enzyme particle will undergo less intrusion by the solvent. Hydrophobicity is frequently related to the $\log P$ parameter, P being the partition coefficient (of the solvent) between n-octanol and water. Solvents with $\log P$ values higher than 4 are adequate, while those with values lower than 2 are not.[97] Unfortunately, many water-immiscible solvents lie between these values, where enzyme behavior is hard to predict. Advances in biocatalyst engineering in the last decade have been impressive, allowing the design of enzyme catalysts which are well suited to perform under these harsh conditions;[98] however, expression of activity is in most cases still low for technological purposes. Reverse micelles represent another microheterogeneous system where micelles are spontaneously formed by the addition under agitation of a small amount of water to a surfactant-containing hydrophobic solvent.[99] The conditions in the inner cavities of the micelles may well resemble a more natural environment for enzymes than aqueous medium and, in fact, increased expression of activity has been reported in some cases.[100] However, reverse micelles are mechanically weak, their behavior is hard to predict and the surfactant interferes with product recovery and purification.[101] Direct solubilization of enzyme aggregates in organic solvents by the addition of minute amounts of water and surfactant has also been studied, and a protective effect against solvent inactivation has been observed in the case of less hydrophobic solvents.[102]

Macroheterogeneous systems are those in which two clearly distinguishable phases are present, one aqueous and one organic. Substrates are dissolved in the organic phase, while the enzyme is in the aqueous phase, so that substrate molecules migrate to the interface and are partitioned into the aqueous phase and are acted upon by the enzyme; the product formed can then be partitioned into the organic phase, which is desirable to reduce possible enzyme inhibition and depress hydrolytic reactions when hydrolases are used as catalysts.[101–103] This is an appealing system in the case of poorly water-soluble substrates and products, its main drawback being mass transfer limitations and substrate and product transport across the interface.[104] Mass transfer rates can be increased by strong agitation, but this will produce enzyme inactivation.[105]

In summary, among the different systems for biocatalysis in organic media, the use of enzymes in low water containing hydrophobic solvents is the most promising. Substantially reduced activity in such media is a major constraint, but by carefully selecting the type of enzyme preparation, solvent and reaction conditions, activities on the same order of magnitude as those in aqueous media are achievable.[4]

3.6 Enzymes as Catalysts for Organic Synthesis

In the last few decades, the focus on enzyme biocatalysis has moved from conventional degradation processes in aqueous media to more sophisticated processes of organic synthesis where the potential added value is much higher, representing now the most promising and challenging area for enzyme biocatalysis.[106,107] Biocatalysis in organic synthesis is closely related to non-conventional media, since most reactions in organic synthesis require non-aqueous media.[21] Enzymes, being labile and costly, are not ideal industrial catalysts, but their exquisite selectivity, specificity and activity under mild conditions are significant attributes, especially for the synthesis of fine-chemicals, pharmaceuticals and other bioactive compounds.[108] After the initial reluctance of the chemical industry to employ biological catalysts in their processes, the above-mentioned qualities are now well appreciated within the framework of green chemistry.[109,110]

Good properties of enzymes as process catalysts have profound technological implications. Chemical precision and mild operation conditions imply a significant reduction in the cost of equipment, energy supply and downstream operations. The production of the non-caloric sweetener aspartame illustrates this: the enzymatic synthesis with the protease thermolysin only requires the protection of the amino group in aspartic acid, being regioselective with respect to the carboxyl group adjacent to the amino group, and enantioselective with respect to L-phenylalanine methyl ester, so that the racemate can be used, these features representing distinctive advantages over the conventional chemical process.[111] Asymmetric organic synthesis is an ample and open field for biocatalysis.[112] In fact, there is increasing pressure on the pharmaceutical industry to produce chiral drugs as pure enantiomers; this is not possible by chemical catalysis, but enzymes are chiral catalysts such that the chirality present in the substrate molecule is recognized upon the formation of the enzyme–substrate complex, and in this way, both enantiomers of a racemic substrate may react at quite different rates, affording kinetic resolution. In this way, a high proportion of the drug can be produced as the required enantiomer (*eutomer*) and the potential adverse effects of the unwanted enantiomer (*distomer*) can be avoided. Selectivity then has a profound influence in process economics by reducing the number of steps and protective reactions required and by reducing the number of downstream operations for product purification.

Complex reactions of organic chemistry like ketone oxidation,[113] ring expansion,[114] hydroxylation, epoxidation and transamination can be done in one-pot enzymatic reactions, which is also a distinctive advantage of enzyme biocatalysis.[115]

There are myriad reactions in organic synthesis that have been studied using enzymes or enzyme-containing cells, many of them of technological relevance.[116,117] However, the entry of enzyme biocatalysts into the organic synthesis industry has been hampered by enzymes being considered expensive, not easily available, unstable, cofactor-requiring, and only acting properly in natural aqueous habitats and on natural substrates.[118] Some of these comments can be argued against: enzymes are indeed expensive catalysts, but the point is their cost relative to product formation and, besides, the cost of enzymes has been continuously dropping over the last few decades; the use of hydrolases in non-conventional media avoids the use of coenzyme-requiring enzymes, and coenzyme regeneration and recycling is technologically feasible for other classes of enzymes;[119] depending on the enzyme and reaction medium, the activity can approximate that in aqueous medium and the stability may be even higher;[4,120] even though enzymes are reputedly highly specific catalysts, the specificity is sometimes broad enough to act on non-natural substrates, a condition that can be obtained by genetic manipulation,[121] and besides, enzymes may be quite promiscuous in the sense of catalyzing reactions sometimes barely related to the one they supposedly evolved to catalyze, so enzyme promiscuity is a concept that is gaining increasing importance in biocatalysis.[122,123]

3.7 Conclusions

Many of the constraints mentioned above have been or are on the way to being overcome. In fact, enzyme prices have been persistently dropping, and advances in biocatalyst and medium engineering have led to acceptable activities and operational stabilities. Sound enzyme performance in non-conventional media, as usually required for organic synthesis, is a reality and alternatives to the use of environmentally objectionable solvents have emerged, which are in compliance with the principles of green chemistry; ionic liquids[124,125] and solvent-free systems[126,127] may represent in the forthcoming years better options than organic solvents as reaction media for biocatalysis in organic synthesis.

References

1. M. R. Kula, in *Enzyme Catalysis in Organic Synthesis*, ed. K. Drauiz and H. Waldmann, Wiley-VCH, Weinhein, 2002, p. 12.
2. A. Zaks and A. M. Klibanov, *Proc. Natl. Acad. Sci. U. S. A.*, 1985, **82**, 3192.
3. A. M. Klibanov, *Trends Biotechnol.*, 1997, **15**, 97.
4. P. Adlercreutz, in *Organic Synthesis with Enzymes in Non-Aqueous Media*, ed. G. Carrea and S. Riva, Wiley-VCH, Weinheim, 2008, p. 3.
5. R. León, P. Fernandes, H. M. Pinheiro and J. M. S. Cabral, *Enzyme Microb. Technol.*, 1998, **23**, 483.
6. B. van den Burg, *Curr. Opin. Microbiol.*, 2003, **6**, 213.
7. M. Ferrer, O. Golyshina, A. Beloqui and P. N. Golyshin, *Curr. Opin. Microbiol.*, 2007, **10**, 207.
8. H. L. Steele, K. E. Jaeger, R. Daniel and W. R. Streit, *J. Mol. Microbiol. Biotechnol.*, 2009, **16**, 25.
9. T. Uchiyama and K. Miyazaki, *Curr. Opin. Biotechnol.*, 2009, **20**, 616.
10. Y. Ijima, K. Matoishi and Y. Terao, *Chem. Commun.*, 2005, **7**, 877.
11. L. Zheng, U. Baumann and J. L. Reymond, *Nucleic Acids Res.*, 2004, **32**, 14.
12. V. G. H. Eijsink, S. Gåseidnes, T. V. Borchert and B. van den Burg, *Biomol. Eng.*, 2005, **22**, 21.
13. P. A. Dalby, *Curr. Opin. Struct. Biol.*, 2011, **21**, 1.
14. B. Davis, *Curr. Opin. Biotechnol.*, 2003, **14**, 379.
15. K. Sangeetha and T. E. Abraham, *J. Mol. Catal. B: Enzym.*, 2006, **38**, 171.
16. J. M. Guisán, *Immobilization of Enzymes and Cells*, Humana Press, New Jersey, 2006.
17. C. Mateo, J. M. Palomo, G. Fernandez-Lorente, J. M. Guisan and R. Fernandez-Lafuente, *Enzyme Microb. Technol.*, 2007, **40**, 1451.
18. J. J. Roy and T. E. Abraham, *Chem. Rev.*, 2004, **104**, 3705.
19. R. Sheldon, *Org. Process Res. Dev.*, 2011, **15**, 213.
20. M. T. Ru, K. C. Wu and J. P. Lindsay, *Biotechnol. Bioeng.*, 2002, **75**, 187.
21. H. L. Yu, L. Ou and J. H. Xu, *Curr. Opin. Biotechnol.*, 2010, **14**, 1424–1432.
22. J. Durand, E. Teuma and M. Gómez, *C. R. Chim.*, 2007, **10**, 152.
23. Z. Yang, Z. Guo and X. Xu, *J. Am. Oil Chem. Soc.*, 2012, **89**, 1049.
24. S. Garcia, N. M. T. Lourenço, D. Lousa, A. F. Sequeira, P. Mimoso, J. M. S. Cabral, C. A. M. Afonso and S. Barreiros, *Green Chem.*, 2004, **6**, 466.
25. K. Rezaei, F. Temellib and E. Jenab, *Biotechnol. Adv.*, 2007, **25**, 272.
26. R. V. Ulijn, L. D. Martin, L. Gardossi and P. J. Halling, *Curr. Org. Chem.*, 2003, **7**, 1333.
27. P. Chaiwuta, P. Kanasawud and P. J. Halling, *Enzyme Microb. Technol.*, 2007, **40**, 954.
28. A. Illanes, in *Comprehensive Biotechnology*, ed. M. Moo-Young, Elsevier, Boston, 2nd edn, 2011, vol. 1, p. 25.
29. S. L. Neidleman, in *Biocatalysts for Industry*, ed. J.S. Dordick, Plenum Press, New York, 1991, p. 21.
30. S. Sánchez and A. L. Demain, *Org. Process Res. Dev.*, 2011, **15**, 224.

31. A. Illanes, A. Cauerhff, L. Wilson and G. Castro, *Bioresour. Technol.*, 2012, **115**, 48.
32. A. Illanes, *Enzyme Biocatalysis: Principles and Applications*, Springer, United Kingdom, 2008.
33. C. L. Walk, M. R. Bedford, T. S. Santos, D. Paiva, J. R. Bradley, H. Wladecki, C. Honaker and A. P. McElroy, *Poult. Sci.*, 2013, **92**, 719.
34. B. Liao, G. A. Hill and W. J. Roesler, *Biochem. Eng. J.*, 2012, **64**, 8.
35. K. Hoyer, M. Galbe and G. Zacchi, *Proc. Biochem.*, 2013, **48**, 289.
36. K. H. Maurer, in *Encyclopedia of Industrial Biotechnology: Bioprocess, Bioseparation, and Cell Technology*, ed. M.C. Flickinger, Wiley Interscience, 2010, p. 1.
37. A. Illanes and L. Wilson, in *Enzyme Biocatalysis: Principles and Applications*, ed. A. Illanes, Springer, United Kingdom, 2008, p. 21.
38. S. H. Bhosale, M. B. Rao and V. V. Deshpande, *Microbiol. Rev.*, 1996, **60**, 280.
39. J. Miller and V. Nagarajan, *Trends Biotechnol.*, 2000, **18**, 190.
40. M. Karabec, A. Lyskowski, K. C. Tauber, G. Stweinkellner, W. Kroutil, G. Grogan and K. Gruber, *Chem. Commun.*, 2010, **46**, 6314.
41. K. Tauber, M. Hall, W. Kroutil, W. M. F. Fabian, K. Faber and S. M. Glueck, *Biotechnol. Bioeng.*, 2011, **108**, 1462.
42. A. Berenguer-Murcia and R. Fernandez-Lafuente, *Curr. Org. Chem.*, 2010, **14**, 1000.
43. A. Zaks and A. M. Klibanov, *Proc. Natl. Acad. Sci. U. S. A.*, 1985, **82**, 3192.
44. A. M. Klibanov, *Chem. Technol.*, 1986, **6**, 354.
45. A. M. Klibanov, *Trends Biotechnol.*, 1989, **14**, 141.
46. A. Zaks and A. M. Klibanov, *Science*, 1984, **224**, 1249.
47. F. Hasan, A. A. Shah and A. Hameed, *Enzyme Microb. Technol.*, 2006, **39**, 235.
48. B. G. Davis and V. Boyer, *Nat. Prod. Rep.*, 2001, **18**, 618.
49. A. M. Klibanov, *Curr. Opin. Biotechnol.*, 2003, **14**, 427.
50. E. Barzana, A. Klibanov and M. Karel, *Appl. Biochem. Biotechnol.*, 1987, **15**, 25.
51. S. Lamare and M. Legoy, *Trends Biotechnol.*, 1993, **11**, 413.
52. B. C. Dravis, K. E. LeJeune, A. D. Hetro and A. J. Russell, *Biotechnol. Bioeng.*, 2000, **69**, 235.
53. A. J. Mesiano, E. Beckman and A. J. Russell, *Chem. Rev.*, 1999, **99**, 623.
54. C. G. Laudani, M. Habulin, Ž. Knez, G. D. Porta and E. Reverchon, *J. Supercrit. Fluids*, 2007, **41**, 92.
55. P. Lozano, E. Garcia-Verdugo, S. V. Luis, M. Pucheault and M. Vaultier, *Curr. Org. Synth.*, 2011, **8**, 810.
56. Ž. Knez, *J. Supercrit. Fluids*, 2009, **47**, 357.
57. F. van Rantwijk, R. M. Lau and R. A. Sheldon, *Trends Biotechnol.*, 2003, **21**, 131.
58. F. van Rantwijk and R. A. Sheldon, *Chem. Rev.*, 2007, **107**, 2757.
59. H. Olivier-Bourbigou, L. Magna and D. Morvan, *Appl. Catal., A*, 2010, **373**, 1.

60. H. Zhao, *J. Chem. Technol. Biotechnol.*, 2010, **85**, 891.
61. S. H. Ha and Y. M. Koo, *Korean J. Chem. Eng.*, 2011, **28**, 2095.
62. J. Durand, E. Teuma and M. Gómez, *C. R. Chim.*, 2007, **10**, 152.
63. N. I. Ruzich and A. S. Bassi, *Can. J. Chem. Eng.*, 2010, **88**, 277.
64. M. Moniruzzaman, K. Nakashima, N. Kamiya and M. Goto, *Biochem. Eng. J.*, 2010, **48**, 295.
65. R. V. Ulijn, L. D. Martin, L. Gardossi, A. E. M. Janssen, B. D. Moore and P. J. Halling, *Biotechnol. Bioeng.*, 2002, **80**, 509.
66. P. Chaiwuta, P. Kanasawud and P. J. Halling, *Enzyme Microb. Technol.*, 2007, **40**, 954.
67. R. V. Ulijn, A. E. M. Janssen, B. D. Moore and P. J. Halling, *Chem.–Eur. J.*, 2001, **7**, 2089.
68. A. Basso, P. Spizzo, M. Toniutti, C. Ebert, P. Linda and L. Gardossi, *J. Mol. Catal. B: Enzym.*, 2006, **39**, 105.
69. M. Erbeldinger, X. Ni and P. J. Halling, *Biotechnol. Bioeng.*, 1998, **59**, 68.
70. M. I. Youshko, H. Moody, A. Bukhanov, V. H. J. Boosten and V. K. Švedas, *Biotechnol. Bioeng.*, 2004, **85**, 323.
71. A. Illanes, C. Altamirano, M. Fuentes, F. Zamorano and C. Aguirre, *J. Mol. Catal. B: Enzym.*, 2005, **35**, 45.
72. C. Vera, C. Guerrero, A. Illanes and R. Conejeros, *Biotechnol. Bioeng.*, 2011, **108**, 2270.
73. C. Vera, C. Guerrero, R. Conejeros and A. Illanes, *Enzyme Microb. Technol.*, 2012, **50**, 188.
74. A. M. P. Koskinen and A. M. Klibanov, *Enzymatic Reactions in Organic Media*, Blackie Academic & Professional, London, 1996.
75. M. N. Gupta and I. Roy, *Eur. J. Biochem.*, 2004, **271**, 2573.
76. R. A. Sheldon, *Green Chem.*, 2005, **7**, 267.
77. P. Anastas and N. Eghbali, *Chem. Soc. Rev.*, 2010, **39**, 301.
78. A. Klibanov, *Nature*, 2001, **409**, 241.
79. F. Secundo, G. Carrea, C. Tarabiono, P. Gatti-Lafranconi, S. Brocca, M. Lotti, K. E. Jaeger, M. Puls and T. Eggert, *J. Mol. Catal. B: Enzym.*, 2006, **39**, 166.
80. K. E. Jaeger and T. Eggert, *Curr. Opin. Biotechnol.*, 2002, **13**, 390.
81. M. Petkar, A. Llai, P. Caimi and M. Daminati, *J. Mol. Catal. B: Enzym.*, 2006, **39**, 83.
82. F. K. Malcata, *Engineering of/with Lipases*, Kluwer Academic Publishers, Dordrecht, 1996.
83. M. T. Reetz, *Curr. Opin. Chem. Biol.*, 2002, **6**, 145.
84. R. Sharma, Y. Chisti and U. C. Banerjee, *Biotechnol. Adv.*, 2001, **19**, 627.
85. S. Torres and G. Castro, *Food Technol. Biotechnol.*, 2004, **42**, 271.
86. O. Abián, C. Mateo, G. Fernández-Lorente, J. M. Palomo, R. Fernández-Lafuente and J. M. Guisán, *Biocatal. Biotransform.*, 2001, **19**, 489.
87. A. Illanes, L. Wilson, E. Caballero, R. Fernández-Lafuente and J. M. Guisán, *Appl. Biochem. Biotechnol.*, 2006, **133**, 189.
88. A. Illanes and A. Fajardo, *J. Mol. Catal. B: Enzym.*, 2001, **11**, 587.
89. S. Tang, C. L. Jones and H. Zhao, *Bioresour. Technol.*, 2013, **129**, 667.

90. M. N. Gupta, Enzyme function in organic solvents, *Eur. J. Biochem.*, 1992, **203**, 17.
91. L. E. S. Brink, J. Tramper, K. C. A. M. Luyben and K. Van't Riet, *Enzyme Microb. Technol.*, 1988, **10**, 736.
92. G. Carrea and S. Riva, *Angew. Chem., Int. Ed.*, 2000, **39**, 2226.
93. H. Ogino and H. Ishikawa, *J. Biosci. Bioeng.*, 2001, **91**, 109.
94. P. J. Halling, *Philos. Trans. R. Soc., B*, 2004, **359**, 1287.
95. D. S. Clark, *Philos. Trans. R. Soc., B*, 2004, **359**, 1299.
96. F. Xu and H. Ding, *Appl. Catal., A*, 2007, **317**, 70.
97. Z. Yang and A. J. Russell, in *Enzymatic Reactions in Organic Media*, ed. A.M.P. Koskinen and A. M. Klibanov, Blackie Academic & Professional, London, 1996, p. 52.
98. M. Adamczak and S. Hari-Krishna, *Food Technol. Biotechnol.*, 2004, **42**, 251.
99. C. M. L. Carvalho and J. M. S. Cabral, *Biochimie*, 2000, **82**, 1063.
100. M. Castro and J. Cabral, *Enzyme Microb. Technol.*, 1989, **11**, 6.
101. F. Bordusa, *Chem. Rev.*, 2002, **102**, 4817.
102. U. Akbar, C. D. Aschenbrenner, M. R. Harper, H. R. Johnson, J. S. Dordick and D. S. Clark, *Biotechnol. Bioeng.*, 2007, **96**, 1030.
103. S. Barberis, E. Quiroga, M. C. Arribére and N. Priolo, *J. Mol. Catal. B: Enzym.*, 2002, **17**, 39.
104. S. Hari Krishna, *Biotechnol. Adv.*, 2002, **20**, 239.
105. S. Colombié, A. Gaunand and B. Lindet, *Enzyme Microb. Technol.*, 2001, **28**, 820.
106. E. García-Junceda, J. F. García-García, A. Bastida and A. Fernández-Mayoralas, *Bioorg. Med. Chem.*, 2004, **12**, 1817.
107. N. Ran, L. Zhao, Z. Chen and J. Tao, *Green Chem.*, 2008, **10**, 361.
108. V. Gotor, I. Alfonso and E. García-Urdiales, *Asymmetric Organic Synthesis with Enzymes*, Wiley-VCH, Weinheim, 2008.
109. R. A. Sheldon, I. Arends, and U. Hanefeld, *Green Chemistry and Catalysis*, Wiley-VCH, Weinheim, 2007.
110. H. P. Meyer and N. J. Turner, *Mini-Rev. Org. Chem.*, 2009, **6**, 300.
111. M. Erbeldinger, A. J. Mesiano and A. J. Russell, *Biotechnol. Prog.*, 2000, **16**, 1129.
112. S. Schätzle, F. Steffen-Munsberg, A. Thontowi, M. Höhne, K. Robins and U. T. Bornscheuer, *Adv. Synth. Catal.*, 2011, **353**, 2439.
113. V. Alphand, G. Carrea, R. Wohlgemuth, R. Furstoss and J. M. Woodley, *Trends Biotechnol.*, 2003, **21**, 318.
114. D. Thakur, M. K. Roy and T. C. Bora, *Biotechnol. Lett.*, 2009, **31**, 1059.
115. D. J. Pollard and J. M. Woodley, *Trends Biotechnol.*, 2007, **25**, 66.
116. K. Faber, *Biotransformations in Organic Chemistry*, Springer, Berlin, Heidelberg, New York, 1997.
117. K. Drauz and H. Waldmann, *Enzyme Catalysis in Organic Synthesis*, Wiley VCH, Weinhein, 2012, vol. 2.
118. A. S. Bommariusand and B. R. Riebel, *Biocatalysis: Fundamentals and Applications*, Wiley-VCH, Weinheim, 2004.

119. W. Liu and P. Wang, *Biotechnol. Adv.*, 2007, **25**, 69.
120. P. V. Iyer and L. Ananthanarayan, *Process Biochem.*, 2008, **43**, 1019.
121. P. A. Dalby, *Curr. Opin. Struct. Biol.*, 2011, **21**, 473.
122. O. Khersonsky and D. S. Tawfik, *Annu. Rev. Biochem.*, 2010, **79**, 471.
123. P. Gatti-Lafranconi and F. Hollfelder, *ChemBioChem*, 2013, **14**, 285.
124. M. P. C. Marques, N. M. T. Lourenço, P. Fernandes and C. C. C. R. de Carvalho, in *Green Solvents I Properties and Applications in Chemistry*, ed. A. Mohammad and Inamuddin, Springer, United Kingdom, 2012, p. 121.
125. P. Domínguez de María, *Angew. Chem., Int. Ed.*, 2008, **47**, 6960.
126. P. Tufvesson, A. Annerling, R. Hatti-Kaul and D. Adlercreutz, *Biotechnol. Bioeng.*, 2007, **97**, 447.
127. S. Shah and M. N. Gupta, *Process Biochem.*, 2007, **42**, 409.

Microwave Assisted Enzyme Catalysis: Practice and Perspective

GANAPATI D. YADAV*[a] AND SARAVANAN DEVENDRAN[a]

[a]Department of Chemical Engineering, Institute of Chemical Technology, India
*E-mail: gdyadav@yahoo.com, gd.yadav@ictmumbai.edu.in

4.1　Introduction

During the twentieth century, many spectacular advances have taken place in different industries which have ultimately led to faster communication, faster transport, higher food production per hectare, improved quality and longevity of life and unbelievable means of entertainment, without much concern, particularly until the 1980s, for the environment and the waste generated during processing and manufacturing. Industry was only concerned with the target of meeting the growing demand for products, without any concern for the drawbacks associated with processes. In the case of the chemical and allied industries, most reaction protocols and industrial manufacture were practiced in the absence of an adequate database on the hazardous and toxic properties of the chemicals and solvents. As a result, industry continued to use excess raw materials, inefficient processing equipment, excess solvents and different materials, leading to a huge quantity of waste. Waste minimization of either the raw materials or energy consumption was not practiced.

RSC Green Chemistry No. 45
White Biotechnology for Sustainable Chemistry
Edited by Maria Alice Z. Coelho and Bernardo D. Ribeiro
© The Royal Society of Chemistry 2016
Published by the Royal Society of Chemistry, www.rsc.org

Due to public pressure and strict environmental laws, industry was forced to adopt clean, green and smart processes. To overcome the shortage of raw materials and to counteract environmental problems, industry has started to make a paradigm shift from customary concepts that center only on the yield of the desired compounds to alternative processes and products with minimal environmental footprints.[1-4] In the case of the pharmaceutical and fine chemical industries, the *E*-factor, the amount of waste generated per unit mass of product, could be as high as 50–100,[2] which could be linked to their multi-step, multi-campaign production facilities with minimum use of reusable heterogeneous catalytic processes.

"Green Chemistry" as a concept was introduced in the early 1990s by the US Environmental Protection Agency (EPA) in order to promote environmentally benign chemical processes or technologies. However, the concept is not new and the genesis of both chemical engineering and green chemistry can be traced[5] back to the famous Solvay soda ash process which came into being during the 1860s, as an alternative to the polluting LeBlanc process which was advocated and practiced then when no synthetic ammonia was manufactured. Green chemistry aims at reducing/eliminating the environmental impact of chemical processes by adopting the so-called dozen principles covering waste minimization strategies and sustainability.[4] Waste minimization embraces the reduction of consumption of raw materials, solvents, *etc.* In addition, green engineering principles[6] have been introduced, some of which are common to green chemistry, to develop inherently safer chemical manufacturing processes that reduce or eliminate the hazards associated with materials and operations used in the processes.[7] Catalysis is one of the foremost principles of green chemistry and both chemical and biocatalysis have been advocated in the manufacture of a variety of chemicals. Biotechnological products can contribute to more sustainable processes in a number of industries. Enzymes are already being applied in industries such as chemicals, pulp and paper, food processing, mining, consumer goods, and textiles. They are also very good biocatalysts to make chemical processes selective.

White biotechnology is a European term used for modern industrial biotechnology, which employs microorganisms such as yeast, molds, bacteria, genetically modified strains and enzymes derived there-from for the production of chemicals.[8] White biotechnology has become much more widely attractive due to newly developed genetic techniques. Numerous enzyme variations can be created rapidly and then screened for catalysis of the desired industrial reaction.[9] Over the last two decades, several studies have shown that isolated enzymes or microbes can be used as catalysts for chemical transformation, either standalone or in conjunction with chemical steps in a multi-process scheme. White biotechnology has since evolved successfully for a variety of processes from food to active pharmaceutical ingredients (APIs) to pharmaceuticals.[10-12] Thus, green chemistry will play an important role in the development of white biotechnology. Among the dozen principles of green chemistry, atom economy, catalysis and the use of renewable resources are the primary principles to reduce waste and develop environmentally benign processes which will also be cost-effective. In this regard, process intensification with the incorporation of innovative approaches in separation and reactor design will aid white biotechnology.

In the cases of pharmaceuticals and fine chemicals, five different types of selectivities – *chemo, regio, stereo, enantio*, and *diastereo* – can interplay in a process, because of the presence of multiple functional groups and chiral centers, and there are limitations on the use of synthetic organic chemistry alone in such practices.[5] White biotechnology thus provides a tremendous opportunity for the manufacture of a spectrum of chemicals, within the ambit of green chemistry, ranging from bulk chemicals to APIs, across various industries.[13] Enzyme catalysis in particular will be of great use when coupled with microwave energy to augment both the reaction rate and selectivity, thereby reducing the reactor volume and processing cost. In this chapter, the focus is on the use of microwave irradiation in enzyme catalysis to intensify processes as a green chemistry tool in white biotechnology.

4.2   Enzyme Catalysis

Enzymes are well known as natural catalysts for various biochemical reactions that occur in living cells. Apart from biochemical reactions, enzymes/whole cells have been employed for the production of cheese, beer, wine, vinegar and the like.[14] However, a major breakthrough in enzyme catalysis was realized after the first determination of the 3-D structure of an enzyme using X-ray diffraction (XRD), followed by the identification of the interactions of enzymes with substrates and inhibitors. It was found that the rate enhancement by enzyme catalysis is due to the non-covalent interaction of enzyme active sites and substrates, which is the same as the conventional chemical catalytic mechanism.[15] It thus became routine to use enzymes as conventional catalysts for organic or bulk chemical synthesis.[16] Enzyme catalysis has several advantages over chemical catalysis. For instance, enzymes work under mild reaction conditions such as low temperature and pH, and are capable of accepting a wide range of chemicals as substrates; there is better selectivity towards the desired product or specific functional group, with minimum formation of by-products or their complete suppression. Thus, enzyme catalysis leads to simple and easy downstream processing. Enzymes can provide stereo-, diastereo- or regio-selectivity in complex syntheses involving multi-functional molecules. Therefore, biocatalysts can be used for the synthesis of highly complex molecules without the necessity of cumbersome blocking and deblocking steps. Conventional chemical catalysts are not capable of providing some of these advantages or would need additional separation stages. Enzymes reduce the energy requirements and operation costs on an industrial scale. Furthermore, they are easily biodegradable and typically lead to reduced or no toxicity when released into the environment after use in industrial production. Because of these attributes, enzymes are viewed as environmental friendly alternatives to conventional chemical catalysts. Enzymes have generated huge interest among diverse industries and academic groups.[17-19] Over the years, thus, thousands of enzymes have been identified from various resources and reported in enzyme databases. All of these enzymes have been classified into six different categories:[20] (i) *oxidoreductases*: transfer electrons from one compound to another (*e.g.* alcohol dehydrogenase, laccase, peroxidases); (ii) *transferases*: transfer functional groups from one compound to another (*e.g.* methyl transferases, acyl transferases); (iii) *hydrolases*: perform addition of water molecules to esters and amides, and epoxidation of C=C bonds (*e.g.* lipase, protease, epoxy hydrolase); (iv) *lyases*: perform non-hydrolytic bond cleavage or addition of HX to double bonds such as C=C, C=N and C=O (*e.g. aldolases, oxynitrilases*); (v) *isomerases*: perform intramolecular rearrangement within molecules (*e.g. cis–trans isomerases, racemases, epimerases*); (vi) *ligases*: form bonds such as C–C, C–O, C–N, C–S or phosphate bonds between two compounds (*e.g. synthetases*).[20]

Notwithstanding the merits of enzymatic processes as potentially attractive catalysts to industry to meet environmental laws, the inherent drawbacks associated with the use of enzymes must be overcome before they are used on a commercial scale and sustainably. The identification of a suitable enzyme, which involves tedious screening from different sources like microorganism, plant and animal sources, or for that matter, genetic modification, is costly and time-consuming. Secondly, the isolation and purification of the required enzyme is also expensive. Most industrial processes operate at high temperature and varying pH, with the use of organic solvents and high substrate and product concentrations, which are in direct contrast to most enzymes which are not tolerant of these severe conditions. Enzymes are effective under mild conditions, low substrate concentrations and aqueous conditions. Sometimes enzymes will lose their stability when used under unnatural conditions. A further handicap of enzymatic processes is the scale up from laboratory to industrial level. Scale up requires not only a systematic understanding of the interaction between the enzyme and substrates, but also the physical environment, type of reactor and the operating conditions like pH, temperature and agitation speed need to be optimized for industrial applications.[21,22] However, there are a number of ways by which the activity and stability of an enzyme can be improved, such as immobilization,[23] medium engineering,[24] the use of non-thermal heating like microwave irradiation,[25] or by molecular level changes using protein engineering or enzyme evolution.[26,27]

4.3 Microwave Irradiation

In the electromagnetic spectrum, the microwave radiation region is located between infrared radiation and radio waves (Figure 4.1).[28,29] Microwaves have frequencies between 0.3 GHz and 300 GHz, corresponding to wavelengths between 1 mm and 1 m, respectively. Virtually all domestic and commercial equipment today uses a frequency of 2.45 GHz (wavelength 12.2 cm)

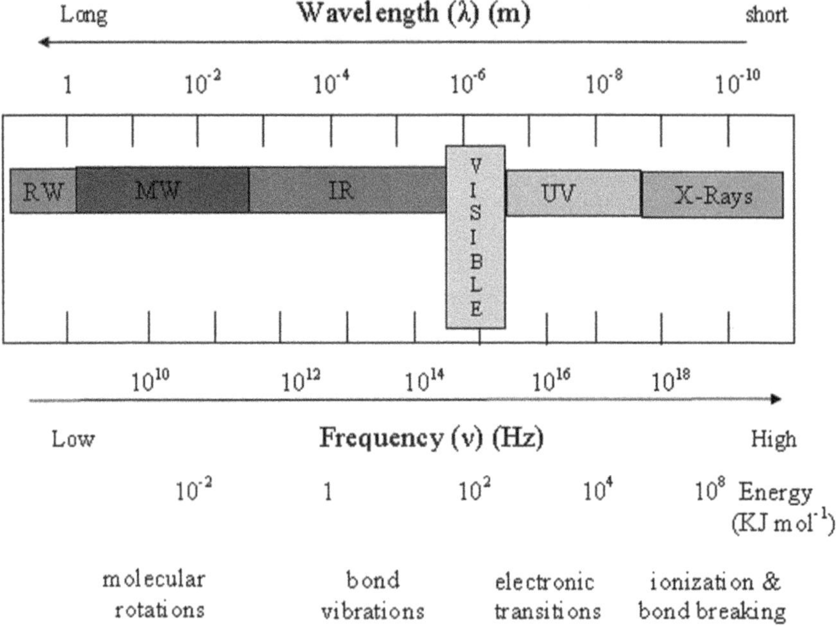

Figure 4.1   Electromagnetic spectrum.

for operation. At this microwave frequency, the electromagnetic waves can induce molecular rotations without breaking/altering the atomic bonds.[29]

4.3.1   Brief History of Microwave Technology

It is pertinent to make reference to the development of microwave technology over the years since it was used in a high power magnetron for radar signaling and communication by Spencer (Table 4.1). CEM developed the batch type microwave reactor for laboratory use in 1978. It has come a long way since then, and its first use for organic synthesis was reported in 1986 by Gedye, Majetich and Giguere, hailing from different universities.[30,31] They observed that under microwave irradiation, the rate of reaction was enhanced greatly, reducing the reaction time by orders of magnitude vis-à-vis conventional thermal heating. Thereafter, sporadic activity commenced with researchers publishing a huge amount of data on reactions in green synthesis using domestic microwave ovens until the advancement of mono- and multi-mode commercial equipment and ever since, the investigation of microwave energy as an alternative heating method for various organic reactions and materials synthesis has continued.[30,31]

The advantages of microwave technology have been extended to other industrial applications; for instance, thermal molding of wood and plastics, drying of medicinal plants and fibers, sintering of ceramics, cancer treatment,

Table 4.1 Brief history of microwave development.[30]

Sr. no.	Year	Development
1	1946	Percy Le Baron Spencer discovered microwave radiation as a method of heating
2	1947	First commercial domestic microwave oven was developed
3	1978	First microwave instrument for laboratory research was developed by CEM Corporation to analyze moisture in solids
4	1980–82	Microwave radiation was developed to dry organic materials
5	1983–85	Microwave radiation was used for chemical analysis
6	1986	Robert Gedye, Laurentian University, Canada, George Majetich, University of Georgia, USA, and Raymond Giguere of Mercer University, USA published papers relating to microwave irradiation in chemical synthesis
7	1990s	Microwave chemistry emerged and was developed as a field of study for applications in chemical reactions
8	1990	Milestone s.r.l. generated the first high pressure vessel (HPV 80) for performing complete digestion of difficult to digest materials like oxides, oils and pharmaceutical compounds
9	1992–1996	CEM developed a batch system(MDS 200) reactor, and a single mode cavity system (Star 2) that were used for performing chemical synthesis
10	1997	Milestone s.r.l. and Prof. H. M. (Skip) Kingston of Duquesne University published a reference book titled "Microwave enhanced chemistry—Fundamentals, sample preparation and applications" edited by H. M. Kingston and S. J. Haswell
11	2000	First commercial microwave synthesizer was introduced to conduct chemical synthesis

waste water treatment, drying and sterilization of foodstuffs, vulcanization of rubber, pretreatment of biomass for bio alcohol production, *etc.*[32]

4.3.2 Microwave Principles

Microwave energy transfer is proposed to occur *via* two fundamental mechanisms, dipole polarization and ionic induction.[33]

4.3.2.1 Dipole Polarization

According to this mechanism, the interaction between the electric field component and polar molecules takes place when polar molecules are exposed to electromagnetic irradiation.[33] Polar molecules try to align themselves with the applied alternating electric field. However, due to inter-molecular forces, polar molecules face some degree of resistance to their motions and are not kept in phase with the applied field. They collide with each other and lose their absorbed energy in the form of heat. Both the frequency of the applied field and the viscosity of the solution influence the intermolecular force between molecules and their ability to align with the applied field.

In high frequency radiation, field oscillation is very rapid and intermolecular forces are strong enough to stop the motion of polar molecules before they respond to the applied field, resulting in inadequate interactions and no energy being transformed into heat. In contrast, at low frequency radiation, the intermolecular forces are weak and polar molecules have enough time to align with the applied field. Although molecules absorb some energy, there is less chance for random interaction or collision of molecules and therefore, low heat formation results. Thus, an appropriate frequency of irradiation is required to maintain sufficient oscillation between molecules. Microwave radiation with a frequency of 2.45 GHz is capable of aligning polar molecules with an applied field. The oscillation of dipoles under an applied alternating field induces collision or friction between polar molecules. The energy losses in this process result in dielectric heating.[33]

4.3.2.2   *Ionic Conduction*

According to the ionic conduction mechanism, strong interactions occur between electric components and ions in the medium of a sample.[33] The sample ions move in solution with the applied electric field. The oscillating electromagnetic field generates an oscillation of electrons or ions in solution. While in oscillation, ions collide with each other or provide friction to the applied field and lose their gained energy which is converted into heat. This property of ionic species can be used to improve the heating ability of non-polar solvents under microwave irradiation. The energy loss in ionic samples is much higher in the ionic conduction mechanism than dipole interaction.[33] A major limitation of this method is that it is not applicable to materials with high conductivity, since such materials reflect most of the energy that falls on them.

4.3.3   Interaction Between Microwave Irradiation and Reaction Medium

Rate enhancement in a chemical reaction is influenced by the thermal effects of microwave irradiation. The thermal effect is the heat energy produced by the selective absorption of microwave irradiation by the reactants/reaction medium or both. The magnitude of the heat energy produced is directly correlated with the dielectric properties of the materials being heated, and the microwave penetration depth at a given frequency and temperature.[34,35] The dielectric constant or complex permittivity ε^* is the amount of microwave absorbance and energy stored in a material and is given by:

$$\varepsilon^* = \varepsilon' + i\varepsilon'' \tag{4.1}$$

where ε' is the dielectric constant of the material and ε'' is the dielectric loss of the material and is defined as the ability of the material to convert absorbed microwave energy into heat. The ratio of the dielectric loss to the dielectric constant is known as the loss tangent (tan δ) or dissipation factor (D) which is given as:[35]

$$D = \tan \delta = \frac{\varepsilon''}{\varepsilon'} \tag{4.2}$$

The tangent loss or dissipation factor is used to describe the ability of a solvent to convert electromagnetic energy into heat at a given frequency and temperature. The efficiency on conversion of microwave energy into thermal energy depends on both the dielectric and thermal properties of the material. The fundamental relationship is:

$$P = \sigma|E|^2 = (\omega\varepsilon_0\varepsilon'')|E|^2 = (\omega\varepsilon_0\varepsilon'D)|E|^2 \tag{4.3}$$

where ε_0 is permittivity of free space, P is the power dissipation per unit volume, σ is the conductivity of the material, E is the strength of the electric field in the sample and ω is the angular frequency.

Assuming negligible heat loss and diffusion, the rate of heating or temperature rise (ΔT) in a time interval can be expressed as:

$$\frac{\Delta T}{t} = \frac{\sigma|E|^2}{\rho C} = \frac{(\omega\varepsilon_0\varepsilon'D)|E|^2}{\rho C} \tag{4.4}$$

where ρ is the density, σ is the conductivity of the molecule and C is the specific heat capacity.

The absorbance capacity of organic solvents used in microwave chemistry can be classified into three categories based on the tangent loss as high- (tan δ > 0.5), medium- (tan δ = 0.1–0.5), and low-absorbing (tan δ < 0.1) (Table 4.2).

Highly polar solvents such as water, methanol, dichloromethane and acetonitrile easily absorb microwave irradiation due their high dielectric constants. In contrast, non-polar solvents such as hexane, heptanes and

Table 4.2 Classification of solvents according to microwave radiation absorbance.[29]

Sr. no.	MW absorbance level	Solvent
1	High	Dimethylsulfoxide, ethanol, methanol, propanol, nitrobenzene, formic acid, ethylene glycol
2	Medium	Water, DMF, NMP, butanol, acetonitrile, HMPA, MEK, acetone and other ketones, nitromethane, *o*-dichlorobenzene, 1,2-dichloroethane, 2-methoxyethanol, acetic acid, trifluoroacetic acid
3	Low	Chloroform, dichloromethane, carbon tetrachloride, 1,4-dioxane, tetrahydrofuran, dimethoxyethane (glyme) and other ethers, ethyl acetate, pyridine, triethylamine, toluene, benzene, chlorobenzene, xylene, pentane, hexane and other hydrocarbons

iso-octane have no dielectric properties or are transparent to microwaves and cannot be employed as solvents in microwave assisted reactions unless the reaction mixture contains other dielectric components, or additives such as ionic liquids or passive heating elements made of strongly microwave absorbing materials are added, which in turn increase the absorbance level of these solvents.[27] Another parameter which affects the microwave heating is the microwave penetration depth (D_p), which is defined as the depth into the material where the power is reduced to ~1/3 of its original intensity. The penetration depth (D_p) of an electromagnetic wave is inversely proportional to the absorption coefficient (α), which is dependent on the dielectric constant and dielectric loss of the given compound and is described as follows:

$$D_p = \frac{1}{\alpha} = \frac{\lambda\sqrt{\varepsilon'}}{2\pi\varepsilon''}. \tag{4.5}$$

Thus, the penetration depth decreases with increasing microwave frequency. In the microwave frequency range, the absorption coefficient is moderate and the depth of penetration is on the order of the cm to mm range, which results in consistent heating of the reaction mass. For example, at 2.45 GHz, the usual frequency used in microwave reactors, the penetration depth of a microwave at room temperature is 13 mm for water.

The rapid increase in the rate of reaction under microwave irradiation is not due to the thermal effect alone, some other effects also play an important role such as hot spots, selective heating and specific microwave effects (due to non-thermal effects). In the case of microwave effects, energy transformation from microwave irradiation is directly coupled to conformational changes in molecules or superheating of the reaction medium to levels above its boiling point. However, this is hotly debated due to the lack of clear data associated with experimental difficulties in discriminating thermal and non-thermal effects.[36] Several reports have appeared on the existence of non-thermal effects of microwaves in both chemical and biocatalysis.[37,38]

4.3.4 Microwave Heating *vs.* Conventional Heating

Conventional heating sources are oil baths or heating mantles for chemical synthesis, wherein heat is transferred through a conductive mechanism in series through the vessel wall to the mixture of solvent and reactants. This mode of heat transfer is dependent on the thermal conductivity of the material of construction. There is a large temperature drop in the reaction mixture compared to the outer surface, which leads to delays in reaching the desired temperature, and therefore control over the reaction becomes difficult, which may lead to the formation of by-products. In contrast, microwave radiation is transparent to the reaction vessel. Microwaves easily enter the reaction vessel and couple directly with polar molecules, leading to a fast rise in temperature. The rate of energy transmission is dependent on the dielectric loss of the reaction components. Thus, it is totally independent of the conductivity of the material of construction. The

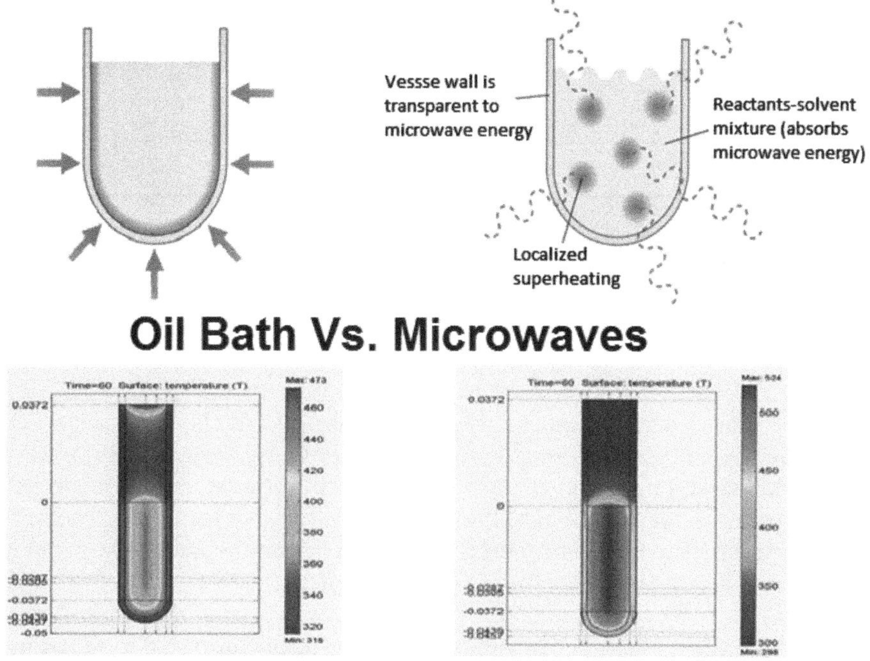

Figure 4.2 Comparison of microwave heating with conventional heating[130] (source: http://lab-plusinternational.com/).

vessel and the molecules of the reaction mass get heated simultaneously. Microwave heating is very fast. The temperature gradient is thus exactly opposite and starts from the reactants and solvent rather than at the reaction vessel (Figure 4.2). Many organic reactions which are slow in nature can be performed under microwave heating to enhance their rates, improve yields and modify selectivities. With the help of microwave irradiation, it is also possible to carry out reactions that do not otherwise occur under conventional heating.[29–32]

4.3.5 Application of Microwaves in Enzymatic Reactions – Green Chemistry Approach

Stringent environmental legislation has necessitated the generation of cleaner methods of chemical production that reduce or, preferably, eliminate the generation of waste of both material and energy, and avoid the use of toxic and/or hazardous reagents and solvents, which all fall under the ambit green chemistry.[4] Catalysis is one of the most important principles of green chemistry. For heterogeneous catalysis, whether supported chemical entities or immobilized, enzymes play a very significant role in the design and development of safer and economical processes of industrial relevance, and give added advantages of ease of separation, mild reaction conditions, better selectivity, waste minimization, less expensive construction materials, *etc.*[1–6]

Temperature plays a dominant role in the conversion, yield and selectivity of any chemical reaction, in the absence of transport effects, and thus, the means of delivery of heat energy and the timing of its introduction into the reactor will be of paramount importance. Microwave irradiation has enormous potential to supply controlled energy directly to the molecules of interest in the reaction mass, and has been established as a green and sustainable alternative to conventional heating. Although biocatalytic transformations are considered as eco-friendly processes, they are still too sluggish to attain the desired conversion and yield.

Low temperature microwave irradiation technology can reduce or eliminate the side reactions caused by severe thermal effects, and is applied widely because of its high speed and safety. There are several ways used to achieve low temperature, such as adjusting the microwave radiation source, use of a cooling medium, changing the physical properties of the reactants and controlling the initial temperature of the reaction medium. A large body of literature exists on the chemistry of this aspect.

Enzymatic reactions conducted under microwave irradiation are faster, cleaner, and offer much better yields in shorter time compared to reactions carried out under conventional heating. Microwave assisted enzymatic reactions have easy work-ups and reduced times, because of which there is a saving of energy, improved enzyme stability, and less solvent consumed. Furthermore, in combination with solvent-free conditions or in neoteric solvents like ionic liquids, microwave methods can offer an efficient and safe technology conforming to green chemistry principles.[29,37] Some proteins exposed to microwave radiation do not undergo protein denaturation and some natural products can be extracted without degradation and oxidation.

4.3.5.1 *Initial Studies on the Application of Microwave Irradiation to Enzymatic Reactions*

The number of instruments based on microwave technology has rapidly increased for a variety of applications. Different types of system ranging from living organisms to purified enzymes have been studied to assess the specific effects (non-thermal effects) of microwave irradiation on biological processes.[38]

Reactions were carried out in aqueous medium (in buffer solutions) to study the effects of microwaves in comparison with conventional heating by keeping all other parameters constant. Enzymatic activity was monitored either during or after the exposure to the particular heating mode. The effect of microwave irradiation has been studied on the hydrolytic activity of acid phosphatase[39] and acetylcholinesterase[40] in aqueous media and there was apparently no change in activity after exposure to microwave irradiation. Homenko *et al.*[41] have studied the effects of microwave irradiation spanning the frequency range 300 MHz–100 GHz and reported that it lacks sufficient energy to break chemical bonds. Whether this range of frequency is detrimental to cellular dysfunction, according to these authors, may be difficult to explain using complex systems such as whole animals, cells, or cell extracts.

Orlando *et al.*[42] studied the activity of ascorbate oxidase in both free and encapsulated form at 25 °C under microwave irradiation and showed that the free enzyme in solution has no effect, but the encapsulated form shows reduced activity. Kamerke *et al.*[44] reported the microwave assisted synthesis of nucleotide activated oligosaccharides by using β-galactosidase isolated from *Bacillus circulans*. All reactions were performed in a temperature controlled instrument at 30 °C. The loss of hydrolytic activity of the enzyme was found to be very rapid and also time dependent. Complete loss of activity occurred within 30 min, compared to conventional heating, where 20% residual activity was retained after 60 min. Similarly Cara *et al.*[45] studied the specific effects of microwave irradiation on the hydrolytic activity of thermo-stable β-galactosidase isolated from *Bacillus acidocaldarius* at 70 °C. This enzyme is highly thermo-stable and it is possible to conduct the reaction at high temperature when a sufficient amount of microwave energy is consumed to reach the desired temperature. The activity of the enzyme decreased as the duration of microwave exposure increased, whereas the activity remained constant in conventional heating. This clearly indicated that the microwave heating induced non-thermal effects on enzyme activity. However, microwave induced enzyme inactivation was observed at low concentration of enzyme loading only, and the same effect was not found at higher enzyme loading.[45] Similar kinds of experiments were performed by Porcelli *et al.*[46] using *S*-adenosylhomocysteine hydrolase and 5'-methylthioadenosine phosphorylase isolated from thermo-stable archaea, *Sulfolobus solfataricus* under 10.4 GHz microwave irradiation. All reactions were performed in the range of 70–90 °C in order to differentiate thermal and non-thermal effects. Both enzymes undergo irreversible and time dependent inactivation. The extent of inactivation is different for both enzymes. *S*-Adenosylhomocysteine hydrolase showed greater sensitivity than 5'-methylthioadenosine phosphorylase at the same temperature and time compared to a control incubated experiment without irradiation. Rutin was subjected to intermolecular transglycosylation assisted by microwave irradiation using cyclodextrin glucanotransferase (CGTase) produced from *Bacillus* sp. SK13.002. Compared with the conventional enzymatic method, microwave assisted reaction was much faster and thus more efficient.[48] Enzyme activity increased due to a sharp increase in collision frequency under microwave heating.

Pavelkic *et al.*[47] studied the influence of microwave heating (2.45 GHz) on enzyme kinetics by using porcine pepsin catalyzed hydrolysis of bovine serum albumin–bromophenol blue as substrate and observed a significant reduction of enzyme activity (45.9%) after exposure for 20 min. Furthermore, they analyzed the conformational changes in pepsin after microwave irradiation using SDS-PAGE. They found a more diffuse, wider band and the presence of small fragments compared to the control experiment. Inactivation and fragments in the band are due to autodigestion of enzyme molecules induced by microwave irradiation. In contrast, amylase and trypsin have shown good catalytic activity for starch and protein hydrolysis, respectively.[49,53]

In all these cases, the specific effect (non-thermal) of microwave radiation has been found to be a negative effect on enzyme catalysis. All studies have been done using free enzyme in aqueous media, in which case microwave irradiation easily penetrates into the aqueous medium due to its high dielectric properties. It is well known that protein's secondary structures are preserved by hydrogen bonding between the carbonyl and amide groups of a peptide. Microwave radiation does not possess sufficient energy to break or modify peptide bonds in enzyme molecules.[45] However, it has sufficient energy to break and reorient the hydrogen bonds between peptide chains and peptide–water. Microwave irradiation can affect hydrogen bonding, van der Waals forces and hydrophobic bonds. The consequent loosening, breakage and recombination of hydrogen bonds alter the orientation of the hydrogen bonds that originally maintained a stable secondary structure. Thus, hydrogen bonds are redistributed, changing the conformation and activity of the enzyme. However, in some cases, this will have a negative influence such as reduction or loss of activity upon microwave irradiation.[45,48] Rearrangement of hydrogen bonds could also allow more flexible and easy access of reactants to the active site, leading to a positive influence on the enzyme activity.[51–53] In the case of amylase, during microwave irradiation, it was observed that the α-helix content gradually decreased and the content of β-sheet turns and non-regular coils increased. These changes made the enzyme structure more flexible and led to improved activity.[49] Unlike free enzymes, those immobilized on a solid support have shown different behavior in aqueous media. Alkaline phosphatase immobilized on a 96-well plate showed no decrease in activity in the hydrolysis of *p*-nitrophenylphosphate.[41] Similarly, β-galactosidase isolated from *Kluyveromyces lactis* immobilized on Duolite A-568 showed improved catalytic activity for the synthesis of galacto-oligosaccharides. The yield improved from 38 g L^{-1} to 44 g L^{-1} under microwave irradiation.[43] Yu *et al.*[54] studied the influence of microwave irradiation on glucose isomerase for isomerization of glucose to fructose and observed it to have a pronounced effect (Scheme 4.1). The microwave assisted reaction generated the equivalent yield of fructose in a relatively shorter time compared with conventional heating. Similarly, different immobilized lipases have shown

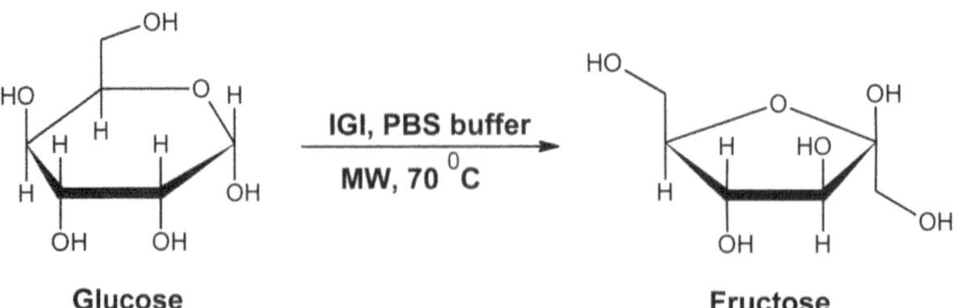

Scheme 4.1   Glucose isomerase catalyzed isomerism of glucose to fructose.[54]

Table 4.3 Microwave assisted enzymatic reactions in aqueous phase.

Sr. no.	Enzyme	Temperature (°C)	Reaction type	Effect	References
1	Acid phosphatase	60	Hydrolysis	No effect	39
2	Acetylcholinesterase	25	Hydrolysis	No effect	40
3	Calf alkaline phosphatase	25	Hydrolysis	Low and significant	41
4	Ascorbate oxidase	25	Oxidation	No or reduced activity	42
5	Beta-galactosidase	40	Hydrolysis/ transglycosylation	No effect/loss of activity	44 and 45
6	*S*-Adenosylhomocysteine hydrolase and 5′-methylthioadenosine phosphorylase	70–90	Hydrolysis	Inactivation	46
7	Porcine pepsin	37	Hydrolysis	Reduced activity	47
8	Cyclodextrin glucanotransferase	40	Transglycosylation	Significant	48
9	Beta-glycosidase	50 and 65	Transglycosylation/ reverse hydrolysis	Significant	50
10	Lactate dehydrogenase	23	Conversion of pyruvate to lactate	Improved activity up to 900 MHz, then activity decreased	51
12	Alpha-amylase, glucoamylase	62	Hydrolysis	Significant	49 and 52
13	Trypsin	60–90	Protein digestion	Fast	53,56–58
14	Glucose isomerase	70	Isomerization	Fast	54
15	Lipase	37	Hydrolysis	Improved activity	55

good conversion and enhanced activity within a short time for biodiesel production and bio-esterification.[55] The results are summarized in Table 4.3.

4.4 Application to Different Industrially Relevant Reactions

Despite the controversy about the effects of microwave irradiation on enzymatic reactions, a significant number of reports have appeared on the application of microwave assisted enzymatic reactions relevant for different industries.

The two important factors that have significant contributions to the effects of microwaves on enzyme catalysts are: (i) dielectric properties and (ii) polarity of the reaction medium. In the former case, as was pointed out earlier, microwaves induce collision between polar substances/reactants which subsequently increase the conversion and rate of reaction; whereas in the latter case, energy released by oscillation of the polar solvent causes changes in interaction between the enzyme structure and its microenvironment. This leads to easy accessibility of reactants to the active sites which further increases the rate of reaction. Lipase catalyzed reactions have been extensively reported and have been widely used for organic synthesis, and the majority of these studies are based on lipase catalyzed reactions in non-aqueous media or ionic liquids.

Here, we will discuss the combined use of microwave irradiation and enzyme catalyzed reactions in different industries (Table 4.4).

4.4.1 Microwave Assisted Enzymatic Hydrolysis for Proteomics

The use of microwaves in enzyme applications began with the use of enzymatic hydrolysis reactions for protein digestion in proteomic analysis. Chen and coworkers[56] used microwave irradiation to enhance the rate of hydrolysis of peptides and proteins in a 6 M HCl solution in a closed system. Pramanik *et al.*[57] successfully demonstrated the complete digestion of several known biologically active proteins within a few minutes, compared to several hours using the conventional method. Microwave assisted protein digestion showed a complete digestion profile even for tightly folded proteins like ubiquitin, which usually require a longer time for enzymatic digestion under normal methods. Lin *et al.*[58] prepared trypsin immobilized magnetic nanoparticles for proteomic analysis. The efficiency of microwave assisted protein digestion was studied under aqueous conditions using three different protein molecules, bovine serum albumin, myoglobin and cytochrome c. The microwave assisted digestion was completed in 15 s in a domestic microwave oven (1100 W) in open system mode, compared to conventional heating, which took 12 h for equivalent digestion. The nanoparticles showed good reusability after the fourth reuse. Furthermore, they showed the same digestion capability for a crude protein mixture isolated from rat liver. Similarly, Lin *et al.*[59] studied the microwave assisted trypsin catalyzed standard known protein digestion in organic solvents like acetonitrile, methanol and a combination of methanol and chloroform with water. Under microwave irradiation, the proportion of digested protein increased with increasing acetonitrile content, indicating that acetonitrile did not deactivate the enzyme. In comparison, the percentage of protein digested decreased as the methanol content increased, suggesting a loss in the enzyme's activity in this solvent. However, the enzyme showed a loss of activity in the medium containing a combination of methanol and acetonitrile under conventional heating. In all cases, microwave irradiation increased the digestion efficiencies compared to

Table 4.4 Microwave assisted enzymatic reactions in non-aqueous phase.

Sr. no.	Enzyme	Temp (°C)	Reaction type[a]	Reaction scheme	Product	Solvent	MW effect	Industrial use	References
1	Trypsin	37	H	protein + H_2O $\xrightarrow[\text{MW, 37°C}]{\text{Trypsin}}$ amino acid	Cytochrome c, lysozyme, myoglobin	Acetonitrile	Yes	Proteomic analysis	59
2	Lipase	40	E	$CH_3(CH_2)_7CH=CH(CH_2)_7COOH + EtOH$ $\xrightarrow[\text{MW, 40°C}]{\text{Lipase}}$ $CH_3(CH_2)_7CH=CH(CH_2)_7COOEt + H_2O$ Ethyl Oleate	Ethyl oleate	Cyclohexane	No effect	Biodiesel	60
3	Lipase	40	T	Macauba oil + EtOH $\xrightarrow[\text{MW, 40°C}]{\text{Lipase}}$ Fatty acid ethyl ester + Glycerol	Fatty acid ethyl ester	Solvent-free	Yes	Biodiesel	61
4	Lipase	40	T	Soybean oil + MeOH $\xrightarrow[\text{MW}]{\text{Lipase}}$ Fatty acid methyl ester + Glycerol	Methyl ester of fatty acid	Solvent-free	Yes	Biodiesel	62
5	Lipase	60	T	Soybean oil + MeOH $\xrightarrow[\text{MW +IL}]{\text{Lipase}}$ Fatty acid methyl ester + Glycerol	Fatty acid methyl ester	Ionic liquid	Yes	Biodiesel	63

(*continued*)

Table 4.4 (*continued*)

Sr. no.	Enzyme	Temp (°C)	Reaction type[a]	Reaction scheme	Product	Solvent	MW effect	Industrial use	References
6	Lipase	50	T	Beef tallow oil + EtOH $\xrightarrow[\text{MW, 50}^{\circ}\text{C}]{\text{Lipase}}$ Fatty acid ethyl ester + Glycerol	Fatty acid ethyl ester	Solvent-free	Yes	Biodiesel	64
7	Lipase	43	T	Palm oil + EtOH $\xrightarrow[\text{MW}]{\text{Lipase}}$ Fatty acid ethyl ester + Glycerol	Fatty acid ethyl ester	Solvent-free	Yes	Biodiesel	65
8	Lipase	40	H	Waste cooking oil + H_2O $\xrightarrow[\text{MW, 40}^{\circ}\text{C}]{\text{Lipase}}$ Fatty acid + Glycerol	Hydrolysis of waste cooking oil	Bi-phasic medium (oil/buffer)	Yes	Biodiesel	66
9	Lipase	86	T	gamma-Caprolactone + H_2O $\xrightarrow[\text{Microwave}]{\text{Lipase, Solvent}}$ Polymer	Caprolactone	Benzene	Yes	Polymer	67
10	Lipase	90	T	gamma-Caprolactone + H_2O $\xrightarrow[\text{Microwave}]{\text{Lipase}}$ Polymer	Caprolactone	Solvent-free	Yes	Polymer	68
11	Lipase	60	E	$HOOC(CH_2)_4COOH$ (Adipic acid) + C_4H_9OH (n-Butanol) $\xrightarrow{\text{Lipase}}$ $C_4H_9OOC(CH_2)_4COOH$ (Monobutyl adipate); $C_4H_9OOC(CH_2)_4COOH$ (Monobutyl adipate) + C_4H_9OH (n-Butanol) $\xrightarrow{\text{Lipase}}$ $C_4H_9OOC(CH_2)_4COOC_4H_9$ (Dibutyl adipate)	Adipic acid esters	1,4-Dioxane	Yes	Chemical	69
12	Lipase	30	E	CH_2OH–$CH{-}OH$–CH_3 + $C_{11}H_{23}COOH$ $\xrightarrow{\text{Lipase}}$ $CH_2OOCC_{11}H_{23}$–$CH{-}OH$–CH_3 + H_2O	Mono 1,2-propyl laurate	1,4-Dioxane	Yes	Chemical	70

No.	Enzyme	Temp (°C)	Type	Reaction scheme	Application	Solvent	Effect		Ref.
13	Lipase	40	E	Lactic acid + EtOH → Ethyl lactate + H_2O (Lipase, Ionic liquid, MW, 40 °C)	Ethyl lactate	Ionic liquid	Yes	Chemical	71
14	Lipase	70	E	Starch (R'-OH) + Fatty acid → Starch ester + H_2O (Lipase, MW, 70 °C)	Starch esters	Dimethyl formamide	Yes	Chemical	72–75
15	Lipase	50	T	Methyl acetoacetate + Alcohol (R—OH) → alkyl acetoacetate + Methanol (Lipase, MW, Toluene, 50 °C)	Alkyl acetoacetate	Toluene	Yes	Chemical	76
16	Lipase	50	T	Keto ester + → + (Lipase, MW, 50 °C)	Keto esters synthesis	Solvent-free	Yes	Chemical	77
17	Lipase	70	T	Methyl acetoacetate + n-Propanol → propyl acetoacetate + Methanol (Lipase, MW, Toluene, 70 °C)	Propyl acetoacetate	Toluene	No effect	Chemical	78
18	Lipase	55	Ep	Styrene → Styrene oxide (Lipase, Toluene, 30% H_2O_2, lauric acid, MW)	Styrene oxide	Toluene	Yes	Chemical	79
19	Lipase	60	1,2 Dihydroxylation	Olefins → 1,2 dihydroxy olefins (Lipase, 50% H2O2, MW)	1,2-Dihydroxyl olefins	Ethyl acetate	Yes	Chemical	80
20	Lipase	60	R	R-N_3 (Azides) → R-NH_2 (Amines) (Lipase, MW, 60 °C)	Amine synthesis	Methanol	Yes	Chemical	81

(continued)

Table 4.4   (*continued*)

Sr. no.	Enzyme	Temp (°C)	Reaction type[a]	Reaction scheme	Product	Solvent	MW effect	Industrial use	References
21	Lipase	60	*N*-Acylation	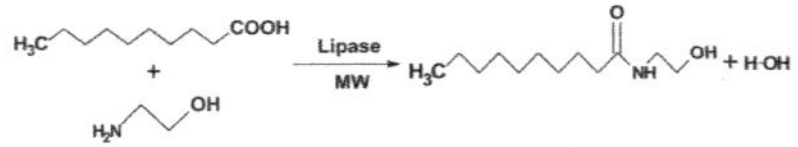	Fatty acid amide synthesis using ethanolamine	1,4-Dioxane	Yes	Chemical	82
22	Xylose isomerase	80	I	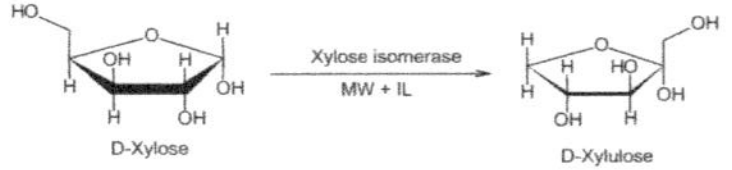	Xylulose	Ionic liquid	Yes	Chemical	83
23	Lipase	80	H	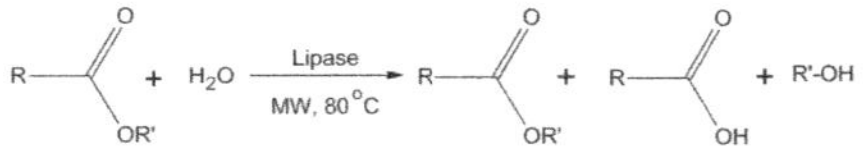	Partial hydrolysis of palm oil	Solvent-free	Yes	Food and cosmetics	84
24	Lipase	95	E	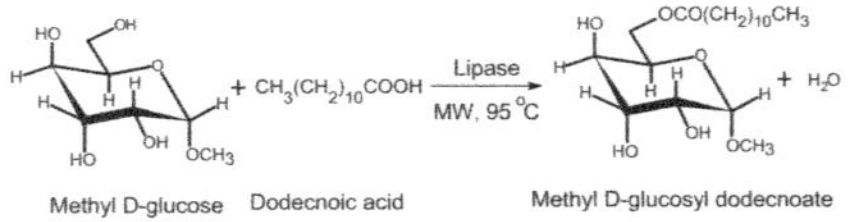	Dodecanoic acid esters of α-D-glucopyranoside	Di-ethyl ether	Yes	Food and cosmetics	85
25	Lipase	50	E	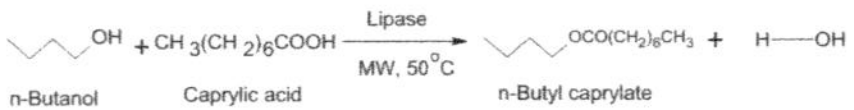	Butyl caprylate	Hexane	Yes	Food and cosmetics	86
26	Lipase	50	E	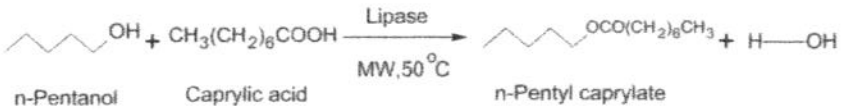	Pentyl caprylate	Octane	MW induce alcohol inhibition	Food and cosmetics	87
27	Lipase	65	E	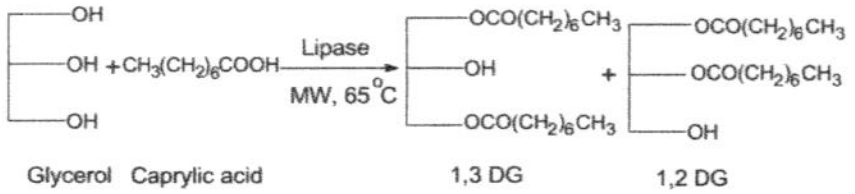	Glyceryl caprylate	Solvent-free	MW induced weak 1,3 specificity	Food and cosmetics	88
28	Lipase	55	E	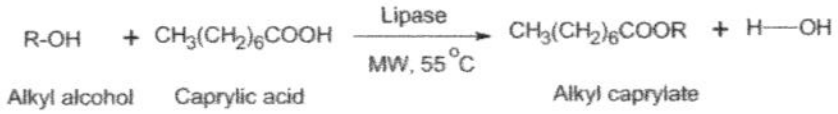	Alkyl caprylate	Solvent-free or hexane	Yes	Food and cosmetics	89

No.	Enzyme	Temp. (°C)		Reaction scheme	Product	Solvent	Effect	Application	Ref.
29	Lipase	55	E	R-OH + CH$_3$(CH$_2$)$_{13}$COOH $\xrightarrow[\text{MW, 55°C}]{\text{Lipase}}$ CH$_3$(CH$_2$)$_{10}$COOR + H—OH; Alkyl alcohol, Lauric acid, Alkyl laurate	Alkyl laurate esters	Hexane	Yes	Food and cosmetics	90
30	Lipase	60	E	Isoamyl alcohol + CH$_3$(CH$_2$)$_{12}$COOH (Myristic acid) $\xrightarrow[\text{MW, 60°C}]{\text{Lipase}}$ OCO(CH$_2$)$_{12}$CH$_3$ (Isoamyl myristate) + H—OH	Isoamyl myristate	Solvent-free	Yes	Food and cosmetics	91
31	Lipase	50	T	Citronellol + Vinyl acetate $\xrightarrow[\text{MW}]{\text{Lipase}}$ Citronellyl acetate + Vinyl alcohol → Acetaldehyde	Citronellyl acetate	Toluene	Yes	Food and cosmetics	92
32	Lipase	60	T	Ethyl cinnamate + Citronellol $\xrightarrow[\text{MW, 60°C}]{\text{Lipase}}$ Citronellyl cinnamate + OH	Citronellyl cinnamate	Heptane	Yes	Food and cosmetics	93
33	Lipase/cutinase	50	T	Ethyl butyrate + n-Butanol $\xrightarrow[\text{MW, 50°C}]{\text{Lipase, Organic media}}$ Butyl butyrate + Ethanol	Butyl butyrate	Organic media	Yes	Food and cosmetics	94
34	Lipase	100	T	Ethyl butyrate + n-Butanol $\underset{\text{MW, 100°C}}{\overset{\text{Lipase, Organic media}}{\rightleftharpoons}}$ Butyl butyrate + Ethanol	Butyl butyrate	Organic solvent	Yes	Food and cosmetics	95
35	Lipase	100	T	Ethyl butyrate + n-Butanol $\underset{\text{MW, 100°C}}{\overset{\text{Lipase}}{\rightleftharpoons}}$ Butyl butyrate + Ethanol	Butyl butyrate	Solvent-free	No effect	Food and cosmetics	96
36	Lipase	40	T	Ethyl butyrate + n-Butanol $\underset{\text{MW, 40°C}}{\overset{\text{Lipase Ionic liquid}}{\rightleftharpoons}}$ Butyl butyrate + Ethanol	Butyl butyrate	Ionic liquid	Yes	Food and cosmetics	97
37	Lipase	Room temperature	H	De-acylation of cephalosporins $\xrightarrow[\text{MW, Basic Alumina}]{\text{FGL}}$ + H$_3$C—COOH	De-acylation of cephalosporins	Aq. ammonia	Yes	Pharma	98

(continued)

Table 4.4 (*continued*)

Sr. no.	Enzyme	Temp (°C)	Reaction type[a]	Reaction scheme	Product	Solvent	MW effect	Industrial use	References
38	Lipase	60	E	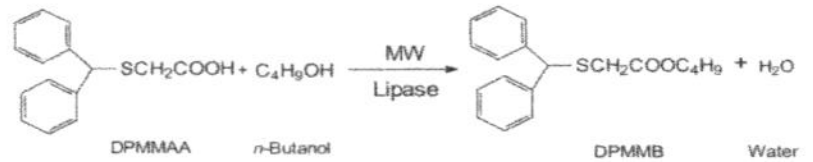	Butyl diphenyl methyl-mercapto acetate	Toluene	Yes	Pharma	99
39	Lipase	60	E	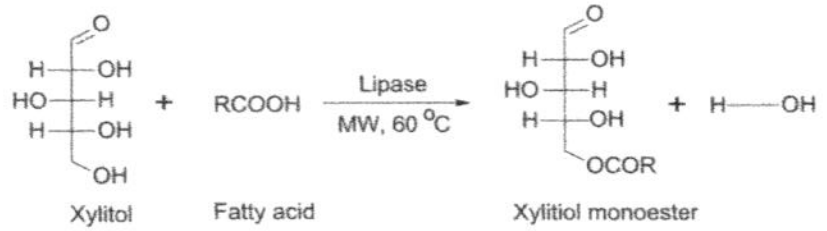	Xylitol mono ester	t-Butanol	Yes	Pharma	100
40	Lipase	30	E	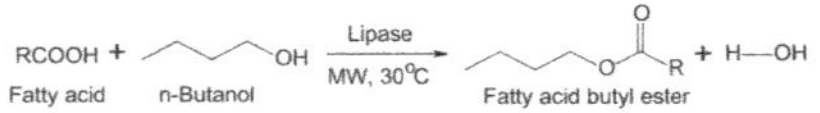	Fatty acid butyl ester	Isooctane	Yes	Pharma	101
41	Lipase	55	E	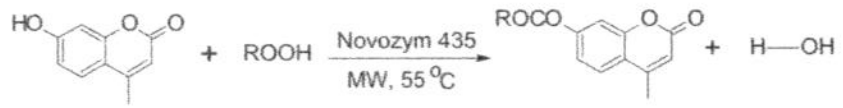	4-Methyl-2-oxo-2H-chromene-7yl acetate	Dry media	Yes	Pharma	102
42	Lipase	80	T	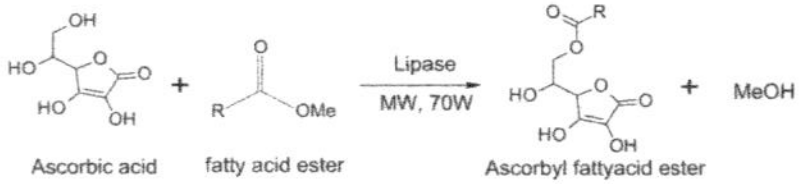	Ascorbyl fatty acid esters	Dry media	Yes	Pharma	103
43	Lipase	60	T	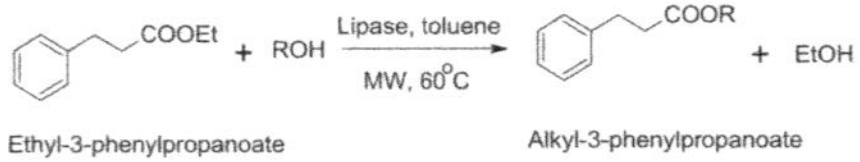	Butyl 3-phenylpropanoate	Toluene	Yes	Pharma	104
44	Lipase	70	E	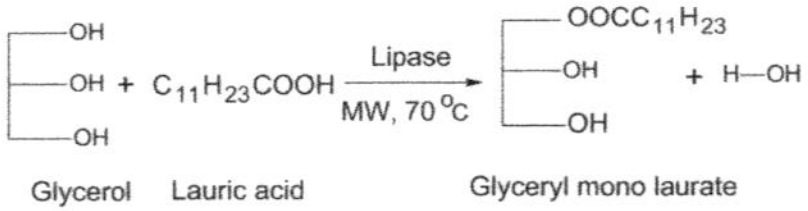	Glycerol mono laurate	Bi-phasic condition	Yes	Pharma	105
45	Lipase	50	N-Acylation	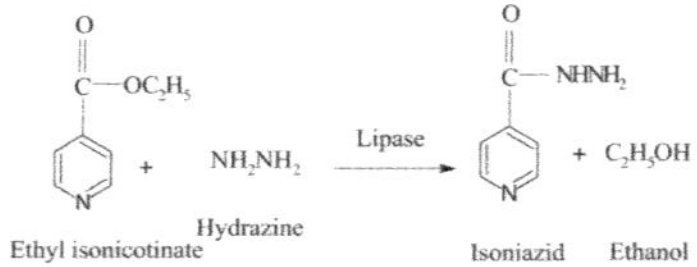	Isoniazid	1,4-Dioxane	Yes	Pharma	106

[a]E – esterification, Ep – epoxidation, H – hydrolysis, R – reduction, T – transesterification.

those obtained under conventional conditions and gave complete digestion within 10 min.[59]

4.4.2 Application of Microwave Irradiation for Enzyme Catalyzed Biodiesel Production

Biodiesel is a mixture of mono-alkyl esters of fatty acids from vegetable oil, animal fat and algae obtained by transesterification reactions. Although well-established chemical catalytic methods are known, they have several drawbacks like the use of high temperature, hazardous chemicals and salt contamination in the product. The use of enzymes for biodiesel production has shown significant advantages compared to chemical catalysis. However, enzyme catalyzed reactions are slow in nature and enzymes are deactivated in the presence of excess water or highly polar compounds in the reaction mixture. In order to reduce the reaction time and improve the stability of enzymes to give high yield, microwave irradiation has been used. Costa *et al.*[60] studied the existence of non-thermal microwave effects in biodiesel synthesis by esterification of oleic acid using ethanol at 40 °C with cyclohexane used as solvent. There was no significant difference in reaction yield between both heating modes and it was concluded that there was no effect of microwave irradiation on biodiesel production.[60] The low reaction temperature and the use of a non-polar solvent could be the reason for this observation. In contrast, several authors have reported that the use of microwaves significantly reduced the reaction time and improved the yield in enzymatic transesterification of various oils with methanol or ethanol. Nogueira *et al.*[61] studied microwave assisted enzymatic transesterification of macauba oil using ethanol under solvent-free conditions at 40 °C and observed an improvement in yield by almost one order of magnitude compared to conventional heating. Similarly, Yu *et al.*[62] studied the influence of microwave irradiation on the production of fatty acid methyl esters by the transesterification of soybean oil with methanol at 40 °C using *t*-amyl alcohol as solvent. Under optimum conditions, 94% of the ester was achieved in 12 h under microwave conditions, whereas conventional heating took almost 24 h to produce the same yield. No loss in enzyme activity was observed after five recycles, showing the stability of the enzyme under microwave irradiation. Furthermore, the same group[63] studied the synergistic effects of ionic liquids and microwave irradiation on this reaction. Different types of ionic liquid were screened to find that 1-ethyl-3-methylimmidazolium hexafluorophosphate gave a significant enhancement in product yield without any loss in enzyme activity under microwave irradiation. Similar results were obtained in the case of lipase catalyzed ethanolysis of beef tallow oil[64] and palm oil.[65] Furthermore, to improve the yield of biodiesel production, Saifuddin *et al.*[66] developed a hybrid method, which used enzymatic hydrolysis followed by acid catalyzed esterification. In enzymatic hydrolysis, the combined three phase partitioning (TPP) method and microwave drying were used to improve the stability and catalytic activity of enzyme particles. Firstly, the enzyme was subjected to TPP and pH tuning, then dried using microwave irradiation (2.45 GHz and power output 100 W) for 10 s and then cooled; this step was repeated 15 times and the tuned enzyme was directly used for hydrolysis of waste cooking oil at 40 °C in bi-phasic medium (oil/buffer). Microwave irradiation was found to increase the initial rate of reaction by at least 1.6 times compared to normal drying.

4.4.3   Application of Microwave Irradiation to Enzyme Catalyzed Polymer Synthesis

Enzymatic synthesis of polymeric esters has been widely reported in recent years. However, the use of microwave heating in enzyme catalyzed polymer synthesis has not yet been thoroughly studied. There is a dearth of literature in this area compared to other industries. Kerep and Ritter[67] studied the effect of microwave irradiation on the *Candida antarctica* lipase catalyzed polymerization of γ-caprolactone in organic media (Scheme 4.2). Highly polar solvents like tetrahydrofuran and 1,4-dioxane showed no conversion, whereas medium range polar solvents like toluene, benzene and diethyl ether were effective under microwave irradiation. The conversion of γ-caprolactone was found to be higher in diethyl ether than in the other two solvents. Both toluene and benzene gave lower conversions than the same reaction performed under thermal heating. The authors concluded that there was a significant effect of microwave irradiation on the polymerization reaction. As compared to thermal heating, microwaves induce accelerating, as well as decelerating effects, based on the boiling solvent.[67] Matos *et al.*[68] studied the influence of different parameters such as temperature, microwave intensity and time for the same reaction under solvent-free conditions and observed a positive influence of temperature on polymer synthesis, whereas an increase in microwave intensity had a negative effect. Under optimum conditions, a four-fold increase was observed at low microwave power (50 W). The authors concluded that an increase in power leads to an increase in temperature of the microenvironment of the catalyst compared to the bulk reaction mixture. This could affect the enzyme stability and activity. A clear trend in this regard has not yet been identified and further study needs to be done. However, microwave irradiation leads to an increase in the rate of reaction and a reduction in time duration in enzyme catalyzed polymerization.

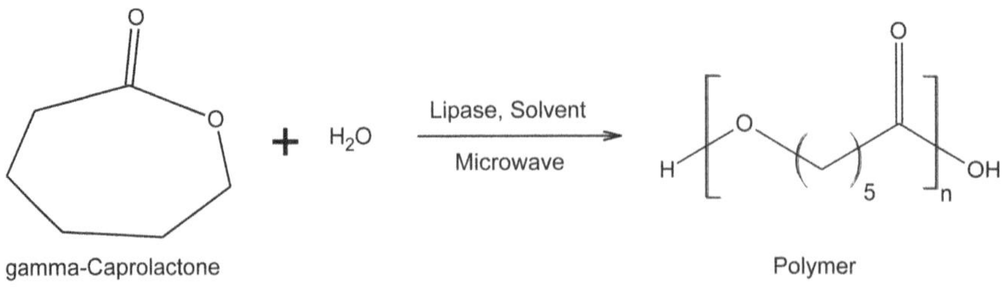

Scheme 4.2   Lipase catalyzed ring-opening polymerization of γ-caprolactone.[67]

4.4.4 Application of Microwave Irradiation for Enzyme Catalyzed Reactions in the Chemical Industry

The synergism between microwave assisted and lipase catalyzed synthesis of fine chemicals has gained much importance in the chemical industries. Yadav and Lathi[69] studied the synergistic effects of low power microwaves on the immobilized lipase catalyzed synthesis of adipic acid esters at 60 °C in 1,4-dioxane. The rate of reaction was improved by ~1.07 to 2.63 fold, depending on the chain length of the alcohol used. However, the selectivity towards the monoester was reduced as the consecutive reaction under microwave irradiation leads to subsequent diester formation. The same authors also reported the esterification of 1,2-propanediol with lauric acid.[70] Unlike the diester of adipic acid, they observed improved selectivity towards the monoester of 1,2-propane diol. The enhanced activity of the enzyme was due to an increase in entropy of the system. The activation energies for both types of heating are identical but ~2.84 times enhancement in the pre-exponential factor in the Arrhenius equation was observed under microwave irradiation vis-à-vis conventional heating. Major *et al.*[71] studied the esterification of lactic acid and ethanol with phosphonium ionic liquid as a medium. Improved ethyl lactate ester production was observed under microwave irradiation. The reaction time was drastically reduced from 24 h to 7 h. They concluded that the improved ester production was due to rapid hydrolysis of lactoyllactic acid (dimer of lactic acid) to the monomer, which subsequently reacted with ethanol under microwave irradiation.

Modified starch has found innumerable applications in different sectors, like the chemical, plastic and biomedical industries. Modification of starch by chemical methods requires strict reaction control, sophisticated instruments, the use of toxic solvents and is not viable for industrial scale production. The enzymatic modification of starch by microwaves is a potentially viable process. A number of researchers have investigated the effect of microwave irradiation on the esterification of starch with high chain length fatty acids or recovered coconut oil using different microbial lipases such as *Burkholderia cepacia*,[72] *Candida rugosa*,[73] *Thermomyces lanuginose*[74] and pig pancreas.[75] All of these lipases gave a good degree of substitution within a few minutes without changes in morphology, compared to hours in the conventional method.

β-Keto esters are important building blocks used in a variety of processes as starting precursors or intermediates. Yadav and Lathi[76] studied the effect of microwaves on immobilized *Candida antarctica* lipase B catalyzed transesterification of keto esters with different alcohols using toluene as solvent. Good to moderate conversion of esters was obtained in a short time. Furthermore, the conversion decreased with increasing chain length of the alcohol. Around 2.2–4.6 fold improvement in the rate of reaction was obtained compared to conventional reactions. Risso *et al.*[77] also studied the same reaction system under solvent-free conditions and found a significant enhancement in the conversion and rate of reaction. In contrast, Leadbeater *et al.*[78] reported that microwave irradiation had no influence on the lipase catalyzed transesterification of methyl acetoacetate with *n*-propanol using toluene or diethyl ether as solvent (Scheme 4.3). However, the reason for the non-existence of microwave effects was not completely discussed. This would also suggest that the micro-environment of the enzyme should be made favorable by the solvent.

Lipase catalyzed epoxidation and 1,2-dihydroxylation are important reactions to produce the respective epoxide and diol from a ketone using hydrogen peroxide or peroxy acids as oxidizing agent. Yadav and Borkar[79] studied the microwave assisted lipase catalyzed epoxidation of styrene in toluene. The first step of the reaction is the conversion of lauric acid into perlauric acid which acts as an oxidizing agent for the second step and donates oxygen to styrene to form styrene oxide. Both the conversion and rate of reaction were higher in microwave than conventional heating. Similar results were obtained for the *Pseudomonas* sp. [PSLG6] lipase catalyzed conversion of ketone to diol in the presence of microwave irradiation. The reaction was completed within a few minutes, whereas conventional heating required more than 20 h.[80]

Generally, the reduction of azide derivatives is carried out by using chemical reducing agents, and some disadvantages come into play like toxicity, high cost, low yield, *etc.* Mazumder *et al.*[81] used the microwave induced lipase catalyzed conversion of azide to amine in organic media and found that the enzyme showed high selectivity in the reduction of the N=N bond compared to other functional groups like C=C or MeO. The microwave assisted process gave good to moderate conversion in less than 10 min compared to a few hours for thermal heating. Kidwai *et al.*[82] observed selective *N*-acylation of fatty acid esters by ethanolamine in organic media under microwave irradiation instead of ethanolysis (Scheme 4.5). Enhanced conversion and rate of

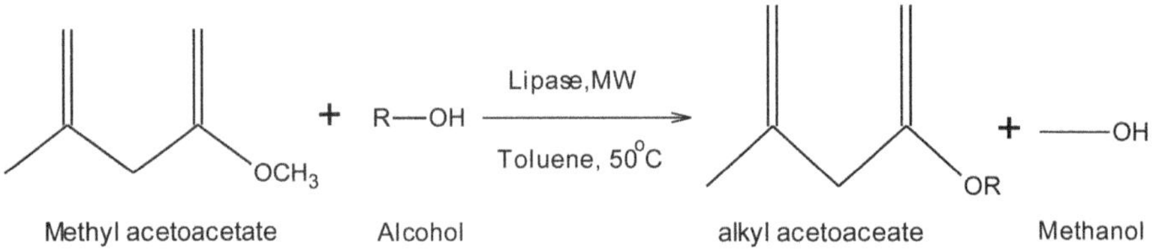

Scheme 4.3   Lipase catalyzed transesterification of methyl acetoacetate with different alcohols.[76]

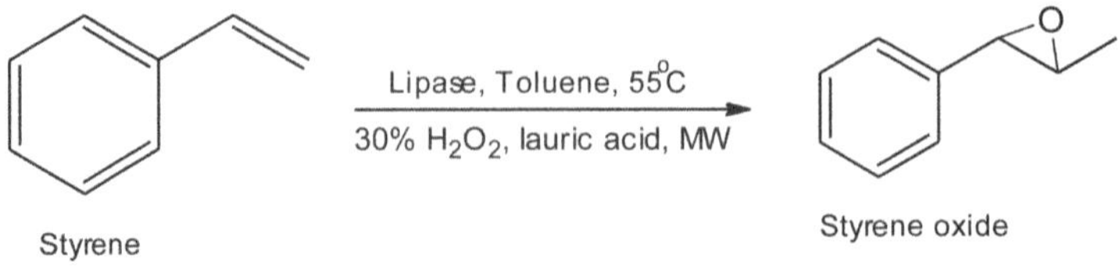

Scheme 4.4   Lipase catalyzed epoxidation of styrene.[79]

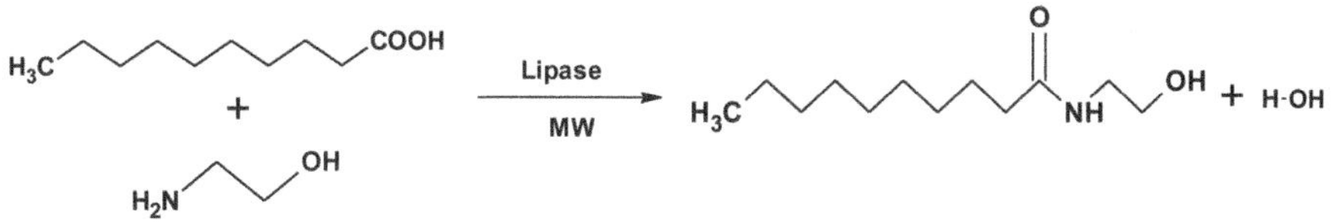

Scheme 4.5 Lipase catalyzed *N*-acylation of ethanolamine with decanoic acid.[82]

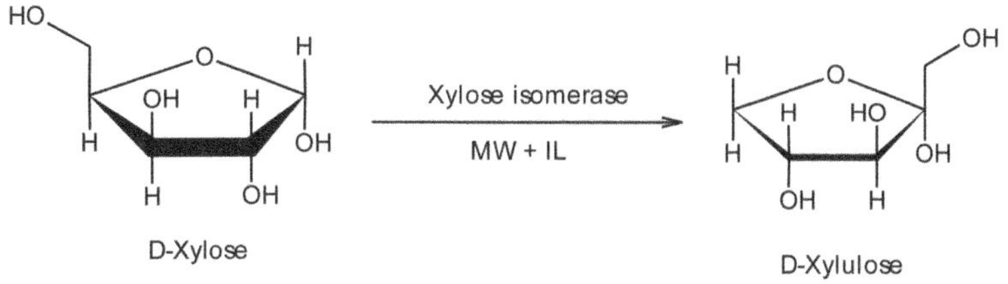

Scheme 4.6 Xylose isomerase catalyzed isomerization of xylose to xylulose in ionic liquid.[83]

reaction were obtained in dry media compared to liquid media under microwave and conventional heating.

Unnatural ketoses such as D-xylulose and ribulose are important precursors for the chemical synthesis of amino sugars or glycosides. These sugars can be obtained by isomerization of the corresponding aldo-sugars by using isomerase as biocatalyst. Although enzymatic synthesis of keto-sugar has been reported, the conversion rate of the aldo-sugar to the keto-sugar is very low and it is not a feasible process at the industrial level. Yu *et al.*[83] believed that with the combination of microwave irradiation and an ionic liquid, it could be possible to accelerate the rate of the reaction catalyzed by the isomerase enzyme. They studied the isomerization of xylose to xylulose by xylose isomerase in the presence and absence of microwaves in both ionic liquid and solvent-free systems (Scheme 4.6). Both the conversion and rate of reaction were improved from 50 to 68% and 133 to 375 (μmol min^{-1} g^{-1}), respectively, with the combination of microwaves and an ionic liquid. Furthermore, the immobilized xylose isomerase showed good thermal stability in ionic liquid under microwave irradiation. Ionic liquids certainly change the micro-environment of the enzyme and induce higher activity.

4.4.5 Application of Microwaves for Enzyme Catalyzed Reactions in the Food and Cosmetics Industries

Diacylglycerols are used as additives and bases for the food and cosmetics industries and can be obtained by either glycerolysis of plant oil or selective hydrolysis of oil. Matos *et al.*[84] studied the *Burkholderia cepacia* (PS "Amano" SD) lipase catalyzed partial hydrolysis of palm oil under microwave irradiation. The reaction was carried out at 80 °C under solvent-free conditions.

Moderate conversion (26%) was obtained in 5 min after which the enzyme was deactivated under microwave irradiation. Agglomeration of the enzyme was observed in the reaction mixture, which may have formed hotspots during microwave irradiation which led to enzyme deactivation.

Fatty acid esters of glycoside derivatives have many applications in the food and cosmetics industries. Initial studies reported that lipase catalyzed esterification of fatty acids with sugar is slow in nature. Pujic *et al.*[85] investigated the microwave assisted lipase catalyzed synthesis of long chain fatty acid esters of sugar derivatives in order to improve the conversion and rate of reaction. The esterification of dodecanoic acid with different sugar derivatives such as α-D-glucopyranoside, α-D-glucose and α,α-trehalose was carried out at 95–110 °C under dry conditions. Under microwave irradiation, enhanced conversion and selectivity were obtained in the esterification of dodecanoic acid with α-D-glucopyranoside and α,α-trehalose. No change in conversion was obtained in both heating modes when α-D-glucose was used.

Alkyl esters of fatty acids are used as ingredients for the synthesis of flavor and fragrance compounds and are widely used in the food and cosmetics industries. Enzymatic syntheses of flavor esters are currently being developed under thermal heating. The effects of microwave on these has not been fully explored. Wan *et al.*[86] studied the influence of the polarity of the organic solvent on the non-thermal effects of microwave using the *Mucor miehei* lipase catalyzed esterification of caprylic acid with butanol as model reaction. They found a significant effect of microwave irradiation which is dependent on the substrate concentration. No effect of microwave was observed when the substrate concentration was above 2 mol L^{-1}. Solvents with a $\log P$ value ranging from 2 to 4 showed positive non-thermal effects. Above $\log P$ 4, the non-thermal effect was found to decline. No significant correlation of the $\log P$ value with the non-thermal effect was found with lower substrate concentrations. On the other hand, Fang *et al.*[87] found a significant effect of consecutive microwave irradiation on the same lipase catalyzed esterification of pentanol with caprylic acid in octane as solvent. Furthermore, these authors extended their study to *Candida antarctica* lipase B (Novozym 435) catalyzed esterification of glycerol with caprylic acid in the presence and absence of solvent, and reported that consecutive microwave irradiation has the ability to alter the active site of the enzyme which weakens the 1, 3 specificity of the enzyme.[88] Huang *et al.*[89] studied the *Mucor miehei* lipase catalyzed esterification of C_2–C_{10} *n*-alcohol with *n*-caprylic acid under both heating methods. Under microwave irradiation, both the conversion and activity were enhanced and the rate of reaction changed with alcohol chain length with a declining zig-zag shape, which represents the ordinary substrate specificity. Osuna and Rivero[90] studied three different commercially available immobilized lipases, *Candida antarctica* lipase B (Novozym 435), and *Pseudomonas cepacia* lipase immobilized on diatomite (PS-DI) and ceramic (PS-CI), in the catalyzed esterification of lauric acid with six different alcohols, C_1–C_5 *n*-alcohol and allyl alcohol, in hexane under both heating modes. Novozym 435 was found to be the most effective among them. Different lipases have shown different rate profiles with respect to alcohol chain length. For Novozym 435, the rate of reaction increases initially up to ethanol then decreases with increasing chain length of the alcohol. The reaction time was drastically reduced from 24 h to 6 min under microwave irradiation.[90] A similar positive effect was found in Novozym 435 catalyzed esterification of myristic acid with isoamyl alcohol under solvent-free conditions.[91]

Yadav and Borkar[92] used microwave irradiation in the lipase catalyzed transesterification of citronellol with vinyl ester at 50 °C in toluene as solvent (Scheme 4.7). Under microwave irradiation, both the conversion and rate of reaction were improved. Similar results were also obtained in enzymatic alcoholysis of ethyl cinnamate with citronellol in organic media.[93] The improved activity of the enzyme is due to increased collision of substrates and enzyme particles, which in turn leads to an increase in the entropy of the system. In both cases, the reported solvents have low $\log P$ values and show low microwave absorbance capacities, but definite synergism was observed between the enzyme and the microwave radiation. Parker *et al.*[94] studied the influence of microwave irradiation on the cutinase catalyzed transesterification of ethyl butyrate with butanol in organic media (Scheme 4.8).

The reaction was carried out using a hydrated enzyme with three different water activities ranging from 0.58 to 0.97 at 50 °C. At a water activity of 0.58, the rate of reaction was almost doubled under the influence of microwaves compared to the rate under conventional heating. The catalytic activity was increased by 2–3 fold. However, further increase in the water activity to 0.97 led to a drastic fall in the rate of reaction under microwave irradiation and

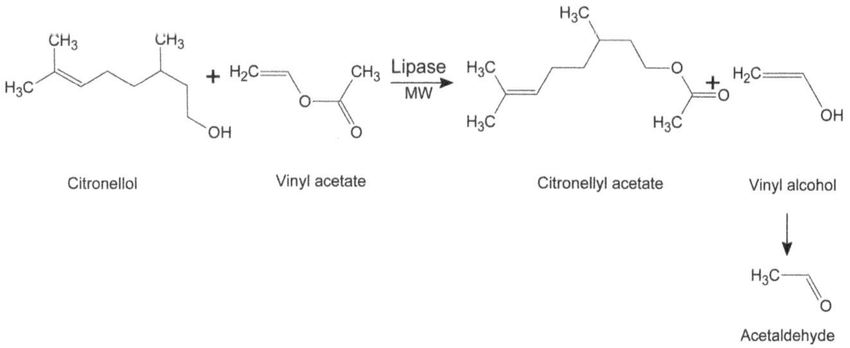

Scheme 4.7 Lipase catalyzed synthesis of citronellyl acetate.[92]

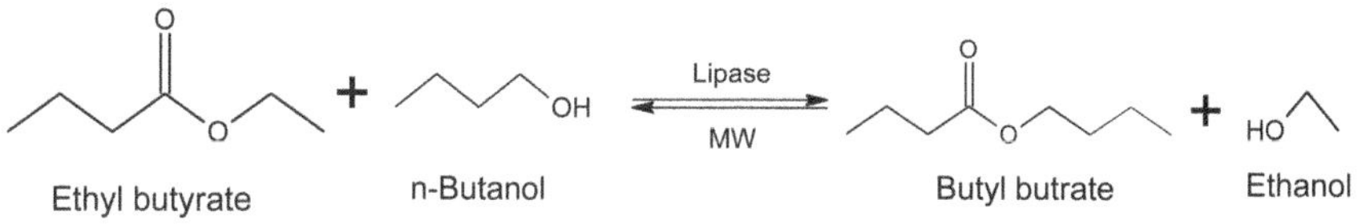

Scheme 4.8 Lipase catalyzed transesterification of ethyl butyrate with *n*-butanol.[94]

it was also observed that an increase in temperature from 50 to 70 °C produced the same results. These authors concluded that the increase in activity is not purely based on thermal effects. Rejasse *et al.*[95] studied the same reaction at 100 °C with immobilized lipase B from *Candida antarctica* (Novozym 435) in different organic solvents. They observed that microwave irradiation improved the stability of the enzyme and it depended on the polarity of the solvent used. A change in temperature does not have any effect on the stability of the enzyme under microwave conditions in agreement with previous reports. Furthermore, they studied the influence of microwave irradiation on free lipase in solvent-free conditions, and the same activity of free enzyme was found under both heating methods with improved stability under microwave conditions. They concluded that microwave radiation has no effect on enzyme activity, but influences the stability of the enzyme.[96] Zhao *et al.*[97] studied the effects of ionic liquid properties such as viscosity, polarity and hydrophobicity on Novozym 435 activity for the same reaction under microwave irradiation. They observed that viscosity and ionic liquid polarity do not have direct correlation with enzyme activity but the highly viscous nature of the ionic liquid might reduce the enzyme activity. The hydrophobic effect showed a weak correlation with enzyme activity. Increase in $\log P$ value led to an increase in enzyme activity up to a certain level, and it then decreased with further increase in $\log P$ value. However, higher enzyme activity was observed under microwave heating with hydrated enzyme compared to the dried one, and higher temperature existed near the enzyme molecules than in the reaction medium. Improved enzyme activity under microwave irradiation is due to superheating of the essential water layer around the enzyme molecules.

4.4.6   Application of Microwave Heating for Enzyme Catalyzed Reactions in the Pharmaceutical Industry

Enzymatic synthesis of active ingredients has gained enormous attention within the pharmaceutical industry for selective transformations. Combination with microwave heating will make enzymatic synthesis a much easier, cleaner and greener process. Kidwai *et al.*[98] studied the possibility of the existence of a synergism between microwave irradiation and *Fusarium globulosum* lipase catalyzed deacylation of cephalosporins in aqueous ammonia (Scheme 4.9). Microwave irradiation reduced the duration of reaction from 10 h to 5 min and also the conversion was enhanced from 78–82% to 90–92% for different cephalosporin derivatives.

Scheme 4.9   Lipase catalyzed deacylation of cephalosporins.[98]

Modafinil is a CNS stimulant drug, which is used in the treatment of hypersomnia and narcolepsy. Chemical synthesis of this compound requires harsh conditions and hazardous chemicals. Yadav and Lathi[99] studied the enzymatic synthesis of key intermediate *n*-butyl diphenyl methyl mercapto acetate (BDMMA) using lipase catalyzed esterification of diphenyl methyl mercapto acid with butanol in toluene as solvent. The conversion obtained under thermal heating was around 12% at 10 h. Due to the low conversion, they conducted the same reaction under microwave irradiation at 60 °C. There was a marginal increase from 12 to 18% and the activity was increased from 0.0013 to 0021 mol L^{-1} h^{-1}.

Xylitol mono esters are used as drugs for the treatment of hemoglobin diseases and malignant tumors. Low solubility of xylitol in low toxicity solvents and low regioselectivity are bottlenecks to enzymatic reaction. Rufino *et al.*[100] investigated the effects of non-thermal heating such as microwave and ultrasound heating on *Penicillium camembertii* lipase (immobilized on epoxy silica-PVA), in the esterification of *t*-butanol protected xylitol with different fatty acids in solvent-free conditions. Maximum conversion was achieved under microwave heating compared to ultrasound and conventional heating in the following order: microwave > ultrasound > conventional heating. Around 60% yield was obtained in 7 h under microwave irradiation, which was realized in 15 h under conventional heating.

γ-Linolenic acid is an important precursor for prostaglandin and leukotriene synthesis. A high content of γ-linolenic acid is present in blackcurrant seed oil. Enrichment of the γ-linolenic acid content is required in order to use it as a dietary supplement. Vacek *et al.*[101] tried to enrich γ-linolenic acid by selective esterification of free fatty acids extracted from blackcurrant seed oil with butanol at 30 °C under microwave irradiation. Reactions were carried out using three different lipases, *Candida cylindracea* lipase, *Mucor miehei* lipase and *Pseudomonas cepacia* lipase with Lipozyme in octane as solvent. In all cases, microwave heating showed increased conversion compared to conventional heating. *Pseudomonas cepacia* lipase produced a high amount of butyl-γ-linolenate in fatty acid butyl esters under microwave irradiation.

7-Hydroxy-4-methyl-2*H*-chromene-2-one is one of the members of the benzopyrone family and has various pharmaceutical applications. Derivatives of this molecule have led to diverse molecules with different pharmaceutical properties. Kidwai *et al.*[102] investigated the synergism between enzyme catalysts and microwave heating for acylation of 7-hydroxy-4-methyl-2*H*-chromene-2-one with different fatty acids. The reaction was carried in both organic media and dry media at 50–55 °C. The maximum effect of microwave irradiation was observed in dry media compared to organic media with microwave and conventional heating. Almost complete conversion was obtained with decanoic acid at 30 min in dry media under microwave irradiation and the enzyme showed no loss of activity. The authors concluded that there exists a definite synergism between enzyme catalysts and microwave heating.

L-Ascorbyl fatty acid esters are widely used as anti-oxidants. Enzymatic synthesis of these compounds requires longer times and the use of solvents. Kidwai *et al.*[103] studied the *Burkholderia multivorans* lipase catalyzed transesterification of L-ascorbic acid with different long chain fatty acid methyl esters in solvent-free media. They observed the enhancement in both the conversion and selectivity under low microwave power. Furthermore, they studied the influence of the substrate ratio and found that maximum conversion was achieved at a ratio of 1:2 (ascorbic acid:ester) compared to 1:9 required under conventional synthesis. This lipase showed good stability and no loss in activity under microwave irradiation. Similarly, Yadav and Pawar[104] studied the influence of microwave heating on the lipase catalyzed transformation of a complex molecule like hydroxy cinnamate with butanol in organic media. Under the optimum conditions, the rate of reaction was almost doubled vis-à-vis conventional heating and the enzyme showed good activity after three repeated uses. Similarly, Chen *et al.*[105] investigated the microwave assisted synthesis of glyceryl monolaurate, which is used as an anti-viral agent. The reaction was carried out using Novozym 435 in a reverse micro emulsion system. Under the optimum conditions, the maximum yield of glyceryl monolaurate was achieved within 30 min under microwave irradiation, which was 1.5 times greater yield than in conventional heating. Furthermore, the authors studied the steady state fluorescence profile of the enzyme after exposure to microwaves and found that the microwaves had induced some structural changes in the enzyme which led to easy access of substrate molecules to the enzyme's active sites.

Isoniazid or isonicotinic acid hydrazide is commonly used to treat tuberculosis. The chemical synthesis of isoniazid requires high temperature. Yadav and Sajgure[106] tried to synthesize this compound through enzymatic transamidation of ethyl isonicotinate with hydrazine in organic media (Scheme 4.10). The yield of the process was very low; after 24 h, around 36% conversion was obtained under thermal heating. Furthermore, to improve the conversion, they tried this reaction at 50 °C under microwave irradiation. The conversion was considerably increased to 54% in 4 h, which shows that the microwaves act synergistically with the enzyme catalysts.

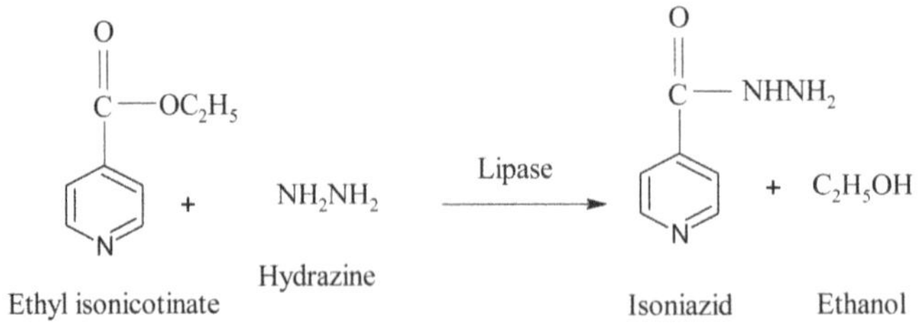

Scheme 4.10   Lipase catalyzed hydrazinolysis of ethyl isonicotinate.[106]

4.4.7 Separation of Racemic Compounds

Enzymatic resolution of racemic compounds has been found to be a better alternative to conventional chemical processes for the synthesis of single enantiomers. Although enzyme catalyzed reactions have gained considerable interest in the synthesis of enantiomeric compounds, they are sluggish in nature and require a long time. They are not economically viable at industrial level. Synergism with microwave heating can enhance the rate of enzymatic reaction without damaging the structure of enzyme molecules and can make it economically viable for chiral molecule synthesis (Table 4.5). Munoz *et al.*[107] studied the lipase catalyzed resolution of (±)-1-phenylethanol under microwave irradiation (Scheme 4.11).

The reactions were carried out by using *Pseudomonas cepacia* lipase and *Candida antarctica* lipase in dry conditions under both conventional heating and microwave irradiation. Under the optimal conditions, both lipases showed an enhanced initial rate and enantioselectivity under microwave irradiation. From the analysis of the effect of acyl donors, a higher enantioselectivity was observed in the case of isopropenyl acetate and octanoic acid compared to ethyl esters. The authors concluded that the improved selectivity was due to the increased polarity of these compounds which led to better absorbance of microwaves. In contrast, de Souza *et al.*[108] studied lipase catalyzed resolution of *rac*-1-phenyl ethanol in organic media at 40 °C with different lipases such as Novozym 435, Amano PS C1, Amano AK and Lipozyme TL IM for 2 h using vinyl acetate as acyl donor under both heating methods. No significant non-thermal effect of microwaves was observed, as both conversion and selectivity were identical for all lipases under both heating methods. The use of low temperature could be responsible for these results. Lin and Lin[109] reported the microwave assisted *Porcine pancreatic* lipase catalyzed acylation of various secondary alcohols in organic media. Both reaction rate and enantioselectivity were improved by 1–14 and 3–9 fold, respectively, under microwave heating compared to conventional heating. The authors concluded that in the former case, alcohol moieties were activated by the absorbance of microwaves, which was not possible with conventional heating. Yu *et al.*[110] studied Novozym 435 lipase catalyzed resolution of (*R,S*)-2-octanol in *n*-heptane as solvent at 60 °C under microwave irradiation. Under the optimum conditions under microwave irradiation, both activity and enantiomeric ratio (*E*-value) were increased from 55 to 203 (μmol min^{-1} mg^{-1}) and 99 to 352, respectively, within 2 h. Similar results were obtained for lipase catalyzed kinetic resolution of various secondary alcohols through transesterification reaction in organic media under microwave irradiation.[111–115] To make the enzymatic process more green and clean, Lundell *et al.*[116] used an ionic liquid as reaction medium and conducted the microwave assisted *Burkholderia cepacia* lipase (PS-C II piase) acylation of *N*-acylated 2-amino-1-phenylethanol and *N*-acylated norphenylephrine using vinyl butanoate as acyl donor. Among different ionic liquids employed, reasonable chemoselectivity was obtained in the presence of a hydrophobic ionic liquid, 1-ethyl-3-methylimidazolium

Table 4.5 Microwave assisted enzymatic resolution.

Sr. no.	Enzyme	T (°C)	Compound name	Reaction Scheme	Solvent	MW effect	References
1	Lipase	70–100	*Rac*-1-Phenyl ethanol		Dry media	Yes	107
2	Lipase	40	*Rac*-1-Phenylethanol		Cyclohexane	No effect	108
3	Lipase	35	1,2,3,4-Tetrahydro-1-naphthol, (±) 1-indanol, (±) menthol		Benzene	Improved rate and enantioselectivity	109
4	Lipase	60	(±) 2-Octanol		*n*-Heptane	Improved rate and enantioselectivity	110

5	Lipase	60	(±)-1-Phenyl-1-propanol, (±)-1-(4-bromophenyl)-propan-1-ol, (±)-1-phenylbut-3-en-1-ol, (±)-3-bromo-2-(2-hydroxypropyl)-1,4-dimethoxynaphthalene	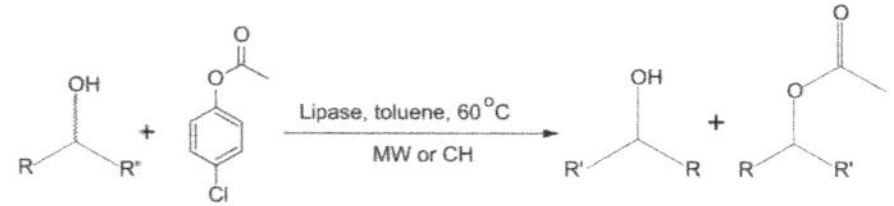	Toluene	Yes	111
6	Lipase	60	(±)-3-Bromo-2-(2-hydroxypropyl)-1,4-dimethoxynaphthalene	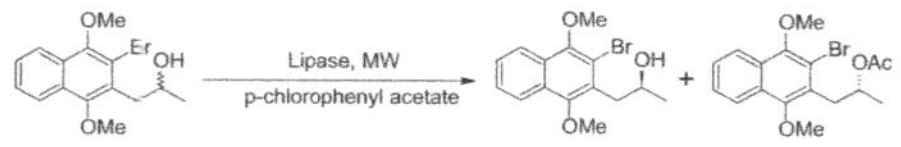	Toluene	Yes	112
7	Lipase	80	(±)-Mandelonitrile	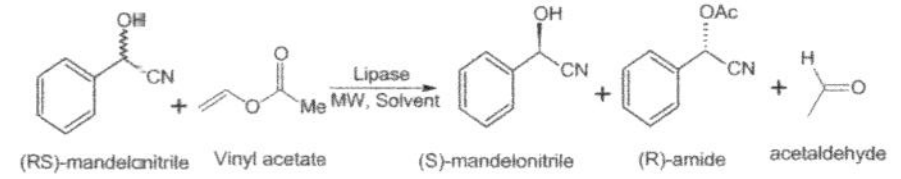	Toluene	Yes	113
8	Lipase	60	DL-(±)-3-Phenyllactic acid	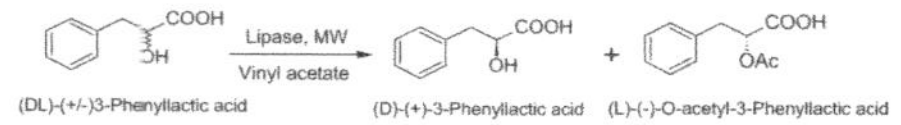	Toluene	Yes	114
9	Lipase	60	(±)-1-(1-Naphthyl)ethanol	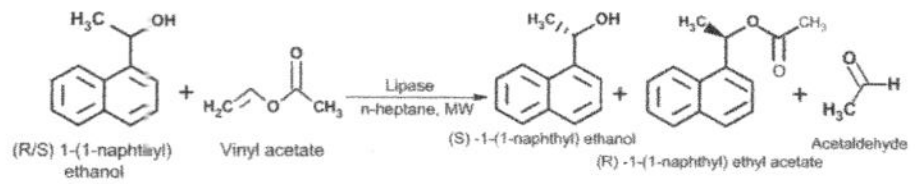	n-Heptane	Yes	115
10	Lipase	60–90	Rac-N-Acylated 2-amino-1-phenylethanol, Rac-N-acylated norphenylephrine	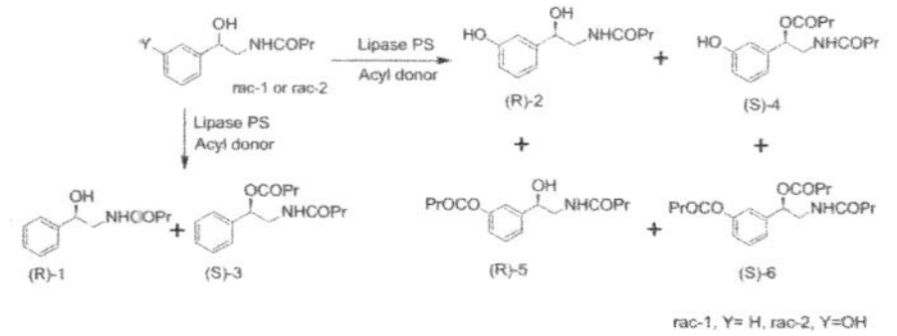	t-Butyl methyl ether or EMI-M·Ntf$_2$	No effect	116

(continued)

Table 4.5 (*continued*)

Sr. no.	Enzyme	T (°C)	Compound name	Reaction Scheme	Solvent	MW effect	References
11	Lipase	73	(±) 2-Octanol		Ionic liquid	Yes	117
12	Lipase	50	(±) 2-Octanol		Dichloro-methane	Yes	118
13	Lipase	50	(±) 2-Butylamine		*n*-Hexane	Yes	119
14	Lipase	50	(±)-1-phenylethyl-amine		*t*-Butyl methyl ether	Yes	120
15	Lipase	100	(±)-1-Phenylethyl-amine		Toluene	No effect	121
16	Lipase	50	Hydrolysis of (±) methyl mandelate		*t*-Butyl alcohol	Yes	122

bis(trifluoromethanesulfonyl) amide (**EMIM·NTf$_2$**), whereas in a hydrophilic ionic liquid, 1-ethyl-3-methylimidazolium tetrafluoroborate (**EMIM·BF$_4$**), the chemoselectivity was low compared to organic solvent. Furthermore, under microwave irradiation, there was no enhancement in the activity or selectivity of the reaction observed in the presence of an ionic liquid compared to those under normal conditions. However, Yu *et al.*[117] compared the enzyme activity and selectivity in the presence and absence of an ionic liquid under both heating methods for the kinetic resolution of (*R,S*)-2-octanol (Scheme 4.12). Novozym 435 showed superior catalytic activity and thermostability in the hydrophobic ionic liquid **EMIM·NTf$_2$** under microwave irradiation compared to those under solvent-free conditions under microwave and conventional heating. The enhanced activity of the enzyme is due to the low viscosity, solvent power and hydrophobic nature of **EMIM·NTf$_2$**, which retained the essential water layer around the enzyme molecule to maintain its structure intact. Furthermore, the ionic liquid interacts with the enzyme's charged group in the active site or at its periphery, which induces flexibility and provides an adequate microenvironment for superior activity and thermostability.

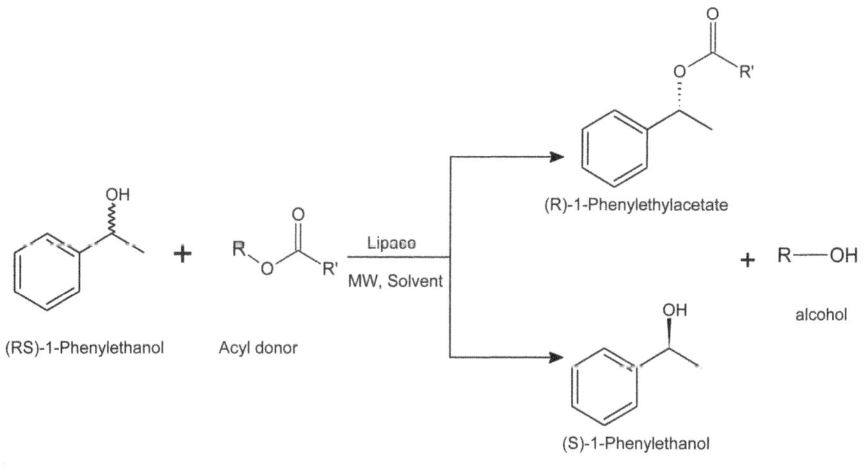

Scheme 4.11   Lipase catalyzed kinetic resolution of (*RS*)-1-phenylethanol.[107]

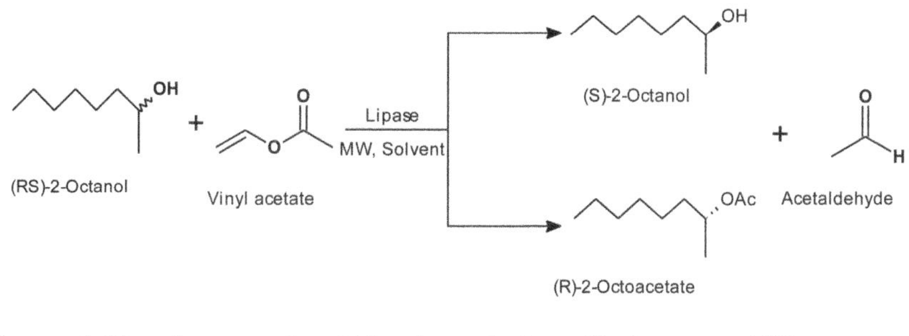

Scheme 4.12   Lipase catalyzed kinetic resolution of (*RS*)-2-octanol.[110]

Yu *et al.*[118] successfully developed a chemo-enzymatic method for sequential kinetic resolution of (*R,S*)-2-octanol under microwave irradiation. Initially they immobilized the enzyme, oxidizing agent (chromium trioxide) and reducing agent (sodium borohydride) on SBA-15, alumina and basic styrene anion exchange resin, respectively. These three immobilized materials were kept in three quartz vessels with connecting tubes and the whole setup with reactants was placed in a three column reactor and then microwave radiation was applied. The reaction was started by pumping the reagents from one column to another. A yield of 84% with an enantiomeric excess of 99% was obtained within 2 h under the optimum conditions. This approach widens the scope of microwave assisted enzymatic resolution.

Pilissao *et al.*[119] studied the effect of microwave radiation on the enzymatic resolution of (*RS*)-*sec*-butylamine in organic media. The reaction was carried out using free and immobilized *Aspergillus niger* lipase using vinyl acetate as acyl donor under microwave heating for 1 min at 35 °C. Around 21% conversion with ee$_p$ of 99% and *E* value > 200 was obtained. There was no reaction with conventional heating. These results showed the existence of an effect of microwaves on both conversion and selectivity. Furthermore, the authors studied the effects of the solvent and acyl donor on enzymatic reaction under microwave heating. *n*-Hexane was found to be a good solvent and was transparent to microwave absorption. As a result, the polar reactant molecules in the reaction mixture were activated by microwave absorption, which causes the enhancement of enzyme activity and selectivity. A similar result was obtained with Novozym 435 catalyzed acylation of (±)-1-phenylethylamine using ethyl acetate as acyl donor under microwave irradiation at 45 °C (Scheme 4.13). Both conversion and selectivity were substantially increased from 35 to 42% and the *E* value from 10 to 42 with shorter reaction time (from 10 h to 4 h) under microwave irradiation. At high concentration, (±)-1-phenylethylamine binds to the enzyme active site and forms a dead end complex which leads to the loss of enzyme activity under microwave irradiation.[120] In contrast, Parvulescu *et al.*[121] investigated the microwave effect on

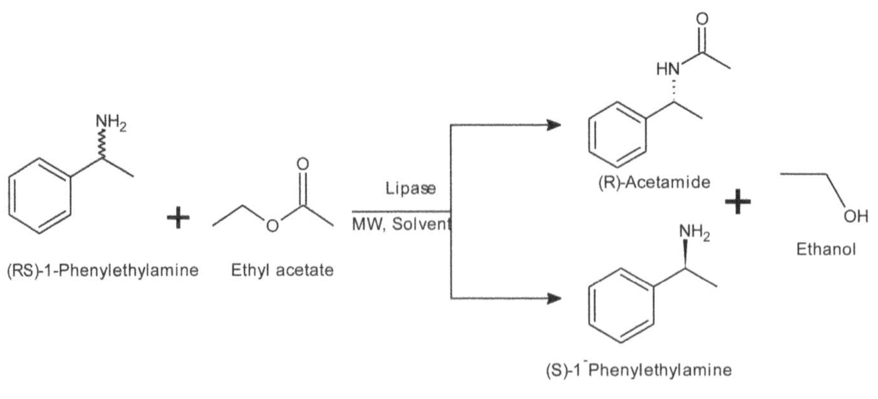

Scheme 4.13　Lipase catalyzed kinetic resolution of (*RS*)-1-phenylethylamine.[120]

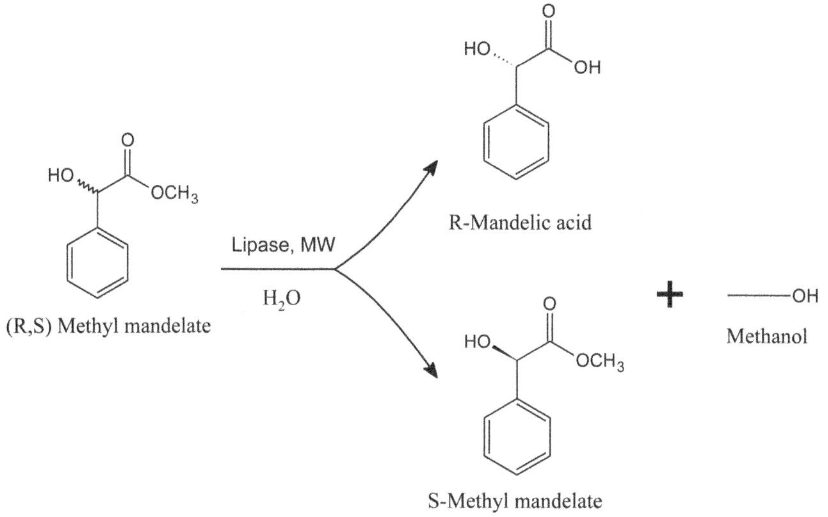

HO
OCH_3
O
(R,S) Methyl mandelate
Lipase, MW
H_2O
HO
OH
O
R-Mandelic acid
HO
OCH_3
O
S-Methyl mandelate
+
OH
Methanol

Scheme 4.14   Lipase catalyzed selective hydrolysis of methyl mandelate.[122]

Novozym 435 catalyzed acylation of 1-phenylethylamine using isopropyl acetate as acylating agent in toluene as solvent at 100 °C for 10 min. There was no significant difference obtained in the activity or selectivity of enzymatic resolution of *rac*-1-phenylethylamine under both heating modes. There could be denaturation of the enzyme at such a high temperature.

Yadav *et al.*[122] studied the effect of microwave irradiation on the lipase catalyzed hydrolysis of a racemic ester in organic media (Scheme 4.14). The reaction was carried out using an equimolar concentration of methyl mandelate and water in *t*-butyl alcohol as solvent at 50 °C for 4 h. The reaction under conventional heating gave only 19% conversion with an *E* value of 78% after 24 h. However, microwave irradiation gave 30% conversion with an *E*-value of 84.2% within 4 h. These results showed that both the conversion and selectivity were improved significantly under microwave irradiation.

4.4.8   Application of Microwaves to Enzyme Immobilization

Immobilization is one of the strategies used to improve enzyme stability and reusability. Traditional immobilization is slow and usually needs 12 to 24 h or longer to give sufficient interaction between the enzyme and support material. Usually enzymes are immobilized by any of the following methods: physical adsorption, entrapment or covalent interaction. Among them, covalent interaction has proven to offer good stability and reusability. The main disadvantage of covalent immobilization is a significant loss in activity during immobilization. In order to reduce the time and retain the enzyme activity, Nahar and Bora[123] studied the microwave mediated covalent immobilization of the enzymes such as horseradish peroxidase, glucose oxidase and urease on an activated polystyrene microtiter plate as support material. Both enzyme solution and activated support were exposed to microwaves, and immobilization started within 10 s with maximum immobilization obtained in 60 s, whereas there was no/less immobilization of enzyme obtained under the conventional method at 37 °C. In the case of horseradish peroxidase, the absorbance value obtained after 60 s of microwave exposure was 0.242 OD at 490 nm. However, it took 45 min to obtain the same value of absorbance under conventional immobilization.

Bezbradica *et al.*[124] compared microwave assisted covalent immobilization of *Candida rugosa* lipase on two different Eupergit supports, Eupergit C and C250 L. The immobilization of an enzyme occurs on these supports *via* covalent interaction between their oxirane groups and the amino group of the enzyme. Eupergit C has larger pore size, whereas Eupergit C250L has a greater number of oxirane groups. Enzyme loading onto the support was found to be 29% lower under microwave irradiation. However, the highest activity of 130 IU g^{-1} was observed with lipase immobilized on Eupergit C after 3 min under microwave exposure, which was 2.45 times higher than conventional immobilization which required 48 h. Lipase immobilized on Eupergit C250L showed the same activity (110 IU g^{-1}) as the conventional method after 3 min exposure to microwaves. The pore size of Eupergit C250L is smaller than that of Eupergit C and *C. rugosa* lipase has a molecular size of $5 \times 5 \times 7$ nm. Microwave irradiation promotes lipase diffusion into the pores of supports that are big in size. Eupergit C250L limits the diffusion rate of the substrates and products into pores and leads to an overall decrease in the activity of lipase immobilized on Eupergit C.

Du *et al.*[125] studied the microwave assisted covalent immobilization of *Candida antarctica* lipase B on mesocellular siliceous foam. Immobilization of the enzyme was carried out over 120 s using 400 W microwave power. A specific activity of 114.4 U mg^{-1} was obtained, which was 1.23 times higher than that for the conventional method, and the turnover number of the immobilized lipase was significantly improved (1.26 fold) compared to the free enzyme. Furthermore, the quenching of excess sites on the activated support by alcohol led to improved thermal stability of the immobilized enzyme. Around 60.64% residual activity was retained after the immobilized lipase was quenched with ethanol at 50 °C for 5 h. This is around 2.36 and 1.99 times higher than free lipase and immobilized lipase without quenching, respectively.[125] Similar results were reported in the covalent immobilization of aldolase[126] and thermolysin[127] on mesocellular siliceous foam under microwave irradiation.

Wang *et al.*[128] studied the effect of microwaves on enzyme immobilization in mesocellular siliceous foam. Simple adsorption immobilization was carried by using two different enzymes, penicillin acylase ($7 \times 5 \times 5.5$ nm) and papain ($3.7 \times 3.7 \times 5$ nm). Microwave irradiation greatly enhanced the rate of immobilization in both cases. The size of penicillin acylase was much bigger than that of papain; however, microwaves still took only 140 s to reach adsorption equilibrium, whereas the conventional method required 15 h. The relative activity of the immobilized penicillin acylase was significantly improved from 127.1% to 178.1% compared to the relative activity of free penicillin acylase (100%), which was 1.38 times higher than the activity obtained for the conventional method. In the case of papain, the activity of the immobilized enzyme was 1.86 times higher under microwave irradiation. The possible reasons for the improved immobilization and enzyme activity are, (i) microwave irradiation accelerates the rate of external mass transfer and intra-particle diffusion, which leads to enhanced transport of enzyme molecules onto the support material. (ii) Microwaves induce uniform distribution of enzyme molecules on support particles and improve the rate of substrate accessibility to the enzyme active site. (iii) Direct exposure of lipase to microwaves during immobilization could change the conformation of the enzyme and improve its activity.[123–128] Microwave assisted immobilization of different enzymes is summarized in Table 4.6. Furthermore, microwave irradiated enzymatic separation of chiral molecules that are useful in the pharmaceutical and fine chemical industries will be of great use.[129]

4.5   Kinetic Models and Their Critical Analysis

The derivation of kinetic models for enzymatic reactions has been performed in a variety of cases and can be found in books and monographs. The usual model can be used by collecting initial rate data and Burk–Lineweaver double inversion plots in the absence of external mass transfer resistance and intra-particle diffusion limitation for immobilized enzymes. However, precisely how microwave irradiation affects the various constants in these equations has rarely been examined barring very few studies. An enzyme includes one or more stereo-specific active sites where the substrate binds and the reaction takes place. The specificity of the enzyme for the substrate will differ from molecule to molecule and is mostly governed by the size of the active site of the enzyme, the size of the molecule, the attached functional groups and the stereo-selectivity of the molecule. Generally, the substrate binds to the enzyme and forms an enzyme–substrate complex or transition complex, much like chemical catalysis, which is then converted into the product. The enzyme activity increases with increasing substrate concentration up to a certain level, then remains the same or decreases after that. Inhibition or poisoning of active sites is a concentration dependent phenomenon arising from strong adsorption of a substrate used in high concentration or a lack of vacant sites for other reactants used in lower concentration. It is essential to recognize the behavior of the enzyme and the sequence of substrate binding to the enzyme active site during the reaction. Kinetic models are very useful for scale up and reactor design. Kinetic models have been developed by conducting studies on the effects of various parameters on rates, including the oft-used Burk–Lineweaver double inversion plots to correlate the affinity of the enzyme towards a particular substrate(s) and to forecast the activity of the enzyme based on substrate concentration.

Single substrate reaction follows the famous Michaelis–Menten kinetic model, whereas in two or multi-substrate based reactions, the order of

Table 4.6 Effect of microwave heating on enzyme immobilization.

Enzyme	Support material	Immobilization method	Microwave		Conventional		References
			Time	Activity	Time	Activity	
HR peroxidase	Polystyrene	Covalent interaction	60 s	0.242 OD	45 min	0.266 OD	123
Glucose oxidase			60 s	0.147 OD	45 min	141 OD	
Urease			60 s	0.263 OD	45 min	0.244 OD	
Candida rugosa lipase	Eupergit resin C	Covalent interaction	3 min	130 ± 1.2 IU g^{-1}	48 h	53 ± 1.1 IU g^{-1}	124
	Eupergit resin C 250 L		3 min	110 ± 1.6 IU g^{-1}	48 h	107 ± 2.1 IU g^{-1}	
Candida antarctica lipase B	Mesocellular siliceous foam	Covalent	2 min	114.4 U mg^{-1}	12 h	92.5 U mg^{-1}	125
Aldolase[a]	Mesocellular siliceous foam	Covalent interaction	5 min	63.4%	20 h	23.2%	126
Thermolysin[a]	Mesocellular siliceous foam	Covalent interaction	3 min	82.9%	20 h	35.4%	127
Penicillin acylase	Mesocellular siliceous foam	Adsorption	140 s	141.76 U mg^{-1}	20 h	101.64 U mg^{-1}	128
Papain			80 s	779.6 U mg^{-1}	20 min	419.1 U mg^{-1}	

[a]Relative activity after immobilization with all other immobilization conditions the same.

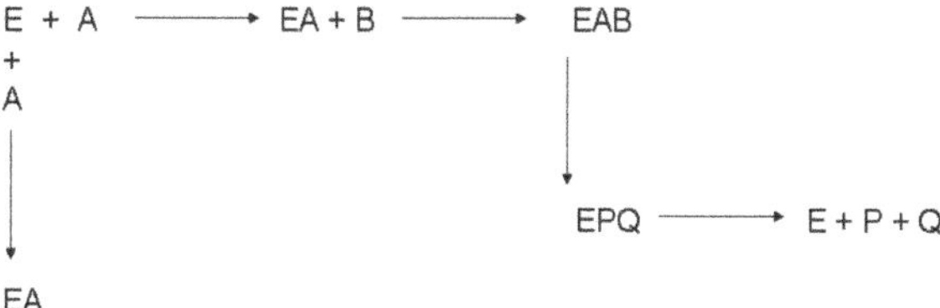

Scheme 4.15　Ordered bi–bi mechanism.

the substrate binding step determines the mechanism of the reaction. For instance, for two substrate based reactions, there are two types of mechanisms, namely, (i) random substrate binding and (ii) ordered substrate binding. In the former mechanism, any one of two substrates can bind to the active site. As the name suggests, the binding of the first substrate is independent of the second substrate. Similarly, either one of the products formed in the reaction is released first, whereas in the latter case, two more mechanisms are possible. One is called the ordered bi–bi mechanism, where successively two substrates bind to the enzyme and form a ternary complex, and then subsequently, the product is released. The second is called the ping-pong bi–bi mechanism, where the first substrate binds to the enzyme and forms an intermediate enzyme–substrate complex and releases the product before binding of the second substrate. Finally, the second product is released.

In the case of lipase catalyzed reactions, it has been established that the lipase first forms an acyl–enzyme complex with the acyl donor, and it follows either an ordered bi–bi mechanism or a ping-pong bi–bi mechanism. For example, lipase catalyzed hydrazinolysis of ethyl nicotinate follows an ordered bi–bi mechanism[106] and the reaction mechanism is shown in Scheme 4.15.

The lipase (E) will react with the acyl donor (A) to form a complex (EA). The second reactant (B) then reacts to form a ternary complex (EAB). At higher concentration of A, a dead end complex is formed.

The appropriate equation for this mechanism is:

$$\frac{r}{r_{\max}} = \frac{[A][B]}{K_{i(A)}K_{m(B)} + K_{m(A)}[B] + K_{m(B)}[A] + [A][B]} \tag{4.6}$$

whereas lipase catalyzed kinetic resolution of (±) 1-(1-naphthyl) ethanol follows a ping-pong bi–bi mechanism[115] and the reaction mechanism is shown in Scheme 4.16.

By analogy to the classical mechanism of lipase, it is assumed that the acyl donor (A) binds first to the free enzyme (E) and forms a non-covalent enzyme–acetate complex (EA), which releases the first product, hydrolyzed product (P) and the modified enzyme (F). The second substrate alcohol (B) reacts with the activated enzyme (F) to give the complex (FB), which gives the product (Q) and the free enzyme (E). Along with this, alcohol (B) also forms the dead end complex (E′B) by binding to the free enzyme (E) and the acyl donor (A) also forms the dead end complex with the modified enzyme (F) to give complex FA.

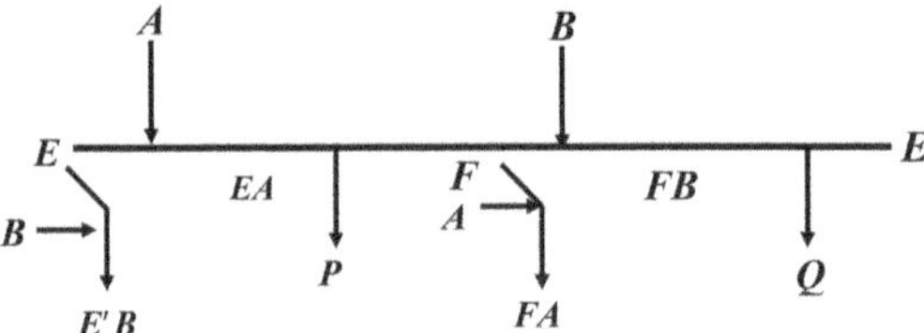

Scheme 4.16   Ping-pong bi–bi mechanism.

The rate equation obtained for the above mechanism is:

$$v = \frac{v_{\mathrm{m}}[A][B]}{K_{\mathrm{mA}}[B]\left(1+\dfrac{[B]}{K_{\mathrm{iB}}}\right)+K_{\mathrm{mB}}[A]\left(1+\dfrac{[A]}{K_{\mathrm{iA}}}\right)+[A][B]} \tag{4.7}$$

Yadav and Lathi[76] studied the kinetic mechanism of lipase catalyzed transesterification of methyl acetoacetate with *n*-butanol under microwave heating. The reaction followed a ping-pong bi–bi mechanism, with *n*-butanol acting as a dead end inhibitor at high concentration. The kinetic parameter values are smaller under microwave irradiation without any change in reaction mechanism. The affinity of *n*-butanol for the enzyme was drastically improved by around 10 times under microwave irradiation as the K_{m} value of *n*-butanol decreased from 1.096 to 0.1061 mol L^{-1}. Yadav and Borkar[79] reported kinetic analysis for the chemoenzymatic epoxidation of styrene oxide. The first step is the generation of perlauric acid from lauric acid and hydrogen peroxide in the presence of *Candida antarctica* lipase B. This generated perlauric acid is allowed to react simultaneously with styrene to give styrene oxide. These authors also conducted a separate kinetic study on the lipase catalyzed formation of perlauric acid from hydrogen peroxide and lauric acid. Increases in the concentration of hydrogen peroxide led to a decrease in the rate of reaction at lower lauric acid concentration, and increasing concentration of hydrogen peroxide led to an increase in the rate of reaction at higher lauric acid concentration. Based on progress curve analysis, the reaction follows an ordered bi–bi mechanism with hydrogen peroxide acting as a dead end inhibitor. The same kinds of profile curve were observed under both heating methods. However, under microwave heating, an improved initial rate and a reduced K_{m} value for both substrates were observed. Furthermore, they developed a kinetic model for *in situ* epoxidation of styrene and observed a similar effect under microwave irradiation (Scheme 4.4). The collisions of molecules are faster under microwave heating than in conventional heating, which is reflected in lower values of dissociation constants under microwave heating.[92]

Fang *et al.*[87] studied the kinetic details of microwave assisted *Mucor miehei* lipase catalyzed esterification of caprylic acid with *n*-pentanol in non-aqueous media. The Michaelis constant for pentanol was drastically reduced from 4.15 to 0.58, whereas there was no significant difference in the K_{m} value of caprylic acid observed under microwave irradiation when compared to conventional heating. This implies that microwave irradiation led to an improved affinity towards pentanol when excess caprylic acid was present. Pentanol has a lower $\log P$ (−1.4) than caprylic acid (2.9) and absorbs most microwave energy and passes it to the enzyme which could lead to a greater affinity towards *n*-pentanol. Furthermore, the authors observed that the reaction follows a ping-pong bi–bi mechanism. Microwave irradiation induced pentanol inhibition when the pentanol concentration was >0.75 mol L^{-1} and when the caprylic acid concentration was >0.25 mol L^{-1}. However, this inhibition was not found under conventional heating. Under microwave conditions, the enzyme already shows high affinity and further increased concentration of alcohol led to competition with acid molecules at active sites and caused alcohol inhibition. It was also observed that the pre-exponential factor in the kinetic rate constants was enhanced due to an increase in the frequency of collision.

The kinetic mechanisms reported for lipase catalyzed reactions are summarized in Table 4.7. However, there are not many studies reported on microwave assisted kinetics and mechanism.

Table 4.7 Kinetic models for microwave irradiated enzyme catalysis.

Sr. no.	Enzyme	Reaction	Kinetic model	Inhibition by	References
1	*Candida antarctica* lipase B	Esterification of 1,2 propanediol with lauric acid	Ping-pong bi–bi mechanism	1,2-Propanediol	70
2	*Candida antarctica* lipase B	Transesterification of methyl acetoacetate with *n*-butanol	Ping-pong bi–bi mechanism	No inhibition	76
3	*Candida antarctica* lipase B	Epoxidation of styrene	Ordered bi–bi mechanism	Hydrogen peroxide	79
4	*Mucor Miehei* lipase	Esterification of pentanol with caprylic acid	Ping-pong bi–bi mechanism	*n*-Pentanol	87
5	*Candida antarctica* lipase B	Esterification of myristic acid with isoamyl alcohol	Ping-pong bi–bi mechanism	Iso-amyl alcohol	91
6	*Candida antarctica* lipase B	Transesterification of citronellol with vinyl acetate	Ping-pong bi–bi mechanism	Both citronellol and vinyl acetate	92
7	*Candida antarctica* lipase B	Transesterification of citronellol with ethyl cinnamate	Ordered bi–bi mechanism	Citronellol	93
8	*Candida antarctica* lipase B	Esterification of *n*-butanol with diphenyl methyl mercapto acetic acid	Ping-pong bi–bi mechanism	*n*-Butanol	99

Table 4.7   (*continued*)

Sr. no.	Enzyme	Reaction	Kinetic model	Inhibition by	References
9	*Candida antarctica* lipase B	Transesterification of ethyl-3-phenylpanoate with *n*-butanol	Ping-pong bi–bi mechanism	*n*-Butanol	104
10	*Candida antarctica* lipase B	Hydrazinolysis of ethyl isonicotinate	Ordered bi–bi mechanism	Ethyl isonicotinate	106
11	*Candida antarctica* lipase B	Kinetic resolution of DL-(±)-3-phenyllactic acid with vinyl acetate	Ordered bi–bi mechanism	DL-(±)-3-Phenyllactic acid	114
12	*Candida antarctica* lipase B	Kinetic resolution of (±)-1-(1-naphthyl) ethanol with vinyl acetate	Ordered bi–bi mechanism	(±)-1-(1-Naphthyl) ethanol and vinyl acetate	115
13	*Candida antarctica* lipase B	Kinetic resolution of (±)-1-phenylethylamine with ethyl acetate	Ping-pong bi–bi mechanism	(±)-1-Phenylethyl amine	120
14	*Candida antarctica* lipase B	Hydrolysis of methyl mandelate	Ordered bi–bi mechanism	*RS*-(±)-Methyl mandelate	122

4.6   Conclusions

White biotechnology will be taken to new heights through green chemistry. Catalysis is of paramount importance in waste minimization, process intensification and selective production of a variety of chemicals. Biocatalysis has emerged as a green technology for synthesizing industrially important chemicals through a sustainable process across different industries and will be the pillar of white biotechnology or modern industrial biotechnology. Although biocatalytic processes are found to be potentially alternative routes, enzyme catalyzed reactions are naturally slow processes and require longer periods to achieve the desired yield. Microwave irradiation has already proven to be a clean and green heating method for organic and materials synthesis. The combination of microwave irradiation with biocatalysis is not fully appreciated as compared to chemical catalysts as controversies about the effects of microwaves on enzyme catalysts have been reported. Some literature has reported the existence of synergetic effects of microwave irradiation on enzyme catalysts in both solvent and solvent-free reactions. Microwave assisted processes have been successfully demonstrated for various industrial applications, such as protein digestion, biodiesel production, polymerization, synthesis of fine chemicals, kinetic resolution of racemic compounds and enzyme immobilization. Immobilized enzymes are better activated by microwaves than by free enzymes. The microenvironment in the particle is greatly influenced by the nature of the solvent and whether it can retain the water layer in the pore space or strip it off. Thus, many non-polar solvents, which do not strip off water, are found to work efficiently to increase the activity; for instance, hexane and toluene. In the case of ionic liquids, there could be changes in the enzyme structure under microwave irradiation, leading to favorable conformations for higher activity and selectivity. In the majority of cases, microwave assistance has demonstrated a remarkable reduction in reaction time and has also induced conformational changes in the enzyme that have led to improved activity and selectivity. The immobilization of enzymes on supports using microwaves is beneficial since it reduces the reaction time by an order of magnitude, leading to superior activity. The influence of microwave irradiation on mass transfer coefficients, effectiveness factors, and adsorption and desorption coefficients must be properly studied in order to remove controversy or contradictory reports. Several researchers in the past have used domestic microwaves to get 'fantastic' results without providing any explanation, since the use of mono-mode or multi-mode microwave reactors, and the position of the reactor in the microwave oven greatly influence the outcome. Properly designed mechanically agitated contactors need to be used to avoid mass transfer resistance, as well as the attrition of enzyme loaded particles. Some of the reports thus may not be reproducible, if proper precautions are not exercised in conducting experiments in well-designed reactors which will ensure the same degree and quality of micro-mixing. Furthermore, microwave irradiated enzymatic separation of chiral molecules in the pharmaceutical and fine chemicals industries will be greatly useful and it is believed that this area will be further probed. The combination of microwave irradiated biocatalysis and heterogeneous catalysis will add to the prowess of white biotechnological processes in the pharmaceutical and fine chemicals industries. However, there are only a limited number of reports in this area. Microwave assistance in biocatalysis is relatively new and the majority of studies have been performed using hydrolytic enzymes. There is scope for studying the kinetics of these reactions so that scale up and reactor design could be reliable. Furthermore, extending microwave effects to other classes of enzymes is timely and a lot of study is required for implementation at an industrial scale. The combination of biocatalysis and microwave heating makes chemical synthesis much greener and more economically competitive than conventional heating. Newer types of support will also be explored in future for microwave assisted enzymatic reactions which will extend nano-biotechnology's frontiers.

Nomenclature

ε'	Dielectric constant
ε''	Dielectric loss
$\tan\delta$	Tangent loss

D	Dissipation factor
P	Power dissipation per unit volume
ε_0	Permittivity of free space
E	Electric field strength
σ	Conductivity
ρ	Density
C	Specific heat capacity
ΔT	Temperature change
t	Time
D_p	Penetration depth
α	Absorption coefficient
λ	Wavelength
A	Initial concentration of acyl donor, mol dm^{-3}
B	Initial concentration of alcohol, mol dm^{-3}
E	Free enzyme
EA	Enzyme–acyl complex with A
EAB	Ternary complex
$E'B$	Enzyme–alcohol dead end complex
F	Modified enzyme
FA	Modified enzyme–acyl donor complex
FB	Modified enzyme–alcohol complex with B
P	Product
Q	Product
K_{iB}	Inhibition constant for alcohol, mol dm^{-3}
K_{mA}	Michaelis constant for acyl donor, mol dm^{-3}
K_{mB}	Michaelis constant for alcohol, mol dm^{-3}
v	Rate of reaction, mol dm^{-3} s^{-1} g^{-1}-enz
v_m	Maximum rate of reaction, mol dm^{-3} s^{-1} g^{-1}-enz

References

1. P. Anastas and N. Eghbali, *Chem. Soc. Rev.*, 2010, **39**, 301.
2. R. A. Sheldon, *CHEMTECH*, 1994, 38.
3. M. Ikunaka, *Catal. Today*, 2004, **96**, 93.
4. P. Anastas and J. C. Warner, *Green Chemistry: Theory and Practice*, Oxford University Press, Oxford, 1998.
5. G. D. Yadav, *Top. Catal.*, 2004, **29**, 145.
6. D. T. Allen and D. R. Shonnard, *Green Engineering: Environmentally Conscious Design of Chemical Processes*, Prentice Hall, Upper Saddle River, NJ, 2002.
7. R. A. Sheldon, *Chem. Soc. Rev.*, 2012, **41**, 1437.
8. A. Liese and M. V. Filho, *Curr. Opin. Biotechnol.*, 1999, **10**, 595.
9. S. Wenda, S. Illner, A. Mell and U. Kragl, *Green Chem.*, 2011, **13**, 3007.
10. U. Stottmeister, A. Aurich, H. Wilde, J. Andersch, S. Schmidt and D. Sicker, *J. Ind. Microbiol. Biotechnol.*, 2005, **32**, 651.
11. R. H. Kaul, U. Tornvall, L. Gustafsson and P. Borjesson, *Trends Biotechnol.*, 2007, **25**, 119.

12. M. Alcalde, M. Ferrer, F. J. Plou and A. Ballesteros, *Trends Biotechnol.*, 2006, **24**, 281.

13. J. Arcil, M. Vincente, M. Martinez and M. Poulina, *J. Biotechnol.*, 2006, **124**, 213.

14. H. E. Schoemaker, D. Mink and M. G. Wubbolts, *Science*, 2003, **299**, 1694.

15. A. Schmid, J. S. Dordick, B. Hauer, A. Kiener, M. Wubbolts and B. Withold, *Nature*, 2001, **409**, 258.

16. K. E. Jaeger, *Curr. Opin. Biotechnol.*, 2004, **15**, 269.

17. T. W. J. Cooper, I. B. Campbell and S. J. F. Macdonald, *Angew. Chem., Int. Ed.*, 2010, **49**, 8082.

18. J. S. Carey, D. Laffan, C. Thomson and M. T. Williams, *Org. Biomol. Chem.*, 2006, **4**, 2337.

19. J. V. Belien and Z. Li, *Curr. Opin. Biotechnol.*, 2002, **13**, 338.

20. S. H. Krishna, *Biotechnol. Adv.*, 2002, **20**, 239.

21. W. A. Loughlin, *Bioresour. Technol.*, 2000, **74**, 49.

22. M. Ayala and E. Torres, *Appl. Catal., A*, 2004, **272**, 1.

23. W. Marconi, *React. Polym.*, 1989, **11**, 1.

24. A. Illanes, A. Cauerhff, L. Wilson and G. R. Castro, *Bioresour. Technol.*, 2012, **115**, 48.

25. I. Roy and M. N. Gupta, *Curr. Sci.*, 2003, **85**, 1685.

26. P. Lorenz and J. Eck, *Eng. Life Sci.*, 2004, **4**, 501.

27. W. L. Tang and H. Zhao, *Biotechnol. J.*, 2009, **4**, 1725.

28. E. D. Neas and M. J. Collins, *Introduction to microwave sample preparation theory and practice*, ed. H. M. Kingston and L. B. Jassie, American Chemical Society, 1998, vol. 2, pp. 7–32.

29. B. L. Hayes, *Microwave Synthesis: Chemistry at the Speed of Light*, CEM Publishing, Matthews, NC, 2002.

30. M. Taylor, S. S. Atri and S. Minhas, *E-valueserve analysis: Developments in microwave chemistry*, 2005, http://www.rsc.org/images/evalueserve_tem18-16758.pdf.

31. D. Stuerga, *Microwaves in organic synthesis*, ed. A. Loupy, Wiley-VCH, Weinheim, 2nd edn, 2006, ch. 1, pp. 1–61.

32. H. K. Solanki, V. D. Prajapati and G. K. Jani, *Int. J. PharmTech Res.*, 2010, **2**, 1754.

33. P. Lidstrom, J. Tierney, B. Wathey and J. Westman, *Tetrahedron*, 2001, **57**, 9225.

34. C. Gabriel, S. Gabriel, E. H. Grant, B. S. J. Halstead and D. M. P. Mingos, *Chem. Soc. Rev.*, 1998, **27**, 213.

35. S. Chandrasekaran, S. Ramanathan and T. Basak, *AIChE J.*, 2012, **58**, 330.

36. R. A. England, *Biotage company brochure*, 2012, http://www.data.biotage.co.jp/pdf/literature/2154.pdf.

37. F. A. Bassyouni, S. M. Abu-Bakr and M. A. Rehim, *Res. Chem. Intermed.*, 2012, **38**, 283.

38. B. Rejasse, S. Lamare, M. D. Legoy and T. Besson, *J. Enzyme Inhib. Med. Chem.*, 2007, **22**, 518.

39. J. G. Greco, L. Gianfreda, G. d'Ambrosio, R. Massa, A. Scaglione and M. R. Scarfi, *Bioelectromagnetics*, 1990, **11**, 57.

40. D. B. Millar, J. P. Christopher, J. Hunter and S. S. Yeandle, *Bioelectromagnetics*, 1984, **5**, 165.

41. A. Homenko, B. Kapilevich, R. Kornstein and M. A. Firer, *Bioelectromagnetics*, 2009, **30**, 167.

42. A. R. Orlando, C. Arcovito, A. Palombo, A. L. Serafino and G. Mossa, *J. Liposome Res.*, 1993, **3**, 7170724.

43. T. Maugard, D. Gaunt, M. D. Legoy and T. Besson, *Biotechnol. Lett.*, 2003, **25**, 623.

44. C. Kamerke, M. Pattky, C. Huhn and L. Elling, *J. Mol. Catal. B: Enzym.*, 2012, **79**, 27.

45. F. La Cara, M. R. Scarffi, S. D'Auria, R. Massa, G. d'Ambrosio, G. Franceschetti, M. Rossi and M. De Rosa, *Bioelectromagnetics*, 1999, **20**, 172.

46. M. Porcelli, G. Cacciapuoti, S. Fusco, R. Massa, G. d'Ambrosio, C. Bertoldo, M. De Rosa and V. Zappia, *FEBS Lett.*, 1997, **402**, 102.

47. V. M. Pavelkic, D. R. Stanisavljev, K. R. Gopcevic and M. V. Beljanski, *Russ. J. Phys. Chem. A*, 2009, **83**, 1473.

48. T. Sun, B. Jiang and B. Pan, *Int. J. Mol. Sci.*, 2011, **12**, 3786.

49. H. Bo-lin and L. Ji-xing, *Theoretics physics and life science*, Shanghai Science and Technology Press, Shanghai, 1997.

50. M. Zarevucka, M. Vacek, Z. Wimmer, C. Brunet and M. D. Legoy, *Biotechnol. Lett.*, 1993, **21**, 785.

51. V. Vojisavljevic, E. Pirogova and I. Cosic, *Med. Biol. Eng. Comput.*, 2011, **49**, 793.

52. Z. Xiao-yun, Q. Wenqiug, T. Xueda and H. Meng, *J. Cent. South Univ. Technol.*, 2011, **18**, 1029.

53. P. M. Reddy, W. Y. Hsu, J. F. Hu and Y. P. Ho, *J. Am. Soc. Mass Spectrom.*, 2010, **21**, 421.

54. D. Yu, H. Wu, A. Zhang, L. Tian, L. Liu, C. Wang and X. Fang, *Process Biochem.*, 2011, **46**, 599.

55. S. Bradoo, P. Rathi, R. K. Saxena and R. Gupta, *J. Biochem. Biophys. Methods*, 2002, **51**, 115.

56. S. T. Chen, S. H. Chiou and K. T. Wang, *J. Chin. Chem. Soc.*, 1991, **38**, 85.

57. B. N. Pramanik, U. A. Mirza, Y. H. Ing, Y.-H. Liu, P. L. Bartner, P. C. Weber and A. K. Bose, *Protein Sci.*, 2002, **11**, 2676.

58. S. Lin, D. Yun, D. Qi, C. Deng, Y. Li and X. Zhang, *J. Proteome Res.*, 2008, **7**, 1297.

59. S. S. Lin, C. H. Wu, M. C. Sun, C. M. Sun and Y. P. Ho, *J. Am. Soc. Mass Spectrom.*, 2005, **16**, 581.

60. I. C. R. Costa, S. G. F. Leite, I. C. R. Leal, L. S. M. Miranda and R. O. M. A. de Souza, *J. Braz. Chem. Soc.*, 2011, **10**, 1993.

61. B. M. Nogueira, C. Carretoni, R. Cruz, S. Freitas, P. A. Melo, R. C. Felix, J. C. Pinto and M. Nele, *J. Mol. Catal. B: Enzym.*, 2010, **67**, 117.

62. D. Yu, L. Tian, D. Ma, H. Wu, Z. Wang, L. Wang and X. Fang, *Green Chem.*, 2010, **12**, 844.

63. D. Yu, C. Wang, Y. Yin, A. Zhang, G. Gao and X. Fang, *Green Chem.*, 2011, **13**, 1869.
64. P. C. M. Da Ros, H. F. de Castro, A. K. F. Carvalho, C. M. F. Soares, F. F. de Moraes and G. M. Zanin, *J. Ind. Microbiol. Biotechnol.*, 2012, **39**, 529.
65. P. C. M. Da Ros, L. Freitas, V. H. Perez and H. F. de Castro, *Bioprocess Biosyst. Eng.*, 2013, **36**, 443.
66. N. Saifuddin, A. Z. Raziah and H. N. Farah, *E-J. Chem.*, 2009, **6**, S485.
67. P. Kerep and H. Ritter, *Macromol. Rapid Commun.*, 2006, **27**, 707.
68. T. D. Matos, N. King, L. Simmons, C. Walker, A. R. McClain, A. Mahapatro, F. J. Rispoli, K. T. McDonnell and V. Shah, *Green Chem. Lett. Rev.*, 2011, **4**, 73.
69. G. D. Yadav and P. S. Lathi, *Synth. Commun.*, 2005, **35**, 1699.
70. G. D. Yadav and P. S. Lathi, *Enzyme Microb. Technol.*, 2006, **38**, 814.
71. B. Major, I. K. Horvath, Z. Csanadi, K. B. Bako and L. Gubicza, *Green Chem.*, 2009, **11**, 614.
72. A. Rajan and T. E. Abraham, *Bioprocess Biosyst. Eng.*, 2006, **29**, 65.
73. A. Rajan, J. D. Sudha and E. Abraham, *Ind. Crops Prod.*, 2008, **27**, 50.
74. A. Rajan, V. S. Prasad and T. E. Abraham, *Int. J. Biol. Macromol.*, 2006, **39**, 265.
75. M. Lukasiewicz and S. Kowalski, *Starch-Starke*, 2012, **64**, 188.
76. G. D. Yadav and P. S. Lathi, *J. Mol. Catal. A: Chem.*, 2004, **223**, 51.
77. M. Risso, M. Mazzini, S. Kroger, P. S. Mendez, G. Seoane and D. Gamenara, *Green Chem. Lett. Rev.*, 2012, **5**, 539.
78. N. E. Leadbeater, L. M. Stencel and E. C. Wood, *Org. Biomol. Chem.*, 2007, **5**, 1052.
79. G. D. Yadav and I. V. Borkar, *AIChE J.*, 2006, **52**, 1235.
80. K. Sarma, N. Borthakur and A. Goswami, *Tetrahedron Lett.*, 2007, **48**, 6776.
81. S. Mazumder, D. D. Laskar, D. Prajapati and M. K. Roy, *Chem. Biodiversity*, 2004, **1**, 925.
82. M. Kidwai, R. Podder and P. Mothsra, *Beilstein J. Org. Chem.*, 2009, **5**, 10.
83. D. Yu, Y. Wang, C. Wang, D. Ma and X. Fang, *J. Mol. Catal. B: Enzym.*, 2012, **79**, 8.
84. L. M. C. Matos, I. C. R. Leal and R. O. M. A. de Souza, *J. Mol. Catal. B: Enzym.*, 2011, **72**, 36.
85. M. G. Pujic, E. G. Jampel, A. Loupy, S. A. Galema and D. Mathe, *J. Chem. Soc. Perkin Trans.*, 1, 1996, 2777.
86. H. Wan, S. Sun, X. Hu and Y. Xia, *Appl. Biochem. Biotechnol.*, 2012, **166**, 1454.
87. Y. Fang, W. Huang and Y. Xia, *Process Biochem.*, 2008, **43**, 306.
88. Y. Fang, S. Sun and Y. Xia, *J. Mol. Catal. B: Enzym.*, 2008, **55**, 6.
89. W. Huang, Y. Xia, H. Gao, Y. Fang, Y. Wang and Y. Fang, *J. Mol. Catal. B: Enzym.*, 2005, **35**, 113.
90. V. Osuna and I. A. Rivero, *J. Mex. Chem. Soc.*, 2012, **56**, 176.
91. G. D. Yadav and P. A. Thorat, *J. Mol. Catal. B: Enzym.*, 2012, **83**, 16.
92. G. D. Yadav and I. V. Borkar, *Ind. Eng. Chem. Res.*, 2009, **48**, 7915.
93. G. D. Yadav and S. D. Shinde, *Int. Rev. Chem. Eng.*, 2012, **4**, 589.

94. M. C. Parker, T. Besson, S. Lamare and M. D. Legoy, *Tetrahedron Lett.*, 1996, **37**, 8383.
95. B. Rejasse, S. Lamare, M. D. Legoy and T. Besson, *Org. Biomol. Chem.*, 2004, **2**, 1086.
96. B. Rejasse, T. Besson, M. D. Legoy and S. Lamare, *Org. Biomol. Chem.*, 2006, **4**, 3703.
97. H. Zhao, G. A. Baker, Z. Song, O. Olubajo, L. Zanders and S. M. Cambell, *J. Mol. Catal. B: Enzym.*, 2009, **57**, 149.
98. M. Kidwai, B. Dave, K. R. Bhushan, P. Misra, R. K. Saxena, R. Gupta, R. Gulati and M. Singh, *Biocatal. Biotransform.*, 2002, **20**, 377.
99. G. D. Yadav and P. S. Lathi, *Clean Technol. Environ. Policy*, 2007, **9**, 281.
100. A. R. Rufino, F. C. Biaggio, J. C. Santos and H. F. de Castro, *Int. J. Biol. Macromol.*, 2010, **47**, 5.
101. M. Vacek, M. Zarevucka, Z. Wimmer, K. Stransky, K. Demnerova and M. D. Legoy, *Biotechnol. Lett.*, 2000, **22**, 1565.
102. M. Kidwai, P. Mothsra and R. Podder, *Ind. J. Chem.*, 2009, **48**, 1307.
103. M. Kidwai, P. Mothsra, N. Gupta, S. S. Kumar and R. P. Gupta, *Synth. Commun.*, 2009, **39**, 1143.
104. G. D. Yadav and S. V. Pawar, *Bioresour. Technol.*, 2012, **109**, 1.
105. Y. Chen, X. Xu, B. Xu, Z. Jin, R. Lim, M. Bashari and N. Yang, *Eur. Food Res. Technol.*, 2010, **231**, 719.
106. G. D. Yadav and A. D. Sajgure, *J. Chem. Technol. Biotechnol.*, 2007, **82**, 964.
107. J. C. Munoz, D. Bouvet, E. G. Jampel, A. Loupy and A. Petit, *J. Org. Chem.*, 1996, **61**, 7746.
108. R. O. M. A. de Souza, O. A. C. Antunes, W. Kroutil and C. O. Kappe, *J. Org. Chem.*, 2009, **74**, 6157.
109. G. Lin and W. Y. Lin, *Tetrahedron Lett.*, 1998, **39**, 4333.
110. D. Yu, Z. Wang, P. Chen, L. Jin, Y. Cheng, J. Zhou and S. Cao, *J. Mol. Catal. B: Enzym.*, 2007, **48**, 51.
111. P. Bachu, J. S. Gibson, J. Sperry and M. A. Brimble, *Tetrahedron: Asymmetry*, 2007, **18**, 1618.
112. P. Bachu, J. Sperry and M. A. Brimble, *Tetrahedron*, 2008, **64**, 4827.
113. S. S. Ribeiro, J. R. de Oliveira and L. M. Porto, *J. Braz. Chem. Soc.*, 2012, **23**, 1395.
114. G. D. Yadav and S. V. Pawar, *Appl. Microbiol. Biotechnol.*, 2012, **96**, 69.
115. G. D. Yadav and S. Devendran, *J. Mol. Catal. B: Enzym.*, 2012, **81**, 58.
116. K. Lundell, T. Kurki, M. Lindroos and L. T. Kanerva, *Adv. Synth. Catal.*, 2005, **347**, 1110.
117. D. Yu, D. Ma, Z. Wang, Y. Wang, Y. Pan and X. Fang, *Process Biochem.*, 2012, **47**, 479.
118. D. Yu, P. Chen, L. Wang, Q. Gu, Y. Li, Z. Wang and S. Cao, *Process Biochem.*, 2007, **42**, 1312.
119. C. Pilissao, P. O. Carvalho and M. G. Nascimento, *J. Braz. Chem. Soc.*, 2012, **23**, 1688.
120. J. B. Sontakke and G. D. Yadav, *J. Chem. Technol. Biotechnol.*, 2011, **86**, 739.

121. A. N. Parvulescu, E. V. Eycken, P. A. Jacobs and D. E. de Vos, *J. Catal.*, 2008, **255**, 206.
122. G. D. Yadav, A. D. Sajgure and S. B. Dhoot, *J. Chem. Technol. Biotechnol.*, 2008, **83**, 1145.
123. P. Nahar and U. Bora, *Anal. Biochem.*, 2004, **328**, 81.
124. D. Bezbradica, D. Mijin, M. Mihailovic and Z. C. Jugovic, *J. Chem. Technol. Biotechnol.*, 2009, **84**, 1642.
125. Z. Du, A. Wang, C. Zhou, S. Zhu and S. Shen, *J. Chem. Eng. Jpn.*, 2009, **42**, 441.
126. A. Wang, M. Wang, Q. Wang, F. Chen, F. Zhang, H. Li, Z. Zeng and T. Xie, *Bioresour. Technol.*, 2011, **102**, 469.
127. F. Chen, F. Zhang, F. Du, A. Wang, W. Gao, Q. Wang, X. Yin and T. Xie, *Bioresour. Technol.*, 2012, **115**, 158.
128. A. Wang, M. Liu, H. Wang, C. Zhou, Z. Du, S. Zhu, S. Shen and P. Quyang, *J. Biosci. Bioeng.*, 2008, **106**, 286.
129. G. D. Yadav, A. D. Sajgure and S. B. Dhoot, in *Enzyme Mixtures and Complex Biosynthesis*, ed. S. K. Bhattacharya, Landes Biosciences, Austin, TX, 2007.
130. R. England, *LabPlus Internat.*, April/May 2003, pp. 1–3, (http://www.biotechmedia.com; http://lab-plusinternational.com).

Lipase-Catalyzed Reactions in Pressurized Fluids

RAQUEL LOSS[a], LINDOMAR LERIN[a], JOSÉ VLADIMIR DE OLIVEIRA[a], AND DÉBORA DE OLIVEIRA*[a]

[a]Federal University of Santa Catarina, Brazil
*E-mail: debora@enq.ufsc.br

5.1 Introduction

The major drawback to the widespread usage of many enzymes compared to chemical catalysts is their relatively low stability in their native state. There is great interest in developing competitive biocatalysts for industrial applications by improvement of their activity, stability, and re-usage capacity. Furthermore, modification of the reaction environment has been explored using alternative solvents, including organic solvents, solvent-free systems, dense or supercritical gases and ionic liquids. Therefore, enzymes suitable for use in industrial biocatalysis may require the application of a combination of these improvement methods, and research into new environmentally benign solvents and catalysts has become an area of significant research in green chemistry.[1]

Non-conventional solvents are receiving more and more attention in biocatalysis, mainly in enzymatic hydrolysis, transesterification, esterification, interesterification, and enantioselective synthesis. Solvents can modify the conformation of an enzyme and hence alter its catalytic efficiency or specificity, which may increase the regio- and enantioselectivity of reactions and also

RSC Green Chemistry No. 45
White Biotechnology for Sustainable Chemistry
Edited by Maria Alice Z. Coelho and Bernardo D. Ribeiro

Published by the Royal Society of Chemistry, www.rsc.org

increase the enzyme's stability.[2] Therefore, it is very important to optimize and alter the course of enzymatic reactions in non-aqueous media, including conventional organic solvents, supercritical fluids and ionic liquids.[3–6]

Supercritical fluids are a rapidly growing alternative to conventional reaction media, as these fluids allow higher rates of mass transfer in view of their favorable transport properties. Also, the solvation power of supercritical and other parameters, such as density, dielectric constant, diffusivity, viscosity, and the solubility affecting the reaction, are easily manipulated by temperature and pressure.[7] Processes in supercritical fluids also tend to have advantages in terms of energy reduction and reduction in side reactions. Another advantage of the use of supercritical/pressurized gases as solvents for enzyme-catalyzed reactions is their simple downstream processing.[8]

The combination of the properties of the liquid and vapor phases characteristic of the supercritical state is extremely advantageous for the use of supercritical fluids as solvents. Supercritical fluids have densities close to the liquids, which strengthens their solvent properties. Moreover, viscosity, surface tension, and diffusivity have values close to the gaseous state, which makes the transport properties highly favorable for processes. All of these unique properties of supercritical fluids make them quite interesting media for chemical reactions.[9]

The most extensively utilized supercritical fluid in biocatalysis is $scCO_2$ due to its low-polarity, useful characteristics, such as its non-toxicity, non-flammability, environmental acceptability, low cost, availability in large quantities, tunable solvent properties, solvation power, high (liquid-like) density, low viscosity and high diffusivity. $scCO_2$ is regarded as a green solvent and has low surface tension, which reduces the substrate diffusion limitations and CO_2 has a moderate critical temperature and pressure (31.1 °C and 7.38 MPa).[7,10] Due to its non-toxicity and non-flammability, it is a suitable solvent for the preparation of food additives.[11] On the other hand, some challenges are present in enzyme catalysis with $scCO_2$, mainly related to carbamate formation on the enzyme and the control of the pH of the reaction due to carbonic acid formation.[4,12] Nevertheless, CO_2 is not the only gas with adequate properties for biocatalysis. For example, methane, ethane, propane, fluoroform and sulfur hexafluoride have also been used as supercritical fluids for biocatalysis.

The enzyme activity at high pressure is a crucial point to develop and understand enzymatic processes in supercritical and pressurized fluids. Enzyme stability and activity may depend on the enzyme species, supercritical fluid, water content of the enzyme/support/reaction mixture, decompression rates, exposure times, and the pressure and temperature of the reaction system.[12–14] In other words, the results depend on the results of pressure/temperature induced changes to the properties of supercritical fluids.[7]

The use of non-aqueous solvents for enzymatic reactions is attractive for several reasons. An enzyme in a non-aqueous solvent may have solvent/enzyme interactions similar to those in its native environment and may thus show increased activity compared to pure water. Substrates may also be more soluble in a non-aqueous solvent, so that reaction rates are higher in such solvents. The thermostability of biomolecules in pressurized fluids is greater than that in water and there is the possibility of recycling the solvent.[9,15]

The phase behavior of the fluid can facilitate a reaction through the use of controlled depressurization, which can allow the separation of substrates and products, without leaving harmful solvent residues. Diffusion is typically faster in supercritical fluids as compared to liquids, and can speed up both homogeneous and heterogeneous reactions.[16] It should also be noted that water in the supercritical state cannot be used with enzymes, as its critical parameters are well above those tolerated by proteins.[12]

In light of the increasing interest in the development of alternative media for lipase-catalyzed reactions to overcome the problems related to the use of organic solvents, and taking into account the possible industrial applications and the concept of "green chemistry", the aim of this section is to perform a review of lipase-catalyzed reactions in supercritical/pressurized fluids, showing the potential of this alternative technique. The first section of this chapter will be dedicated to the behavior of lipases in supercritical/pressurized fluids. The knowledge of this is of fundamental importance for further application of these biocatalysts in these alternative media. A review of the application of supercritical/pressurized fluids as solvents for lipase-catalyzed processes will also be presented.

5.2   Behavior of Lipases in Supercritical and Compressed Fluids

Significant progress has been made over the last two decades in terms of supercritical/pressurized fluid processing of fats and oils, with some applications reaching commercial level.[17] However, as in any field, there are still some challenges ahead and more work is needed to reach the full potential of possibilities. Currently, one can infer that the most relevant challenge is related to understanding the real influence of the pressurized/supercritical fluid on the lipase behavior. The objectives of this section are to reflect on the challenges in fat and oil catalyzed reactions in these alternative fluids from a fundamental perspective. Attention will be devoted to the main hypotheses presented in the literature in an attempt to understand the influence of supercritical fluids on lipase activity and stability. Special attention will be given to the use of scCO$_2$ as an alternative pressurized solvent, as it has been mostly used for this purpose. Whenever possible, discussion about the use of other pressurized fluids will be also presented.

There are a lot of studies that have reported on enzyme stability and activity in different supercritical fluids or in compressed gases. Table 5.1 summarizes some relevant studies on this subject recently presented in the literature. The enzyme behavior in compressed fluids is of primary importance as the

Table 5.1 Summary of studies on the behavior of lipases in supercritical/pressurized fluids.

Catalyst	Solvent	Reaction conditions	Behavior	References
M. miehei lipase	CO_2	110.4 bar; 35 °C	0.4% Activity loss	18
	Propane	250 bar; 75 °C	1.1% Activity loss	
	Butane	10 bar; 35 °C	0.0% Activity loss	
C. antarctica lipase B	CO_2	71.5 bar; 35 °C	1.3% Activity loss	
	Propane	30 bar; 35 °C	1.7% Activity loss	
	Butane	250 bar; 75 °C	21.5% Activity gain	
Y. lipolytica	CO_2	71.5 bar; 35 °C	10.2 Activity loss	25
	Propane	30 bar; 35 °C	0.0% Activity loss	
	Butane	250 bar; 75 °C	0.5% Activity loss	
P. simplicissimum	Propane	30 bar; 35 °C	427% Residual activity	29
C. antarctica lipase B	$scCO_2$	80 bar; 60 °C	923^a mU cm^{-2}	32
C. viscosum lipase	$scCO_2$/	110 bar; 40 °C	ee$_p$ 65%; $E = 11^b$	36
Porcine pancreas lipase	buffer biphasic		ee$_p$ 90%; $E = 21$	
P. cepacia lipase	system		ee$_p$ 83%; $E = 40$	
M. miehei lipase	$scCO_2$	59 bar; 50 °C	15×10^{-6} mmol min^{-1} g$^{-1\,c}$	39
C. antarctica lipase B	$scCO_2$	90 bar; 40 °C	ee$_p$ 99.8%^{b} ee$_s$ 90.6% $E > 100$	40
		130 bar; 40 °C	ee$_p$ 99.7%^{b} ee$_s$ 89.6% $E = 1850$	
C. antarctica lipase B	$ncCO_2$	300 bar; 40 °C	98%^{d}	41
C. antarctica lipase B	$scCO_2$	100 bar; 35 °C	200.7 mmol min^{-1} g$^{-1\,c}$	44
C. antarctica lipase B	$scCO_2$	150 bar; 95 °C	>99%^{b}; 48%^{d}	47
C. antarctica lipase B	$scCO_2$	260 bar; 60 °C	79	48
M. miehei lipase			58	
P. cepacia lipase			41	
Porcine pancreas lipase	C_3H_8	100 bar; 50 °C	43%^{d}	50
Porcine pancreas lipase immobilized		100 bar; 40 °C	70%^{d}	
M. miehei lipase	$scCO_2$	140 bar; 40 °C	30%^{d}	55
P. cepacia lipase	$scCO_2$	120 bar; 40 °C	ee$_s$ 98%^{b} ee$_p$ 18%	56
C. antarctica lipase B	$scCO_2$	100 bar; 40 °C	73%^{d}	57
	sc-Ethane		98%^{d}	
M. miehei lipase	$scCO_2$	180 bar; 60 °C	1600 mmol min^{-1} g$^{-1\,c}$	61
C. cylindracea lipase	$scCO_2$	110 bar; 45 °C	0.05 mmol h^{-1} mg$^{-1\,c}$	65
	$scSF_6$		5.5 mmol h^{-1} mg$^{-1\,c}$	
	scC_3H_8		0.5 mmol h^{-1} mg$^{-1\,c}$	
	scC_2H_6		0.3 mmol h^{-1} mg$^{-1\,c}$	
	C_2H_4		0.3 mmol h^{-1} mg$^{-1\,c}$	
	$scCHF_3$		0.3 mmol h^{-1} mg$^{-1\,c}$	

(continued)

Table 5.1   (*continued*)

Catalyst	Solvent	Reaction conditions	Behavior	References
M. miehei lipase	scCO$_2$	100 bar; 50 °C	70%[b]	67
C. antarctica lipase B	scCO$_2$	110 bar; 40 °C	0.41 mol mol^{-1} s^{-1}[c]	68
Hog pancreas lipase	scCO$_2$	83 bar; 65 °C	22.5%[d]	70
P. roqueforti lipase		83 bar; 50 °C	15.3%[d]	
R. arrhizus lipase	scCO$_2$	83–110 bar; 35 °C	15 × 10^{-7} mol L^{-1} s^{-1}[c]	71

[a]Enzymatic activity.
[b]Stereoselectivity.
[c]Initial rate.
[d]Conversion.

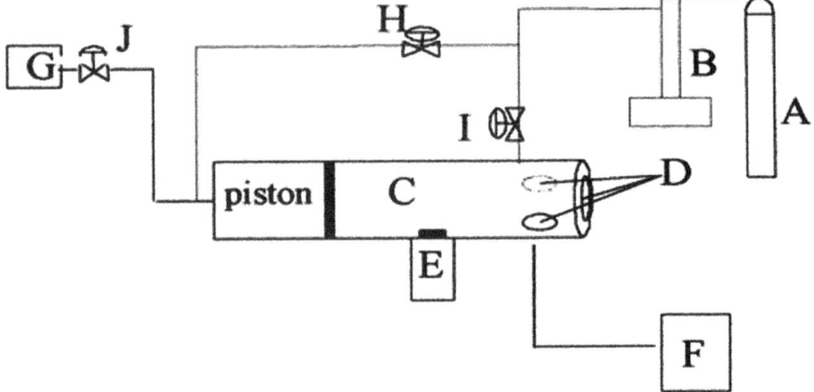

Figure 5.1   Schematic diagram of the high-pressure apparatus for enzyme activity tests. A, solvent reservoir; B, syringe pump; C, equilibrium cell; D, sapphire windows; E, magnetic stirrer; F, white light source; G, pressure transducer; H, ball valve; I, micrometering valve; J, relief valve. From: Oliveira *et al.* (2006).

loss of enzyme activity may lead to undesirably poor reaction rates and low yields of target products. Enzyme stability and activity may depend on the enzyme species, the characteristics of the compressed fluid, the water content of the enzyme/carrier/reaction mixture and the process variables manipulated.[18] Some of these aspects will be addressed briefly in this section. Figure 5.1 presents a schematic diagram of the high-pressure apparatus used for enzyme activity tests.

5.2.1   Effect of Nature of Solvent

Oliveira *et al.*[18] investigated the behavior of two commercial immobilized lipases subjected to compressed carbon dioxide, propane and *n*-butane under different reaction conditions (35 to 75 °C, 10 to 280 bar, exposure times from 1 to 6 h, with distinct depressurization rates). The lipase Lipozyme RM IM from *Mucor miehei* showed activity losses in all compressed solvents, markedly in carbon dioxide. For Novozym 435, treatment in carbon dioxide also led to activity losses, while the use of propane and *n*-butane promoted enhancement of the enzyme activity. A non-commercial lipase from *Yarrowia lipolytica* also presented losses of activity for all tested fluids, markedly again for carbon dioxide (Figure 5.2).

A compilation of these results is presented in Figure 5.3, for a determined experimental condition. The treatment in carbon dioxide also had a deleterious effect on the activity of the commercial Novozym 435 and Lipozyme IM, though to a lesser degree compared to immobilized YLL. On the other hand, it is worth noticing that treatment of Novozym 435 in compressed propane and *n*-butane improved the enzyme activity for this experimental condition, with resulting activity gains of around 10%. For these two compressed gases, however, activity losses were also observed for Lipozyme IM.

Thermogravimetric analyses showed that the thermal profiles of Novozym 435 treated in *n*-butane and in carbon dioxide were similar to that of the untreated enzyme. Scanning electron micrographs (SEM) of Novozym 435 indicated that the material subjected to carbon dioxide presented morphological alterations when compared to the untreated enzyme.

Housaindokht and Monhemi[19] and Housaindokht, Bozorgmehr and Monhemi[20] performed a very interesting study in an attempt to explain the different behaviors presented by an enzyme subjected to different pressurized fluids. The authors assessed the structure of *Candida antarctica* lipase B (CALB) by molecular dynamics simulations. In this tool, a common way to monitor the structural stability of a macromolecule is to calculate the root mean square deviation (rmsd) from the initial structure during the simulation. Following this procedure, the authors explained the changes that occur in the conformation of the enzyme when subjected to high pressure, using $scCO_2$ and near-critical propane. The rmsd values show that the enzyme has a more native-like structure in near-critical propane compared to water. Although the overall structural integrity is needed for enzymatic activity, more effects could be related to the conformation of residues near the active site. Minor structural changes in these regions may lead to significant alterations in the activity and specificity.

Analyzing the structural alignments of the crystal and the simulated structures of the enzyme in $scCO_2$ and near-critical propane, it can be seen that the overall structure of the enzyme is very close to the crystal form in near-critical propane, but there are significant structural deviations in the case of $scCO_2$. In particular, the conformations of regions, including the active site funnel entrance (α_5 and α_{10}) are partially decomposed in $scCO_2$ and deviate from those in the crystal. However, in near-critical propane the helical arrangements of α_5 and α_{10} remain very close to those in the crystal. These data stated that the enzyme activity in near-critical propane can be related to the native-like structure of the active site entrance in this solvent.

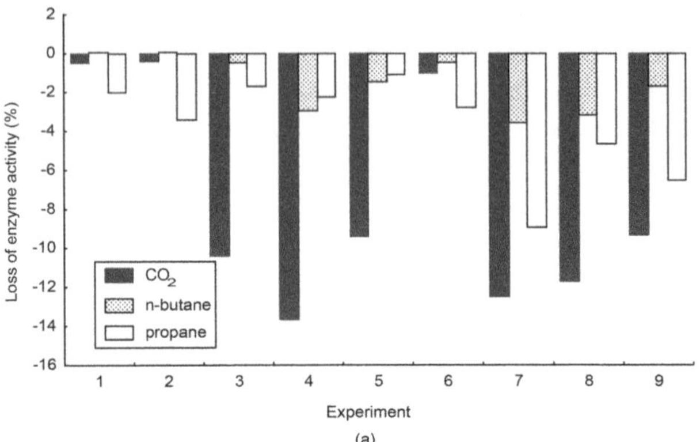

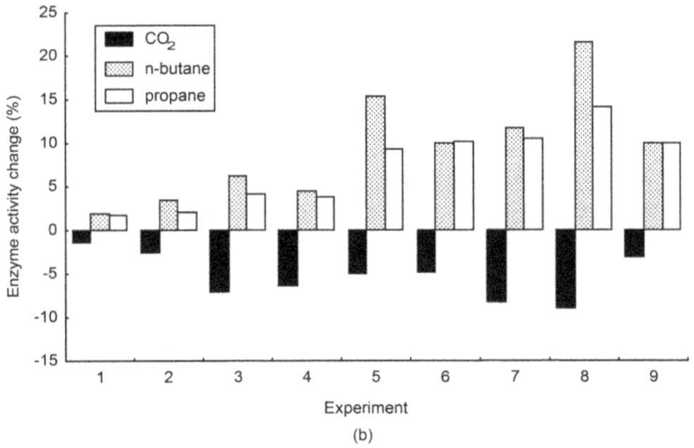

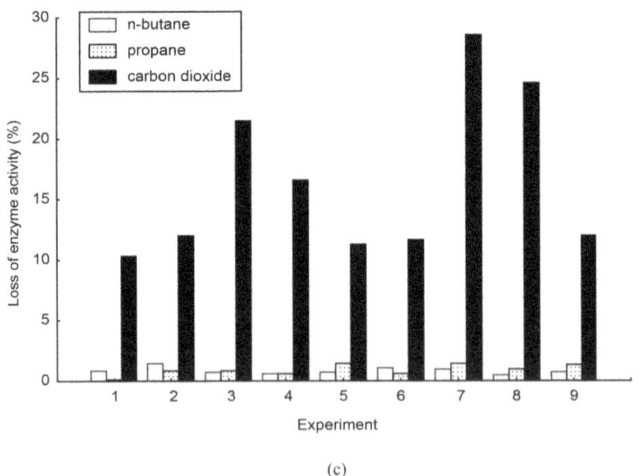

Figure 5.2 Enzyme activity change after treatment in compressed carbon dioxide, propane and *n*-butane: (a) Lipozyme IM, (b) Novozym 435 and (c) lipase from *Yarrowia lipolytica*. From: Oliveira *et al.* (2006).

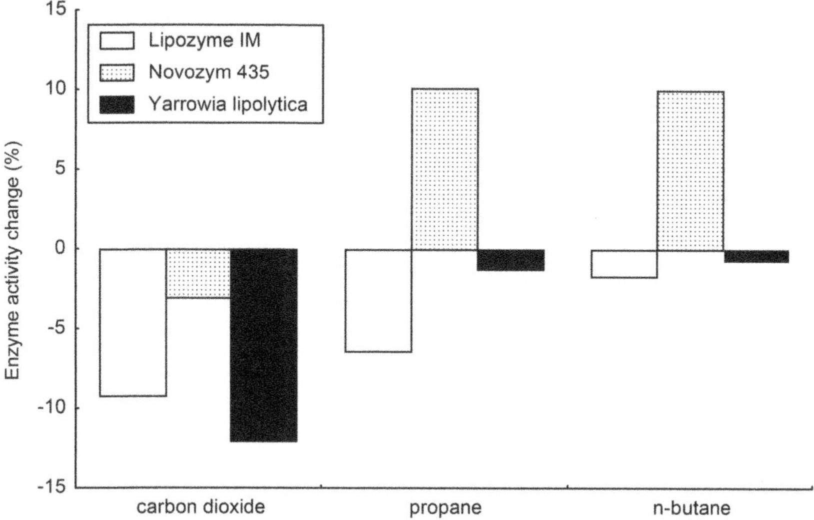

Figure 5.3 Comparison of activity changes between lipase from *Yarrowia lipolytica* with two commercial lipases after treatment in compressed carbon dioxide, propane and *n*-butane at experimental condition 9 of Tables 5.2–5.4.

Near-critical propane stabilizes the α_5 residues. Inversely, in scCO$_2$, the α_5 residues have a steep rmsd. All of these results suggest that the hydrophobic nature of the solvents can stabilize important regions of the enzyme, such as α_5 in CALB. α_5 is mainly composed of residues with hydrophobic side chains. Hydrophobic interactions of propane molecules with these residues rigidify α_5 and thus, this important element remains native in near-critical propane.[19]

The conformational deviation of the enzyme over the course of the simulation can be further elucidated by examining its time-averaged secondary structure content. Then, by analyzing the secondary structures of CALB in crystal form, in near-critical propane and in scCO$_2$ through molecular dynamics simulations, it can be observed that the helix structures of residues P12-G19, G44-S50, P119-K124, V139-S150 (α_5) and T267-A276 (α_{10}) of CALB are partially decomposed in scCO$_2$, while in near-critical propane, these structures are almost native and close to those in the crystal. All of these helices are surface elements and their structural deviations are directly related to solvent effects. The β-sheet contents of CALB in all of the solvents under study are almost intact because they are internal secondary structures in CALB and the solvent is not able to induce their conformations.

The radius of gyration (R_g) is another parameter that describes the equilibrium conformation of a total system. It is an indicator of the protein structure compactness of the conformation[21] Monhemi and Housaindokht[22] analyzed the values of R_g obtained by molecular dynamics simulations of CALB in each solvent and observed a more compact and, thus, more active conformation of the enzyme in near-critical propane than in $scCO_2$. In near-critical propane, the hydrophobic residues of the active site entrance interact advantageously with propane molecules. As a result, these regions are stable in this condition and the cavity between α_5 and α_{10} remains native and open.

There is experimental evidence that confirms the greater stability of the enzymes in near-critical propane than in $scCO_2$. Habulin and Knez[23] reported the stability and activity of lipases from *Pseudomonas fluorescens*, *Rhizopus javanicus*, *Rhizopus niveus*, porcine pancreas and *Candida rugosa* in a non-solvent system at atmospheric pressure, in $scCO_2$ and near-critical propane at 100 bar and 40 °C. The authors observed that the enzymes are more stable in near-critical propane than in $scCO_2$. The large loss in enzyme activity in $scCO_2$ was attributed to interactions between CO_2 and enzyme molecules since this loss was not observed in near-critical propane. In $scCO_2$ there was a great loss in activity of the examined lipases. In contrast, the use of near-critical propane improved the activity of lipases in comparison to the non-solvent system by four times (porcine pancreas lipase) to nine times (*R. javanicus* lipase).

Housaindokht, Bozorgmehr and Monhemi[20] used molecular dynamics simulations to evaluate the structure of *Candida antarctica* lipase B (CALB) in $scCO_2$ and water. The authors observed that the values of rmsd in water increased gradually within 1 ns of simulation and then reached equilibrium values of about 0.6 nm. These high values show that, in comparison with the wild-type enzyme, a significant conformational change occurred for CALB. However, the rmsd of the enzyme is still considerably larger in $scCO_2$ than in water. In $scCO_2$, the structural variations of the enzyme are greater than those in the aqueous solution and therefore, dissolving the enzyme in $scCO_2$ imposes undesirable instability on its structure, once high values indicate a conformationally labile model. Despite the many advantages of $scCO_2$ in enzymatic reactions, it is necessary to develop enzyme stabilization strategies in this solvent.

These results are in agreement with those reported by Ishikawa *et al.*[24] These authors showed that $scCO_2$ treatment at 35 °C and 25 MPa could reduce the residual α-helix content of lipase, alkaline protease, acid protease, and glucoamylase to 62.9, 31.3, 37.6 and 12.4%, respectively. They also found a linear relationship between residual activities against the residual α-helix contents. The reduction of the ordered secondary structure is effective in both the activity and stability of proteins. Oliveira *et al.*[25] observed that significant activity losses were obtained when the treatment of *Yarrowia lipolytica* lipase was conducted in $scCO_2$.

Housaindokht and Monhemi[19] used molecular dynamics simulations to evaluate the structure of *Burkholderia cepacia* lipase (BCL, formerly known as *Pseudomonas cepacia* lipase). To detect the possibility of lid movements in the open conformation of BCL, the rmsd of the αC carbon for the overall enzyme structure and also for the α_5 helix in water and in near-critical propane were computed. The profile obtained in water indicates that the rmsd of the overall structure gradually increased and reached an equilibrium value of about 0.4 nm, while the rmsd of the α_5 helix increased quickly and reached high values of about 1 nm. A different story occurred in near-critical propane. The rmsd values have similar profiles for the overall enzyme structure and the α_5 helix with only slight differences in the final times of simulation. This shows that lid movement in the open conformation of BCL most likely occurs only in water. Moreover, the rmsd values for the overall enzyme structure in near-critical propane reach about 0.3 nm which is lower than those values obtained in water. Thus, the enzyme shows higher conformational stability in near-critical propane than in water.

To get a better picture of lid movement, the authors measured the distance between α_5 and α_9 as a criterion for lid closure in the open conformation of BCL during simulation. The enzyme conformation is shifted to the inactive form whenever this distance is reduced. By simulation of the open conformation of BCL in water, the distance between α_5 and α_9 decreased from 2.3 nm to about 1 nm. Thus, the enzyme conformation changes from open to closed in water. However, in near-critical propane, reduction of the distance between α_5 and α_9 does not occur. In the compressed gas, the enzyme not only remains active, but a more open conformation was observed in comparison with the crystal.

BCL has so far only been crystallized in an open conformation. So, for the simulation of the closed model of the enzyme, the average final structure which was obtained after the simulation of open BCL in water was used. Monitoring the distance between α_5 and α_9 from the initially closed BCL, it was found that in water, these two helices tended to be closer to each other, while in near-critical propane the distance increased and the conformation shifted from closed to open to some extent.

Interfacial activation may be the major reason for the high activity of lipases in compressed propane. It is difficult to experimentally monitor the conformational transition of the enzymes, especially in the high pressure conditions of a compressed gas. To get evidence about the possibility of the interfacial activation of lipases in compressed gases, multiple molecular dynamics simulations were performed for open and closed BCL. When open BCL was simulated in water, the α_5 helix (lid region) showed large deviations from the crystal structure in comparison with the overall enzyme structure. In near-critical propane, however, deviations of the α_5 helix were very close to those of other regions. The measured distance between the α_5 and α_9 helices as a criterion for lid closure in BCL clearly showed that in water, the conformation changed from open to closed, but in near-critical propane, the conformation remained open. By simulation of the closed conformation of BCL, other interesting results were obtained. The closed conformation remained closed in water and shifted to open conformation in near-critical propane. Exposing BCL to near-critical propane leads to opening of the lid, while in water lid closure occurred.

The results on the lid closure of BCL in aqueous solvents are in complete agreement with recently reported data in the theoretical investigations of Barbe *et al.*[26] and Trodler, Schmid and Pleiss.[27] They also found that the open conformation of BCL dominates in organic solvents such as octane and toluene. Similar results have been obtained by James *et al.*[28] for *Candida rugosa* lipase at alkane/aqueous interfaces. Thus, the conformational stability of the enzyme may be a common reason for the activeness of different enzymes in compressed propane, but it is not the major reason for the high activity of lipases. This is confirmed in the experimental observations which argued that in compressed propane, the activity is even higher than that in the non-solvent system. When two non-commercial lipases from *Penicillium simplicissimum* and *Aspergillus parasiticus* were pretreated in compressed propane, an enhancement in the residual activities after incubation was observed under several experimental conditions.[29] The increment in activity even reached 427% for *P. simplicissimum*. These results can be explained by the existence of an important event in the active site region (especially the lid) by exposing the enzyme to compressed gas. It is most likely that the major factor which determines the lipase activity in compressed propane is lid opening.

In addition to the nature of the solvent, some other parameters can affect the lipase activity under the effect of a pressurized fluid. Some of these will be discussed below.

5.2.2 Effects of Changing Pressure

Pressure is likely to affect the reaction performance indirectly by changing either the rate constant or the reactants solubility. At higher pressures more solute–solvent interactions take place, resulting in better solvent capacity. Regarding the effects of changing pressure, several suggestions have been made in the literature.[14]

Aaltonen[30] reported that apart from the direct conformational changes in enzymes, which may occur at very high pressures, pressure affects enzymatic reaction rates in supercritical fluids in two ways. First, the reaction rate constant changes with pressure according to transition state theory and standard thermodynamics. Theoretically, one can predict the effect of pressure on reaction rate if the reaction mechanism, the activation volumes and the compressibility factors are known. Second, the reaction rates may change with the density of SCFs because physical parameters, such as the dielectric constant, change with density. These changes may indirectly influence enzyme activity.

The impact of operating conditions on the enzymatic esterification of *n*-octyl oleate catalyzed by Lipozyme TL IM was investigated by Laudani *et al.*[31] The experimental evidence was that changing the pressure actually changed the enzymatic reaction rate at constant substrate concentration. A series of tests at various pressures was performed in a constant volume reactor, keeping the substrate concentration constant. The results showed that in all the reported cases, with the increase in pressure, the reaction rate decreased. The reason for this particular behavior could be explained by taking into account that with increasing pressure, at constant volume, the molar fraction of substrates decreased. Consecutively, reduction in the initial reaction rate could be observed.

Lozano *et al.*[32] evaluated the synthesis of butyl butyrate from vinyl butyrate and 1-butanol using Novozym 435 and reported that, at any fixed temperature between 40 and 60 °C, an increase in pressure (80–150 bar) resulted in a decrease in the synthetic activity of the enzyme. This effect was attributed to a pressure-related increase in density of $scCO_2$.

Ikushima *et al.*[33] and Ikushima *et al.*[34] suggested that changes in enantioselectivity with pressure are due to the interaction of CO_2 and enzyme molecules. Studying the catalytic activity of *Candida cylindracea* lipase, the authors found that the enzyme was activated near the critical point of CO_2. The authors suggested that CO_2, in the near-critical region, triggered this activation by causing movement of the α-helical lid from a closed conformation (enzyme inactive) to an open conformation (enzyme active).

Mase *et al.*[35] reported enhanced enantioselectivity for the lipase-catalyzed desymmetrization of 1,3-propanediacetate in $scCO_2$ as compared to that in organic solvents. This was also attributed to the transformation of lysine groups by reaction with CO_2 to form carbamates. However, Hartmann, Meyer and Scheper[36] described the peak in enantioselectivity (*E* value) at 103 bar for the hydrolysis of 3-hydroxy-5-phenyl-4-pentenoic acid ethyl ester catalyzed by *Pseudomonas cepacia* lipase in a biphasic buffer/$scCO_2$ system. Above and below 103 bar, the *E* value decreased, and this effect was attributed to a direct inactivation of the biocatalyst by formation of carbamates.

Some reports have suggested that reactions are enhanced near the critical point. Albrycht *et al.*[37] demonstrated that the reactivity and selectivity of the kinetic resolution of *P*-chiral hydroxymethanephosphinates, catalyzed by Novozym 435, can be controlled by tuning the pressure at values high above the critical point. The reaction was reported to be fastest at 130 bar, way above the critical pressure. Erickson, Schyns and Cooney[38] reported a drop in reaction rate as the pressure increased, particularly as the critical pressure is approached, for the reaction between trilaurin and palmitic acid catalyzed by *Rhizopus arrhizus* lipase. Nakaya, Miyawaki and Nakamura[39] described the transesterification of triolein and stearic acid catalyzed by Lipozyme TL IM and classified the reaction into three regions according to the pressure. Below 50 bar, the reaction rate was very slow and limited to the liquid triolein phase; in the near-critical region (50–100 bar), the rate was maximal at 59 bar, possibly due to stabilization of the enzyme–substrate complex; in the supercritical region (>100 bar), the reaction rate increased with increasing pressure, reflecting an increase in substrate solubility.

Knez and Habulin[8] exposed immobilized lipase from *R. miehei* for 24 and 46 h to CO_2, nitrogen, *n*-butane and a mixture of *n*-propane : *n*-butane (70 : 30) at 100 bar and 35 °C, and observed that there was no change in enzyme activity in $scCO_2$, butane and the mixture of propane–butane. Some deactivation of the enzyme could be determined only with longer exposure to nitrogen. In the same work, non-immobilized lipases *Pseudomonas fluorescens*, *Rhizopus javanicus*, *Rhizopus niveus* and porcine pancreas, were exposed to $scCO_2$ and near-critical propane at a pressure of 300 bar and at 40 °C for 24 h and the results showed that there was no activity change for any of the examined lipases, when exposed to high pressure pure CO_2 and propane.

However, Matsuda *et al.*[40] demonstrated that changes in pressure for the continuous kinetic resolution of *rac*-1-phenylethanol catalyzed by Novozym 435 in $scCO_2$ did not greatly affect either conversion or *E* values. Steytler, Moulson and Reynolds[41] stated that on increasing the pressure of $scCO_2$ to 500 bar, the synthesis of butyl laurate catalyzed by *Candida antarctica* B lipase was not significantly affected. Also, nonyl acetate synthesis catalyzed by *Mucor miehei* lipase was not affected significantly by changes in pressure[42] and the effect of changes in pressure (100–250 bar) was small in the hydrolysis of blackcurrant oil catalyzed by Lipozyme TL IM.[43]

In fact, there seems to be no "rule of thumb" for predicting enzyme activity and enantioselectivity in $scCO_2$. Some authors say that working near the critical point is advantageous for good selectivity; some say that it is sufficient that the conditions are at or above the critical point, and yet still others report that it does not make any difference. It may be that the effect of CO_2 on enzyme activity is very dependent on the specific enzyme, substrates, and reaction studied.

5.2.3   Effects of Changing Temperature

Oliveira *et al.*[44] studied the synthesis of decyl acetate by the transesterification of vinyl acetate with decanol using $scCO_2$ as solvent and Novozym 435 as catalyst and observed that the increase of temperature has a slightly positive effect on the reaction rate. Similar behavior was found by Habulin *et al.*,[45] evaluating the esterification of citronellol with lauric acid in $scCO_2$–ethyl methylketone media, where the authors reported that an enhancement of the reaction rate was observed when the temperature was increased from 50 to 60 °C, but at 70 °C, the reaction rate decreased. When the effect of pressure was evaluated, they observed that on increasing the pressure from 102 to 130 bar, 160 and 190 bar, the conversion after 1 min of reaction decreased by only 3%, 11% and 15%, respectively. This effect could be related to changes in the density-dependent properties of $scCO_2$ (*e.g.* partition coefficient, dielectric constant, Hildebrand solubility parameters) that indirectly regulate the activity, specificity and stability of enzymes.

Nakaoki, Kitoh and Gross[46] have shown that Novozym 435 is still active even after heating to 140 °C in $scCO_2$. Overmeyer *et al.*[47] also observed good Novozym 435 activity and enantioselectivity at temperatures above 95 °C for the kinetic resolution of ibuprofen with *rac*-1-phenylethanol, and this is supported by the work of Turner *et al.*[48] for the hydrolysis of retinyl palmitate acetate by the same enzyme. It was suggested that dry compressed CO_2 stabilizes the protein structure of Novozym 435 or that there is a faster mass transfer of the substrate to the active site of the enzyme plus higher reaction rates at elevated temperatures.[47,48]

In contrast, Primožič, Habulin and Knez[49] demonstrated the deactivation of Lipolase 100T in scCO$_2$ at 200 bar and 50 °C. They suggested that this is due to the denaturation of the enzyme. Other reports suggested an increase in the enzyme thermal stability in nc-propane as compared to that in water; for example, the optimum reaction temperature for porcine pancreas lipase in water is 40 °C, but in near-critical propane, the optimum temperature is 50 °C.[50] It is suggested that this is probably a consequence of protein structural and conformational rigidity in propane and this may give better substrate specificity for the reaction studied.[51]

Various hypotheses and considerations have underlined the fact that the pressure leads to a more compact, rigid enzyme structure, reducing the thermal effects caused by high temperatures. For instance, one of the considerations was related to the water exchange between the interior part of the enzyme and the bulk solvent, which was reported to increase below a pressure of 1000 bar due to conformational fluctuations.[52] Pressure was also reported to increase the density of the first hydration shell at the protein surface, which induces a constraint on lateral chain motion.[53,54]

An optimum temperature of 40 °C in scCO$_2$ was reported for the synthesis of butyl laurate catalyzed by crude *Candida antarctica* lipase B for the synthesis of geranyl acetate catalyzed by Lipozyme TL IM and for the resolution of 3-hydroxyoctanoic acid methyl esters catalyzed by *Pseudomonas cepacia* lipase.[55,56]

However, 62 °C has been reported as the optimum temperature for the hydrolysis of 3-hydroxy-5-phenyl-4-pentenoic acid ethyl ester catalyzed by *Pseudomonas cepacia* lipase.[36] Conversely, Peres, Silva and Barreiros[57] reported that changes in temperatures between 40 and 60 °C have little effect on geranyl acetate synthesis by Novozym 435. Sovová and Zarevucka[43] demonstrated that changes in temperature between 30 and 40 °C also have little effect on the catalytic activity of Lipozyme TL IM.

The temperature dependence of the enantioselectivity of enzyme-catalyzed reactions and the importance of both entropic and enthalpic factors were first systematically studied in the late 1980s.[58] A number of detailed studies on the temperature dependence of the enantioselectivity of lipase-catalyzed reactions in organic solvents were conducted by Sakai[59] who exploiting temperature to improve the enantioselectivity of a lipase-catalyzed reaction in an organic solvent. An initial experiment demonstrated that 1-azirine methanol could be esterified using *Burkholderia cepacia* lipase in diethyl ether with an *E* value of 99 at −40 °C in diethyl ether, but an *E* value of only 17 was observed at room temperature.

In scCO$_2$, Matsuda *et al.*[60] looked at the Novozym 435 catalyzed enantioselective acetylation of *rac*-1-(*p*-chlorophenyl)-2,2,2-trifluoroethane with vinyl acetate at 31, 40, 55, and 60 °C. A rapid change in *E* value was observed between 31 and 40 °C, and a more gradual change was observed at higher temperatures. The authors noted that these changes correlate well with the changes in CO$_2$ density and went on to evaluate the *E* values at various temperatures and pressures but at the same density. They reported that the

E values were affected by temperature, with higher temperatures resulting in lower enantioselectivity, in line with the observations for enzyme-catalyzed reactions in either aqueous or organic solvents.

Al-Duri, Goddard and Bosley[61] investigated lipase (Lipozyme TL IM) in scCO$_2$. Their results suggested that increasing the pressure from 13 to 18 MPa did not significantly affect the lipase activity. The results also showed that the lipase activity increased when the temperature increased from 40 to 60 °C, and hence the rate constant increased too. Besides, pressure changes had an insignificant effect on the stability, while temperature had a prominent effect. At 60 °C the enzyme lost 30% of its activity over 8 h, compared to a 15% loss at 40 °C. This confirms the negative effect of higher temperature on the enzyme structure leading to denaturation.

The activity of Lipozyme TL IM exhibits an optimum operating temperature for *n*-octyl oleate esterification in the 70–80 °C range.[31] At a pressure level of 10 MPa, an increase in temperature also resulted in higher solubility of substances in supercritical fluids because the increase in the vapor-pressure of the compounds to be dissolved overcomes the reduction in density. In this case, the temperature effect, in addition to its effect on the enzyme activity, was directly related positively to the supercritical fluid's solvating power. Thus, operating in supercritical fluids at 10 MPa, a good compromise among optimum solvating power and enzyme activity, on one hand, and enzyme thermal stability, on the other hand could be found.

5.2.4   Effect of Changing Water Content

Water is crucial for enzymes and affects their action by influencing the structure *via* non-covalent binding and disruption of hydrogen bonds, by facilitating reagent diffusion, and by influencing the reaction equilibrium. In the complete absence of water, enzymes are catalytically inactive. The most common explanation for this is that a minimum of a single layer of water molecules is required at the critical points on the enzyme's surface to maintain the native protein structure. Use of an enzyme in pure scCO$_2$ may lead to the removal of water, which is included in or bonded to the enzyme. The quantity of removed water is temperature and pressure dependent. To avoid enzyme deactivation as a consequence, water could be added to the system at the start of the reaction. The optimal initial water concentration should be determined for each reaction system, because even small differences in initial water concentration may cause big differences in enzyme action.[62,63]

Zaks and Klibanov[64] were the first to note that enzymes are more active in hydrophobic rather than in hydrophilic organic solvents, and they suggested that this was due to differences in water partitioning between the enzyme and the bulk solvent. In essentially non-aqueous systems, any water present will partition between the enzyme and the solvent. On considering hydrophilic solvents, water will partition preferably into the solvent, and this will tend to strip the essential water off the enzyme, hence destroying the native structure and any enzyme activity. In contrast, hydrophobic solvents will not strip the essential layer of water from the enzyme, as these solvents become saturated with water at much lower concentrations; hence, the activity of the enzyme is maintained.

Kamat *et al.*,[65] in a comparison of reaction rates and hydrophobicity of supercritical fluids, observed that reaction rates increase with increasing hydrophobicity of supercritical fluids due to reduced stripping of the essential water molecules surrounding the enzyme. Then, in terms of hydrophobicity of the supercritical fluids evaluated, the authors obtained the trend $CO_2 < CHF_3 < C_2H_4 < C_2H_6 < C_3H_8 < SF_6$.

A similar trend in conversion was obtained for the synthesis of octyl palmitate[12] and for the transesterification reaction between 2-ethylhexanol and methylmethacrylate,[65] wherein the highest activity was observed in supercritical methane. However, it is imperative to note that the initial rates in the four supercritical fluids cannot be directly compared, because the densities of the fluids relative to their critical density are different. Furthermore, it is important to note that, although the initial rates of the reaction are different, the conversion at the end of 20 h is nearly the same in all systems.

Experiments in supercritical fluids have demonstrated that $scCO_2$ can strip water off enzymes, reducing their activity. Kamat *et al.*[65] reported the *Candida cylindracea* lipase-catalyzed transesterification of methylmethacrylate in several supercritical fluids and observed a marked decrease in enzyme activity in $scCO_2$. Reaction rate increases were found to correlate with increasing hydrophobicity of the supercritical fluids. Hence, it appears that the loss of activity was a result of the enzyme losing essential water. This is surprising since CO_2 is generally considered to be a hydrophobic solvent. However, Jackson *et al.*[66] explained that CO_2 is more hydrophilic than fluoroform or hexane and is therefore capable of stripping essential water from an enzyme, thereby inactivating it.

Habulin *et al.*[51] demonstrated increased enzyme activity in near-critical propane as compared with $scCO_2$ for the lipase-catalyzed esterification of butyric acid with ethanol, and they suggest that this is due to the stripping of water from the enzyme into CO_2. Rantakyla and Aaltonen[67] investigated the esterification catalyzed by *Mucor miehei* lipase in $scCO_2$ and showed that a pressure increase from 10 to 25 MPa reduced the initial reaction rates. It was presumed that at higher pressure, a more considerable water amount was extracted from the enzyme beads, which resulted in lower reaction yields.

Steytler, Moulson and Reynolds[41] studied the synthesis of butyl laurate using crude *Candida antarctica* B lipase in $scCO_2$ and demonstrated that the reaction was enhanced on addition of water. Three experiments were reported as follows: (i) dry enzymes in the absence of water: the performance of the enzyme in $scCO_2$ was comparable with that in toluene under equivalent conditions of temperature and pressure; (ii) water-saturated enzymes: the reaction was severely retarded and hydrolysis was forced, and (iii) water-saturated $scCO_2$ was added above the enzyme contained in the water phase. In this case, the transfer of water between the two phases was minimized since both enzyme and solvent were hydrated; therefore, the reaction rate was enhanced.

Dijkstra *et al.*[68] evaluated the enantioselective esterification of *rac*-1-phenylethanol by vinyl acetate catalyzed by cross-linked enzyme crystals of *Candida antarctica* B lipase and observed that the reaction is very sensitive to the amount of water present. A concentration of 0.05 g L^{-1} resulted in optimum enzymatic activity, while the enzyme was (reversibly) deactivated at lower water concentrations. This was attributed to the stripping of catalytically important water molecules from the surface of the enzyme. However, Kmecz *et al.*[69] reported that the use of dry or humid CO_2 made little difference to the activity of Amano lipase from *Pseudomonas fluorescens* for the acylation of 3-benzyloxypropane-1,2-diol.

Alternative studies have looked at the effect of varying the water content in the system. Vermue *et al.*[42] described the decrease in transesterification of nonanol and ethyl acetate by Lipozyme TL IM in $ncCO_2$ on increasing the water content from 0.05 to 0.2% (v/v). Srivastava and Madras[70] studied the hydrolysis of *p*-nitrophenyl laurate to *p*-nitrophenyl catalyzed by hog pancreas lipase or *Penicillium roqueforti* lipase in $scCO_2$ and reported that both enzymes were hindered on increasing the water content, due to either the inactivation of the enzyme or the formation of an aqueous layer around the enzyme that contributes to mass transfer resistance. Still others have reported that changes in water content do not affect the intrinsic activity of the enzyme, although it is generally agreed that the higher the water content is, the greater the degree of unwanted substrate/product hydrolysis observed.[71,43]

5.2.5   Water Activity (a_w)

Halling[72] has suggested that the thermodynamic activity of water, rather than water concentration is the key parameter in understanding the effect of water on enzymatic reactions. The term water activity (a_w) describes the amount of water available for hydration of materials. A value of one indicates pure water while zero indicates the total absence of "free" water molecules; the addition of solutes always lowers a_w which is defined as the product of the activity coefficient of water in the solvent and the molar fraction of water in the solvent.

A low a_w can be achieved and fine-tuned in $scCO_2$ using zeolite molecular sieves, such as NaA[73] or salt hydrates, $Na_2CO_3 \cdot H_2O/Na_2CO_3 \cdot 10H_2O$.[74] The effect of these solid state buffers has been extensively studied, and it was found that an acid–base effect was actually occurring. A transesterification reaction, catalyzed by subtilisin cross-linked enzyme crystals, was noted to increase in rate up to 10-fold with increasing amount of zeolite and therefore the corresponding a_w in $scCO_2$. The initial hypothesis was that a_w was low enough to decrease carbonic acid formation (hence minimize changes in pH), but still adequate for the function of subtilisin; from the same observations, it was also noted that subtilisin requires a formal negative charge on the catalytic triad for full activity. This would require removal of a proton and replacement by a counterion such as Na^+ for electroneutrality. This was tested by performing the reaction under three conditions in sc-ethane: (i) with zeolite only, (ii) with both zeolite and CAPSO [3-(cyclohexylamine)-2-hydroxy-1-propanesulfonic acid (sodium salt), a sodium/proton acid–base buffer], and (iii) with CAPSO only. The initial rate in the presence of buffer, regardless of the presence of zeolite, was reasonably similar; therefore, it was concluded that the zeolite effect must be of an acid–base nature.[73,75]

Other such reports have been made, including the investigation into the best solid state acid–base buffer to use in supercritical fluids. The buffer Na_2CO_3–$NaHCO_3$ was shown to increase enzyme activity up to 54-fold, probably due to its high basicity and capacity to counteract the deleterious effect of carbonic acid.[74] Six zwitterionic proton/sodium buffers were tested, and it was concluded that the higher the basicity of the buffer is, the higher the catalytic activity obtained.[76]

Hence, this work highlights the need for the evaluation of the acid–base behavior of an extensive set of salt hydrates to identify one that is able to optimize the activity of an enzyme in $scCO_2$. Overall, Fontes, Halling and Barreiros[76] strongly recommend the use of acid–base buffers in enzymatic reactions in non-aqueous solvents, especially in supercritical fluids where the use of salt hydrates still remains the most practical technique for setting and controlling a_w.

5.2.6   Effect of Pressurization and Depressurization

Habulin *et al.*[50,51] exposed crude lipases from *Pseudomonas fluorescens*, *Rhizopus javanicus*, *Rhizopus niveus* and porcine pancreas to $scCO_2$, and also to nc-propane, and reported no activity change for the esterification of butyric acid following a depressurization step. The ability to perform the reaction, catalyzed by porcine pancreatic lipase in nc-propane, numerous times with the same batch of enzyme was also demonstrated. The conversion level only decreased to half its initial value after 10 reaction cycles, and the decrease was shown to be due to the increase in water released during the esterification reaction at the enzyme surface and not inactivation due to the pressurization and depressurization steps. Lanza *et al.*[13] investigated the influence of temperature, pressure, exposure time, and decompression rate on lipase activity in high-pressure CO_2 media, using Novozym 435. The results showed that an increase in temperature and density led to an enhancement of enzyme activity losses, while the decompression rates had a weak influence on enzyme inactivation.

5.3   Lipase-Catalyzed Reactions in Supercritical and Compressed Fluids

The first reports on lipase-catalyzed reactions in supercritical fluids were those by Randolph *et al.*[77] and Hammond *et al.*[78] Lipases now constitute the most important group of biocatalysts for the synthesis of many interesting compounds for the pharmaceutical, cosmetics and food industries. Among the various enzymatic methods in which supercritical/pressurized fluids find applications are esterification, transesterification, synthesis of biodiesel and hydrolysis of oil and fat. Below, we show some application examples reported in the literature. A compilation of these data is presented in Tables 5.2–5.4.

5.3.1   Esterification

The performance of lipase from *Mucor miehei* in the myristic acid esterification in hexane was compared to that in $scCO_2$ and higher activity was found in supercritical media. The maximum velocity appears 1.5-fold higher in $scCO_2$ than in hexane.[79]

Lipase from *Candida cylindracea* was used for the esterification between *n*-valeric acid and citronellol. The results showed a dependence between the reaction rate and the pressure, which increased significantly until reaching a maximum at 7.55 MPa, near the critical point of CO_2.[80]

Knez and Habulin[8] evaluated the esterification of oleic acid with oleyl alcohol, catalyzed by immobilized lipase from *R. miehei*. Reactions were performed at 20–350 bar and at 20–50 °C, using as reaction media $scCO_2$, butane and a mixture of propane–butane. The highest reaction rates were obtained using $scCO_2$ at 250 MPa and 50 °C. However, when the same authors evaluated the esterification between butyric acid and ethanol catalyzed by non-immobilized lipases from *Pseudomonas fluorescens*, *Rhizopus javanicus*, *Rhizopus niveus*, *Candida Rugosa* and porcine in $scCO_2$ and propane at high pressure (100 bar and 40 °C), they observed that $scCO_2$ used as reaction medium deactivated all of the lipases studied, while in propane, just *C. rugosa* was deactivated.

Srivastava, Modak and Madras[81] reported the synthesis of commercially important flavor esters of isoamyl alcohol using crude hog pancreas lipase and observed that the overall conversion under solvent-free conditions was higher than that in $scCO_2$, but under conditions of low enzyme loading (5 mg), the conversions obtained in $scCO_2$ were higher than the conversions obtained under solvent-free conditions. This indicates that $scCO_2$ might be more commercially viable at low enzyme concentrations.

Novak *et al.*[82] reported similar findings: porcine pancreas lipase immobilized as a sol–gel demonstrated much improved conversion for the esterification of butyric acid with isoamyl alcohol in nc-propane as compared to $scCO_2$.

Peres *et al.*[57] reported that *Candida antarctica* lipase B is more active in sc-ethane compared with $scCO_2$ for the esterification of geraniol with acetic acid. Besides, the conversion was higher in sc-ethane (98%) than in $scCO_2$ (73%) at 100 bar and 40 °C.

Madras, Kumar and Modak[83] suggested that sc-methane is the supercritical fluid of choice for the esterification of octyl palmitate catalyzed by *Candida antarctica* lipase B, possibly due to the high solubility of substrates or a more favorable enzyme conformation in this medium. The highest conversion

Table 5.2 Lipase-catalyzed esterification reactions in supercritical fluids.

Catalyst	Solvent	Substrate + products	Reaction conditions	Yield or initial rate	References
M. miehei lipase	scCO$_2$ Hexane	Myristic acid + ethanol → ethyl myristate	12.5 MPa; 50 °C Patm; 50 °C	0.833 µmol min^{-1} mg^{-1} 0.532 µmol min^{-1} mg^{-1}	79
C. cylindracea lipase	scCO$_2$	Valeric acid + citronellol → citronellol valerate	7.55 MPa; 35 °C	282.1 µmol h^{-1} g^{-1}	80
M. miehei lipase	scCO$_2$ Butane Butane–propane	Oleic acid + oleyl alcohol → oleyl oleate	250 MPa; 50 °C	13 mmol h^{-1} g^{-1} 5 mmol h^{-1} g^{-1} 8 mmol h^{-1} g^{-1}	8
P. fluoresces lipase *R. javanicus* lipase *R. niveus* lipase *C. rugosa* lipase Porcine pancreas lipase	Propane	Butyric acid + ethanol → ethyl butyrate	100 bar; 40 °C	29 mmol h^{-1} g^{-1} 39 mmol h^{-1} g^{-1} 18 mmol h^{-1} g^{-1} — 14 mmol h^{-1} g^{-1}	
Crude hog pancreas lipase	scCO$_2$	Acid + isoamyl alcohol → flavor esters	90 bar; 45 °C	3–70%	81
Porcine pancreas lipase	scCO$_2$ nc-Propane	Butyric acid + isoamyl alcohol → isoamyl butyrate	100 bar; 40 °C	31% 79%	82
C. antarctica lipase B	scCO$_2$ sc-Ethane	Acetic acid + geraniol → geranyl acetate	100 bar; 40 °C	73% 98%	57
C. antarctica lipase B	scCO$_2$ sc-Ethane sc-Methane	Palmitic acid + octanol → octyl palmitate	100 bar; 55 °C 46 bar; 55 °C 130 bar; 55 °C	76% 80% 85%	83
M. miehei lipase	scCO$_2$	Ibuprofen + propanol	10 MPa; 45 °C	70%	67

(continued)

Table 5.2 (*continued*)

Catalyst	Solvent	Substrate + products	Reaction conditions	Yield or initial rate	References
C. antarctica lipase B	$scCO_2$	Acetic anhydride + iso-amyl alcohol → isoamyl acetate	15 MPa; 40 °C	100%	84
C. antarctica lipase B	$scCO_2$	1-(*p*-Chlorophenyl)-2,2,2-trifluoroethanol (*RS*)-1 + vinyl acetate	9.1 MPa; 40 °C	25%	85
C. antarctica lipase	$scCO_2$	Lauric acid + 1-propanol → propyl-laurate	110 bar; 35 °C	36.5%	86
C. antarctica lipase B	$scCO_2$	Acetic acid + lavandulol → lavandulyl acetate	10 MPa; 60 °C	86%	87
M. miehei lipase	$scCO_2$	Oleic acid + 1-dodecanol → lauryl oleate	10 MPa; 50 °C	80%	88
C. antarctica lipase B	$scCO_2$	Sugar + fatty acid → sugar fatty acid esters	10 MPa; 80 °C	67%	11
T. lanuginosus lipase	$scCO_2$	Camel hump fat + tristearin → cocoa butter analog	10 MPa; 40 °C	74.45%	89
C. rugosa lipase	$scCO_2$	α-Terpineol + acetic anhydride → terpinyl acetate	10 MPa; 50 °C	95.1%	90
C. antarctica lipase B	$scCO_2$	Lactic acid + butanol → lactate esters	40 MPa; 55 °C	100%	91

Table 5.3 Lipase-catalyzed transesterification reactions in supercritical fluids.

Catalyst	Solvent	Substrate + products	Reaction conditions	Yield or initial rate	References
M. miehei lipase	$ncCO_2$	Nonanol + ethyl acetate → nonyl acetate + ethanol	12 MPa; 60 °C	8 μmol s^{-1} kg^{-1}	42
P. cepacea lipase	$scSF_6$	1-Phenylethanol + vinyl acetate	10 MPa; 50 °C	50%	92
	$scCO_2$		20 MPa; 50 °C	45%	
C. antarctica lipase B	$scCO_2$	Ethylene glycol + ethyl acetate	10 MPa; 50 °C	65%	93
C. antarctica lipase B	$scCO_2$	Butyl butyrate + geranyol → geranyl butyrate	100 bar; 50 °C	230 mmol g^{-1} h^{-1}	94
	scC_2H_6		60 bar; 50 °C	290 mmol g^{-1} h^{-1}	
	scC_2H_4		160 bar; 50 °C	293 mmol g^{-1} h^{-1}	
	$scCH_4$		130 bar; 50 °C	408 mmol g^{-1} h^{-1}	
R. oryzae lipase	$scCO_2$	Citronellol + vinyl acetate → citronellol acetate	8 MPa; 45 °C	91%	95
		Citronellol + vinyl butyrate → citronellol butyrate		98%	
		Citronellol + vinyl laurate → citronellol laurate		99%	
C. antarctica lipase B	$scCO_2$	*Jatropha curcas* oil + ethanol/methanol → FAEE/FAME	68 bar; 45 °C	51; 44%	96
		Pongamia pinnata oil + ethanol/methanol → FAEE/FAME		50; 45%	
		Groundnut oil + ethanol/methanol → FAEE/FAME		66; 57%	
		Palm oil + ethanol/methanol → FAEE/FAME		75; 61%	
C. antarctica lipase B	Propane	Soybean + ethanol → FAEE	50 bar; 65 °C	75%	97
C. antarctica lipase B	$scCO_2$	Sesame oil + ethanol/methanol → FAEE/FAME	100 bar; 50 °C	8.49; 10.23 mol g^{-1} h^{-1}	98
		Mustard oil + ethanol/methanol → FAEE/FAME		11.37; 14.49 mol g^{-1} h^{-1}	
C. antarctica lipase B	$scCO_2$	Castor oil + ethanol/methanol → FAEE/FAME	68 bar; 50 °C	35; 28%	99
C. antarctica lipase B	Propane	Soybean + ethanol → FAEE	50 bar; 65 °C	92%	100
T. lanuginosus lipase	$ncCO_2$	Canola oil + methanol → FAME	100 bar; 30 °C	99%	101

Table 5.4   Lipase-catalyzed hydrolysis reactions involving oils and fats in supercritical fluids.

Catalyst	Solvent	Substrate + products	Reaction conditions	Yield or initial rate	References
M. miehei lipase	$scCO_2$	Canola oil + H_2O → FFA + MG, DG, TG	10–38 MPa; 35 and 55 °C	63–67% TG conversion	106
M. miehei lipase	$scCO_2$	Canola oil + H_2O → glycerol + FFA	24 MPa; 35 °C	90%	107
M. miehei lipase	$scCO_2$	Canola oil + H_2O → glycerol + FFA	24 MPa; 35 °C	97% TG conversion	108
M. miehei lipase	$scCO_2$	Blackcurrant oil + H_2O → products, specific toward linoleic acids	10–25 MPa; 30–40 °C	100%	43
Lipolase 100T	$scCO_2$	Sunflower oil + water → oleic acid + linoleic acid	20 MPa; 50 °C	$0.300\ g_{linoleic\ acid}\ g^{-1}_{oil\ phase}$	49
Lipolase 100T	$scCO_2$	Sunflower oil + H_2O → oleic acid + linoleic acid	20 MPa; 50 °C	$0.193\ g_{oleic\ acid}\ g^{-1}_{oil\ phase}$; $0.586\ g_{linoleic\ acid}\ g^{-1}_{oil\ phase}$	109
T. lanuginosus lipase	$scCO_2$	Milk fat + H_2O → FFA + MG, DG, TG	30 MPa; 55 °C	98% Conversion of CLA in TG form to FFA form	110

was obtained when sc-methane (85%) was used as an esterification medium, while lower conversion was obtained with $scCO_2$ (76%).

Rantakyla and Aaltonen[67] evaluated the enantioselective esterification of (*R,S*)-ibuprofen with propanol catalyzed by Lipozyme TL IM. The initial reaction rates increased with pressure, but enantioselectivity was not affected by pressure changes. The reaction rates for the esterification of ibuprofen were similar in both $scCO_2$ and *n*-hexane.

Immobilized *Candida antarctica* lipase B (CALB) was successfully used as a catalyst for esterification of butyl butyrate in $scCO_2$. All supercritical conditions essayed enhanced the activity by 84-fold with respect to synthesis in organic solvents, while maintaining stability, showing a 360 cycle half-life. The best results were achieved at 60 °C and 8 MPa and were explained by improved micro-environments around the enzyme.[32] CALB was also effective at synthesizing isoamyl acetate in $scCO_2$ at higher initial reaction rates when compared to hexane, while maintaining activity from 8 to 30 MPa.[84]

The enantioselective acetylation of racemic 1-(*p*-chlorophenyl)-2,2,2-trifluoroethanol (*RS*)-1 with lipases and vinyl acetate in $scCO_2$ was studied by Matsuda *et al.*,[85] and they found that the enantioselectivity of the reaction catalyzed by *Candida antarctica* lipase B can be controlled by adjusting the pressure and temperature of $scCO_2$. Enzymatic esterification at pressures ranging from 8 and 19 MPa and for different reaction times at 55 °C showed that the E value decreased from 50 to 10 continuously when the pressure was changed from 8 to 19 MPa, regardless of the reaction time. The highest enantioselectivity ($E = 38$) was obtained at 9.1 MPa.

Lipases from *Candida antarctica* and *Mucor miehei* encapsulated in lecithin water-in-oil (w/o) microemulsion-based organogels (MBGs) were used as catalysts in the esterification of lauric acid and 1-propanol in $scCO_2$ (35 °C, 110 bar) with isooctane as solvent. The initial rates in $scCO_2$ were higher than those observed in isooctane.[86]

The monoterpene lavandulol has been successfully converted to lavandulyl acetate by enzymatic esterification in $scCO_2$ using immobilized *Candida antarctica* lipase B. Conversions of up to 86% were observed at substrate concentrations of 60 mM at 60 °C and 10 MPa.[87]

Mucor miehei lipase was used in the esterification of lauryl oleate from oleic acid and 1-dodecanol using $scCO_2$ as a reaction medium. The catalytic efficiency increased up to 10 MPa which was attributed to either substrate–solvent clustering, the stabilization effect of $scCO_2$ treatment or to the stabilization of lipase in the "open" form by hydrophobic interactions.[88]

Habulin, Šabeder and Knez[11] evaluated the enzymatic synthesis of sugar fatty acid esters in organic solvent and in $scCO_2$ at 10 MPa. The optimal temperature for lipase-catalyzed synthesis in 2-methyl-2-butanol was 60 °C which resulted in 65% conversion, while in $scCO_2$ it was 80 °C which resulted in 67% conversion after 24 h.

Shekarchizadeh *et al.*[89] studied $scCO_2$ as a medium for esterification of camel hump fat and tristearin in producing a cocoa butter analog using immobilized *Thermomyces lanuginosus* lipase (Lipozyme TL IM) as a biocatalyst. The optimum conditions to achieve the maximum yield of the cocoa butter analog were found to be 10 MPa, 40 °C, substrate molar ratio 1:1, water content 10% (w/w) and 3 h of incubation in $scCO_2$.

Liu and Huang[90] evaluated the direct esterification of α-terpineol and acetic anhydride catalyzed by *Candida rugosa* lipase in $scCO_2$ with an organic solvent serving as co-solvent. The highest yield of terpinyl acetate of 95.1% was obtained after 1.5 h of reaction in $scCO_2$ with heptane serving as co-solvent at 50 °C and 10 MPa.

Kuhn *et al.*[29] studied the influence of propane pre-treatment on the esterification activities of two non-commercial lipases from *Penicillium simplicissimum* and *Aspergillus parasiticus* in the lyophilized and immobilized forms. For both lyophilized and immobilized lipases, an enhancement in residual activities after incubation in pressurized propane was observed, under several experimental conditions. The highest increment (427%) occurred with the lyophilized enzyme from *P. simplicissimum*, pressurized at 30 bar for 1 h and then depressurized at the fastest rate (20 bar min^{-1}).

Knez *et al.*[91] evaluated the direct esterification of butanol and lactic acid, catalyzed by immobilized *Candida antarctica* lipase B in $scCO_2$ with or without co-solvent. Experiments were carried out in the pressure range from 7.5 to 40 MPa and at temperatures of 35 and 55 °C. The highest conversion was obtained in $scCO_2$ with hexane serving as a co-solvent at 40 MPa and 55 °C.

5.3.2 Transesterification

Vermue *et al.*[42] studied the enzymatic synthesis of nonanyl acetate *via* transesterification reaction using nonanol and ethyl acetate. The results showed that the transesterification rate in near-critical CO_2 proved to be much lower than that in hexane at comparable conditions of temperature, water content, substrate and enzyme concentration.

Celia *et al.*[92] evaluated the catalytic efficiency of immobilized *Pseudomonas cepacia* lipase in the transesterification of 1-phenylethanol and vinyl acetate using supercritical fluids as reaction media. The enzyme was exposed for 6 h to $scCO_2$ and $scSF_6$ at 50 °C from 4.5 to 25 MPa, and showed high stability in both supercritical fluids with a recovery of enzyme activity of up to 89% after incubation in the supercritical media.

Yasmin *et al.*[93] evaluated the transesterification reaction between a substrate having two functional groups (ethylene glycol) with ethyl acetate in the presence of lipase Novozym 435 in $scCO_2$. The results showed that reaction equilibrium can be reached after 60 min in the presence of CO_2; the equilibrium conversion is higher (65%) than that in the absence of CO_2 (58%). $scCO_2$ could also enhance the selectivity of ethylene glycol monoacetate (EGMA) and suppress the formation of ethylene glycol diacetate (EGDA), and the selectivity could be tuned by the CO_2 pressure.

Varma and Madras[94] investigated the transesterification of butyl butyrate to geranyl butyrate in various supercritical fluids. The initial rate of transesterification of butyl butyrate in different supercritical fluids followed the order: $ScCO_2 < ScC_2H_6 < ScC_2H_4 < ScCH_4$.

Dhake *et al.*[95] investigated the synthesis of citronellol esters with $scCO_2$ as a reaction medium. The optimized conditions of the transesterification reaction for citronellol ester synthesis are: molar ratio of citronellol to vinyl acetate: 1:6, biocatalyst loading: 1.5% (w/v), temperature: 45 °C, pressure: 8 MPa, time: 12 h, 10% hexane as a co-solvent, and $scCO_2$ as solvent.

Biodiesel was enzymatically synthesized using *Candida antarctica* lipase B lipase in the presence of $scCO_2$ using edible oils like palm oil and groundnut oil and from crude non-edible oils like *Pongamia pinnata* and *Jatropha curcas*. The authors observed lower conversion of methyl esters in comparison to ethyl esters for all oils studied, and the highest conversion was observed when the substrate was palm oil.[96]

Dalla Rosa *et al.*[97] investigated the production of fatty acid ethyl ester (FAEE) from soybean oil in compressed propane at 35–65 °C and 50–150 bar using *Candida antarctica* lipase B as catalyst. Complete FAEE conversion was achieved at 65 °C and 50 bar, in 6 h of reaction.

Varma, Deshpande and Madras[98] studied the enzymatic production of biodiesel in $scCO_2$, catalyzed by *Candida antarctica* lipase B using sesame and mustard oil, and sesame oil gave maximum conversions of 51% and 60% with ethanol and methanol, respectively, while mustard oil gave a maximum conversion of 71% in methanol and ethanol at 50 °C and 100 bar. Similar results were obtained by Varma and Madras.[99] The authors evaluated the synthesis of biodiesel catalyzed by *Candida antarctica* lipase B from castor oil using scCO$_2$ as solvent. Conversions of 35 and 28% were achieved in the first 4 h for the transesterification in methanol and ethanol, respectively.

Brusamarelo *et al.*[100] reported catalyzed biodiesel production using soybean oil and ethanol as substrates, using pressurized propane at 50 bar as solvent. The highest content of FAEE, about 92 wt%, was obtained at 65 °C and 50 bar.

Lee *et al.*[101] evaluated the enzymatic synthesis of biodiesel in near-critical CO$_2$ (ncCO$_2$) using different lipases as catalysts. Biodiesel conversion from several edible and non-edible oil feedstocks (canola, soybean, *Jatropha* oils and waste cooking oil) reached 92%. Lipozyme TL IM was found to be most efficient catalyst and the highest conversion was obtained with canola oil. Higher conversion (99.0%) was obtained in a shorter time by employing repeated batch processes with optimized conditions. The enzyme maintained 80.2% of its initial stability after being reused eight times.

5.3.3 Interesterification

Liang, Chen and Liang[102] evaluated the enzymatic interesterification of palm oil by stearic acid in scCO$_2$. The authors used five triglycerides (POP, POS, POO, OOO, and SOO) and two free fatty acids (stearic and palmitic acids) as indicators to monitor the interesterification. The results showed that *Mucor miehei* dominantly catalyzes the interesterification of POP+S«POS and POO+S«SOO when the stearic acid content in the extraction solution is abundant. A very limited amount of SOS is also irregularly found in the product samples. POS and SOO are rarely produced when the loaded stearic acid is completely elutriated, and the weight fraction of POO is significantly increased by the depletion of POP. It is presumed that the palmitoyl group in the 1,3-position is substituted much more readily, and that the stearic acid is a reactive acyl donor for the interesterification in scCO$_2$ when catalyzed by *Mucor miehei*. That large amounts of palmitic acid are found in the transesterified oil confirms this presumption.

A new route for biodiesel production using methyl acetate instead of methanol as the acyl acceptor was proposed by Xu, Du and Liu.[103] The kinetics of lipase-catalyzed interesterification of triglycerides for biodiesel production with methyl acetate as the acyl acceptor was further studied. The authors observed that three consecutive and reversible reactions occurred in the interesterification of triglycerides and methyl acetate. The results showed that k_{DG-MG} (0.1124) and k_{MG-TA} (0.1129) were much higher than k_{TG-DG} (0.0311), which indicated that the first step reaction was the limiting step for the overall interesterification.

Liu, Chang and Liu[104] investigated the substrate oil composition, reaction time, acyl donor, temperature, and pressure effects on the triacylglycerol (TG) content of a cocoa butter analog during the interesterification reaction catalyzed by *Mucor miehei* lipase in $scCO_2$. Among the oil sources used to interact with tristearin, the content of 1(3)-palmitoyl-3(1)-stearoyl-2-monoolein (POS) (P, palmitate; O, oleate; S, stearate) and 1-palmitoyl-2,3-dioleoylglycerol (POO) analogs was most similar to the corresponding TG content of cocoa butter when the analog was prepared with lard. The optimized interesterification reaction was at 17 MPa, 50 °C, and pH 9 for 3 h.

A potential cocoa butter analog was prepared from camel hump fat (CHF) and tristearin (SSS) by enzymatic interesterification in $scCO_2$ using immobilized *Thermomyces lanuginosus* lipase as a biocatalyst. A pressure of 10 MPa, temperature of 42 °C, SSS/CHF ratio of 1.15:1, water content of 10% (w/w), and incubation time of 3 h were found to be the optimum conditions to achieve the most similar cocoa butter analog to the corresponding cocoa butter.[105]

5.3.4 Hydrolysis

Rezaei and Temelli[106] investigated the effects of pressure, temperature, and CO_2 flow rate on the extent of conversion and product composition in the hydrolysis of canola oil in $scCO_2$ catalyzed by *Mucor miehei* immobilized on macroporous anionic resin. A conversion of 63–67% (triglyceride disappearance) was obtained at 24–38 MPa. Mono- and diglyceride production was minimum at 10 MPa and 35 °C. Monoglyceride production was favored at 24 MPa. The amount of product obtained was higher at 24–38 MPa due to enhanced solubility in $scCO_2$.

Hydrolysis of canola oil in $scCO_2$ (24 MPa and 35 °C) catalyzed by *Mucor miehei* immobilized on macroporous anionic resin was studied as a model reaction to develop an on-line extraction–reaction process to extract oil from oilseeds and convert the oil to other valuable products. After a 6 h run at a CO_2 flow rate of 3.9 L min^{-1} (measured at ambient conditions) at 24 MPa and 35 °C, ~90% of the oil in the flakes was recovered. Part of the non-recovered oil was lost during the depressurization step.[107]

The effects of the water flow rate (0.002–0.050 mL min^{-1}), the amount of canola flakes (3.0 and 15.0 g) and enzyme loading (1.0 and 5.0 g of immobilized lipase from *Mucor miehei*, Lipozyme IM), and the reaction cell size were studied in the continuous enzymatic hydrolysis of canola oil in $scCO_2$ at 24 MPa and 33 °C using an online extraction–reaction system by Martinez, Rezaei and Temelli.[108] At a water flow rate of 0.002 mL min^{-1}, 97% triglyceride conversion was achieved, *i.e.*, almost complete hydrolysis. High levels of FFA in the product were obtained with 3 g of canola load, using a smaller reaction cell, and 5 g of enzyme load, indicating a greater extent of hydrolysis.

Sovová and Zarevucka[43] investigated the effect of reaction conditions on the extent of conversion in the hydrolysis of blackcurrant oil in CO_2 saturated with oil and water (55–100%) catalyzed by Lipozyme, a lipase from *Mucor miehei* immobilized on macroporous anionic resin. Complete hydrolysis of the oil was achieved in the experiments carried out with a CO_2 flow rate of

0.4–0.9 g min^{-1}. The effects of pressure (10–25 MPa) and temperature (30–40 °C) on the reaction rate were small, and the effects of CO_2 saturation with water and of enzyme distribution in the reactor were negligible.

Primožič, Habulin, and Knez[49] studied the hydrolysis of sunflower oil at high pressure using Lipolase 100T (*Aspergillus niger* lipase) as catalyst and observed that the activity of Lipolase 100T increased between 35 and 50 °C. The same authors, in another report, studied the hydrolysis of sunflower oil in the presence of the lipase preparation Lipolase 100T (*Aspergillus niger* lipase). scCO$_2$ was used as a solvent for this reaction and the optimal reaction rate and conversion were determined for the hydrolysis process: the concentration of lipase was 0.0714 g per milliliter of CO_2 free reaction mixture, and the highest conversions of oleic acid (0.193 g per gram of oil phase) and linoleic acid (0.586 g per gram of oil phase) were obtained at 50 °C, 200 bar, pH = 7, and an oil/buffer ratio of $1:1$ (w/w).[109]

Conjugated linoleic acid (CLA) in its free form can be obtained by enzymatic hydrolysis of milk fat. Prado *et al.*[110] studied the enzymatic hydrolysis of CLA-enriched anhydrous milk fat (AMF) using scCO$_2$. The maximum level of free fatty acids (FFA) (86.79%, w/w) was achieved using Lipozyme TL IM at 23 MPa, a $1:5$ fat to water ratio (mol mol^{-1}) and 55 °C. The maximum CLA content in the FFA form (6.81 mg g_{fat}^{-1}) was obtained using Lipozyme TL IM at 30 MPa, a $1:30$ fat to water ratio (mol mol^{-1}) and at 55 °C, which corresponds to 98% conversion of CLA in triglyceride form to FFA form.

5.4 Conclusions

A review on lipase-catalyzed reactions in supercritical/pressurized fluids, supported by data in the literature about the behavior of lipases in these alternative solvents was presented in this chapter. The knowledge of this is of fundamental importance for further application of these biocatalysts in these media. The increasing interest in the development of alternative media for lipase-catalyzed reactions to overcome the problems related to the use of organic solvents, taking into account the possible industrial applications and the concept of "green chemistry", permits us to assert the great potential of this technique for reactions catalyzed by lipases. Other enzymes can also be tested in these media, increasing the potential for industrial applications.

References

1. C. Mateo, J. M. Palomo, G. Fernandez-Lorente, J. M. Guisan and R. Fernandez-Lafuente, *Enzyme Microb. Technol.*, 2007, **40**, 1451.
2. S. L. Wells and J. DeSimone, *Angew. Chem., Int. Ed.*, 2001, **40**, 518.
3. C.-S. Chen and C. J. Sih, *Angew. Chem., Int. Ed.*, 1989, **28**, 695.
4. A. J. Mesiano, E. J. Beckman and A. J. Russell, *Chem. Rev.*, 1999, **99**, 623.
5. A. M. Klibanov, *Nature*, 2001, **409**, 241.
6. S. H. Schofer, N. Kaftzik, U. Kragl and P. Wasserscheid, *Chem. Commun.*, 2001, **37**, 425.

7. K. A. Rezaei, F. Temelli and E. Jenab, *Biotechnol. Adv.*, 2008, **25**, 272.

8. Ž. Knez and M. J. Habulin, *J. Supercrit. Fluids*, 2002, **23**, 29.

9. D. Oliveira and J. V. Oliveira, *J. Supercrit. Fluids*, 2001, **19**, 141.

10. D. Senyay-Oncel and O. J. Yesil-Celiktas, *Biosci. Bioeng.*, 2011, **112**, 435.

11. M. Habulin, S. Šabeder and Ž. Knez, *J. Supercrit. Fluids*, 2008, **45**, 338.

12. H. R. Hobbs and N. R. Thomas, *Chem. Rev.*, 2007, **107**, 2786.

13. M. Lanza, W. L. Priamo, J. V. Oliveira, C. Dariva and D. Oliveira, *Appl. Biochem. Biotechnol.*, 2004, **113**, 181.

14. M. Habulin, M. Primožič and Ž. Knez, *Acta Chim. Slov.*, 2007, **54**, 667.

15. S. Kamat, G. Critchley, E. J. Beckman and A. J. Russell, *Biotechnol. Bioeng.*, 1995, **46**, 610.

16. P. G. Jessop and W. E. Leitner, *Chemical Synthesis Using Supercritical Fluids*, Wiley-VCH, Weinheim, 1999.

17. F. Temelli, *J. Supercrit. Fluids*, 2009, **47**, 583.

18. D. Oliveira, A. C. Feihrmann, A. F. Rubira, M. H. Kunita, C. Dariva and J. V. Oliveira, *J. Supercrit. Fluids*, 2006, **38**, 373.

19. M. R. Housaindokht and H. Monhemi, *J. Mol. Catal. B: Enzym.*, 2013, **87**, 135.

20. M. R. Housaindokht, M. R. Bozorgmehr and H. Monhemi, *J. Supercrit. Fluids*, 2012, **63**, 180.

21. M. Y. Lobanov, N. S. Bogatyreva and O. V. Galzitskaya, *Mol. Biol.*, 2008, **42**, 623.

22. H. Monhemi and M. R. Housaindokht, *J. Supercrit. Fluids*, 2012, **72**, 161.

23. M. Habulin and Ž. Knez, *J. Chem. Technol. Biotechnol.*, 2001, **76**, 1260.

24. H. Ishikawa, M. Shimoda, A. Yonekura and Y. Osajima, *J. Agric. Food Chem.*, 1996, **44**, 2646.

25. D. Oliveira, A. C. Feihrmann, C. Dariva, A. G. Cunha, J. V. Bevilaqua, J. Destain and J. V. Oliveira, *J. Mol. Catal. B: Enzym.*, 2006, **39**, 117.

26. S. Barbe, V. Lafaquière, D. Guieysse, P. Monsan, M. Remaud-Simeon and I. Andre, *Proteins: Struct., Funct., Bioinf.*, 2009, **77**, 509.

27. P. Trodler, R. D. Schmid and J. Pleiss, *BMC Struct. Biol.*, 2008, **9**, 1.

28. J. J. James, B. S. Lakshmi, A. S. N. Seshasayee and P. Gautam, *FEBS Lett.*, 2007, **581**, 4377.

29. G. Kuhn, M. Marangoni, D. M. G. Freire, V. F. Soares, M. G. Godoy, A. M. Castro, M. DiLuccio, H. Treichel, M. A. Mazutti, D. Oliveira and J. V. Oliveira, *J. Chem. Technol. Biotechnol.*, 2010, **85**, 839.

30. O. Aaltonen, in *Chemical Synthesis using Supercritical Fluids*, ed. P. G. Jessop and W. Leitner, Wiley-VCH, Weinheim, 1999, p. 414.

31. C. G. Laudani, M. Habulin, G. Della Porta, E. Reverchon and Ž. Knez, in *7th Italian Conference on Chemical and Process Engineering, ICheaP-7*, ed. S. Pierucci, AIDIC, Milano, 2005, p. 843.

32. P. Lozano, G. Víllora, D. Gómez, A. B. Gayo, J. A. Sánchez-Conesa, M. Rubio and J. L. Iborra, *J. Supercrit. Fluids*, 2004, **29**, 121.

33. Y. Ikushima, N. Saito, T. Yokoyama, K. Hatakeda, S. Ito, M. Arai and H. W. Blanch, *Chem. Lett.*, 1993, **22**, 109.

34. Y. Ikushima, N. Saito, M. Arai and H. W. Blanch, *J. Phys. Chem.*, 1995, **99**, 8941.

35. N. Mase, T. Sako, Y. Horikawa and K. Takabe, *Tetrahedron Lett.*, 2003, **44**, 5175.
36. T. Hartmann, H. H. Meyer and T. Scheper, *Enzyme Microb. Technol.*, 2001, **28**, 653.
37. M. Albrycht, P. Kielbasinski, J. Drabowicz, M. Mikolajczyk, T. Matsuda, T. Harada and K. Nakamura, *Tetrahedron: Asymmetry*, 2005, **16**, 2015.
38. J. C. Erickson, P. Schyns and C. L. Cooney, *AIChE J.*, 1990, **36**, 299.
39. H. Nakaya, O. Miyawaki and K. Nakamura, *Biotechnol. Tech.*, 1998, **12**, 881.
40. T. Matsuda, K. Watanabe, T. Harada, K. Nakamura, Y. Arita, Y. Misumi, S. Ichikawa and T. Ikariya, *Chem. Commun.*, 2004, **40**, 2286.
41. D. C. Steytler, P. S. Moulson and J. Reynolds, *Enzyme Microb. Technol.*, 1991, **13**, 221.
42. M. H. Vermue, J. Tramper, J. P. J. Dejong and W. H. M. Oostrom, *Enzyme Microb. Technol.*, 1992, **14**, 649.
43. H. Sovová and M. Zarevucka, *Chem. Eng. Sci.*, 2003, **58**, 2339.
44. M. V. Oliveira, S. F. Rebocho, A. S. Ribeiro, E. A. Macedo and J. M. Loureiro, *J. Supercrit. Fluids*, 2009, **50**, 138.
45. M. Habulin, S. Šabeder, M. Paljevac, M. P. rimožič and Ž. Knez, *J. Supercrit. Fluids*, 2007, **43**, 199.
46. T. Nakaoki, M. Kitoh and R. A. Gross, *ACS Symp. Ser.*, 2005, **900**, 393.
47. A. Overmeyer, S. Schrader-Lippelt, V. Kasche and G. Brunner, *Biotechnol. Lett.*, 1999, **21**, 65.
48. C. Turner, M. Persson, L. Mathiasson, P. Adlercreutz and J. W. King, *Enzyme Microbiol. Technol.*, 2001, **29**, 111.
49. M. Primožič, M. Habulin and Ž. Knez, *J. Am. Oil Chem. Soc.*, 2003, **80**, 643.
50. M. Habulin and Ž. Knez, *Acta Chem. Slov.*, 2001, **48**, 521.
51. M. Habulin and Ž. Knez, *J. Chem. Technol. Biotechnol.*, 2001, **76**, 1260.
52. N. Tanaka, C. Ikeda, K. Kanaori, K. Hiraga, T. Konno and S. Kunugi, *Biochemistry*, 2000, **39**, 12063.
53. P. Mentre' and G. H. B. Hoa, *Int. Rev. Cytol.*, 2000, **201**, 1.
54. N. Smolin and R. Winter, *Biochim. Biophys. Acta*, 2006, **1764**, 522.
55. W. Chulalaksananukul, J.-S. Condoret and D. Combes, *Enzyme Microb. Technol.*, 1993, **15**, 691.
56. A. Capewell, V. Wendel, U. Bornscheuer, H. H. Meyer and T. Scheper, *Enzyme Microb. Technol.*, 1996, **19**, 181.
57. C. Peres, D. R. G. Silva and S. Barreiros, *J. Agric. Food Chem.*, 2003, **51**, 1884.
58. R. S. Phillips, *Trends Biochem. Sci.*, 1996, **14**, 13.
59. T. Sakai, *Tetrahedron: Asymmetry*, 2004, **15**, 2749.
60. T. Matsuda, R. Kanamaru, K. Watanabe, T. Kamitanaka, T. Harada and K. Nakamura, *Tetrahedron: Asymmetry*, 2003, **16**, 909.
61. B. Al-Duri, R. Goddard and J. Bosley, *J. Mol. Catal. B: Enzym.*, 2001, **11**, 825.
62. A. Zaks and A. M. Klibanov, *J. Biol. Chem.*, 1988, **263**, 3194.
63. M. Habulin, M. Primožič and Ž. Knez, *Acta Chim. Slov.*, 2007, **54**, 667.

64. A. Zaks and A. M. Klibanov, *Proc. Natl. Acad. Sci. U. S. A.*, 1985, **82**, 3192.
65. S. Kamat, J. Barrera, E. J. Beckman and A. J. Russell, *Biotechnol. Bioeng.*, 1992, **40**, 158.
66. K. Jackson, L. E. Bowman and J. L. Fulton, *Anal. Chem.*, 1995, **67**, 2368.
67. M. Rantakyla and O. Aaltonen, *Biotechnol. Lett.*, 1994, **16**, 825.
68. Z. J. Dijkstra, H. Weyten, L. Willems and J. T. F. Keurentjes, *J. Mol. Catal. B: Enzym.*, 2006, **39**, 112.
69. I. Kmecz, B. Simandi, L. Poppe, Z. Juvancz, K. Renner, V. Bodai, E. R. Toke, C. Csajagi and J. Sawinsky, *Biochem. Eng. J.*, 2006, **28**, 275.
70. S. Srivastava and G. J. Madras, *Chem. Technol. Biotechnol.*, 2001, **76**, 890.
71. D. A. Miller, H. W. Blanch and J. M. Prausnitz, *Ind. Eng. Chem. Res.*, 1991, **30**, 939.
72. P. J. Halling, *Enzyme Microb. Technol.*, 1994, **16**, 178.
73. N. Fontes, J. Partridge, P. J. Halling and S. Barreiros, *Biotechnol. Bioeng.*, 2002, **77**, 296.
74. N. Harper and S. Barreiros, *Biotechnol. Prog.*, 2002, **18**, 1451.
75. N. Fontes, N. Harper, P. J. Halling and S. Barreiros, *Biotechnol. Bioeng.*, 2003, **82**, 802.
76. N. Fontes, P. J. Halling and S. Barreiros, *Enzyme Microb. Technol.*, 2003, **33**, 938.
77. T. W. Randolph, H. W. Blanch, J. M. Prausnitz and C. R. Wilke, *Biotechnol. Lett.*, 1985, **7**, 325.
78. D. A. Hammond, M. Karel, A. M. Klibanov and V. J. Krukonis, *Appl. Biochem. Biotechnol.*, 1985, **11**, 393.
79. T. Dumont, D. Barth, C. Corbier, G. Branlant and M. Perrut, *Biotechnol. Bioeng.*, 1992, **40**, 329.
80. Y. Ikushima, N. Saito, K. Hatakeda and O. Sato, *Chem. Eng. Sci.*, 1996, **51**, 2817.
81. S. Srivastava, J. Modak and G. Madras, *Ind. Eng. Chem. Res.*, 2002, **41**, 1940.
82. Z. Novak, M. Habulin, V. Krmelj and Ž. Knez, *J. Supercrit. Fluids*, 2003, **27**, 169.
83. G. Madras, R. Kumar and J. Modak, *Ind. Eng. Chem. Res.*, 2004, **43**, 7697.
84. M. D. Romero, L. Calvo, C. Alba, M. Habulin, M. Primosie and Ž. Knez, *J. Supercrit. Fluids*, 2005, **33**, 77.
85. T. Matsuda, T. Harada, K. Nakamura and T. Ikariya, *Tetrahedron: Asymmetry*, 2005, **16**, 909.
86. C. Blattner, M. Zoumpanioti, J. Kröner, G. Schmeer, A. Xenakis and W. J. Kunz, *J. Supercrit. Fluids*, 2006, **36**, 182.
87. T. Olsen, F. Kerton, R. Marriott and G. Grogan, *Enzyme Microb. Technol.*, 2006, **39**, 621.
88. Ž. Knez, C. G. Laudani, M. Habulin and E. Reverchon, *Biotechnol. Bioeng.*, 2007, **97**, 1366.
89. H. Shekarchizadeh, M. Kadivar, S. Hasan, H. S. Ghaziaskar and M. Rezayat, *J. Supercrit. Fluids*, 2009, **49**, 209.
90. K.-J. Liu and Y.-R. Huang, *J. Biotechnol.*, 2010, **146**, 215.

91. Ž. Knez, S. Kavčič, L. Gubicza, K. Bélafi-Bakó, G. Németh, M. Primožič and M. Habulin, *J. Supercrit. Fluids*, 2012, **66**, 192.
92. E. Celia, E. Cernia, C. Palocci, S. Soro and T. Turchet, *J. Supercrit. Fluids*, 2005, **33**, 193.
93. T. Yasmin, T. Jiang, B. Han, J. Zhang and X. Ma, *J. Mol. Catal. B: Enzym.*, 2006, **41**, 27.
94. M. N. Varma and G. Madras, *Biochem. Eng. J.*, 2010, **49**, 250.
95. K. P. Dhake, K. M. Deshmukh, Y. P. Patil, R. S. Singhal and B. M. Bhanage, *J. Biotechnol.*, 2011, **156**, 46.
96. V. Rathore and G. Madras, *Fuel*, 2007, **86**, 2650.
97. C. Dalla Rosa, M. B. Morandim, J. L. Ninow, D. Oliveira, H. Treichel and J. V. Oliveira, *J. Supercrit. Fluids*, 2008, **47**, 49.
98. M. N. Varma, P. A. Deshpande and G. Madras, *Fuel*, 2010, **89**, 1641.
99. M. N. Varma and G. Madras, *Ind. Eng. Chem. Res.*, 2007, **46**, 1.
100. C. Z. Brusamarelo, E. Rosset, A. Césaro, H. Treichel, D. Oliveira, M. A. Mazutti, M. Di Luccio and J. V. Oliveira, *J. Biotechnol.*, 2010, **147**, 108.
101. M. Lee, D. Lee, J. K. Cho, J. Cho, J. Han, C. Park and S. Kim, *Bioprocess Biosyst. Eng.*, 2012, **35**, 105.
102. M.-T. Liang, C.-H. Chen and R.-C. Liang, *J. Supercrit. Fluids*, 1998, **13**, 211.
103. Y. Xu, W. Du and D. Liu, *J. Mol. Catal. B: Enzym.*, 2005, **32**, 241.
104. K.-J. Liu, H.-M. Chang and K.-M. Liu, *Food Chem.*, 2007, **100**, 1303.
105. H. Shekarchizadeh and M. Kadivar, *Food Chem.*, 2012, **135**, 155.
106. K. A. Rezaei and F. Temelli, *J. Supercrit. Fluids*, 2000, **17**, 35.
107. K. A. Rezaei and F. Temelli, *J. Supercrit. Fluids*, 2000, **19**, 263.
108. J. L. Martinez, K. A. Rezaei and F. Temelli, *Ind. Eng. Chem. Res.*, 2002, **41**, 6475.
109. M. Primožič, M. Habulin and Ž. Knez, *J. Am. Oil Chem. Soc.*, 2003, **80**, 643.
110. G. H. Prado, M. Khan, M. D. A. Saldaña and F. Temelli, *J. Supercrit. Fluids*, 2012, **66**, 198.

Biocatalysis in Ionic Liquids

BERNARDO DIAS RIBEIRO[a], ARIANE GASPAR SANTOS[a], AND ISABEL M. MARRUCHO*[b]

[a]Escola de Química, Universidade Federal do Rio de Janeiro, Rio de Janeiro, RJ 21941-598, Brazil; [b]Instituto de Tecnologia Química e Biologica, Universidade Nova de Lisboa, Av. Republica, 2780-157, Oeiras, Portugal
*E-mail: imarrucho@itqb.unl.pt

6.1 Ionic Liquids

Ionic liquids are liquid organic salts that represent a promising development toward environmentally friendly non-aqueous solvents with tailor-made physical properties. In this sense, ionic liquids represent a major break-through in modern chemistry since they permit new synthetic pathways and the development of innovative solutions to well-established chemical processes.

This new class of neoteric solvents is entirely composed of ions and is fluid below the conventional temperature of 100 °C. A wide variety of cations and anions have been proposed, combined and studied. In 1914, Paul Walden discovered ethylammonium nitrate, and in 1948 Hurley and Wier used fluids based on *N*-ethyl pyridinium cations, however, it was only in 1975 when Hussey and co-workers proposed the use of the 1-ethyl-3-methyl imidazolium cation that ionic liquid chemistry was given a significant boost.[1] Since then, distinct families of cations based on dialkyl imidazolium, alkyl pyridinium, dialkyl pyrrolidinium, ammoniums, phosphoniums and morpholiniums have been combined with a wide panoply of anions, ranging from simple

RSC Green Chemistry No. 45
White Biotechnology for Sustainable Chemistry
Edited by Maria Alice Z. Coelho and Bernardo D. Ribeiro

Published by the Royal Society of Chemistry, www.rsc.org

halogenated ions, such as Cl and Br anions, to fluorinated anions such as tetrafluoroborate, hexafluorophosphate and bistrifluorosulfonylimide. Recently, advanced ionic liquids made of biodegradable, less expensive and less toxic ions, such as carboxylic acids, amino acids and other biorefinery platform compounds, have been proposed.[2,3] In principle, at least one million ionic liquids can be readily prepared in the laboratory. It is thus entirely possible to prepare an ionic liquid with the desired combination of properties such as reactivity, solubility, viscosity and toxicity. Moreover, the use of microwave radiation in the synthesis of ionic liquids represents an important step toward the sustainability of these fluids.[4]

Rogers and co-workers[5] described in a very elegant manner, the fast advances in the ionic liquids arena by proposing their classification into three distinct generations. Initially, the first generation of ILs were regarded as solvents and thus, their unique combination of thermophysical properties was studied and understood at the molecular level. The aim of the second generation was to develop materials for targeted purposes, and thus, their chemical properties, such as electrochemistry, energy density and fluidity to mention but a few, were carefully adjusted by appropriate choice of both the anion and the cation. More recently, the third generation of ILs have focused on providing fluids with biological properties such as low toxicity, and antibacterial and antifungal properties.

Some years ago, most authors wrote about the general properties of ionic liquids, such as their wide electrochemical windows, non-flammability, high thermal and chemical stability, extremely low volatility and low melting temperature. Today, due to the broad range of ionic liquids that have been made recently available, authors instead focus on the specific properties of one member or a family of ionic liquids' properties. This structural diversity, which has translated into a wide range of thermophysical properties, can be appreciated in the open literature[6] (see for example Zhang *et al.*'s[7] compilation of ILs' properties) and in good databases such as ILThermo, organized by the US National Institute of Standards and Technology. Some examples of typical property ranges for common ionic liquids are density ($1.1-1.6$ g L^{-1}), viscosity ($40-800$ mPa s), surface tension ($30-50$ mNm^{-1}), water miscibility (from totally miscible to almost completely immiscible).[8] Ionic liquids are frequently misquoted as green solvents, due to their null vapor pressure at ambient conditions. However, it should be stressed that ionic liquids are not intrinsically green. The toxicity and environmental persistence of the most used ILs are now being noted as important green factors.

The most attractive property of ionic liquids is the possibility of design, tailoring their properties according to the chemical processes of interest, and they are often described as *designer solvents* for this reason. These fluids provide a unique architectural synthetic platform on which the properties of both the cation and the anion can be independently tuned to provide the design of new functional materials. The determination and compilation of a large amount of data for a wide variety of ionic liquids allows the establishment of simple general rules regarding the role of both the cation and the anion. It is now common knowledge that *the anion defines the chemistry and the cation controls the thermophysical properties.* However, the plethora of available ions that can be combined into pairs "to make" ionic liquids clearly indicates that there will always be a lack of thermophysical properties for these fluids. In this way, a molecular-based understanding of ILs' properties is vital for the development of predictive models for the quick screening of ILs' thermophysical properties. However, a molecular approach for these fluids is a great challenge since they display a complex interplay of interactions, such as Coulombic, dipole–dipole and van der Waals[9] forces, and it is necessary to understand the effect of changing the cation or the anion on the above mentioned properties. Nonetheless, a wide variety of models have been used to describe the behavior of ionic liquid-containing systems, from molecular dynamics,[10] where appropriate force fields have to be developed and implemented, to accurate quantitative property–structure relationships[11] to classic thermodynamic models such as NRTL, UNIQUAC and UNIFAC. However, due to charge delocalization, these methods require much caution in their application in order to yield accurate results. Another versatile predictive method is the conductor-like screening model for realistic solvation, COSMO-RS,[12,13] which has been implemented to predict and screen the thermophysical properties and phase equilibria of ionic liquid-containing systems,[13] generally giving reliable qualitative results. The usefulness of COSMO-RS in the screening of ILs' properties is well illustrated by its use in the prediction of the polarity of ionic liquids and their mixtures with organic solvents.[14]

The applications of ionic liquids in biotechnology have been mainly focused on enzymatic biocatalysis. The impact of the use of ionic liquids on biocatalysis is very high and excellent reviews have been recently published.[15-18] The widely tunable properties of ionic liquids have enabled the optimization of the enzyme specificity and activity and the enhancement of the product recovery. In some cases, the special solvation capacity of substrates, enzymes and co-factors allows the one-pot progression of usually difficult reactions.[19] However, whole-cell processes[20] are common in diverse biotechnological topics, such as environmental biotechnology, biosynthesis of high value chemicals and recovery of metabolites. For these applications, several IL properties have been discussed as those which have the highest impact on the final outcome, which are the following:

1. Hydrophobicity—It is well known that a good solvent for biocatalysis has to be able to fully solvate the substrate and enzyme without stripping their protective layer of water. The tailoring of an IL's properties needs to take into account these aspects. Despite the fact that the hydrophobicity is mainly defined by the anion, it can be adjusted by changing the alkyl side chain of the cation or by the introduction of functional groups. For example, ILs with highly coordinated anions (such as Cl^-, CH_3COO^-, NO_3^-), can dissolve many compounds which are insoluble or poorly soluble in water, such as cellulose[21] and nucleic acid bases,[22] for example.

2. Viscosity—This is an important parameter since it controls the activity of the enzyme by affecting the mass transfer. Compared to organic solvents, ionic liquids are more viscous (35–500 cP for common ILs compared to 0.6 and 0.9 cP for toluene and water, respectively). The high viscosity of ionic liquids is due to their strong tendency for hydrogen bonding and van der Waals interaction. It is well known that the increase of the alkyl chain length increases the viscosity, while the temperature and small amounts of water or an organic solvent greatly reduce the value of this property. Also, the fluorination effect is also very marked on the viscosity since it reduces the hydrogen bonding effect and thus reduces the viscosity.

3. Polarity—One of the special properties of ionic liquids is their high polarity. The solvent polarity is usually determined based on the shift of a charge transfer absorption band of a solvatochromic probe, such as Reichardt's dye,[23] in the presence of a solvent. On the normalized polarity scale, E^{N}_{T}, using tetrameylsilane at 0.0 and water at 1.0, common ionic liquids normally fall in the region 0.6–0.7, together with small alcohols and formamide. Because of their high polarity, ionic liquids are an ideal reaction milieu for chemical and biochemical reactions due to their ability to dissolve a wide range of different substances, including polar, nonpolar organic and inorganic and polymeric compounds. Despite their high polarity, most ionic liquids are hydrophobic, but they can dissolve up to 1% water, which has a tremendous impact on the ionic liquids' thermophysical properties, such as viscosity. However, due to the complex interplay of interactions existent in ILs, the multiparameter approach of Kamlet, Abboud and Taft,[24] which uses a combination of dyes to describe polarity, acidity and basicity has also been applied to ILs.[25]

4. Structure and organization—Ionic liquids are known to have a wide variety of interactions; despite the fact that the Coulombic forces are the dominant, hydrogen bonding and dispersive forces between the cation and anion also need to be taken into account since they determine the tightness of the interaction in the ion pair. On top of that, ionic liquids exhibit nanostructure association for alkyl side chains greater than four carbon atoms. In this way, different domains with different properties (polar, nonpolar, fluorinated) can exist in the same pure fluid. This organization of the liquid phase determines the interaction and thus the solvation in ionic liquid media.

6.2 Enzymes in Ionic Liquids

Throughout the history of biocatalysis, alternative reaction conditions have been investigated to overcome problems such as substrate solubility, selectivity, yield or catalyst stability. Some progress has been made by the use of organic solvents, the addition of high salt concentrations, and the use of microemulsions, supercritical fluids or ionic liquids.[26-28] Generally, there are three ways to use organic solvents or ionic liquids in a biocatalytic process: as pure solvents, as co-solvents in aqueous systems or in biphasic systems,[3,27–29] depending on the miscibility of the ILs with classical molecular solvents, which is a function of the ions involved, as seen in Table 6.1.[26]

As with organic solvents, proteins are not soluble in most ionic liquids when they are used as pure solvents. As a result, the enzyme is either applied as an immobilized enzyme coupled to a support, such as a polymer, nanoparticles or carbon nanotubes, and encapsulated in hydrogels or as a suspension in its native form. Other options to enhance enzyme stability could include the addition of a small amount of water to the ionic liquids, water-in-IL microemulsions, covalent modification of enzymes with linear polyethylene glycol (PEG) chains, use of an ionic liquid enzyme coating or the use of biocompatible anions, such as saccharinate or dihydrogen phosphate and cations, such as choline and 1-butyl-1-methylpyrrolidinium.[16,28,30–32] For production processes, the majority of enzymes are used as immobilized catalysts in order to facilitate handling and to improve their operational stability. Furthermore, other factors could affect the stability and solubility of the biocatalysts in ionic liquids:[26,28,31–36]

→ **Water:** An essential amount of water may activate the enzyme by increasing the polarity and structural flexibility of the enzyme active site, but too much water is harmful to the enzyme by facilitating enzyme aggregation, thus diminishing the substrate diffusion and eventually leading to enzyme inactivation; whereas water present in the reaction system may cause hydrolysis of some ionic liquids. Besides, as nonaqueous solvents, ILs are assumed to affect enzyme performance *via* three interactions in the same way as do organic solvents:

(1) Retention of water allows them to strip off the essential water that is associated with the enzyme, leading to enzyme deactivation.
(2) Penetration into the microaqueous phase surrounding the enzyme molecules leads to direct contact with the enzyme, thereby changing

Table 6.1 Miscibility of ILs in organic solvents.[a,26]

		Miscibility				
Cation	Anion	Water	Acetonitrile	Isopropanol	Hexane	Toluene
[C$_2$mim]	[BF$_4$]	+	+	–	+/–	–
[C$_4$mim]	[BF$_4$]	+	+	–	–	–
[C$_6$mim]	[BF$_4$]	+/–	+/–	+	–	–
[C$_2$mim]	[PF$_6$]	+	+	–	–	–
[C$_4$mim]	[PF$_6$]	–	+	–	–	–
[C$_6$mim]	[PF$_6$]	–	+	–	–	–
[C$_4$mim]	[NTf$_2$]	–	+	+	+/–	–
[C$_6$mim]	[NTf$_2$]	–	+	+	–	–
[C$_8$mim]	[NTf$_2$]	–	+	–	–	–

[a]+, Totally miscible; +/–, partially miscible; –, immiscible.

the protein dynamics, the protein conformation, and/or the enzyme's active center.

(3) Interaction with the substrates and products either by direct reaction with them or by altering their partitioning between the bulk IL phase and the microaqueous phase that surrounds the enzyme molecule.

→ **Hofmeister Effects:** An ion may affect the enzyme performance by playing the role of a substrate, a cofactor, or an inhibitor to the enzyme. But more generally, the specific ion effects can be better understood by considering the ability of the ion to alter the bulk water structure, to affect the protein–water interactions, and to directly interact with the enzyme molecules because at low salt concentrations (up to 0.01 M), ions affect enzyme performance predominantly *via* electrostatic interactions. However, the Hofmeister ion effects become important when the electrostatic forces are screened by higher salt concentrations, with an enzyme/protein solution normally stabilized by kosmotropic anions and chaotropic cations, but destabilized by chaotropic anions and kosmotropic cations, as seen in Figure 6.1. (With a high charge water density, a kosmotropic ion interacts more strongly with water than water with itself and tends to strengthen the water structure by shifting the water equilibrium to low-density water. The situation is reversed in the case of a chaotropic ion.) In the case of IL cations, 1,3-dimethylimidazolium [C_1mim], 1-ethyl-3-methylimidazolium [C_2mim] and *N*–butylpyridinium are chaotropes, while larger imidazolium and ammonium cations with longer alkyl chains are more kosmotropic due to their stronger hydrophobic hydration. As regards the anion effect, normally IL anions follow the Hofmeister series to affect the enzymes and proteins, however, in the presence of a kosmotropic cation [C_4mim], some enzymes can present their activities in reverse Hofmeister order of the anions.

Normally, Hofmeister effects explain the impact of ILs on biocatalysis, especially when they are used as a co-solvent or as an additive in the aqueous

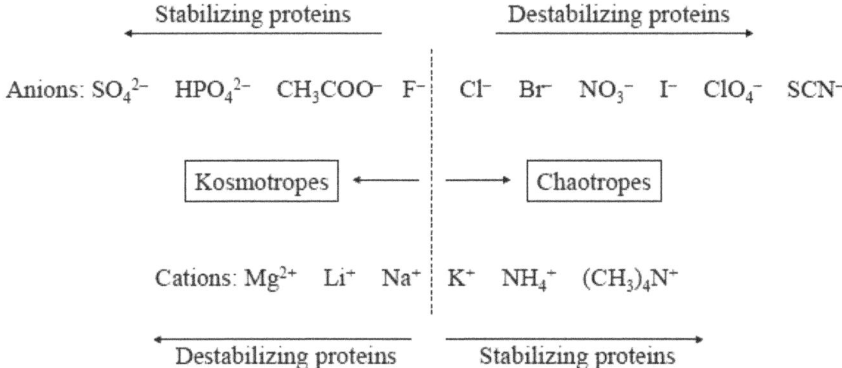

Figure 6.1   The Hofmeister series.[31]

solution, but there are also some other interactions that overlap with these effects:

(1) *Pure IL*: An IL in its pure liquid form presents a polymeric supramolecular structure with a H-bonding network and is hardly dissociated into cations and anions in the presence of the small amount of water added to the system or within the microaqueous phase surrounding the enzyme molecules. This may be the major reason why some ILs do not effectively exhibit Hofmeister effects on the enzyme function.

(2) *Surface pH of the enzyme*: The surface pH determines the ionization state of the amino acid residues of the enzyme's active site, which is important in modulating the enzyme's catalytic activity. Using a modified ion-specific double-layer model, the surface pH of proteins is dependent on the salt concentration and on the ionic species following the Hofmeister series. Holding a higher polarizability than Cl^-, SCN^- has a stronger interaction with the protein surface, resulting in the accumulation of H^+ around the protein's surface and in turn, a reduction in the surface pH, affecting the action mechanism of the enzyme's active site.

(3) *Net charge of the enzyme*: Addition of salts (or any impurities) to the buffer solution may also induce a change in the buffer pH, following the Hofmeister series, which could affect the activity and stability of the enzyme in the buffer solution accordingly. Part of the reason for this variation may be the change in the net charge of the enzyme that is caused. However, it is not certain whether the ion-induced change in the net charge of an enzyme is critical in determining the enzyme performance. On the other hand, the Hofmeister series can be direct or reverse depending on the charge and hydrophobicity/hydrophilicity of the surface in a colloidal system.

(4) *Active site and catalytic mechanism of the enzyme*: Ions may have strong interactions with the functional groups on the surface of the enzyme, especially those in the enzyme's active site, which are crucial for catalysis. This will trigger a change in the enzyme's active site both chemically and physically, resulting in a modification in the enzyme's catalytic activity and even its catalytic mechanism.

→ **Hydrophobicity:** This factor can be quantified in terms of $\log P$, the logarithm of the partition coefficient of the solvent in an octanol–water mixture. Solvents with high $\log P$, such as hexane ($\log P = 3.5$), are usually found to be more hydrophobic and more favorable for enzymatic reactions than those with a low $\log P$, such as ethanol ($\log P = -0.24$), simply because more hydrophobic solvents have a lower tendency to strip off the essential water from the enzyme, thus favoring the maintenance of its native structure. A direct correlation between the hydrophobicity of ILs and their $\log P$ values exists, although ILs present extremely low $\log P$ values (-2.90 to -2.39 in imidazolium ILs, for example); $[C_4mim][NTf_2]$ and

[C$_4$mim][PF$_6$] are immiscible with water (hydrophobic) while others are miscible (hydrophilic), such as [C$_4$mim][BF$_4$] and most ammonium ILs. Furthermore, the increase in hydrophobicity of an IL can be achieved by extending the length of the alkyl group from the cation, promoting a higher enzyme activity, but a lower selectivity, since more free water molecules can act as nucleophile acceptors in transesterification reaction. However, longer alkyl chains may behave as surfactants and destabilize the enzyme, while terminal hydroxylation of alkyl chains could make the IL more compatible with protein stability.

→ **Viscosity:** The viscosity of an IL is usually higher than that of molecular solvents and may control the enzyme activity by affecting the mass transfer limitations in the reaction systems. On the other hand, the high viscosity of ILs may also offer a stabilizing effect, slowing down the migration of protein domains from the active conformation to the inactive one.

→ **Nucleophilicity:** More nucleophilic IL anions such as NO$_3^-$, CF$_3$CO$_2^-$, CH$_3$CO$_2^-$, CF$_3$SO$_3^-$ and CH$_3$SO$_3^-$ can coordinate more strongly to the positively charged sites in the enzyme structure, causing conformational changes. Therefore, the use of ILs with low anion nucleophilicity is essential for enzyme activity.

→ **H-bond Basicity:** IL anions with a strong H-bond basicity, such as alkylsulfate, nitrate and lactate, may cause dissociation of the hydrogen bonds that maintain the structural integrity of the α-helices and β-sheets, which in turn will cause the protein to unfold, dissolving the enzyme.

→ **Others:**

(1) *IL–Buffer Interaction*: Enzymatic reactions are often performed in aqueous buffer solutions; the addition of increasing amounts of ionic liquids sometimes causes precipitates of unknown composition.

(2) *IL–Metal ion Complexes*: Some enzymes require metal ions, such as cobalt, manganese or zinc, for their activity; if these are removed by the ionic liquid by complexation, enzyme inactivation may occur.

(3) *Impurities*: Unlike conventional organic solvents, most research groups prepare the ionic liquids themselves. This may be the reason why, with nominally the same ionic liquid, different results are sometimes obtained, such as for example through the presence of halides and unreacted organic salts.

6.2.1 Lipases, Proteases and Esterases

One of the first studies using enzymes in the presence of ionic liquids was the evaluation of the influence of ethylammonium nitrate–water mixtures on alkaline phosphatase activity and stability, and the production of aspartame using thermolysin in the presence of [C$_4$mim][PF$_6$].[28,29,37] Since this, many investigations have been reported, in which the enzymes lipase and protease were the most cited.

Lipases, known for acting at aqueous/organic interfaces for the hydrolysis of oils and fats, and in other reactions, such as esterification, chiral resolution (more detailed in chapter 10), alcoholysis, ammoniolysis, perhydrolysis and even Baeyer–Villiger oxidation,[38] are the most reported enzymes used in ionic liquids, mainly [C$_4$mim][PF$_6$] and [C$_4$mim][BF$_4$], altering such properties as stability, specificity, regioselectivity and enantioselectivity, and also allowing solubilization of hydrophobic and hydrophilic compounds.[26,28,34,39,40] To preserve or enhance lipase properties, some authors have invested research in immobilization, complexation or coating of enzymes with ionic liquids.[41-50]

On the other hand, proteases such as trypsin, papain and subtilisin can be used for hydrolysis and formation of esters and amides, whether in enantioselective reaction or not.[3,34,51] Normally, esterases like cutinase and feruloyl esterase, enzymes which have similar action in relation to lipases but with no need for an emulsionized substrate, are much less tolerant to ionic liquids than lipases.[37,52]

In comparison with organic molecular solvents, ionic liquids normally increase reaction conversions. Nara *et al.*[53] compared dichloromethane and [C$_4$mim][PF$_6$] as solvents in the transesterification of 2-hydroxymethyl-1,4-benzodioxane using free lipase from *Pseudomonas cepacia*, and achieved almost three times greater conversion with ionic liquid in one hour of reaction. Using the same enzyme, Pan *et al.*[54] showed that a mixture of ionic liquids, mainly [C$_4$mim][NTf$_2$] and molecular solvents such as benzene, *t*-butanol and *n*-hexane, could activate lipase 40% more, and Vidya and Chadha[55] presented similar results when compared with *n*-hexane in the transesterification of 3-(furan-2-yl) propanoic acid. Akbari *et al.*[56] reported a great increase in lipase activity from *Pseudomonas stutzeri* in the presence of 30% v/v [C$_6$mim]Cl (700%), 45% v/v [C$_4$mim]Cl (650%) and 50% v/v [C$_2$mim]Cl (450%), while better results were obtained with 25% v/v molecular solvents acetone and *n*-hexane with 130–140% relative activity. Karbalaei-Heidari *et al.*[57] realized similar results with protease from *Salinivibrio* sp., obtaining 190% relative activity using 50% v/v [C$_6$mim][PF$_6$], whereas with the same concentration of ethyl acetate, chloroform, *n*-hexane and toluene, the activity only reached 120%.

In enzymatic reactions, ionic liquids can also be associated with other green chemical processes, such as supercritical CO$_2$,[58] microwave irradiation,[59] micellar systems,[60,61] aqueous two-phase systems[62-64] and supported liquid membranes, which can improve the mass transfer of solutes and decrease the viscosity of the reaction medium.[27,34] Another alternative is the use of a deep eutectic solvent (DES), a mixture of compounds which presents a lower melting point than each component alone, such as choline chloride (ChCl) and urea, or ChCl–glycerol, which have similar properties to ionic liquids, but are more biocompatible and cheaper.[65-68]

In Table 6.2, a wide variety of products is shown using lipases, proteases and esterases in the presence of ionic liquids, such as active pharmaceutical ingredients (APIs), biodiesel, polymers, flavour esters, solvents and surfactants.

Table 6.2 Applications of lipases, proteases and esterases in ionic liquids.

Application	Products	Enzyme	Ionic liquid[a]	References
API	Ascorbyl oleate	Lipase B from *Candida antartica* (Novozym 435)	$[sec\text{-}C_4mim][BF_4]$ $[sec\text{-}C_5mim][BF_4]$	Park *et al.*[69]
API	Acetylated polysaccharide from lily (*Lilium lancifolium*)	Lipase from *Burkholderia cepacia* (PSL-C "Amano")	$[C_4mim][BF_4]$	Chen *et al.*[70]
API	1-Octyl dihydrocaffeate	Novozym 435	$[N_{1888}][CF_3COO]$	Yang *et al.*[71]
API	Hesperidin fatty acid ester	Novozym 435	$[C_4mim][BF_4]$	Araújo *et al.*[72]
API	Chromene acetate	Novozym 435	$[C_4mim][BF_4]$ $[C_4mim][PF_6]$	Kidwai and Poddar[73]
API	Caffeic acid phenethyl ester	Novozym 435, lipases from *Thermomyces lanuginosa* (Lipozyme TL IM), and *Rhizomucor miehei* (Lipozyme RM IM) and lipase from *B. cepacia* (PS-CI)	$[C_4mim][NTf_2]$ $[N_{1113}][NTf_2]$ $[C_3mpip][NTf_2]$ $[C_3mpyr][NTf_2]$ $[C_3m(3)pz][NTf_2]$	Kurata *et al.*[74]
API	3-Cyclohexylpropyl caffeate	Novozym 435 + chlorogenate hydrolase from *A. japonicus*	$[C_4mim][NTf_2]$	Kurata *et al.*[75]
API	Propyl caffeate	Novozym 435, Lipozyme TL IM, and Lipozyme RM IM	$[C_4mim][NTf_2]$ $[C_4mim][PF_6]$ $[C_4mim][CF_3SO_3]$ $[N_{1888}][NTf_2]$	Pang *et al.*[76]
API	Rutin and esculin fatty acid esters	Novozym 435	$[C_4mim][BF_4]$ $[C_4mim][PF_6]$ $[C_8mim][BF_4]$ $[C_8mim][PF_6]$ $[N_{1888}][NTf_2]$	Lue *et al.*[77]
API	Starch palmitate	Lipase from *C. rugosa*, type VII	$[C_4mim][BF_4]^{+}$ $[C_4mim][CH_3COO]$	Lu *et al.*[78]
API	Ribavirin adipate	Novozym 435	$[C_4mim][BF_4]$	Liu *et al.*[79]
API	Oleyl ferulate	Novozym 435	$[C_6mim][PF_6]$ $[C_8mim][PF_6]$	Chen *et al.*[80]
API	Glyceryl diferulate	Novozym 435	$[C_2mim][PF_6]$ $[C_2mim][NTf_2]$	Sun *et al.*[81]

(*continued*)

Table 6.2   (*continued*)

Application	Products	Enzyme	Ionic liquid[a]	References
API	Ethyl *N*-acetyl-phenylalanine	Subtilisin Carlsberg, Type XIII	$[C_2mim][CF_3SO_3]$	Noritomi *et al.*[82]
API	Propyl *N*-acetyl-phenylalanine	Subtilisin Carlsberg, Type VIII	$[C_4mim][PF_6]$	Shah and Gupta[83]
API	Butyl *N*-acetyl-phenylalanine	α-Chymotrypsin	$[C_4mim][NTf_2]$	Eckstein *et al.*[84]
API	*N*-Acetyl-tryptophan glycyl glycinamide	α-Chymotrypsin	$[C_2mim][FSI]$	Noritomi *et al.*[85]
API	Peptides	α-Chymotrypsin	$[C_1OC_2mim][PF_6]$	Xing *et al.*[86]
API	Propyl *N*-acetyl-phenylalanine	Calbiochem® protease from *Bacillus licheniformis*	ChCl–glycerol Ch[CH_3COO]–Glycerol, both molar ratio 1 : 2	Zhao *et al.*[87]
API	Phenylalanine	Protease from *Bacillus licheniformis* (subtilisin Carlsberg)	$[C_2mim][Gly]$ $[C_2mim][Ala]$ $[C_2mim][Lys]$ $[C_2mim][Glu]$ $[C_2mim][Pro]$ $[C_2mim][4\text{-}ABA]$ $[C_2mim][5\text{-}APA]$ $[C_2mim][6\text{-}AHA]$	Zhao *et al.*[88]
API	Amoxicillin	Penicillin G acylase from *Escherichia coli*	$[C_4mim][BF_4]$ $[C_4mim][NTf_2]$ $[C_4mim][PF_6]$	Pereira *et al.*[89]
API	6-Aminopenicillanic acid	Penicillin acylase	$[C_4mim][BF_4]^+$ $[C_4mim][PF_6]$	Jiang *et al.*[90]
API	ω3-Fatty acids	Lipozyme TL IM	$[Tea][Ms]$	Akanbi *et al.*[91]
API	Acetylated pyridoxine	Novozym 435	$[C_4mim][PF_6]$	Bai *et al.*[92]
Chiral resolution	DL-Phenylalanine	Subtilisin Carlsberg, Alcalase, Protease P "Amano", Novozym 435, porcine pancreas lipase	$[C_2mim][CH_3COO]$ $[C_2mim][Ts]$	Zhao *et al.*[93]
Chiral resolution	DL-Phenylglycine methyl ester	Novozym 435	$[C_4mim][BF_4]$	Lou *et al.*[94]
Chiral resolution	(*R,S*) 1-Phenyl ethanol and its derivatives	Lipase from *Pseudomonas aeruginosa*	$[C_4mim][PF_6]$	Singh *et al.*[95]
Chiral resolution	(*R,S*) 1-Phenyl ethanol	Lipase from *Yarrowia lipolytica*	$[C_4mim][PF_6]$	Li *et al.*[96]

Chiral resolution	(*RS*)-Methyl mandelate	Novozym 435, Lipozyme TL IM, Lipozyme RM IM and lipases from *B. cepacia* (PSL, PSL-D and PSL-C, Amano)	$[C_4mim][BF_4]$ $[C_4mim][PF_6]$	Pilissão and Nascimento[97]
Chiral resolution	1-Phenylethanol, 2-amino-1-phenylethanol and 1-(2-furyl)ethanol	Lipase PS "Amano" SD from *Burkholderia cepacia*	$[C_2mim][NTf_2]$ $[C_2mim][BF_4]$ $[C_4mim][PF_6]$	Hara *et al.*[98]
Chiral resolution	*N*-Acetyl amino acid esters	Porcine pancreas lipase	$[C_2py][CF_3COO]$	Malhotra and Zao[99]
Chiral resolution	D,L-*p*-Hydroxyphenyl-glycine methyl ester	Novozym 435, Lipozyme TL IM, Lipozyme RM IM, lipase from *Mucor miehei* (Lipozyme MM IM), from *Candida cylindracea* and from *C. rugosa* type VII	$[C_6mim][BF_4]$	Lou *et al.*[100]
Chiral resolution	DL-Menthol	Lipase from *C. rugosa*, type VII	$[C_4mim][PF_6]$	Ren *et al.*[101]
Chiral resolution	(*R,S*)-2-Octanol, (*R,S*)-2-butanol and (*R,S*)-1-phenylethanol	Novozym 435, lipase from *P. cepacia* (Amano)	$[C_4mim][PF_6]$ $[C_8mim][dca]$	Bogel-Łukasik *et al.*[102]
Chiral resolution	(*R,S*)-1-Chloro-3-(3,4-difluorophenoxy)-2-propanol	Lipase from *Pseudomonas aeruginosa*	$[C_4mim][BF_4]$ $[C_4mim][PF_6]$	Singh *et al.*[103]
Chiral resolution	D,L-5-Phenyl-1-penten-3-ol acetate and (*RS*)-methyl mandelate	Novozym 435	$[C_4dmim][BF_4]$ $[C_4dmim][PF_6]$	Itoh *et al.*[104]
Chiral resolution	1-Phenylethane-1,2-diol	Lipases PS-C from *P. cepacia* (Amano)	$[C_4mim][PF_6]$	Kamal and Chounan[105]
Fuel	Biodiesel from sunflower and waste cooking oils	Novozym 435	$[C_8mim][PF_6]$ $[C_8mim][NTf_2]$	De los Ríos *et al.*[106]
Fuel	Biodiesel from soybean oil	Lipase from *B. cepacia* (PS "Amano")	$[C_8mpy][BF_4]$	Liu *et al.*[107]
Fuel	Biodiesel from soybean oil	Free lipase B from *C. antarctica*, Novozym 435, porcine pancreas lipase type II, Amano lipase A from *Aspergillus niger*, lipase AK from *Pseudomonas cepacia*, lipase AK from *C. cylindracea*, lipases PS, PS-D I and PS-C I from *B. cepacia*, lipase AK 20 from *P. fluorescens*, and Newlase F (lipase + protease) from *Rhizopus niveus*	ChCl/glycerol molar ratio 1/2	Zhao *et al.*[108]

(*continued*)

Table 6.2   (*continued*)

Application	Products	Enzyme	Ionic liquid[a]	References
Fuel	Microalgal biodiesel	Novozym 435 and lipase from *Penicillium expansum*	[C$_4$mim][PF$_6$]	Lai *et al.*[109]
Fuel	Methyl oleate	Lipase B from *Candida antartica* (Novozym 435 and 525 L) and lipase AK "Amano" from *Pseudomonas fluorescens*	[C$_{16}$mim][NTf$_2$]	De Diego *et al.*[110]
Fuel	Biodiesel from rapeseed oil	Novozym 435, lipases from *Thermomyces lanuginosa* (Lipozyme TL IM), and *Rhizomucor miehei* (Lipozyme RM IM)	Ammoeng 102 Ammoeng 120	Devi *et al.*[111]
Fuel	Biodiesel from Miglyol oil 812	Novozym 435	[Ch][CH$_3$COO]/ glycerol molar ratio 1/1.5	Zhao *et al.*[87]
Polymers	Poly-L-lactide and poly-L-lactide-*co*-glycolide	Novozym 435	[C$_6$mim][PF$_6$]	Chanfreau *et al.*[112]
Polymers	Poly-L-lactide-*co*-glycolide	Novozym 435	[C$_4$mim][PF$_6$]	Mena *et al.*[113]
Polymers	Polyhydroxyalkanoates	Lipase A and B from *C. antarctica* (Roche Chirazyme L-5 and L-2, respectively), esterase and lipase from *Candida rugosa*, lipase from *Thermomyces lanuginosa* (Roche Chirazyme L-8), lipase from *Pseudomonas cepacia* (Amano Lipase PS), esterase from *P. fluorescens*, porcine pancreatic lipase, protease from *Bacillus lentus*, α-chymotrypsin, pepsin, subtilisin Carlsberg, Novozym 435	[C$_4$mim][NTf$_2$]	Gorke *et al.*[114]
Polymers	Polyesters	Novozym 435	[C$_4$mim][BF$_4$] [C$_4$mim][PF$_6$] [C$_4$mim][NTf$_2$]	Marcilla *et al.*[115]
Polymers	Poly(butylene sebacate)	Lipase from *P. cepacia* PS-C "Amano"	[C$_4$mim][PF$_6$]	Nara *et al.*[116]
Polymers	Poly(ε-caprolactone)	Novozym 435	[C$_2$mim][NTf$_2$] [C$_6$mim][NTf$_2$] [C$_{12}$mim][NTf$_2$]	Wu *et al.*[117]
Polymers	Hyperbranched poly-L-lactide	Lipase B from *Candida antartica*	[C$_4$mim][PF$_6$]	Mena *et al.*[118]
Polymers	Polycaprolactone and poly(5, 5-dimethyl-1,3-dioxan-2-one)	Porcine pancreas lipase	[C$_4$mim][PF$_6$]	Zhang *et al.*[119]

Flavours	Citronellyl esters	Novozym 435	[C$_4$mim][PF$_6$] [C$_6$mim][PF$_6$] [C$_8$mim][PF$_6$]	Lozano et al.[120]
Flavours	Geranyl acetate, citronellyl acetate, neryl acetate and isoamyl acetate	Novozym 435	[C$_{12}$tma][NTf$_2$] [C$_{14}$tma][NTf$_2$] [C$_{16}$tma][NTf$_2$] [C$_{18}$tma][NTf$_2$]	Lozano et al.[121]
Flavours	Isoamyl acetate	Lipase B from *C. antartica* (Lipozyme CALB L)	[C$_4$mpy][dca]	Pohar et al.[122]
Flavours	Geranyl acetate	Novozym 435	[C$_4$mim][PF$_6$]	Barahona et al.[123]
Surfactants	Fructose palmitate	Novozym 435	[C$_4$mim][CF$_3$SO$_3$]$^+$ [C$_8$mim][NTf$_2$]	Ha et al.[124]
Surfactants	Glucose laurate	Novozym 435	[C$_4$mim][CF$_3$SO$_3$]$^+$ [C$_4$mim][PF$_6$]	Lee et al.[125]
Surfactants	Phosphatidylserine	Phospholipase D from *Streptomyces* PMF	[C$_4$mim][PF$_6$]	D'Arrigo et al.[126]
Surfactants	Monoglycerides	Novozym 435	Ecoeng 500^a	Guo and Xu[127]
Surfactants	Diglycerides	Novozym 435, Lipozyme TL IM, and Lipozyme RM IM	[C$_4$mim][NTf$_2$] [C$_4$mim][PF$_6$] [N$_{1888}$][NTf$_2$] Ammoeng 102 Ammoeng 120	Kahveci et al.[128] and Guo et al.[129]
Sweeteners	Z-Aspartame	Thermolysin from *Bacillus thermoproteolyticus rokko*	[P$_{4444}$][Z-Asp]$^+$ [mPhe]^s[MeSO$_4$]	Furukawa et al.[130]
Solvents	Butyl acetate	Novozym 435	[C$_4$mim][PF$_6$] [C$_4$mim][NTf$_2$]	Park et al.[131]
Solvents	Ethyl lactate	Novozym 435	Cyphos 104 Cyphos 201	Findrik et al.[132]

^{a}Abbreviations for some ionic liquids: [C$_3$mpip] = *N*-methyl-*N*-propylpiperidinium; [C$_3$mpyr] = *N*-methyl-*N*-propylpyrrolidinium; [C$_3$m(3)pz] = 1-propyl-2,3,5-trimethylpyrazolium; [FSI] = bis(fluorosulfonyl)imide; [Ala] = alanine; [Glu] = glutamic acid; [Gly] = glycine; [Lys] = lysine; [Pro] = proline; [4-ABA] = 4-aminobutanoic acid; [5-APA] = 5-aminopentanoic acid; [6-AHA] = 6-aminohexanoic acid; [Tea][Ms] = triethylammonium mesylate; [Ts] = tosylate; [dca] = dicyanamide; [C$_4$dmim] = 1-butyl-2,3-dimethylimidazolium; [C$_8$mpy] = 1-octyl-3-methylpyridinium; Ammoeng 102 = ethyloctanodecanoyl oligoethyleneglycol ammonium ethylsulfate; Ammoeng 120 = methyloctanodecanoyl oligoethyleneglycolate ammonium methylsulfate; [tma] = trimethylammonium; Ecoeng 500 = cocosalkyl pentaethoxymethyl ammonium methosulfate; [Z-Asp] = Z-aspartic acid; [mPhe] = phenylalanine methyl ester; Cyphos 104 = trihexyl(tetradecyl)phosphonium bis 2,4,4-(trimethylpentyl)phosphinate; Cyphos 201 = tributyl(tetradecyl)phosphonium dodecylbenzenesulfonate.

6.2.2 Glycosidases

In their natural role, glycosidases act in glycosidic bonds, hydrolyzing polysaccharides or glycosylated compounds. A recent application associating ionic liquids with these enzymes, such as cellulase and xylanase, is in lignocellulosic residue treatment. Ionic liquids can dissolve polysaccharides ([C$_2$mim][CH$_3$COO] and [C$_4$mim]Cl, for example), altering their crystalline form to amorphous when regenerated. Glycosidases can act in both cases, but their activities decrease significantly in the presence of polysaccharide-dissolving ILs.[16,17] Another use for glycosidases is for carbohydrate synthesis *in vitro*, by applying two methodologies, condensation (reverse hydrolysis) and transglycosylation.[3,34,37] In Table 6.3, some examples of the applications of glycosidases in ILs are shown, while in Table 6.4 some data on glycosidase stability in ILs are given, such as storage and operational stabilities.

6.2.3 Other Enzymes

Oxidoreductases, isomerases and lyases are other classes of enzymes that are utilized in the presence of ionic liquids. Oxidoreductases, such as dehydrogenases, peroxidases and laccases, catalyze *in vivo* oxidation and reduction reactions that are engaged in essential roles in living cell metabolism, with applications ranging from environmental goals, such as biodegradation and bioremediation, to synthetic performances involving the formation of chiral centers, or C–O (or other heteroatoms such as S or halogens) bond formation in organic substrates.[152] Ionic liquids have been used with these enzymes in some applications, including phenol degradation,[153] sulfoxidation of thioanisole,[154,155] oxidation of methyl-parathion,[156] chlorination of monochlorodimedone,[155] in nanocomposites as supports for choline oxidase,[157] oxygen reduction with bilirubin oxidase immobilized in a gold nanoparticle,[158] polymerization of aniline,[159] delignification of Hinoki cypress (*Chamaecyparis obtusa*) chips,[48] and asymmetric reduction of ketones.[160]

In some cases, ionic liquids can act as inhibitors, for example Hong *et al.*[161] and Park *et al.*[162] reported [C$_4$mim][BF$_4$] and [C$_4$mpy][BF$_4$], respectively, which exhibited noncompetitive inhibition of horseradish peroxidases, and Park *et al.*[163] cited [C$_4$mim][MeSO$_4$] as an uncompetitive inhibitor, and inhibition of other enzymes such as catalase,[164] tyrosinase[165] and laccase[166] by ionic liquids has also been reported. Another example is alcohol dehydrogenase, reported by Dabirmanesh *et al.*,[167,168] stabilized by [Hmim]Cl. Some strategies have been utilized to enhance or maintain enzyme activity, as reported by Zhou *et al.*[169] who used a [C$_4$mim][PF$_6$] microemulsion to maintain the activity of the enzymes laccase from *Trametes versicolor* and lignin peroxidase from *Phanerochaete chrysosporium*; Mohidem and Mat[170] performed a sol–gel immobilization of laccase assayed using [C$_4$mim][CF$_3$SO$_3$] and [C$_4$py][CF$_3$SO$_3$]; an aqueous biphasic system has been reported by Cao *et al.*[171] who explored the system [C$_4$mim]Cl – K$_2$HPO$_4$ with horseradish peroxidase; a strategy of applying more biocompatible ionic liquids, such as hydrated choline dihydrogen phosphate, can also be used.[172]

Table 6.3 Applications of glycosidases in ionic liquids.

Enzyme	Ionic liquid[a]	Reaction	Conditions	Yield	References
Cellulases (Celluclast 1.5 L + Novozym 188)	$[C_2mim][CH_3COO]$	Hydrolysis of (1) microcrystalline cellulose (Avicel) and (2) yellow poplar (*Liriodendron tulipifera*)	180 rpm, pH 4.8, 50 °C, 24 h, 15% v/v IL, enzyme loadings: 75 FPU cellulase and 80 CBU β-glucosidase per gram of substrate, (1) 0.24% (m/v) Avicel, (2) 0.6% (m/v) yellow poplar	91.0% (1) 45.3% (2)	Wang *et al.*[133]
Cellulases (Celluclast 1.5 L + Novozym 188)	$[C_1mim][DMP]$	Hydrolysis of α-cellulose	pH 4.8, 45 °C, 48 h, 10% v/v IL, 1% (m/v) substrate. Enzyme quantity: 0.01% m/v Novozym 188 + 0.05% (m/v) from purified Celluclast (60% endoglucanase and 40% cellobiohydrolase)	69.2%	Engel *et al.*[134]
Endoglucanase from *Thermatoga maritima*	$[C_2mim][CH_3COO]$	Hydrolysis of corn stover	900 rpm, pH 4.8, 80 °C, 15 h, 10% v/v IL, 2% (m/v) substrate. Enzyme quantity: 6% m/v	35%	Datta *et al.*[135]
β-Galactosidase from *Thermus thermophilus*	$[C_8mim][PF_6]$	Synthesis of N-acetyl-D-lactosamine	pH 6.0, 65 °C, *p*-nitrophenyl-β-D-galactopyranoside (5.12% m/v), *N*-acetyl-D-glucosamine (18.3% m/v), 30% v/v IL. Enzyme loading: 1.8 U	79%	Sandoval *et al.*[136]
α-Amylases from *Bacillus amyloliquefaciens* (BAA) and *B. lichiniformis* (BLA)	$[C_6mim]Cl$	Hydrolysis of starch	pH 7.4, 37 °C, 3 min, 1.8% m/v starch, 0.2 mM CaCl$_2$, 20% v/v IL. Enzymes: 3.12 µg mL^{-1} BLA and 0.2 µg mL^{-1} BAA	90% (BLA) 58% (BAA)	Dabirmanesh *et al.*[137]

(*continued*)

Table 6.3 (*continued*)

Enzyme	Ionic liquid[a]	Reaction	Conditions	Yield	References
Cellulase from *Trichoderma reesei*	[C$_2$mim][DEP]	Hydrolysis of microcrystalline cellulose (Avicel)	pH 5.0, 40 °C, 24 h, Avicel 1% m/v, 20% v/v IL, enzyme: 0.2% m/v	53%	Kamiya *et al.*[138]
Cellulases (Celluclast 1.5 L)	[C$_1$mim][DMP] (1) [C$_2$mim][CH$_3$COO] (2)	Hydrolysis of *Miscanthus giganteus*	pH 4.8, 50 °C, 16 h, *Miscanthus* 0.1% m/v, 10% m/m IL	57% (1) 65% (2)	Wolski *et al.*[139]
β-Galactosidase from *Bacillus circulans*	[C$_1$mim][MeSO$_4$]	Synthesis of *N*-acetyl-D-lactosamine	pH 7.3, 23 °C. *N*-Acetyl-D-glucosamine 600 mM, lactose 62.5 mM, 20% v/v IL. Enzyme solution, 0.2% v/v	32%	Kaftzik *et al.*[140]
Cellulase from *Trichoderma reesei*	[C$_1$mim][DMP]	Hydrolysis of microcrystalline cellulose	pH 4.8, 50 °C, 24 h, Avicel 1% m/v, 20% v/v IL, enzyme: 0.8% m/v. Cellulose pretreatment: 30 min, 60 °C (1); (1) + sonication 45 kHz, 100W (2)	76% (1) 95% (2)	Yang *et al.*[141]
Hesperidinase from *Aspergillus niger*	[C$_2$mim][BF$_4$] (1) [C$_4$mim][BF$_4$] (2)	Hydrolysis of rutin	120 rpm, pH 9.0, 40 °C, 24 h, rutin 0.02% m/v, 10% v/v IL, enzyme: 1% m/v	87% (1) 94% (2)	Wang *et al.*[142]
β-Glycosidase CelB from *Pyrococcus furiosus*	[C$_1$mim][MeSO$_4$]	Synthesis of galactosyl glycerol	pH 5.5, 80 °C. Glycerol 513 mM, lactose 75 mM, 45% v/v IL. Enzyme solution, 0.2% v/v	+10%[b]	Lang *et al.*[143]

[a]Abbreviations for some ionic liquids: [DMP] = dimethylphosphate; [DEP] = diethylphosphate.
[b]Increased the yield in 10% with IL.

Table 6.4 Relative activity of glycosidases in ionic liquids.

Enzyme	Ionic liquid[a]	Substrate	Conditions	Relative activity	References
Cellulase from *Trichoderma reesei*	[C$_4$mim]Cl	Cellulose Azure	60 rpm, 50 °C, pH 4.8, 90 min, enzyme: 0.034% m/v, 21.80 mM IL (1), 39.70 mM IL (2)	94.5% (1) 72.7% (2)	Turner *et al.*[144]
Celluclast 1.5 L	[Amim]Cl (1) [C$_2$mim][CH$_3$COO] (2) [C$_4$mim]Cl (3)	α-Cellulose	1000 rpm, pH 4.8, 45 °C, 30 min, 10% v/v IL, enzyme: 0.05% (m/v)	23% (1) 15% (2) 18% (3)	Engel *et al.*[145]
Cellulase from *Aspergillus niger*	[C$_4$mim]Cl	Sodium carboxymethylcellulose	pH 5.0, 30 °C, 20 min, 10% v/v IL, enzyme: 0.25% m/v. Pressure: 0 MPa (1), 200 MPa (2), 400 MPa (3)	50% (1) 100% (2) 56% (3)	Salvador *et al.*[146]
β-Galactosidase from *Aspergillus oryzae*	[C$_4$mim][BF$_4$] (1) [C$_4$mim][PF$_6$] (2) [C$_4$mim][dca] (3) [C$_4$mim][MeSO$_4$] (4)	O-Nitrophenyl-β-D-galactopyranoside	pH 7.0, 25 °C, 40 min, 12% v/v IL. Enzyme: 1.6×10^{-4}% m/v	12.5% (1) 23.9% (2) 10.9% (3) 25.8% (4)	Singh *et al.*[147]
Cellulase Cel A2 M1	ChCl/glycerol	4-Methylumbelliferyl-β-D-cellobioside	pH 7.2, 30 °C, 15 min, 5% v/v DES (1), 30% v/v DES (2)	75% (1) 30% (2)	Lehmann *et al.*[148]
Cellobiohydrolase from *Halorhabdus utahensis*	[C$_2$mim][CH$_3$COO] (1) [C$_2$mim]Cl (2) [C$_4$mim]Cl (3) [Amim]Cl (4)	Sodium carboxymethylcellulose	pH 7.0, 37 °C, 60 min, 20% m/m IL. Enzyme: 8×10^{-4}% m/v	98% (1) 115% (2) 105% (3) 102% (4)	Zhang *et al.*[149]
Cellulase from *Aspergillus niger*	[HEMA][MeSO$_4$]	Cellulose Azure	pH 4.8, 2 h, 50% IL w/w. Enzyme: 4.76×10^{-6}% m/v. T (°C): 55 (1), 65 (2), 75 (3)	12% (1) 22% (2) 112% (3)	Bose *et al.*[150]
β-Glucosidase from *Volvariella volvacea*	[C$_1$mim][DMP] (1) [C$_2$mim][DMP] (2) [C$_2$mim][CH$_3$COO] (3) [C$_2$mim][DEP] (4)	*p*-Nitrophenyl β-D-glucopyranoside	12h. 15% v/v IL. Enzyme: 5×10^{-4}% v/v	96.1% (1) 86.2% (2) 80.1% (3) 9.5% (4)	Thomas *et al.*[151]
Xylanase E2	[C$_1$mim][DMP] (1) [C$_2$mim][DMP] (2) [C$_2$mim][CH$_3$COO] (3) [C$_2$mim][DEP] (4)	*p*-Nitrophenyl β-D-xylopyranoside	12h. 15% v/v IL. Enzyme: 5×10^{-4}% v/v	104.6% (1) 109.6% (2) 86.3% (3) 58.9% (4)	Thomas *et al.*[151]
Arabinofuranosidase F1	[C$_1$mim][DMP] (1) [C$_2$mim][DMP] (2) [C$_2$mim][CH$_3$COO] (3) [C$_2$mim][DEP] (4)	*p*-Nitrophenyl α-D-arabinofuranoside	12h. 15% v/v IL. Enzyme: 5×10^{-4}% v/v	100.3% (1) 109.0% (2) 87.9% (3) 95.3% (4)	Thomas *et al.*[151]

[a]Abbreviations for some ionic liquids: [Amim] = 1-alkyl-3-methylimidazolium; [dca] = dicyanamide; [DEP] = diethylphosphate; [HEMA] = tris-(2-hydroxyethyl)-methylammonium.

Isomerases act in reactions of conversion of a substrate into an isomer, that is, a substance with the same number and types of atoms. Most isomerases are intracellular and some of them require cofactors, but not organic coenzymes. Glucose isomerase is the most exploited technologically, and is applied in the production of high fructose syrups (HFSs), mostly from corn starch.[173] Wang *et al.*[174] tested this enzyme in the isomerization of glucose to fructose in the presence of 3% v/v [C$_2$mim]Cl associated with an ultrasound bath, obtaining 40% higher enzymatic activity, with a yield of 45.3% fructose in 10 h of reaction. Another common reaction is the isomerization of xylose into xylulose, which was reported by Ståhlberg *et al.*[175] using *N,N*-dibutylethanolammonium octanoate and by Yu *et al.*[176] with [C$_2$mim]Cl and microwave irradiation.

Lyases, such as aldolases, carboxylases and hydratases, catalyze reactions of non-hydrolytic and non-oxidative cleavage of chemical bonds, like C–C, C–O, C–N, C–S, C–X (halides), P–O and other bonds. Enzymes belonging to this family perform different metabolic functions associated not only with cell catabolism, but also with biosynthesis by acting in reverse, and have been also studied for asymmetric synthesis of optically active organic compounds.[173] Kifazume[177] reported the use of aldolase antibody 38C2 promoting aldol and Michael addition reactions in the presence of [C$_4$mim][PF$_6$] and [C$_2$mim][CF$_3$SO$_3$], respectively.

Other enzymes, such as firefly luciferase,[178] D-amino acid oxidase[179] and epoxide hydrolases[3,28] have also been evaluated for the effects of ionic liquids on their stability and activity, with the solubility of organic substrates being enhanced in aqueous solutions when water-miscible ILs were used.

6.3   Whole-Cell Processes in Ionic Liquids

The production of chemicals by whole-cell processes can be more advantageous than enzymatic biocatalysis, when, for example, cofactor regeneration (NADPH) is required. The interest of ionic liquids application in whole-cell processes is due to the possibility of enhancing the efficiency of the processes to give high yields and productivities. These improvements can be achieved through the increase of the substrates' solubility in monophasic systems where ionic liquids are co-solvents, and also in biphasic systems where ionic liquids act as a reservoir of toxic products and substrates. In order to use these solvents in whole-cell biocatalysis processes, microorganisms commonly applied in this area have been widely studied for their tolerance toward ionic liquids and the efficiency of biocatalysis in their presence.

6.3.1   Toxicity Toward Microorganisms

Non-volatility is the main attribute that confers to ionic liquids their benign character. However, other aspects need to be taken into account, such as their effect on living organisms once released as industrial effluents, and their persistence in the environment. As they are very innovative, knowledge about the environmental effects of these new solvents is still recent and limited. In addition, due to their wide range of compositions, given the wide number of possible combinations of cations and anions generating new ionic liquids, it becomes necessary to have prudence to generally consider them environmentally friendly. Among the principles of green chemistry is reduction or non-toxicity of chemical compounds toward living organisms,[180,181] therefore the toxicity of ionic liquids toward living organisms is an important issue to be investigated from the point of view of ecotoxicology. This notion includes the tolerance evaluation of diverse live organisms, such as the luminescent marine bacteria *Vibrio fischeri*,[182,183] algae (*e.g.* the green algae *Pseudokirchneriella subcapitata*),[184] invertebrates (*e.g.* the freshwater snail *Physa acuta* and the freshwater cladoceran *Daphnia magna*),[185,186] vertebrate animals (*e.g.* the zebrafish *Danio rerio*[187]) and mammalian cells IPC-81.[183]

In the above ecotoxicological model, the organisms used are quite diverse and are mainly focused on aquatic organisms, with microbial organisms having only a few representatives. Despite the importance of investigation of the toxic effects of ionic liquids toward environmental microorganisms, the evaluation of the tolerance of microorganisms commonly used in whole-cell biotransformation processes is an important issue to be investigated, due to the increased interest in the use of ionic liquids in this kind of process. Biocompatibility is an essential requirement for the successful application of these solvents in a biocatalytic process involving live microorganisms. Once in contact with the solvent, any interaction that may cause damage to the cell and affect its activity may compromise the process efficiency.

6.3.1.1 Methods

A wide range of methods and criteria have been used to evaluate the toxic effects of ionic liquids toward microorganisms. When in contact with the compound to be tested, the tolerance of the organism may be related to growth inhibition, cell viability, growth rate change or modification of the metabolic activity measured by the amount of product formed or substrate uptake.

One of the simplest methods to evaluate the tolerance of microorganisms in the presence of ionic liquids is the test of agar diffusion that has been used for many years as an antibiotic susceptibility test in clinical laboratories. This assay consists of a filter-paper disc impregnated with the test substance, which is deposited on a solid culture with the microorganism swabbed across the plate. The toxic effect is measured through the zone of growth inhibition around the disc. This type of test has been applied as a screening method for the detection of biocompatible ionic liquids, as it is a method of easy implementation and gives results in a short time.[188–191] Despite its advantages, this method may have some limitations, such as the possibility of ionic liquid interaction with the cellulose filter paper or solid medium, and heterogeneous distribution of the ionic liquid over the agar plate, producing irregular inhibition zones, and thus erroneous measurements.[191]

Another simple method that can provide results about the toxicity of ionic liquids toward microorganisms is through measuring cell viability. The viability of a microorganism's cells can be assayed after cell incubation for a certain time with a determined concentration of ionic liquid. At the end of this period, the toxic effect is detected using a cell viability dye or by plate counting, and the growth is expressed as the number of colony-forming units (CFUs).[192–197] Indirect methods which provide information about the metabolic activity of microorganisms are also applied to assess the toxic effects of these solvents. For example, the production of lactic acid, a typical metabolite generated by lactic acid-producing bacteria acts as an indicator of cellular activity.[193]

Information about the growth inhibition can be obtained by measuring a microorganism's growth rate in the presence of ionic liquids. Quantitative data on culture are collected at regular intervals over the time exposure and can be expressed using the optical density (OD),[180,191,198–201] colony forming units per milliliter (CFU mL^{-1})[182] or cell dry weight[202,203] of the microorganism culture. Therefore, the growth of live cells can be followed during the entire exposure time and compared to the growth rate of cultures free from ionic liquids.

In addition, the standard assay currently applied to assess the susceptibility of clinical microorganisms, which determine the minimum inhibitory concentration (MIC) or minimum biocidal concentration (MBC) has also been applied to analyze the toxicity of ionic liquids toward microorganisms.[204–212] The guidelines for this method can be found at the documents of the Clinical Laboratory Standards Institute (CLSI) and depending on the microorganism subjected to the assay, a specific norm should be followed. The MIC value is the lowest concentration of the tested substance for which no visible growth was detectable after a period of 24 h or 48 h, depending on the microorganism. The MBC is the lowest concentration at which no viable cell is detected after an aliquot of the assay is cultured in an agar medium, in other words the substance tested prevents colony formation. This test is strong and gives more accurate results, since it scans a broad range of concentrations. Furthermore, the application of these standard tests facilitates the comparison between results from different work. The choice of method does not seem to have very clear criteria and the application of various experimental conditions, including exposure time, concentration range, amount of inoculum and culture medium, among other factors, makes it difficult to make comparisons between the results obtained in various studies regarding the toxicity of these compounds.

6.3.1.2 General Trends

With the data currently available, finding a relationship between the type of ionic liquid and the toxicity toward a microorganism is a very complex task. Due to the diversity of ionic liquids available and the emergence of new combinations it is difficult to perform this kind of systematization. Additionally, each organism may have variable responses toward different ionic liquids. Knowledge about the mechanism of action of these solvents when in contact with microbial cells is also scarce.

Nevertheless, some observations are becoming frequent, such as the relationship between the alkyl chain length and the toxic effect of the ionic liquid. This trend, which seems to be well established, shows that the toxicity of ionic liquids increases with increasing length of the alkyl chain attached to the cation moiety. This behavior has been extensively observed in studies which evaluated the toxic effects of pyridinium, imidazolium, benzilimidazolium and ammonium-based ionic liquids.[204–206,208,211] All of these studies observed this effect in a variety of microorganisms, among them cocci, rods, bacilli and fungi. Other reports have also confirmed this trend, including the work performed by Matsumoto *et al.*[193] with lactic acid-producing bacteria, that of Yang *et al.*[213] with immobilized yeast cells, and that of Wood *et al.*[191] with *Escherichia coli* and a wide range of cations and anions, as well as reports by other authors.[182,198,212] More effective biological activity or toxicity toward microbial cells with increasing alkyl chain is associated with increasing of the compound's hydrophobicity, thus enabling greater interaction with the cell membrane. However, at a certain alkyl side chain length, this effect can no longer be increased, as observed by Łuczak *et al.*,[212] who evaluated the antibacterial and antifungal activity of 1-alkyl-3-methylimidazolium chlorides with alkyl side chains ranging between 2 and 18 carbon atoms. The authors noticed that the highest antifungal and antibacterial activities were similar to compounds with 16 or 18 carbon atoms in their alkyl side chains, in other words, the ionic liquid containing 16 carbons was the most toxic. In this case, the cut-off effect was attributed to the tendency to self-assemble, in other words the micellization process limits the rate of diffusion to the surface of the cell, decreasing the free compounds and their ability to act on the cell membrane.[212] Pernak *et al.*[205] also noticed a cut-off effect with 3-alkoxymethyl-1-methylimidazolium ionic liquids, where the alkyl chain containing 12 carbons was the most toxic toward the microorganisms evaluated within a range between 3 and 16 carbon atoms. Following the hypothesis of increasing toxicity concomitant with increasing hydrophobicity, in addition to the structural resemblance between quaternary ammonium based surfactants (QACS) and long chain pyridinium and imidazolium ionic liquids, was noticed an interesting linear relationship between the values of critical micelle concentration (CMC) and minimum inhibitory concentration (MIC).[211] When log CMC increased, an increase in MIC value was seen, and therefore lower toxicity toward microorganisms, thus the CMC can act as an index to estimate the toxicity of ionic liquids.

Concerning the nature of the cationic head group, this factor seems to have no significant effect on the toxicity of ionic liquids. Cornellas *et al.*[211] analyzed the toxicity of a series of 1-alkyl-3-methylimidazolium bromides and 1-alkylpyridinium bromides toward gram-positive cocci, gram-negative rods, bacilli and fungi and no significant differences were observed in the results between two cationic head groups. Other authors also investigated the effects of similar ionic liquids, 1-alkyl-3-methylimidazolium bromides and 1-alkyl-3-methylpyridinium bromide toward *Escherichia coli*, *Staphylococcus aureus*, *Bacillus subtilis*, *Pseudomonas fluorescens* and *Saccharomyces cerevisiae* and the type of cationic group seemed to have a secondary influence on the ionic liquid's toxic effects.[182]

The influence of the anions is a more complex issue; in some cases, the anions can play an insignificant role and in others their effect on microorganisms is more marked. Pernak *et al.*[205] evaluated the toxicity of 3-alkoxymethyl-1-methylimidazolium ionic liquids associated with anions, Cl^-, BF_4^- and PF_6^- toward a wide range of microorganisms, including cocci, rods and fungi and noticed that the anion effect was less pronounced compared to the cation alkyl chain effect. An interesting analysis of the secondary role played by the anion with regard to toxicity was realized by Łuczak *et al.*[212] who compare the impact of increased anion and cation alkyl chain length on the minimum inhibitory concentration (MIC) values for *Candida albicans*. The introduction of seven carbon atoms in the anion, from $MeSO_4^-$ to $OctSO_4^-$, resulted in a 4–6 fold increase of the MIC values. On the other hand, when the carbon atoms were introduced in the cation, from $[C_2mim]$ to $[C_8mim]$, a 15 and 130 fold increase in the MIC values of these cations associated with $OctSO_4^-$ and Cl^-, respectively, was observed.

However, in general, this secondary role can be crucial in a whole-cell process and diverse studies have demonstrated the effects of the more common ionic liquids involving anions such as BF_4^-, PF_6^- and NTf_2^-, which are most widely used in whole-cell processes. Several studies have considered PF_6^- a biocompatible anion and a good choice to form biphasic systems in whole-cell processes. Different microorganisms, with distinct physiological characters, such as the gram negative bacteria *Escherichia coli*, the gram positive *Bacillus cereus* and the yeast *Pichia pastoris*, had their growth investigated in the presence of $[C_4mim][PF_6]$ and $[C_4mim][BF_4]$ in different concentrations.[180] The authors noticed that *E. coli* and *B. cereus* tolerated concentrations higher than 1% $[C_4mim][PF_6]$ and *P. pastoris* growth was observed in the presence of 10% ionic liquid. In contrast, $[C_4mim][BF_4]$ inhibited the growth of all microorganisms at concentrations higher than 1%. This trend was also observed by other authors with other microorganisms. Immobilized *Saccharomyces cerevisiae* cells also showed a better tolerance toward $[C_4mim][PF_6]$ than $[C_4mim][BF_4]$ in a volume ratio of around 3% IL.[202] The metabolic activity of the fungus *Aureobasidium pullulans* was higher in the presence of $[C_4mim][PF_6]$ than with $[C_4mim][BF_4]$ in 50% volume ratio, demonstrating one more time that the PF_6^- anion seems to be more biocompatible.[214]

Contrary to these findings, some studies verified the opposite behavior, for example, in an investigation realized by Wang *et al.*,[201] who evaluated the kinetic growth and gas production of *Clostridium* sp., an anaerobic bacterium, in a wide range concentrations of ionic liquids with the cation

1-methoxyethyl-3-methyl imidazolium ([MeOC$_2$mim]) associated with BF$_4^-$, CF$_3$COO$^-$, MeSO$_3^-$, PF$_6^-$, and NTf$_2^-$ anions. The authors reported that PF$_6^-$ was more toxic toward *Clostridium* sp. than BF$_4^-$. In another case, Lee *et al.* (2005)[198] assessed the growth inhibition of *E. coli* bacteria, which were exposed to four ionic liquids, among them [C$_2$mim][BF$_4$] and [C$_4$mim] [PF$_6$], and the latter was more toxic than [C$_2$mim][BF$_4$]. Possibly, the difference in the number of carbon atoms in the alkyl chain could be the reason for this variation, as in the other cases the cation was exactly the same. The negative effects of BF$_4^-$, when compared with other anions, were also observed by an assay realized with immobilized cells of *Trigonopsis variabilis* AS2.1611[194] and *Rhodotorula* sp. AS2.2241.[215] Both studies used the same range of imidazolium-based ionic liquids with the anions NO$_3^-$, PF$_6^-$, Cl$^-$, CF$_3$SO$_3^-$ and Br$^-$, and the results showed that BF$_4^-$ was the most toxic anion, causing the death of 100% of cells after 24 h exposure, and NO$_3^-$ was considered the most biocompatible anion.

In studies of biphasic systems, where the ionic liquid has to be hydrophobic, PF$_6^-$ and NTf$_2^-$ are the most used anions. Some reports have noted NTf$_2^-$ anions as biocompatible, as observed for *Lactobacillus kefir*,[192] *Escherichia coli* and *Saccharomyces cerevisiae*[196] which were exposed to [C$_4$mim][PF$_6$], [C$_4$mim][NTf$_2$] and [N$_{1888}$][NTf$_2$] at a volume ratio of 20%. The cell viability of *E. coli* and *L. kefir* was measured after 5 hours of exposure and for *S. cerevisiae* the measurement was performed after 20 hours. Both studies showed that the microorganisms assessed were tolerant toward these ionic liquids. The membrane integrity of *L. kefir* was 89% in the presence of [C$_4$mim][NTf$_2$], and around 30% for the others. However, in general, NTf$_2^-$ seems to produce toxic ionic liquids, as will be seen from the following results. Wang *et al.*[195] demonstrated that [C$_4$mim][PF$_6$] showed more biocompatibility with immobilized *Rhodotorula* sp. AS2.2241 cells than [C$_4$mim][NTf$_2$] at a volume ratio of 20%. Immobilized cells of *Candida parapsilosis* CCTC M203011[216] and *E. coli* cells[198] also followed the same trend. In a study realized by Wang *et al.*,[201] as cited previously, *Clostridium* sp. showed the following order of inhibition from highest to lowest: NTf$_2^-$ ≥ PF$_6^-$ > BF$_4^-$ > CF$_3$COO$^-$ > MeSO$_3^-$, and this toxicity sequence was attributed to the number of fluorine atoms present in the molecule. The authors showed a strong linear relationship between the amount of fluorine and the anion toxicity. This hypothesis can explain the results obtained by Bräutigam *et al.*,[217] where the NTf$_2^-$ was slightly more biocompatible than PF$_6^-$ toward *E. coli*, and tris(pentafluoroethyl)trifluorophosphate (E3FAP$^-$), which contains 18 fluorine atoms, was the most toxic anion. Wood *et al.*[191] also reported the high toxicity of NTf$_2^-$, which completely inhibited the growth of *E. coli* MG 1655 combined with several cations.

Besides the inherent particularities of microorganisms, which confer singular results to biocompatibility assays, these controversial data can be influenced by a series of other variables, such as different times of exposure, the concentration of ionic liquids, inoculum concentration, the use of immobilized cells, and impurities presents in the ionic liquids from their synthesis, among others. Therefore, the conclusion and choices have to consider all of these details and the available data have to be extrapolated with some caution. Furthermore, most of the current literature includes only a small range of ionic liquids, mainly imidazolium-based ionic liquids, and the trend is for the types and number of possible combinations of cations and anions to increase more and more.

6.3.2   Whole-Cell Biocatalysis

The first whole-cell process involving ionic liquids was evaluated by Cull *et al.*,[218] in the biotransformation of 1,3-dicyanobenzene to 3-cyanobenzamide and then to 3-cyanobenzoic acid by *Rhodococcus* R312, where [C$_4$mim][PF$_6$] was used as a second phase replacing organic solvents, in this case toluene, acting as a reservoir for the hydrophobic substrate and decreasing the toxic effects of it on the microorganism. The water–[C$_4$mim][PF$_6$] system showed a slower initial rate of amide production than the water–toluene system, however, *Rhodococcus* R312 showed greater tolerance toward the ionic liquid. Thus, the authors attributed the lower initial rate in the presence of [C$_4$mim][PF$_6$] to its high viscosity, which limited substrate mass transfer.

Following the same approach, a series of other studies have been realized since. Most of them applied ionic liquids in reactions of asymmetric reduction, more specifically in the reduction of ketones to chiral alcohols, as shown in Table 6.5. The enzymes that catalyze this kind of transformation are the oxidoreductases, alcohol dehydrogenase (ADH), which need cofactor regeneration (NAD$^+$/NADP$^+$), thus the use of isolated enzymes requires the addition of exogenous cofactors, representing an increase in the process cost. Whole-cell processes enable *in situ* regeneration of cofactors, and therefore are more attractive than isolated enzymes. On the other hand, some factors can interfere in the efficiency of this process, such as limited substrate solubility in aqueous medium, and toxic or inhibitory effects of the substrate and product toward microorganisms. In order to overcome these problems, solvents can be applied as co-solvents in a monophasic system, where water miscible ionic liquids are used to improve the solubility of substrates, and as a second phase, where water immiscible ionic liquids act as a reservoir of substrate and product, reducing their negative effects. Therefore, the use of ionic liquids can improve the yield and enantioselectivity of a biotransformation, as seen in Table 6.5.

In addition to the asymmetric reduction reaction, other types of reaction can be found in the literature. Dipeolu *et al.*[199] investigated the reduction of nitrobenzene to aniline by the bacteria *Clostridium sporogenes* in the presence of water-miscible ionic liquids, in order to enhance the substrate solubility in aqueous medium. The ionic liquids used were choline dimethyl phosphate ([Ch][C$_2$PO$_4$]), *N,N*-dimethylethanolammonium acetate (DMEAA), cocosalkyl-pentaethoxy-methyl-ammonium-methosulfate   (AMMOENG 100), 1-butyl-3-methylimidazolium tetrafluoroborate ([C$_4$mim][BF$_4$]) and

Table 6.5 Application of ionic liquids in asymmetric reduction by whole-cell biocatalysis.

Microorganism	Ionic liquids[a]	Reaction	Conditions	Yield/ee[b]	References
Immobilized baker's yeast	[C$_4$mim][PF$_6$]	Cyclohexanone to cyclohexanol (1) Cyclopentanone to cyclopentanol (2) Hexan-2-one to (*S*)-hexan-2-ol (3) Pentan-2,4-dione to (*S*)-4-hydroxy-pentan-2-one (4) Ethyl acetoacetate to (*S*)-ethyl-3-hydroxybutanoate (5) Ethyl-2-oxocyclopentanecarboxylate to ethyl(1*R*,2*S*)-2-hydroxycyclo-pentanecarboxylate (6) Ethyl 2-oxopropanoate to ethyl (*S*)-2-hydroxypropanoate (7)	91% v/v IL 2 mL Methanol 10 g Calcium alginate beads containing yeast 10 mmol Ketone 33 °C, 72 h	35% (1) 20% (2) 40%/79% (3) 22%/95% (4) 70%/95% (5) 75%/84% (6) 60%/76% (7)	Howarth *et al.*[220]
Lactobacillus kefir	[C$_4$mim][PF$_6$] (1) [C$_4$mim][NTf$_2$] (2) [N$_{1888}$][NTf$_2$] (3)	Asymmetric reduction of 4-chloro-acetophenone to (*R*)-1-(4-chloro-phenyl)ethanol	20% v/v IL 600 mM 4-Chloroacetophenone 50 g L^{-1} *L. kefir*; 3h	88%/99.8% (1) 92.8%/99.7% (2) 88.4%/99.4% (3)	Pfruender *et al.*[192]
Saccharomyces cerevisiae	[C$_4$mim][PF$_6$] (1) [C$_4$mim][NTf$_2$] (2)	Asymmetric reduction of 4-chloro acetoacetate to (*S*)-4-chloro-3-hydroxybutanoate	20% v/v IL; 0.68 mM 4-chloro acetoacetate; 20 g$_{DCW}$ L^{-1} *S. cerevisiae*; 200 mM glucose; 27 °C; 300 rpm; 20 h 20% v/v IL; 0.68 mM 4-chloro acetoacetate; 50 g$_{DCW}$ L^{-1} *S. cerevisiae*; 200 mM glucose; 100 µM NADP$^+$; 27 °C; 300 rpm; 20 h	28.7% (1) 12.4% (2) 80.2%/83.9% (1)	Pfruender *et al.*[196]
Lactobacillus kefir	[C$_4$mim][NTf$_2$]	Asymmetric reduction of *tert*-butyl 6-chloro-3,5-dioxohexanoate to *tert*-butyl (3*R*,5*S*)-6-chlorodihy-droxyhexanoate	20% v/v IL; 0.4 M *tert*-butyl 6-chloro-3,5-dioxohexa-noate; 50 g$_{DCW}$ L^{-1} *L. kefir*; 200 mM glucose; 27 °C; 300 rpm; 20 h	81.3%	Pfruender *et al.*[196]

(continued)

Table 6.5 (*continued*)

Microorganism	Ionic liquids[a]	Reaction	Conditions	Yield/ee[b]	References
Immobilized *Saccharomyces cerevisiae* cells	$[C_4mim][PF_6]$ (1)	Asymmetric reduction of acetyltrimethylsilane to (*S*)-1-trimethylsilylethanol	14% v/v IL; pH 7.3; 30 °C; 84 mM acetyltrimethylsilane (1)	99.9%/>99.9% (1)	Lou *et al.*[216]
	$[C_4mim][BF_4]$ (2)		10% v/v IL; pH 7.5; 30 °C; 77 mM acetyltrimethylsilane (2)	99.2%–>99.9% (2)	
Immobilized *Trigonopsis variabilis* cells	$[C_2OHmim][NO_3]$	Asymmetric reduction of 4'-methoxyacetophenone to (*R*)-1-(4-methoxyphenyl)ethanol	2.5% v/v IL; pH 8.5; 15 mM 4'-methoxyacetophenone and 200 rpm	97.2%/>99%	Lou *et al.*[194]
	$[C_2OHmim][NO_3]$ (1)		5% v/v IL; pH 8.5; 30 °C; 50 mM 4'-methoxyacetophenone; 200 rpm	95%/99% (1)	
	$[C_2OHmim][PF_6]$ (2)			30.5%/60% (2)	
	$[C_2OHmim][BF_4]$ (3)			4.5%/5% (3)	
	$[C_2OHmim]Cl$ (4)			41.4%/73% (4)	
	$[C_2OHmim][CF_3SO_3]$ (5)			34.6%/60% (5)	
	$[C_2mim][NO_3]$ (6)			71.5%/87% (6)	
	$[C_4mim][NO_3]$ (7)			39.2%/84% (7)	
	$[C_4mmim][NO_3]$ (8)			34.3%/70% (8)	
	$[C_2mim][BF_4]$ (9)			8.6%/22% (9)	
	$[C_3mim][BF_4]$ (10)			7.7%/15% (10)	
	$[C_4mim][BF_4]$ (11)			4.5%/11% (11)	
	$[C_5mim][BF_4]$ (12)			2.3%/8% (12)	
	$[C_4mim]Cl$ (13)			18.9%/33% (13)	
	$[C_4mim]Br$ (14)			24.6%/12% (14)	
	$[C_5mim]Br$ (15)			17.1%/18% (15)	
	$[C_6mim]Br$ (16)			9.8%/25% (16)	
	$[C_7mim]Br$ (17)			6.8%/28% (17)	
Pichia membranaefaciens	$[C_4mim][BF_4]$	Asymmetric reduction of ethyl acetoacetate to ethyl (*R*)-3-hydroxybutyrate	2.5% v/v IL; pH 8.0; 30 °C; 0.55 M ethyl acetoacetate; 240 g L^{-1}; 200 rpm; 12 h	77.8%/73.0%	He *et al.*[221]

Biocatalyst	Ionic liquid	Reaction	Conditions	Conversion/ee	Reference
Immobilized *Rhodotorula* sp.	$[C_4mim][PF_6]$ (1) $[C_5mim][PF_6]$ (2) $[C_6mim][PF_6]$ (3) $[C_7mim][PF_6]$ (4) $[C_2mim][NTf_2]$ (5) $[C_4mim][NTf_2]$ (6)	Asymmetric reduction of 4'-methoxyactophenone to (*S*)-1-(4-methoxyphenyl)ethanol	20% v/v IL; pH 7.5; 25 °C; 10 mM 4'-methoxyactophenone; 0.32 g mL^{-1} immobilized cells; 180 rpm; 30 min Optimal condition of (1): 20% v/v IL; pH 7.5; 25 °C; 40 mM 4'-methoxyactophenone; 0.32 g mL^{-1} immobilized cells; 180 rpm; 30 min	69.5%/>99% (1) 55.1%/>99% (2) 51.6%/>99% (3) 38.2%/>99% (4) 66.4%/>99% (5) 58.3%/>99% (6) Optimal condition: 95.5%/>99% (1)	Wang *et al.*[195]
Immobilized *Rhodotorula* sp.	$[C_2OHmim][NO_3]$ (1) $[C_2OHmim][PF_6]$ (2) $[C_2OHmim][BF_4]$ (3) $[C_2OHmim]Cl$ (4) $[C_2OHmim][CF_3SO_3]$ (5) $[C_2mim][NO_3]$ (6) $[C_4mim][NO_3]$ (7) $[C_4mmim]^b[NO_3]$ (8) $[C_2mim][BF_4]$ (9) $[C_3mim][BF_4]$ (10) $[C_4mim][BF_4]$ (11) $[C_5mim][BF_4]$ (12) $[C_4mim]Cl$ (13) $[C_4mim]Br$ (14) $[C_5mim]Br$ (15) $[C_6mim]Br$ (16) $[C_7mim]Br$ (17)	Asymmetric reduction of 4'-methoxyactophenone to (*S*)-1-(4-methoxyphenyl)ethanol	10% v/v; pH 8.0; 30 °C; 10 mM 4'-methoxyactophenone; 0.32 g mL^{-1} immobilized cells; 180 rpm Optimal conditions of (1): 5% v/v IL; pH 8.5; 25 °C; 12 mM 4'-methoxyactophenone; 0.32 g mL^{-1} immobilized cells; 180 rpm	82.3%/>99% (1) 72.3%/>99% (2) 5.3%/8% (3) 80.9%/92% (4) 37.2%/73% (5) 52.2%/86% (6) 48.5%/81% (7) 69.9%/96% (8) 9.6%/25% (9) 8.1%/14% (10) 5.6%/12% (11) 3.1%/9% (12) 57.2%/86% (13) 59.5%/91% (14) 44.2%/90% (15) 42.6%/88% (16) 29.7%/83% (17) Optimal condition: 98.3%/>99% (1)	Lou *et al.*[215]

(continued)

Table 6.5 (*continued*)

Microorganism	Ionic liquids[a]	Reaction	Conditions	Yield/ee[b]	References
Escherichia coli	[C$_6$mpyr][NTf$_2$]	Asymmetric reduction of 2-octanone to (*R*)-2-octanol	20% v/v IL; pH 6.5; 20 °C; 600 mM 2-octanone; 50 g$_{cdw}$ L^{-1} cells; 4 h	95%/99.7%	Bräutigam *et al.*[222]
Escherichia coli	[C$_6$mpyr][NTf$_2$] (1) [C$_4$mim][NTf$_2$] (2)	Asymmetric reduction of 4-chloro-acetophenone to (*R*)-1-(4-chloro-phenyl)ethanol	20% v/v IL; pH 6.5; 20 °C; 600 mM 4-chloroacetophenone; 50 g$_{cdw}$ L^{-1} cells; 3 h	96%/>99.6% (1) (2)	Bräutigam *et al.*[222]
Escherichia coli	[C$_6$mpyr]^a[NTf$_2$]	Asymmetric reduction of 2-octanone to (*R*)-2-octanol	20% v/v IL; pH 6.5; 20 °C; 300 mM 2-octanone; 50 g$_{cdw}$ L^{-1} cells; 600 rpm. 6 h	98.5%/>99.5%	Den-newald *et al.*[223]
Escherichia coli	[C$_4$mim][PF$_6$] (1) [C$_6$mim][PF$_6$] (2) [C$_4$mim][NTf$_2$] (3) [C$_6$mim][NTf$_2$] (4) [C$_4$mpyr] [NTf$_2$] (5) [C$_4$mpyr][NTf$_2$] (6)	Asymmetric reduction of ethyl 4-chloroacetoacetate to ethyl (*S*)-4-chloro-3-hydroxybutyrate	20% v/v IL; 600 mM ethyl 4-chloroacetoacetate; 50 g$_{cdw}$ L^{-1} cells; 1 h	95.8%/99.7% (1) 98.2%/99.6% (2) 97.1%/99.7% (3) 97.5%/99.7% (4) 99.3%/99.7% (5) 99.1%/99.6% (6)	Bräutigam *et al.*[217]
		Asymmetric reduction of phenacyl chloride to (*S*)-α-chloro-1-phenylethanol	20% v/v IL; 600 mM phenacyl chloride; 50 g$_{cdw}$ L^{-1} cells; 1 h	69.5%/99.7% (1) 89%/99.6% (2) 51,1%/99.7% (3) 69.8%/99.7% (4) 66,6%/99.7% (5) 74.2%/99.7% (6)	
Immobilized *Geotrichum candidum* on water-absorb-ing polymer	[C$_4$mim][PF$_6$] (1) [C$_4$mim][BF$_4$] (2) [C$_4$mim][NTf$_2$] (3) [C$_2$mim][BF$_4$] (4) [C$_2$mim][NTf$_2$] (5) [C$_8$mim][PF$_6$] (6)	Asymmetric reduction of *o*-fluoroacetophenone to (*S*)-1-(*o*-fluorophenyl)ethanol	2 mL IL; 0.041 mmol *o*-fluoro-acetophenone; 1 g immobi-lized cell; 30 °C; 16 h	92%/>99 (1) 82%/>99 (2) 49%/>99 (3) 96%/>99 (4) 61%/>99 (5) 88%/>99 (6)	Matsuda *et al.*[224]

Aureobasidium pullulans	[C$_4$mim][PF$_6$]	Asymmetric reduction of ethyl 4-chloro-3-oxobutanoate to ethyl (*S*)-4-chloro-3-hydroxybutanoate	50% v/v IL; pH 6.6; 30 °C; 1.46 mmol 4-chloro-3-oxobutanoate; 50 g$_{cdw}$ L^{-1} cells; 180 rpm; 8 h	95.6%/98.5%	Fan *et al.*[214]
Cell lysate from *Escherichia coli*	[C$_4$mim][PF$_6$]	Asymmetric reduction of ethyl 4-cyanobenzoylformate to (*R*)-4-cyanomandelate	50% v/v IL; 500 mM ethyl 4-cyanobenzoylformate; 40 g$_{cdw}$ L^{-1} cells; 8.5 h (1)	15%/>99% (1)	Kratzer *et al.*[225]
			50% v/v IL; 500 mM ethyl 4-cyanobenzoylformate; 40 g$_{cdw}$ L^{-1} cells; 8.5 h; 490 μM [NAD$^+$] (2)	22%/>99% (2)	
Immobilized *Candida parapsilosis*	[C$_4$mim][PF$_6$]	Asymmetric reduction of 4-(trimethylsilyl)-3-butyn-2-one to (*S*)-4-(trimethylsilyl)-3-butyn-2-ol	20% v/v IL; pH 5.5; 30 °C; 24 mM 4-(trimethylsilyl)-3-butyn-2-one; 0.15 g mL^{-1} immobilized cells; 12 h	97.7%/>99%	Lou *et al.*[216]

[a]Abbreviations for some ionic liquids: [C$_4$mmim] = 1-buthyl-2,3-dimethylimidazolium; [C$_6$mpyr] = 1-hexyl-1-methylpyrrolidinium; [C$_4$mpyr] = 1-butyl-1-methylpyrrolidinium.
[b]ee = enantiomeric excess.

1-ethyl-3-methylimidazolium ethylsulfate ([C_2mim][C_2SO_4]). DMEAA and [Ch][C_2PO_4] were most biocompatible, increasing the growth rate of *C. sporogenes*, [C_2mim][C_2SO_4] (2% v/v) inhibited growth by 58% and the others were totally inhibitory. In the presence of [C_2mim][C_2SO_4], the product yield reached values up to 79% and was higher than in conventional solvents, such as ethanol (4% v/v) and heptane (phase ratio 0.33) where the yield was 8% and 45%, respectively.

Sendovski *et al.*[203] studied the bioproduction of 2-phenylethanol by *Saccharomyces cerevisiae*, which is associated with cell growth, in contrast to the more common asymmetric reduction biotransformation, which is restricted to resting cells. The production of 2-phenylethanol occurs in 3 steps from L-phenylalanine *via* an Ehrlich pathway. In order to improve the productivity by reducing the product inhibition with *in situ* product removal, a biphasic ionic liquid aqueous system was used. As second phase were used the ionic liquids, 1-methyl-1-propylpiperidinium bis(trifluoromethylsulfonyl)imide ([C_3pyr][NTf$_2$]), methyltrioctylammonium bis(trifluoromethylsulfonyl)imide ([N_{1888}][NTf$_2$]) and 1-butyl-3-methylimidazolium bis(trifluoromethylsulfonyl)imide ([C_4mim][NTf$_2$]) in a volume ratio of 20% (v/v). These ionic liquids increased the final concentration of 2-phenylethanol by 3–5 fold, in an initial stress condition with 2.5 g L^{-1} 2-phenylethanol. Other applications of ionic liquids in whole-cell processes associated with the growth of microorganisms are still scarce and only have preliminary results, such as the work with lactic acid-producing bacteria.[193,219] The authors investigated the toxicity of some imidazolium-based ionic liquids toward bacteria of the genus *Lactobacillus*, *Leuconostoc*, *Pediococcus* and *Bacillus*, and the distribution coefficient of some organic acids, mainly lactic acid, in ionic liquids, in order to apply an *in situ* product removal system for this fermentation.

A recent study performed by Chen *et al.*[226] investigated the hydrolysis of glycyrrhizin to glycyrrhetic acid 3-*O*-mono-β-D-glucuronide, used in the pharmaceutical and food industries, by a whole-cell system containing ionic liquids. The authors evaluated the performance of *Penicillium purpurogenum* Li-3 and recombinant strains, *Escherichia coli* BL21 and *Pichia pastoris* GS115 in the presence of some imidazolium-based ionic liquids. The fungus *P. purpurogenum* was the most tolerant toward ionic liquids, mostly to [C_4mim][PF$_6$], and in optimized reaction conditions reached a product yield of 87.63% after 60 h. Before optimization, yields with [C_4mim][PF$_6$] were 13% higher than those obtained with the commonly used organic solvent, *t*-butyl alcohol.

Other studies also applied ionic liquids in different whole-cell systems, such as the use of this solvents in the production of biodiesel fuel from soybean oil[227] and for the degradation of phenol compounds.[228] Arai *et al.*[227] applied three ionic liquids, namely [C_2mim][BF$_4$], [C_4mim][BF$_4$] and [C_2mim][CF$_3SO_3$], as an extracting phase for glycerol, a by-product of biodiesel production, and also a reservoir of methanol, used in biocatalysis in the transesterification reaction of soybean oil triglycerides. Both compounds often have a negative effect on microorganisms. The reaction was performed using immobilized cells of wild-type *Rhizopus oryzae* producing triacylglycerol lipase, recombinant *Aspergillus oryzae* expressing *Fusarium heterosporum* lipase, *Candida antarctica* lipase B, and mono- and diacylglycerol lipase from *A. oryzae*. Among the fungi tested, *R. oryzae* showed the highest activity in the presence of ionic liquids, producing about 45% methyl ester with 50% (v/v) of [C_2mim][BF_4] and [C_4mim][BF_4], in contrast to around 5% methyl ester content in an ionic liquid-free reaction medium. Varying the methanol ratio content in the reaction medium, the authors observed that the efficiency of conversion was improved with ionic liquids, demonstrating that they can mitigate the inactivation of microorganism activity by acting as a reservoir phase.

Baumann *et al.*[228] evaluated phenol degradation by three xenobiotic-degrading bacteria, namely *Pseudomonas putida*, *Achromobacter xylosoxidans*, and *Sphingomonas aromaticivorans* in the presence of six phosphonium-based ionic liquids as a second phase. Among the six ionic liquids investigated, only trihexyl(tetradecyl)phosphonium bis(trifluoromethylsulfonyl)amide ([$P_{6,6,6,14}$][NTf_2]) was biocompatible. In an assay of phenol degradation by *P. putida*, the authors observed that the total consumption of phenol compounds occurred after 12 hours of reaction, which was similar to other results previously obtained with two-phase bioreactors involving organic–aqueous systems.

Finding out the best ionic liquid composition to obtain good performance in whole-cell biocatalysis can be a laborious process, and in general is realized without well-marked criteria. Therefore, one of the challenges in this regard is to understand the driving force which acts in each particular process with a particular microorganism, and then select the ion combination for the synthesis of a tuned ionic liquid. Another very important issue to discuss regarding the industrial use of these solvents in whole-cell systems is the ionic liquid recovery and recyclability. Due to the current high cost of ionic liquids, although in some cases demonstrating highly efficient results, economic viability cannot be achieved if the ionic liquid cannot be reused. Furthermore, the environmental aspect have to be considered, and the disposal of ionic liquids as effluent has to be avoided. Only a few studies have been realized experimentally to investigate the effectiveness of the complete process, evaluating the possibility and the impact of ionic liquid recyclability in whole-cell processes.[214,223,226,229] When the ionic liquid has a good thermal and chemical stability and the product is volatile, the ionic liquid phase containing the product can be subjected to distillation and after product removal, the ionic liquid can be reused successfully in the process.[223] But in other cases, ionic liquids are extracted with traditional organic solvents, such as ethanol[229] and isopropanol,[214] followed by vacuum distillation. These processes can be efficient from an economic standpoint, but are opposed to the main reason that leads researchers to try to replace volatile organic solvents by more environmentally friendly solvents, such as ionic liquids.

References

1. P. Walden, *Bull. Russian Acad. Sci.*, 1914, 405.
2. G. Imperato, B. König and C. Chiappe, *Eur. J. Org. Chem.*, 2007, **2007**, 1049.
3. J. Gorke, F. Srienc and R. Kazlauskas, *Biotechnol. Bioprocess Eng.*, 2010, **15**, 40.
4. V. Polshettiwar and R. S. Varma, *Acc. Chem. Res.*, 2008, **41**, 629.
5. W. L. Hough, M. Smiglak, H. Rodríguez, R. P. Swatloski, S. K. Spear, D. T. Daly, J. Pernak, J. E. Grisel, R. D. Carliss, M. D. Soutullo, J. H. Davis, Jr. and R. D. Rogers, *New J. Chem.*, 2007, **31**, 1429.
6. J. D. Holbrey and R. D. Rogers, in *Ionic Liquids in Synthesis*, ed. P. Wasserscheid and T. Welton, WILEY-VCH Verlag GmbH & Co. KGaA, Weinheim, 2nd edn, 2008, vol. 1, ch. 3.1, pp. 57–71.
7. S. Zhang, N. Sun, X. He, X. Lu and X. Zhang, *J. Phys. Chem. Ref. Data*, 2006, **35**, 1475.
8. S. Werner, M. Haumann and P. Wasserscheid, *Annu. Rev. Chem. Biomol. Eng.*, 2010, **1**, 203.
9. H. Weingärtner, *Angew. Chem., Int. Ed. Engl.*, 2008, **47**, 654.
10. J. N. Canongia Lopes, J. Deschamps and A. A. H. Pádua, *J. Phys. Chem. B*, 2004, **108**, 2038.
11. R. L. Gardas and J. A. P. Coutinho, *Fluid Phase Equilib.*, 2008, **263**, 26.
12. A. Klamt, F. Eckert and W. Arlt, *Annu. Rev. Chem. Biomol. Eng.*, 2010, **1**, 101.
13. M. Diedenhofen and A. Klamt, *Fluid Phase Equilib.*, 2010, **294**, 31.
14. J. Palomar, J. S. Torrecilla, J. Lemus, V. R. Ferro and F. Rodríguez, *Phys. Chem. Chem. Phys.*, 2010, **12**, 1991.
15. Z. Yang and W. Pan, *Enzyme Microb. Technol.*, 2005, **37**, 19.
16. M. Moniruzzaman, K. Nakashima, N. Kamiya and M. Goto, *Biochem. Eng. J.*, 2010, **48**, 295.
17. H. Olivier-Bourbigou, L. Magna and D. Morvan, *Appl. Catal., A*, 2010, **373**, 1.
18. C. Roosen, P. Müller and L. Greiner, *Appl. Microbiol. Biotechnol.*, 2008, **81**, 607.
19. E. P. Hudson, R. K. Eppler and D. S. Clark, *Curr. Opin. Biotechnol.*, 2005, **16**, 637.
20. G. Quijano, A. Couvert, A. Amrane, G. Darracq, C. Couriol, P. Le Cloirec, L. Paquin and D. Carrié, *Chem. Eng. J.*, 2011, **174**, 27.
21. R. P. Swatloski, S. K. Spear, J. D. Holbrey and R. D. Rogers, *J. Am. Chem. Soc.*, 2002, **124**, 4974.
22. J. M. M. Araújo, R. Ferreira, I. M. Marrucho and L. P. N. Rebelo, *J. Phys. Chem. B*, 2011, **115**, 10739.
23. C. Reichardt, *Green Chem.*, 2005, 7, 339.
24. M. J. Kamlet, J. L. Abboud and R. W. Taft, *J. Am. Chem. Soc.*, 1977, **99**, 6027.
25. M. A. Ab Rani, A. Brant, L. Crowhurst, A. Dolan, M. Lui, N. H. Hassan, J. P. Hallett, P. A. Hunt, H. Niedermeyer, J. M. Perez-Arlandis,

M. Schrems, T. Welton and R. Wilding, *Phys. Chem. Chem. Phys.*, 2011, **13**, 16831.

26. P. Lozano, T. De Diego and J. L. Iborra, in *Green Catalysis, vol. 3: Biocatalysis. in Anastas, P.T., Handbook of Green Chemistry Series*, ed. R. H. Crabtree, WILEY-VCH Verlag GmbH & Co. KGaA, Weinheim, 2009, vol. 3, ch. 3, pp. 51–73.

27. M. P. C. Marques, N. M. T. Lourenço, P. Fernander and C. C. C. R. de Carvalho, in *Green Solvents I: Properties and Applications in Chemistry*, ed. A. Mohammad and Inamuddin, Springer, Dordrecht, 2012, vol. 1, ch. 3, pp. 121–146.

28. S. Klembt, S. Dreyer, M. Eckstein and U. Kragl, in *Ionic Liquids in Synthesis*, ed. P. Wasserscheid and T. Welton, WILEY-VCH Verlag GmbH & Co. KGaA, Weinheim, 2nd edn, 2008, vol. 1, ch. 8, pp. 641–661.

29. J. J. Jodry and K. Mikami, in *Green Reaction Media in Organic Synthesis*, ed. K. Mikami, Blackwell Publishing Ltd, Oxford, 2005, ch. 7, pp. 9–58.

30. F. Fischer, J. Mutschler and D. Zufferey, *J. Ind. Microbiol. Biotechnol.*, 2011, **38**, 477.

31. Z. Yang, in *Ionic Liquids in Biotransformations and Organocatalysis: Solvents and Beyond*, ed. P. D. de María, John Wiley & Sons Publication, Hoboken, 2012, ch. 2, pp. 15–71.

32. N. J. M. Sanghamitra and T. Ueno, in *Green Solvents II: Properties and Applications of Ionic Liquids*, ed. A. Mohammad and Inamuddin, Springer, Dordrecht, 2012, vol. 2, ch. 10, pp. 253–273.

33. H. Zhao, O. Olubajo, Z. Song, A. L. Sims, T. E. Person, R. A. Lawal and L. A. Holley, *Bioorg. Chem.*, 2006, **34**, 15.

34. F. van Rantwijk and R. A. Sheldon, *Chem. Rev.*, 2007, **107**, 2757.

35. Z. Yang, *J. Biotechnol.*, 2009, **144**, 12.

36. M. Naushad, Z. A. Alothman, A. B. Khan and M. Ali, *Int. J. Biol. Macromol.*, 2012, **51**, 555.

37. A. P. de los Ríos, F. J. Hernández-Fernández, L. J. Lozano and C. Godínez, in *Green Solvents II: Properties and Applications of Ionic Liquids*, ed. A. Mohammad and Inamuddin, Springer, Dordrecht, 2012, vol. 2, ch. 7, pp. 169–188.

38. A. J. Kotlewska, F. van Rantwijk, R. A. Sheldon and I. W. C. E. Arends, *Green Chem.*, 2011, **13**, 2154.

39. A. M. Gumel, M. S. M. Annuar, T. Heidelberg and Y. Chisti, *Process Biochem.*, 2011, **46**, 2079.

40. T. Kobayashi, *Biotechnol. Lett.*, 2011, **33**, 1911.

41. J. K. Lee and M. Kim, *J. Org. Chem.*, 2002, **67**, 6845.

42. J. K. Lee and M.-J. Kim, *J. Mol. Catal. B: Enzym.*, 2011, **68**, 275.

43. T. Maruyama, H. Yamamura, T. Kotani, N. Kamiya and M. Goto, *Org. Biomol. Chem.*, 2004, **2**, 1239.

44. T. Itoh, Y. Matsushita, Y. Abe, S.-H. Han, S. Wada, S. Hayase, M. Kawatsura, S. Takai, M. Morimoto and Y. Hirose, *Chem.–Eur. J.*, 2006, **12**, 9228.

45. J. Mutschler, T. Rausis, J.-M. Bourgeois, C. Bastian, D. Zufferey, I. V. Mohrenz and F. Fischer, *Green Chem.*, 2009, **11**, 1793.

46. Y. Abe, K. Yoshiyama, Y. Yagi, S. Hayase, M. Kawatsura and T. Itoh, *Green Chem.*, 2010, **12**, 1976.
47. F.-X. Dong, L. Zhang, X.-Z. Tong, H.-B. Chen, X.-L. Wang and Y.-Z. Wang, *J. Mol. Catal. B: Enzym.*, 2012, **77**, 46.
48. M. Moniruzzaman and T. Ono, *Biochem. Eng. J.*, 2012, **60**, 156.
49. M. B. Abdul Rahman, K. Jumbri, N. A. Mohd Ali Hanafiah, E. Abdulmalek, B. A. Tejo, M. Basri and A. B. Salleh, *J. Mol. Catal. B: Enzym.*, 2012, **79**, 61.
50. R. L. de Souza, E. L. P. de Faria, R. T. Figueiredo, L. D. S. Freitas, M. Iglesias, S. Mattedi, G. M. Zanin, O. A. A. dos Santos, J. A. P. Coutinho, Á. S. Lima and C. M. F. Soares, *Enzyme Microb. Technol.*, 2013, **52**, 141.
51. P. C. A. G. Pinto, S. P. F. Costa, A. D. F. Costa, M. L. Passos, J. L. F. C. Lima and M. L. M. F. S. Saraiva, *J. Mol. Liq.*, 2012, **171**, 16.
52. B. Zeuner, G. M. Kontogeorgis, A. Riisager and A. S. Meyer, *New Biotechnol.*, 2012, **29**, 255.
53. S. J. Nara, J. R. Harjani and M. M. Salunkhe, *Tetrahedron Lett.*, 2002, **43**, 2979.
54. S. Pan, X. Liu, Y. Xie, Y. Yi, C. Li, Y. Yan and Y. Liu, *Bioresour. Technol.*, 2010, **101**, 9822.
55. P. Vidya and A. Chadha, *J. Mol. Catal. B: Enzym.*, 2010, **65**, 68.
56. N. Akbari, S. Daneshjoo, J. Akbari and K. Khajeh, *Appl. Biochem. Biotechnol.*, 2011, **165**, 785.
57. H. R. Karbalaei-Heidari, M. Shahbazi and G. Absalan, *Appl. Biochem. Biotechnol.*, 2013, **170**, 573.
58. O. Miyawaki and M. Tatsuno, *J. Biosci. Bioeng.*, 2008, **105**, 61.
59. H. Zhao, G. A. Baker, Z. Song, O. Olubajo, L. Zanders and S. M. Campbell, *J. Mol. Catal. B: Enzym.*, 2009, **57**, 149.
60. S. P. M. Ventura, L. D. F. Santos, J. A. Saraiva and J. A. P. Coutinho, *Green Chem.*, 2012, **14**, 1620.
61. L. Xue, Y. Zhao, L. Yu, Y. Sun, K. Yan, Y. Li, X. Huang and Y. Qu, *Colloids Surf., B*, 2013, **105**, 81.
62. Z. Li, X. Liu, Y. Pei, J. Wang and M. He, *Green Chem.*, 2012, **14**, 2941.
63. Y. Pei, J. Wang, K. Wu, X. Xuan and X. Lu, *Sep. Purif. Technol.*, 2009, **64**, 288.
64. F. J. Deive, A. Rodríguez, L. P. N. Rebelo and I. M. Marrucho, *Sep. Purif. Technol.*, 2012, **97**, 205.
65. E. Durand, J. Lecomte, B. Baréa, G. Piombo, E. Dubreucq and P. Villeneuve, *Process Biochem.*, 2012, **47**, 2081.
66. E. Durand, J. Lecomte and P. Villeneuve, *Eur. J. Lipid Sci. Technol.*, 2013, **115**, 379.
67. C. Ruß and B. König, *Green Chem.*, 2012, **14**, 2969.
68. Q. Zhang, K. De Oliveira Vigier, S. Royer and F. Jérôme, *Chem. Soc. Rev.*, 2012, **41**, 7108.
69. S. Park, F. Viklund, K. Hult and R. J. Kazlauskas, *Green Chem.*, 2003, **5**, 715.
70. Z.-G. Chen, D.-N. Zhang and Y.-B. Han, *Process Biochem.*, 2013, **48**, 620.
71. Z. Yang, Z. Guo and X. Xu, *J. Am. Oil Chem. Soc.*, 2012, **89**, 1049.

72. M. E. M. B. de Araújo, F. J. Contesini, Y. E. M. Franco, A. C. H. F. Sawaya, T. G. Alberto, N. Dalfré and P. D. O. Carvalho, *Molecules*, 2011, **16**, 7171.

73. M. Kidwai and R. Poddar, *Catal. Lett.*, 2008, **124**, 311.

74. A. Kurata, Y. Kitamura, S. Irie, S. Takemoto, Y. Akai, Y. Hirota, T. Fujita, K. Iwai, M. Furusawa and N. Kishimoto, *J. Biotechnol.*, 2010, **148**, 133.

75. A. Kurata, S. Takemoto, T. Fujita, K. Iwai, M. Furusawa and N. Kishimoto, *J. Mol. Catal. B: Enzym.*, 2011, **69**, 161.

76. N. Pang, S.-S. Gu, J. Wang, H.-S. Cui, F.-Q. Wang, X. Liu, X.-Y. Zhao and F.-A. Wu, *Bioresour. Technol.*, 2013, **139**, 337.

77. B.-M. Lue, Z. Guo and X. Xu, *Process Biochem.*, 2010, **45**, 1375.

78. X. Lu, Z. Luo, S. Yu and X. Fu, *J. Agric. Food Chem.*, 2012, **60**, 9273.

79. B.-K. Liu, Q. Wu, J.-M. Xu and X.-F. Lin, *Chem. Commun.*, 2007, 295.

80. B. Chen, H. Liu, Z. Guo, J. Huang, M. Wang, X. Xu and L. Zheng, *J. Agric. Food Chem.*, 2011, **59**, 1256.

81. S. Sun, F. Qin, Y. Bi, J. Chen, G. Yang and W. Liu, *Biotechnol. Lett.*, 2013, **35**, 1449.

82. H. Noritomi, S. Nishida and S. Kato, *Biotechnol. Lett.*, 2007, **29**, 1509.

83. S. Shah and M. N. Gupta, *Biochim. Biophys. Acta*, 2007, **1770**, 94.

84. M. Eckstein, M. Sesing, U. Kragl and P. Adlercreutz, *Biotechnol. Lett.*, 2002, **24**, 867.

85. H. Noritomi, K. Suzuki, M. Kikuta and S. Kato, *Biochem. Eng. J.*, 2009, **47**, 27.

86. G. Xing, F. Li, C. Ming and L. Ran, *Tetrahedron Lett.*, 2007, **48**, 4271.

87. H. Zhao, G. A. Baker and S. Holmes, *Org. Biomol. Chem.*, 2011, **9**, 1908.

88. H. Zhao, L. Jackson, Z. Song and O. Olubajo, *Tetrahedron: Asymmetry*, 2006, **17**, 1549.

89. S. Cerqueira Pereira, R. Bussamara, G. Marin, R. Lima Camargo Giordano, J. Dupont and R. de Campos Giordano, *Green Chem.*, 2012, **14**, 3146.

90. Y. Jiang, H. Xia, C. Guo, I. Mahmood and H. Liu, *Biotechnol. Prog.*, 2007, **23**, 829.

91. T. O. Akanbi, C. J. Barrow and N. Byrne, *Catal. Sci. Technol.*, 2012, **2**, 1839.

92. S. Bai, M. Ren, L. Wang and Y. Sun, *Front. Chem. Eng. China*, 2008, **2**, 301.

93. H. Zhao, L. Jackson, Z. Song and O. Olubajo, *Tetrahedron: Asymmetry*, 2006, **17**, 2491.

94. W.-Y. Lou, M.-H. Zong, Y.-Y. Liu and J.-F. Wang, *J. Biotechnol.*, 2006, **125**, 64.

95. M. Singh, R. S. Singh and U. C. Banerjee, *Process Biochem.*, 2010, **45**, 25.

96. L. Li, W. Feng and K. Pan, *Colloids Surf., B*, 2013, **102**, 124.

97. C. Pilissão and M. D. G. Nascimento, *Tetrahedron: Asymmetry*, 2006, **17**, 428.

98. P. Hara, U. Hanefeld and L. T. Kanerva, *Green Chem.*, 2009, **11**, 250.

99. S. V. Malhotra and H. Zhao, *Chirality*, 2005, **17**, S240–S242.

100. W.-Y. Lou, M.-H. Zong, H. Wu, R. Xu and J.-F. Wang, *Green Chem.*, 2005, **7**, 500.

101. M.-Y. Ren, S. Bai, D.-H. Zhang and Y. Sun, *J. Agric. Food Chem.*, 2008, **56**, 2388.

102. R. Bogel-Łukasik, N. M. T. Lourenço, P. Vidinha, M. D. R. G. da Silva, C. A. M. Afonso, M. N. da Ponte and S. Barreiros, *Green Chem.*, 2008, **10**, 243.

103. M. Singh, R. S. Singh and U. C. Banerjee, *J. Mol. Catal. B: Enzym.*, 2009, **56**, 294.

104. T. Itoh, Y. Nishimura, N. Ouchi and S. Hayase, *J. Mol. Catal. B: Enzym.*, 2003, **26**, 41.

105. A. Kamal and G. Chouhan, *Tetrahedron Lett.*, 2004, **45**, 8801.

106. A. P. de los Ríos, F. J. Hernández Fernández, D. Gómez, M. Rubio and G. Víllora, *Process Biochem.*, 2011, **46**, 1475.

107. Y. Liu, D. Chen, Y. Yan, C. Peng and L. Xu, *Bioresour. Technol.*, 2011, **102**, 10414.

108. H. Zhao, C. Zhang and T. D. Crittle, *J. Mol. Catal. B: Enzym.*, 2013, **85–86**, 243.

109. J.-Q. Lai, Z.-L. Hu, P.-W. Wang and Z. Yang, *Fuel*, 2012, **95**, 329.

110. T. De Diego, A. Manjón, P. Lozano, M. Vaultier and J. L. Iborra, *Green Chem.*, 2011, **13**, 444.

111. B. L. A. P. Devi, G. Zheng and X. Xuebing, *AIChE J.*, 2011, **57**, 1628.

112. S. Chanfreau, M. Mena, J. R. Porras-Domínguez, M. Ramírez-Gilly, M. Gimeno, P. Roquero, A. Tecante and E. Bárzana, *Bioprocess Biosyst. Eng.*, 2010, **33**, 629.

113. M. Mena, S. Chanfreau, M. Gimeno and E. Bárzana, *Bioprocess Biosyst. Eng.*, 2010, **33**, 1095.

114. J. T. Gorke, K. Okrasa, A. Louwagie, R. J. Kazlauskas and F. Srienc, *J. Biotechnol.*, 2007, **132**, 306.

115. R. Marcilla, M. de Geus, D. Mecerreyes, C. J. Duxbury, C. E. Koning and A. Heise, *Eur. Polym. J.*, 2006, **42**, 1215.

116. S. J. Nara, J. R. Harjani, M. M. Salunkhe, A. T. Mane and P. P. Wadgaonkar, *Tetrahedron Lett.*, 2003, **44**, 1371.

117. C. Wu, Z. Zhang, C. Chen, F. He and R. Zhuo, *Biotechnol. Lett.*, 2013, **35**, 1623.

118. M. Mena, A. López-Luna, K. Shirai, A. Tecante, M. Gimeno and E. Bárzana, *Bioprocess Biosyst. Eng.*, 2013, **36**, 383.

119. Z. Zhang, F. He and R. Zhuo, *J. Mol. Catal. B: Enzym.*, 2013, **94**, 129.

120. P. Lozano, R. Piamtongkam, K. Kohns, T. De Diego, M. Vaultier and J. L. Iborra, *Green Chem.*, 2007, **9**, 780.

121. P. Lozano, J. M. Bernal and A. Navarro, *Green Chem.*, 2012, **14**, 3026.

122. A. Pohar, I. Plazl and P. Žnidaršič-Plazl, *Lab Chip*, 2009, **9**, 3385.

123. D. Barahona, P. H. Pfromm and M. E. Rezac, *Biotechnol. Bioeng.*, 2006, **93**, 318.

124. S. H. Ha, N. M. Hiep and Y.-M. Koo, *Biotechnol. Bioprocess Eng.*, 2010, **15**, 126.

125. S. H. Lee, H. M. Nguyen, Y.-M. Koo and S. H. Ha, *Process Biochem.*, 2008, **43**, 1009.

126. P. D'Arrigo, L. Cerioli, C. Chiappe, W. Panzeri, D. Tessaro and A. Mele, *J. Mol. Catal. B: Enzym.*, 2012, **84**, 132.

127. Z. Guo and X. Xu, *Green Chem.*, 2006, **8**, 54.

128. D. Kahveci, Z. Guo, B. Özçelik and X. Xu, *Process Biochem.*, 2009, **44**, 1358.

129. Z. Guo, D. Kahveci, B. Ozçelik and X. Xu, *New Biotechnol.*, 2009, **26**, 37.

130. S. Furukawa, K. Hasegawa, I. Fuke, K. Kittaka, T. Nakakoba, M. Goto and N. Kamiya, *Biochem. Eng. J.*, 2013, **70**, 84.

131. S. C. Park, W. J. Chang, S. M. Lee, Y. J. Kim and Y. M. Koo, *Biotechnol. Bioprocess Eng.*, 2005, **10**, 99.

132. Z. Findrik, G. Németh, L. Gubicza, K. Bélafi-Bakó and D. Vasić-Rački, *Bioprocess Biosyst. Eng.*, 2012, **35**, 625.

133. Y. Wang, M. Radosevich, D. Hayes and N. Labbé, *Biotechnol. Bioeng.*, 2011, **108**, 1042.

134. P. Engel, S. Krull, B. Seiferheld and A. C. Spiess, *Bioresour. Technol.*, 2012, **115**, 27.

135. S. Datta, B. Holmes, J. I. Park, Z. Chen, D. C. Dibble, M. Hadi, H. W. Blanch, B. A. Simmons and R. Sapra, *Green Chem.*, 2010, **12**, 338.

136. M. Sandoval, Á. Cortés, C. Civera, J. Treviño, E. Ferreras, M. Vaultier, J. Berenguer, P. Lozano and M. J. Hernáiz, *RSC Adv.*, 2012, **2**, 6306.

137. B. Dabirmanesh, S. Daneshjou, A. A. Sepahi, B. Ranjbar, R. A. Khavari-Nejad, P. Gill, A. Heydari and K. Khajeh, *Int. J. Biol. Macromol.*, 2011, **48**, 93.

138. N. Kamiya, Y. Matsushita, M. Hanaki, K. Nakashima, M. Narita, M. Goto and H. Takahashi, *Biotechnol. Lett.*, 2008, **30**, 1037.

139. P. W. Wolski, D. S. Clark and H. W. Blanch, *Green Chem.*, 2011, **13**, 3107.

140. N. Kaftzik, P. Wasserscheid and U. Kragl, *Org. Process Res. Dev.*, 2002, **6**, 553.

141. F. Yang, L. Li, Q. Li, W. Tan, W. Liu and M. Xian, *Carbohydr. Polym.*, 2010, **81**, 311.

142. J. Wang, G.-X. Sun, L. Yu, F.-A. Wu and X.-J. Guo, *Bioresour. Technol.*, 2013, **128**, 156.

143. M. Lang, T. Kamrat and B. Nidetzky, *Biotechnol. Bioeng.*, 2006, **95**, 1093.

144. M. B. Turner, S. K. Spear, J. G. Huddleston, J. D. Holbrey and R. D. Rogers, *Green Chem.*, 2003, **5**, 443.

145. P. Engel, R. Mladenov, H. Wulfhorst, G. Jäger and A. C. Spiess, *Green Chem.*, 2010, **12**, 1959.

146. Â. C. Salvador, M. D. C. Santos and J. A. Saraiva, *Green Chem.*, 2010, **12**, 632.

147. N. R. Singh, D. Narinesingh and G. Singh, *J. Mol. Liq.*, 2010, **152**, 19.

148. C. Lehmann, F. Sibilla, Z. Maugeri, W. R. Streit, P. Domínguez de María, R. Martinez and U. Schwaneberg, *Green Chem.*, 2012, **14**, 2719.

149. T. Zhang, S. Datta, J. Eichler, N. Ivanova, S. D. Axen, C. A. Kerfeld, F. Chen, N. Kyrpides, P. Hugenholtz, J.-F. Cheng, K. L. Sale, B. Simmons and E. Rubin, *Green Chem.*, 2011, **13**, 2083.

150. S. Bose, C. A. Barnes and J. W. Petrich, *Biotechnol. Bioeng.*, 2012, **109**, 434.

151. M. F. Thomas, L.-L. Li, J. M. Handley-Pendleton, D. van der Lelie, J. J. Dunn and J. F. Wishart, *Bioresour. Technol.*, 2011, **102**, 11200.

152. D. Gamenara, P. S. Méndez, G. Seoane and P. D. De María, in *Ionic Liquids in Biotransformations and Organocatalysis: Solvents and Beyond*, ed. P. D. De María, John Wiley & Sons Publication, Hoboken, 2012, ch. 3, pp. 229–259.

153. A. P. M. Tavares, B. Pinho, O. Rodriguez and E. A. Macedo, *Procedia Eng.*, 2012, **42**, 226.

154. K. Okrasa, E. Guibé-Jampel and T. Michel, *Tetrahedron: Asymmetry*, 2003, **14**, 2487.

155. C. Chiappe, L. Neri and D. Pieraccini, *Tetrahedron Lett.*, 2006, **47**, 5089.

156. A. Boškin, C. D. Tran and M. Franko, *Environ. Chem. Lett.*, 2009, **7**, 267.

157. P. Rahimi, H. Ghourchian and S. Sajjadi, *Analyst*, 2012, **137**, 471.

158. J. Kuwahara, R. Ikari, K. Murata, N. Nakamura and H. Ohno, *Catal. Today*, 2013, **200**, 49.

159. Y. Gu and J.-Y. Tsai, *Synth. Met.*, 2012, **161**, 2743.

160. G. De Gonzalo, I. Lavandera, K. Durchschein, D. Wurm, K. Faber and W. Kroutil, *Tetrahedron: Asymmetry*, 2007, **18**, 2541.

161. E. S. Hong, O. Y. Kwon and K. Ryu, *Biotechnol. Lett.*, 2008, **30**, 529.

162. J. H. Park, O. Y. Kwon and K. Ryu, *Biotechnol. Bioprocess Eng.*, 2010, **15**, 993.

163. J. H. Park, I. K. Yoo, O. Y. Kwon and K. Ryu, *Biotechnol. Lett.*, 2011, **33**, 1657.

164. P. C. A. G. Pinto, A. D. F. Costa, J. L. F. C. Lima and M. L. M. F. S. Saraiva, *Chemosphere*, 2011, **82**, 1620.

165. Z. Yang, Y.-J. Yue and M. Xing, *Biotechnol. Lett.*, 2008, **30**, 153.

166. G. Hinckley, V. V. Mozhaev, C. Budde and Y. L. Khmelnitsky, *Biotechnol. Lett.*, 2002, **24**, 2083.

167. B. Dabirmanesh, K. Khajeh, J. Akbari, H. Falahati, S. Daneshjoo and A. Heydari, *J. Mol. Liq.*, 2011, **161**, 139.

168. B. Dabirmanesh, K. Khajeh, B. Ranjbar, F. Ghazi and A. Heydari, *J. Mol. Liq.*, 2012, **170**, 66.

169. G.-P. Zhou, Y. Zhang, X.-R. Huang, C.-H. Shi, W.-F. Liu, Y.-Z. Li, Y.-B. Qu and P.-J. Gao, *Colloids Surf., B*, 2008, **66**, 146.

170. N. A. Mohidem and H. Bin Mat, *Bioresour. Technol.*, 2012, **114**, 472.

171. Q. Cao, L. Quan, C. He, N. Li, K. Li and F. Liu, *Talanta*, 2008, 77, 160.

172. K. Fujita and H. Ohno, *Biopolymers*, 2010, **93**, 1093.

173. A. Illanes, in *Enzyme Biocatalysis: Principles and Applications*, ed. A. Illanes, Springer, Dordrecht, 2008, ch. 1, pp. 1–56.

174. Y. Wang, Y. Pan, Z. Zhang, R. Sun, X. Fang and D. Yu, *Process Biochem.*, 2012, **47**, 976.

175. T. Ståhlberg, J. M. Woodley and A. Riisager, *Catal. Sci. Technol.*, 2012, **2**, 291.

176. D. Yu, Y. Wang, C. Wang, D. Ma and X. Fang, *J. Mol. Catal. B: Enzym.*, 2012, **79**, 8.

177. T. Kifazume, in *Electrochemical Aspects of Ionic Liquids*, ed. H. Ohno, Wiley-Interscience, Hoboken, 2005, ch. 10, pp. 135–142.

178. M. Ebrahimi, S. Hosseinkhani, A. Heydari, R. A. Khavari-Nejad and J. Akbari, *Appl. Biochem. Biotechnol.*, 2012, **168**, 604.

179. S. Lutz-Wahl, E.-M. Trost, B. Wagner, A. Manns and L. Fischer, *J. Biotechnol.*, 2006, **124**, 163.

180. F. Ganske and U. T. Bornscheuer, *Biotechnol. Lett.*, 2006, **28**, 465.

181. G. Quijano, A. Couvert and A. Amrane, *Bioresour. Technol.*, 2010, **101**, 8923.

182. K. M. Docherty and C. F. Kulpa, Jr., *Green Chem.*, 2005, **7**, 185.

183. J. Ranke, K. Mölter, F. Stock, U. Bottin-Weber, J. Poczobutt, J. Hoffmann, B. Ondruschka, J. Filser and B. Jastorff, *Ecotoxicol. Environ. Saf.*, 2004, **58**, 396.

184. T. P. T. Pham, C.-W. Cho, J. Min and Y.-S. Yeoung, *J. Biosci. Bioeng.*, 2008, **105**, 425.

185. A. S. Wells and V. T. Coombe, *Org. Process Res. Dev.*, 2006, **10**, 794.

186. R. J. Bernot, E. E. Kennedy and G. A. Lamberti, *Environ. Toxicol. Chem.*, 2005, **24**, 1759.

187. C. Pretti, C. Chiappe, D. Pieraccini, M. Gregori, F. Abramo, G. Monni and L. Intorre, *Green Chem.*, 2006, **8**, 238.

188. M. Rebros, H. Q. N. Gunaratne, J. Ferguson, R. Seddon and G. Stephens, *Green Chem.*, 2009, **11**, 402.

189. S. M. Saadeh, Z. Yasseen, F. A. Sharif and H. M. Abu Shawish, *Ecotoxicol. Environ. Saf.*, 2009, **72**, 1805.

190. M. I. Hossain, M. El-Harbawi, Y. A. Noaman, M. A. B. Bustam, N. B. M. Alitheen, N. A. Affandi, G. Hefter and C.-Y. Yin, *Chemosphere*, 2011, **84**, 101.

191. N. Wood, J. L. Ferguson, H. Q. N. Gunaratne, K. R. Seddon and G. M. Stephens, *Green Chem.*, 2011, **13**, 1843.

192. H. Pfruender, M. Amidjojo, U. Kragl and D. Weuster-botz, *Angew. Chem., Int. Ed.*, 2004, **43**, 4529.

193. M. Matsumoto, K. Mochiduki and K. Kondo, *J. Biosci. Bioeng.*, 2004, **98**, 344.

194. W. Lou, W. Wang, T. J. Smith and M.-H. Zong, *Green Chem.*, 2009, **11**, 1377.

195. W. Wang, M.-H. Zong and W.-Y. Lou, *J. Mol. Catal. B: Enzym.*, 2009, **56**, 70.

196. H. Pfruender, R. Jones and D. Weuster-Botz, *J. Biotechnol.*, 2006, **124**, 182.

197. D. Torres-martínez, R. Melgarejo-torres, M. Gutiérrez-rojas, L. Aguilera-vázquez, M. Micheletti, G. J. Lye and S. Huerta-Ochoa, *Biochem. Eng. J.*, 2009, **45**, 209.

198. S.-M. Lee, W.-J. Chang, A.-R. Choi and Y.-M. Koo, *Korean J. Chem. Eng.*, 2005, **22**, 687.

199. O. Dipeolu, E. Green and G. Stephens, *Green Chem.*, 2009, **11**, 397.

200. M. Ouellet, S. Datta, D. C. Dibble, P. R. Tamrakar, P. I. Benke, C. Li, S. Singh, K. L. Sale, P. D. Adams, J. D. Keasling, B. A. Simmons, B. M. Holmes and A. Mukhopadhyay, *Green Chem.*, 2011, **13**, 2743.
201. H. Wang, S. V. Malhotra and A. J. Francis, *Chemosphere*, 2011, **82**, 1597.
202. W. Lou, M. Zong and T. J. Smith, *Green Chem.*, 2006, **8**, 147.
203. M. Sendovski, N. Nir and A. Fishman, *J. Agric. Food Chem.*, 2010, **58**, 2260.
204. J. Pernak, J. Rogoza and I. Mirska, *Eur. J. Med. Chem.*, 2001, **36**, 313.
205. J. Pernak, K. Sobaszkiewicz and I. Mirska, *Green Chem.*, 2003, **5**, 52.
206. J. Pernak, I. Goc and I. Mirska, *Green Chem.*, 2004, **6**, 323.
207. J. Pernak and J. Feder-Kubis, *Chem.–Eur. J.*, 2005, **11**, 4441.
208. J. Pernak, M. Smiglak, S. T. Griffin, W. L. Hough, T. B. Wilson, A. Pernak, J. Zabielska-Matejuk, A. Fojutowski, K. Kita and R. D. Rogers, *Green Chem.*, 2006, **8**, 798.
209. J. Pernak, A. Syguda, I. Mirska, A. Pernak, J. Nawrot, A. Pradzyńska, S. T. Griffin and R. D. Rogers, *Chem.–Eur. J.*, 2007, **13**, 6817.
210. A. Busetti, D. E. Crawford, M. J. Earle, M. A. Gilea, B. F. Gilmore, S. P. Gorman, G. Laverty, A. F. Lowry, M. McLaughlin and K. R. Seddon, *Green Chem.*, 2010, **12**, 420.
211. A. Cornellas, L. Perez, F. Comelles, I. Ribosa, A. Manresa and M. T. Garcia, *J. Colloid Interface Sci.*, 2011, **355**, 164.
212. J. Łuczak, C. Jungnickel, I. Łącka, S. Stolte and J. Hupka, *Green Chem.*, 2010, **12**, 593.
213. Z. Yang, R. Zeng, Y. Wang, X. Li, Z. Lv and B. Lai, *Food Technol. Biotechnol.*, 2009, **9862**, 62.
214. Z. Fan, N. I. Ye, S. U. N. Zhihao, Z. Pu, L. I. N. Wenqing, Z. H. U. Po and J. U. Nianfeng, *Chin. J. Catal.*, 2008, **29**, 577.
215. W.-Y. Lou, W. Wang, R.-F. Li and M.-H. Zong, *J. Biotechnol.*, 2009, **143**, 190.
216. W.-Y. Lou, L. Chen, B.-B. Zhang, T. J. Smith and M.-H. Zong, *BMC Biotechnol.*, 2009, **9**, 1.
217. S. Bräutigam, S. Bringer-Meyer and D. Weuster-Botz, *Tetrahedron: Asymmetry*, 2007, **18**, 1883.
218. S. G. Cull, J. D. Holbrey, V. Vargas-Mora, K. R. Seddon and G. J. Lye, *Biotechnol. Bioeng.*, 2000, **69**, 227.
219. M. Matsumoto, K. Mochiduki, K. Fukunishi and K. Kondo, *Sep. Purif. Technol.*, 2004, **40**, 97.
220. J. Howarth, P. James and J. Dai, *Tetrahedron Lett.*, 2001, **42**, 7517.
221. J.-Y. He, L.-M. Zhou, P. Wang and L. Zu, *Process Biochem.*, 2009, **44**, 316.
222. S. Bräutigam, D. Dennewald, M. Schürmann, J. Lutje-Spelberg, W.-R. Pitner and D. Weuster-Botz, *Enzyme Microb. Technol.*, 2009, **45**, 310.
223. D. Dennewald, W.-R. Pitner and D. Weuster-Botz, *Process Biochem.*, 2011, **46**, 1132.
224. T. Matsuda, Y. Yamagishi, S. Koguchi, N. Iwai and T. Kitazume, *Tetrahedron Lett.*, 2006, **47**, 4619.

225. R. Kratzer, M. Pukl, S. Egger and B. Nidetzky, *Microb. Cell Fact.*, 2008, 7, 37.
226. J.-Y. Chen, I. Kaleem, D.-M. He, G.-Y. Liu and C. Li, *Process Biochem.*, 2012, **47**, 908.
227. S. Arai, K. Nakashima, T. Tanino, C. Ogino, A. Kondo and H. Fukuda, *Enzyme Microb. Technol.*, 2010, **46**, 51.
228. M. D. Baumann, A. J. Daugulis and P. G. Jessop, *Appl. Microbiol. Biotechnol.*, 2005, **67**, 131.
229. D.-X. Wu, Y.-X. Guan, H.-Q. Wang and S.-J. Yao, *Bioresour. Technol.*, 2011, **102**, 9368.

Biocatalysis in Micellar Systems

ADELAIDE BRAGA[a] AND ISABEL BELO[*a]

[a]Center of Biological Engineering, University of Minho, Campus de Gualtar, 4710-057, Braga, Portugal
*E-mail: ibelo@deb.uminho.pt

7.1 Introduction

7.1.1 Biocatalysis

Biocatalysis or biotransformation involves the use of biological systems to catalyze the conversion of one compound to another. The catalyst part of the biological system can therefore consist of whole cells, cellular extracts, or isolated enzyme(s).[1] Enzymes and whole-cell biocatalysts have several attractive properties, which make them privileged catalysts for organic synthesis, because they have high chemo-, regio-, and stereo-selectivities and require mild reaction conditions. Compared to chemical technologies, biocatalysis produces fewer by-products, consumes less energy, and generates less pollution to the environment.

Since the 1960s, isolated and immobilized enzymes have been used in large-scale industrial reactors. Nowadays, products resulting from bio-catalytic processes vary from commodity products to high-value pharmaceuticals.[2] In most cases, natively evolved biocatalysts, either in the form of enzymes or microbes, are not optimized for use in industrial reactors. Advances in genetic engineering have made it possible to design and produce

RSC Green Chemistry No. 45
White Biotechnology for Sustainable Chemistry
Edited by Maria Alice Z. Coelho and Bernardo D. Ribeiro
© The Royal Society of Chemistry 2016
Published by the Royal Society of Chemistry, www.rsc.org

more efficient industrial biocatalysts. The use of isolated enzymes provides the chance to further improve the performance of biocatalysts through various post-isolation chemical and physical manipulations.[1] One method currently used for the production of industrial enzymes is immobilization. Several approaches have been pursued for the improvement of the preparation and use of biocatalysts, such as developing enzyme-based assemblies, microscale structures and new types of bioreactors.

Biocatalysis is normally performed in an aqueous environment but can, in many cases, also be conducted in solvent mixtures, liquid–liquid two-phase systems, and even in pure organic solvents.

Biocatalysis performed in monophasic aqueous media using hydrophobic reagents, *i.e.* low-water solubility, can lead to low volumetric productivities and poor enzyme stability in mono-phase aqueous systems. Thus, new techniques to improve biocatalytic performance under unfavorable experimental conditions, which involve solvent engineering and multiphase systems, have been reported.[3]

Biocatalysis in organic solvent media is very useful when one or more components of the enzymatic reaction are poorly water soluble. Several approaches have been used to investigate enzymatic behavior in water-restricted environments, including the use of water miscible organic solvents, biphasic aqueous–organic solvents, reverse micelles and monophasic organic solvent systems.[4]

7.1.2 Micellar Systems

Emulsions are dispersions of oil-in-water (o/w) or water-in-oil (w/o) that can be more or less stable depending on the presence of amphiphilic compounds. Amphiphiles are molecules consisting of a hydrophilic head group and a hydrophobic (lipophilic) tail, and are thus able to interact with both polar and nonpolar compounds. Amphiphilic compounds allow the decrease of the interfacial oil–water tension, leading to the spontaneous formation of a dispersion of one phase in the other.

The presence of amphiphiles (or surfactants) in a mixture of oil and water is determinant for the formation of microemulsions. Depending on the ratio between the components in the mixture, at the two extremes, the microstructure of the microemulsions can vary from very tiny water droplets dispersed in the oil phase (water-in-oil microemulsion) to oil droplets dispersed in the water phase (oil-in-water microemulsion).[5]

Emulsions and microemulsions are different in terms of structure and stability. In contrast to microemulsions, emulsions are unstable systems and without agitation, phase separation will occur. The other difference is that the size of droplets in emulsions is in the range of micrometers, while in microemulsions droplet size is in the range of nanometers, depending on some parameters, such as the surfactant type and concentration, and the extent of the dispersed phase.[6] In microemulsions, simple spherical aggregates, called micelles, are formed. A micellar system appears to be

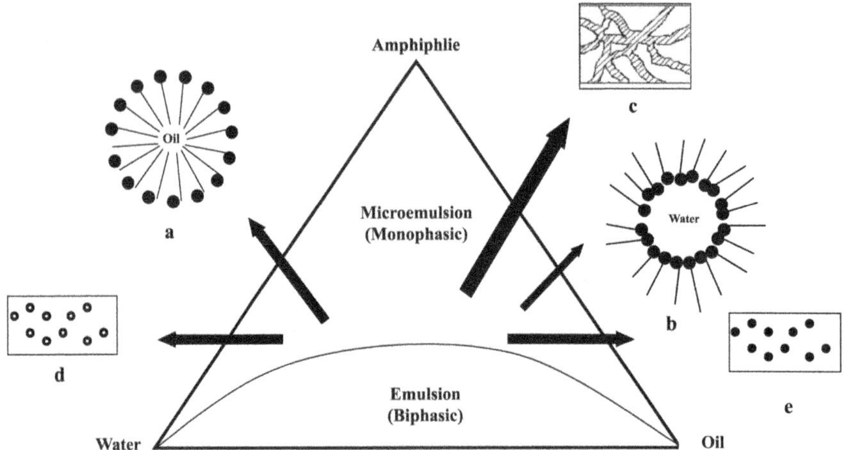

Figure 7.1 A comprehensive ternary phase diagram depicting various structures: (a) o/w microemulsion, (b) w/o microemulsion, (c) bicontinuous microemulsion, (d) and (e) various dispersions. (Adapted from ref. 12.)

homogeneous (monophasic) since these aggregates are of colloidal size. However, the absorbed reactants are in a microheterogeneous two-phase system.[7]

Mixtures of oil, water and amphiphiles can have different structures depending on their composition, and other factors, such as for instance the presence of a co-surfactant and the temperature. Several reports are available in the literature describing with less or more complexity the different types of structure formed, according to the ternary phase diagram of the system.[8-11] Moulik and Rakshit[12] proposed a simplified example of a ternary phase diagram that indicates the principal structures formed according to the mixture composition (Figure 7.1).

Microemulsions can differ in structure and complexity and were first described by Winsor[13] and were reviewed by Moulik and Rakshit.[12] The nature of the amphiphile and the proportion of oil and water are determinant factors for the formation of spherical micelles in oil-in-water systems or reverse micelles in water-in-oil systems.

7.2 Oil-in-Water Systems

Biphasic systems in biotransformation encompass the use of two immiscible liquid phases where one of the phases (aqueous-based) provides a protective environment for the biocatalyst, whereas the second phase is a substrate and/or product pool. The production of relevant compounds using enzymes and whole cells, with this approach, was clearly highlighted in the late 70s and early 80s, and since then the range of applications has constantly increased and diversified.[14]

Originally, the second phase in a biphasic system was based on a water immiscible organic solvent with high solubilization capability for the substrate and/or product. A biphasic system may be alternatively formed by two aqueous phase systems, the organic solvent being replaced by a second aqueous phase which is polymer-based. In a third option, which has been gaining relevance in recent years, the organic solvent is replaced by an ionic liquid.[14]

Recently, biotransformations have gained much attention as alternatives to conventional chemistry for the production of fine chemicals and active intermediate compounds of industrial interest. In particular, synthesis of chiral molecules, enantiomers, and single isomers in the pharmaceutical industry produced by enzymes is becoming increasingly important.[15]

7.2.1 Emulsion Characterization

The physicochemical properties of emulsions are strongly influenced by the characteristics of the droplets that they contain, *i.e.*, their concentration, size, charge, interfacial properties and interactions.[16–18] The droplet concentration in an emulsion influences its texture, stability, appearance, sensory attributes and nutritional quality.[18] Droplet concentration is usually characterized in terms of the dispersed phase volume fraction, which is equal to the volume of emulsion droplets divided by the total volume of the emulsion. The size of the droplets in an emulsion has a strong impact on its stability (*e.g.*, towards gravitational separation, flocculation, and coalescence), its optical properties (*e.g.*, lightness and color), its rheology (*e.g.*, viscosity or modulus), and its sensory attributes (*e.g.*, creaminess).[18] When all droplets in an emulsion have the same size, the emulsion is referred to as "monodisperse". A polydisperse emulsion is characterized by its "particle size distribution", which defines the concentration of droplets in different size classes.[18] The droplets in most emulsions have an electrical charge because of the adsorption of molecules on their surfaces that are ionized or ionizable, *e.g.*, proteins, certain polysaccharides, ionic surfactants, phospholipids and some small ions.[18] The electrical characteristics of a droplet's surface depend on the type and concentration of ionized charge species present at the surface, as well as the ionic composition and physical properties of the surrounding liquid. The charge on an emulsion droplet is important because it determines the nature of its interactions with other charged species (*e.g.*, small ions, macromolecules or colloidal particles) or its behavior in the presence of an electrical field (*e.g.*, electrophoresis). The electrical characteristics of a droplet are usually characterized in terms of its surface charge density (σ) and/or ζ-potential (ζ).[19] The *zeta potential* (ζ) is the electrical potential at the "shear plane", which is defined as the distance away from the droplet surface below which the counter regions remain strongly attached to the droplet when it moves in an electrical field. Practically, the ζ-potential is often a better representation of the electrical characteristics of an emulsion droplet because it inherently accounts for the adsorption of any charged counter ions. In addition, the ζ-potential is much easier to measure than the electrical potential or the surface charge density, and therefore, droplet charges are usually characterized in terms of ζ-potential.[19]

The droplet interface consists of a narrow region (≈1 to 50 nm thick) that surrounds each emulsion droplet, and contains a mixture of oil, water, and emulsifier molecules, as well as possibly other types of molecules and ions (*e.g.*, mineral ions, hydrophilic polyelectrolytes, amphiphilic components). The interfacial region only makes up a significant fraction of the total volume of an emulsion when the droplet size is less than about 1 μm.[18]

A wide variety of different analytical instruments have been developed that can be used to characterize the properties of emulsions (Table 7.1).

For instance, oil-in-water emulsions (castor oil and methyl ricinoleate) stabilized by a non-ionic surfactant, Tween 80 were characterized by Gomes *et al.*[24] by determining the oil droplets' size distribution through a laser granulometry technique. The results obtained indicated the existence of two distinct droplet populations (Figure 7.2). For the population with higher size, the number and diameter of droplets increased with the increase in oil concentration.

Table 7.1 Analytical techniques used to characterize emulsion properties.

Technique		Description	References
Microscopy–Laser scanning confocal microscopy (LSCM)		-Three-dimensional images of samples -Determine the spatial location of the droplets in an emulsion -Powerful at examining the location and transport of specific components within emulsions	20,21
Particle size analyzers	Static light scattering	-The scattering pattern produced when a laser beam is directed through a dilute emulsion depends on the particle size distribution	22
	Dynamic light scattering	-Measurements of the intensity fluctuations that occur over time when light is scattered by particles that change their relative spatial location due to Brownian motion	22
	Electrical pulse counting	-Measurement of changes in the electrical conductivity across a small orifice when a dilute emulsion is pulled through it -The particle size distribution is determined by measuring the height of each individual pulse, since the height of an electrical pulse is proportional to the volume of a particle	18
	NMR techniques	-Interactions between radio waves and the nuclei of hydrogen atoms to obtain information about the microstructure of emulsions -Pulses cause some of the hydrogen nuclei in the sample to be excited to higher energy levels, which leads to the generation of a detectable NMR signal	23

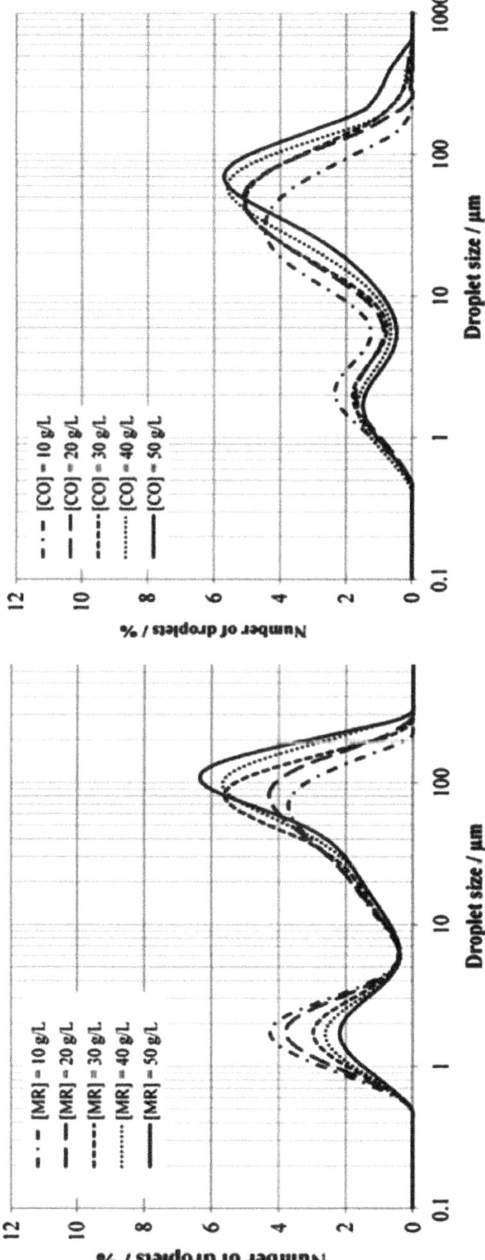

Figure 7.2　Methyl ricinoleate (MR) and castor oil (CO) droplet size distribution in oil-in-water emulsions with Tween 80. (Reproduced with permission from ref. 24.)

7.2.2 Enzyme Catalysis

The use of enzymes in biphasic systems, instead of aqueous media, offers several important advantages, such as a reduction in enzyme substrate and/or product inhibition, the solubilization of hydrophobic compounds, the possibility of shifting the thermodynamic equilibrium towards the desired reaction,[25] and the possibility of re-using the separate phases to increase the enzyme turnover number.

7.2.2.1 Oil and Alkane Systems

Due to the characteristics of lipases, it is of great interest to study reactions catalyzed by these enzymes in biphasic systems. Oliveira *et al.*[26] have studied the reaction of oleic acid with ethanol using a free *R. miehei* lipase and an immobilized *R. miehei* lipase on Accurel EP700 in a hexane and phosphate buffer system. Kraai *et al.*[27] studied the *R. miehei* lipase esterification of oleic acid with various alcohols in a biphasic system consisting of an organic phase with a hydrocarbon diluent and a buffered aqueous phase. Among the solvents used, decane showed the highest reaction rate, however due to cost considerations *n*-heptane was chosen as the most appropriate.

For the splitting of fats and oils, *Candida rugosa* lipase (CRL) is the most recommended because of its random specificity and somewhat inexpensive price. Researchers have found some problems in performing lipolysis in aqueous media. First, the hydrolysis rates are low due to the limitation of interfacial area. Second, the hydrolysis of lipids is subject to substrate inhibition by more than 3–5% of the emulsion.[28] Some approaches, *e.g.* looking for more thermostable lipases or adding a water-immiscible organic solvent to form a biphasic system, have been proposed to eliminate the above obstacles. Also, immobilized lipases with or without organic solvents have been suggested. Biphasic systems seem very attractive where there is no surfactant contamination, as in microemulsions, and where there is no diffusional limitation as in immobilized or membrane reactors. However, some particular problems, like the energy consumption due to agitation and deactivation of an enzyme in the presence of organic solvents, may offset the merits of using biphasic systems for enzymatic reactions.

It has been found that isooctane is the most suitable organic solvent for olive oil splitting by CRL in biphasic systems. Moreover, the reaction rate and stability of lipase in these systems were much greater than those in organic solvent-free systems.[29]

Synthesis of oleic acid L-menthyl ester was analyzed from the viewpoint of reaction equilibrium in an oil–aqueous biphasic system with CRL. The results showed that excess amounts of L-menthol increased the solubility of water in the oil phase, which promoted hydrolysis of the L-menthyl ester.[30]

The transesterification of sunflower oil with methanol was catalyzed by free or immobilized lipases from *Rhizomucor miehei* (Palatase 20000 L) and *Humicola insolens* (Lipozyme TL100L) and the effects of protein amount, temperature, pH, and the molar ratio of methanol to sunflower oil on the enzymatic reaction using free lipase were evaluated.[31]

7.2.2.2 Other Solvents

Enzymatic reactions based on ionic liquids (ILs) and supercritical carbon dioxide (scCO$_2$) as biphasic systems are interesting alternatives to organic solvents for designing clean synthetic chemical processes that provide pure products directly. Because of the unique properties of IL/scCO$_2$ biphasic systems, a solute dissolved in an IL can be easily recovered with scCO$_2$ without any cross contamination. In this context, a new concept for continuous biphasic biocatalysis, where a homogeneous enzyme solution is immobilized in a liquid phase (working phase), while substrates and products reside largely in a supercritical phase (extractive phase), has been proposed.[32] The system was tested for two different enzymatic reactions: the synthesis of butyl butyrate from vinyl butyrate and 1-butanol, and the kinetic resolution of *rac*-1-phenylethanol at 150 bar and in a range of temperatures (40–100 °C). In both cases, an exceptional level of activity, enantioselectivity and operational stability was obtained, even after 11 cycles of 4 h of work. Furthermore, for these IL/scCO$_2$ systems, the substrates must be transported from the supercritical to the enzyme–IL phase, and *vice versa* in the case of the products, their solubility in this liquid phase being the key parameter for controlling the efficiency of the reaction system.[33–35]

Biphasic systems based on ILs and scCO$_2$ have also been used for the enzymatic synthesis of octyl acetate in batch processes, with the IL acting as a solvent rather than as a protective agent of the enzyme.[36] The kinetic resolutions of both *rac*-2-octanol[37] and *rac*-2-phenyl-1-propanol[38] catalyzed by lipase and cutinase, respectively, as well as the dynamic kinetic resolution of *rac*-1-phenylethanol,[39] are additional examples of how IL/scCO$_2$ biphasic systems can be successfully used to develop continuous enzymatic processes for the synthesis of fine chemicals.

7.2.3 Whole-Cell Biotransformations

Although the majority of work on bioconversions in biphasic systems has been performed using enzymes,[40,41] the use of whole cells as biocatalysts is gaining more importance and is becoming a very promising field since a whole complex multi-enzymatic metabolic pathway can be unveiled.

The organic phase in a biphasic system's biotransformation process is normally an hydrophobic substrate however, in most cases, such systems also include an emulsifying agent that may be added to the medium or produced by the cells. As an example, in natural or induced adapted conditions, many microbial systems (including fungi, bacteria and yeasts) are able to produce excellent emulsifying agents which enhance hydrophobic substrate utilization. Mathur *et al.*[42] used a fungal strain, *Syncephalastrum racemosum*, with great emulsification activity to study this activity in the lipolysis of soybean oil. They observed that the degradation of oils to free fatty acids present a positive correlation with the emulsification activity. Therefore, this is a great alternative to use in systems for treatment of hazardous substances in hydrophobic media.

Alternatively, bioremediation in systems with hydrophobic compounds can also be considered as biphasic systems. Smith *et al.*[43] studied the biodegradation kinetics of two hydrophobic organic compounds (phenanthrene and fluoranthene) by the bacterium *Sphingomonas paucimobilis* EPA505. The fungus *Cunninghamella elegans* was used for tributyltin chloride (TBT) degradation, which is of great importance in environmental studies, due to its high toxicity.[44,45] Recently, studies have been performed in order to overcome the low volumetric productivity of aqueous bioconversion systems involving water soluble hydrophobic compounds. Processes are being developed and designed to incorporate green solvent alternatives such as supercritical fluids, ionic liquids, natural oils, and liquid polymers, among others, as an alternative to organic solvents for hydrophobic substrate solubilization. Marques *et al.*[46] studied the cleavage of phytosterols (β-sitosterol), performed by free resting cells of *Mycobacterium* sp. NRRL B-3805, using different green solvents, such as polyethylene glycol (PEG), polypropylene glycol (PPG), ionic liquids (1-ethyl-3-methylimidazolium ethyl sulfate, 1-butyl-3-methylimidazolium dicyanamide, 1-butyl-3-methylimidazolium hexafluorophosphate and 1-octyl-3-methyl-imidazolium) and dioctyl phthalate (DOP) in order to increase the β-sitosterol availability for cells.

As previously stated, in many cases the organic phase is the substrate of the biotransformation in biphasic media, such as in bioprocesses based in *Yarrowia lipolytica* cultures using hydrophobic substrates, such as alkanes or lipids, and sophisticated mechanisms have been developed for the efficient use of hydrophobic substrates as the sole carbon source due to their very efficient mechanisms for breaking down lipids[47] and also due to their capability to produce bioemulsifiers.[48] The biotransformation medium for γ-decalactone production by *Y. lipolytica* is a biphasic system, consisting of an aqueous–liquid phase and an organic–liquid phase of castor oil or its derivatives, forming an emulsion stabilized by the emulsifier Tween 80.[49] In this system, Aguedo *et al.*[50] tested several surfactants (Tween 80, Triton X-100 and Saponin with neutral character; SDS, with anionic character, and CTAB with cationic character) to determine their effects on the viability of yeasts, on the emulsion, on the yeast surface hydrophobicity and on the biotransformation itself, concluding that Tween 80 was the ideal surfactant for this system since it was the compound that allowed the achievement of higher γ-decalactone concentrations. In addition, it had no effect on the viability of cells and did not interact with their membranes, providing a greater interfacial surface to the medium and the largest relative surface hydrophobicity to the cells among all compounds tested.

Later, Gómez-Díaz *et al.*[51] studied the effect of surfactant and oil concentrations in emulsions of water and methyl ricinoleate (MR), stabilized by Tween 80. The presence of the surfactant in the liquid aqueous phase

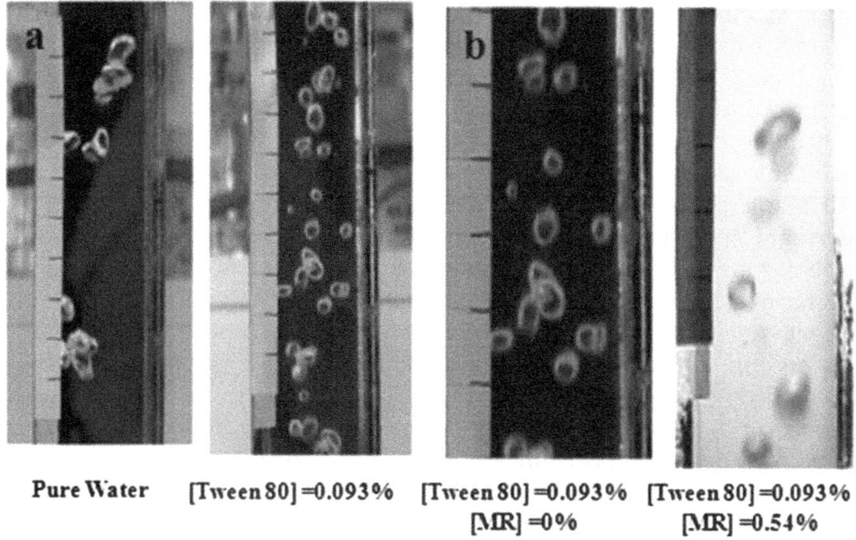

Figure 7.3 Effect of Tween in a three-phase system upon the bubble size distribution in a column bubble bioreactor: (a) effect of the presence of Tween 80 in the liquid phase; (b) influence of methyl ricinoleate (MR) concentration. (Reproduced with permission from ref. 50.)

produced a clear increase in the specific interfacial area for the mass transfer of oxygen from gas to liquid phase due to the decrease of the air bubbles' size (Figure 7.3a). However, when oil was added to the aqueous phase, a continuous increase in the bubbles' diameter was observed due to the partitioning of surfactant molecules in both liquid phases (Figure 7.3b).

On the other hand, it was proven that oil concentration increase has a great impact in γ-decalactone production by the biotransformation of RM with whole cells of *Y. lipolytica*, due to the increase in oil droplets' size. It was proven that larger oil droplets favored aroma production, suggesting that the access of cells to the substrate occurs by their adhesion around larger oil droplets.

7.3 Water-in-Oil Systems

Water-in-oil microemulsions, normally denominated reverse micelles (RMs), are thermodynamically stable single-phase dispersions of water and surfactant within a continuous oil phase.[52] In w/o microemulsions the fraction of the sum of the water and surfactant volumes is not too high, and the microemulsion is oil-continuous and, for many systems, the water is dispersed in the form of tiny droplets which diffuse randomly and independently within the continuous oil medium.[51] These systems are normally optically transparent since, due to their small size, the droplets do not scatter visible light.

Recent interest in w/o microemulsions stems from the discovery that enzymes can be dissolved in them, with retention of their activity and stability.[52] Enzyme-containing water droplets as microreactors stably dispersed within a continuous oil medium have been a greatly attractive concept, especially in bioconversions involving apolar reactancts.[53-55]

The entrapment of enzymes in reverse micelles avoids direct contact with the organic medium, thus protecting the enzymes from denaturation. The interior of the reverse micelles acts like a microreactor that provides a favorable aqueous microenvironment for enzyme activity, as well as an enormous interfacial area through which the conversion of hydrophobic substrates can be catalyzed.[56]

In order to develop suitable synthesis processes, the activity and stability of biocatalysts in these media is of great interest. Therefore, describing the interactions between the enzymes and micelles, which depend on the composition of the microemulsion, has great importance.

7.3.1   Influence of Phase Composition

Several parameters influence the kinetics of enzyme catalyzed reactions in w/o microemulsions, for example, the substrate and enzyme distribution in the different parts of a one-phase microemulsion with different concentrations. Additionally, the oil and surfactant chosen, the water concentration, and the structure of the interfacial layer can influence the activity and stability of biocatalysts.[57]

7.3.1.1   Solvents

Nonpolar and hydrophobic oils are normally used in microemulsion systems and the oil hydrophobicity has a strong influence on the resulting enzyme activity. This may be explained by the interactions of the oil with surfactants.[58] Laane *et al.*[59] have shown that the solubility of the oil in the water pool has great influence on the enzyme activity, independent of the choice of surfactant.[59] They also described the correlation between the hydrophobicity of the oil and the resulting enzyme activity, called the "$\log P$"-concept. P is the distribution coefficient of the oil in the mixture of water and 1-octanol. In general, very hydrophilic oils ($\log P < 2$) are not suitable for enzyme catalysis in microemulsions, since the activity and stability of biocatalysts in these mixtures is very low. In contrast, the use of hydrophobic oils ($\log P > 4$) or aliphatic alkanes ($\log P > 7$) results in higher activities and stabilities, as the enzymes are located in the water pool and are not in contact with the hydrophobic compartments within the microemulsion.[60] In fact, most studies on RMs have been performed with solvents like C_7–C_{15} n-alkanes, isooctane, n-decane, carbon tetrachloride, cyclohexane, octane, n-heptane, n-hexane, hexadecane, *etc.*, as reviewed by Krishna *et al.*[61]

Other nonpolar solvents have been also used to replace alkanes in RM formation, like for instance scCO$_2$. In these systems, the selection of the surfactant is of extreme importance and fluorinated surfactants have been proven to be more effective than hydrocarbon ones to form water-in-scCO$_2$ microemulsions.[62]

The main goal of enzyme catalysis in w/o microemulsions is the replacement of the nonpolar phase by pure hydrophobic substrates, when the substrates used are liquids. Several syntheses have been performed with enzyme catalysts in pure substrates, such as lipase catalyzed esterifications of sugars or lipase catalyzed ring-opening polymerizations.[63]

7.3.1.2 Amphiphiles

Amphiphiles interact with both polar and nonpolar compounds due to their molecular structures. When the hydrophobic tail reaches a certain chain length, amphiphiles reduce the unusually high surface tension of water and are referred to as surfactants. The hydrophilic head group of surfactants can be cationic (*e.g.* benzalkonium chloride; cetyltrimethylammonium bromide), anionic (*e.g.* sodium dodecyl sulfate; sodium-bis-(2-ethylhexyl)sulfosuccinate), zwitterionic (*e.g.* cocamidopropyl betaine), or non-ionic (long chain alcohols such as octanol).[64]

The choice of surfactant, which is mostly constrained by the selection of the oil and the resulting phase behavior of the microemulsion, can have different effects on the enzyme stability and activity. Nevertheless, the surfactant choice has to match the oil in order to obtain a microemulsion.

The critical micelle concentration (CMC) of a surfactant, that is the lowest concentration at which spherical micelles form, is an important parameter in surfactant selection and in the definition of micelle systems formulation. The formation of micelles is characterized by an often sharp discontinuity in system properties, such as surface tension, conductivity, light scattering, self-diffusion, and the molality of dissolved compounds.[62] Sufficient solubility of the amphiphile in water, the medium in which the micelles are formed, is also an important prerequisite.[64] Krishna *et al.*[61] presented an exhaustive list of the most commonly used surfactants, solvents and co-surfactants in reverse micelle formation.

7.3.1.3 Aqueous Phase

Many investigations have been undertaken regarding the effect of the water concentration in the microemulsion on the catalytic behavior of enzymes. The surfactant concentration in a microemulsion defines the size of the internal interface, but has no measurable influence on the enzyme kinetics. The size of a reverse micelle depends on the water-to-surfactant molar ratio (w_0).[53] In the aqueous core, there are different water regions (Figure 7.4). When the w_0 value increases, the concentration of free water also increases. Different models consider that the enzyme activity is determined by the location of the enzyme in a particular water region.[66]

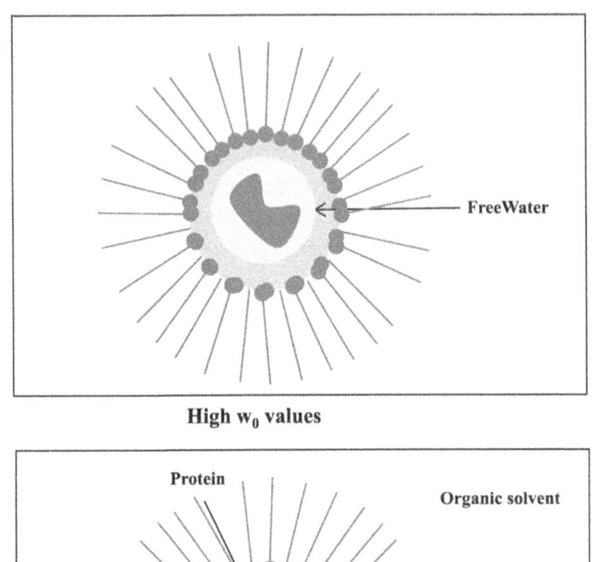

Figure 7.4   Influence of w_0 in reverse micelle structure (adapted from ref. 65).

The physical properties of the water located inside reverse micelles differ from those of bulk water, and the difference becomes progressively smaller as the water concentration increases. Two distinct aqueous regions are present in a reverse micelle. One is located in the inner part of the reverse micelle and has the same physical properties as bulk water; the other is attached to the polar head groups of the surfactant and differs in its physical properties strongly from bulk water.[67] These differences are a lowered melting point and a lower dielectric constant.[68]

The differences in enzyme localization inside reverse micelles depend on their chemical structure. The degree of hydrophobicity of the biomolecule plays an important role in the preferential localization among various microenvironments (water core, bound water, surfactant membrane, and organic solvent).[69] Hydrophilic enzymes are entrapped in the water core, whereas others with an amphiphilic character can be anchored to the surfactant layer, penetrating even into the continuous organic phase. In fact, the water structure in reverse micelles may resemble that of water adjacent to biological membranes. It has been suggested that this system reliably mimics the microenvironment that enzymes encounter in the intracellular medium.[55]

The aqueous phase in micellar systems is of great importance to biocatalytic reaction, and often contains buffers to retain an adequate pH of reaction. Buffers can influence the surfactant CMC, the number of molecules in an aggregate, or the shape of the micelle. Studies have proven that buffered aqueous micellar systems are highly complex.[60] Moreover, the presence of salts in the aqueous phase generally increases the ionic strength, which can decrease the protein solubilization capacity of RMs.[61]

Recent developments in RMs have focused on the addition of ionic liquids to the aqueous phase or even on replacement of water by these liquids, leading to the formation of ionic liquid-in-oil (IL/o) microemulsions.[70] Particular interest is in room temperature ILs that are often associated with green chemistry, because they possess certain advantageous properties typically linked to environmentally friendly solvents.[71]

7.3.2 Enzymatic Reactions

A large variety of enzymes, including cofactor dependent enzymes, have been successfully encapsulated in reverse micellar systems, retaining their catalytic activity.

Enzymatic catalysis in microemulsions has been used for a variety of reactions, such as the synthesis of esters, peptides and sugar acetals, transesterifications, various hydrolysis reactions, and steroid transformations. Orlich and Schomäcker[57] extensively reviewed the published work on enzyme catalysis in w/o-microemulsions. In this review the aerosol-OT (AOT) surfactant was the most applied one. This surfactant can form reverse micelles without the need of a cosurfactant and when compared to most other surfactants it has the ability to solubilize large amounts of water.[61]

A particular case of enzymatic studies in microemulsions is that of lipases.[72] Among the enzymes so far investigated, lipase is one of the most advantageous because it is stable, inexpensive, and widely used in the development of various applications in the detergents, oils and fats, dairy, and pharmaceutical industries. In addition to hydrolysis of triglycerides to glycerol and free fatty acids, lipases can be used in esterification and transesterification reactions in low water content media. This catalytic process is heterogeneous and can be favored by the use of a microemulsion. Lipases are active almost exclusively near interfaces in a classical heterogeneous procedure. Lipase catalyzes the hydrolysis of long-chain aliphatic esters to glycerol and free fatty acids. Several hydrolytic reactions performed by lipases in microemulsion systems have been reported and reviewed by Carvalho and Cabral.[73]

7.3.3   Immobilization of Reverse Micelles

Reverse micelles present some problems of product recovery and of enzyme reuse. Thus, the immobilization of RMs is a very interesting approach to solve these problems. The addition of an aqueous gelatin to a w/o microemulsion can yield a matrix suitable for enzyme immobilization. Gelatin-containing microemulsion-based organogels (MBGs) were first described in 1986,[74] and their physical–structural characterization has since been the subject of a number of studies.[75,76] MBGs offer considerable advantages over w/o microemulsions, such as higher enzyme stabilities, product isolation and reusability, and they enable the use of enzymes in a continuous mode of operation.[77]

The esterification ability of lipase VII from *C. rugosa* was studied in quaternary w/o microemulsion Triton-X100/water/1-hexanol/*n*-hexane-based organogels.[78] Zoumpanioti *et al.*[79] reviewed several applications of MBGs as immobilization matrices for lipases, dated from 1986 to 2008. In the work reviewed by these authors, several biopolymers were used, such as gelatin, agar, alginate, κ-carrageenan and HPMC (hydroxypropylmethyl cellulose), with the most used surfactants used in reverse micelle formation being AOT and lecithin. Recently Zhang *et al.*[80] immobilized lipase from *Candida rugosa* in MBGs and subsequently applied them in large scale synthesis of arylethyl acetate in organic solvents, proving that these MBGs were stable and efficient catalysts. Besides organic solvents, $scCO_2$ has also been used in reverse micelle biocatalysis with MBGs, as is the case for the esterification of lauric acid and 1-propanol catalyzed by MBGs of lipases from *C. antarctica* and *Mucor miehei* in $scCO_2$ (35 °C, 110 bar) as solvent.[81]

7.4   Concluding Remarks

Micellar systems play an important role in biocatalysis and biotransformation. Generally, microemulsions present the advantage of saving operating costs, due to their increased emulsion stability compared to biphasic systems. Moreover, water-in-oil microemulsions increase enzymes' stability and conversion rates. Reverse micelles have been studied by many authors, showing that the use of these systems can have considerable biotechnological relevance and can be extended to a wide range of applications, such as in downstream processing, in drug delivery systems and also in nanoscience and nanotechnology.

An auspicious strategy to perform biocatalysis in RMs is the usage of whole cells instead of purified enzymes, but for this only a few surfactant–oil systems are suitable.

Advances in the application of RM systems have been achieved with the development of immobilization methods, like microemulsion-based organogels. MBGs improve enzymes' stability and reusability. Scaling-up biocatalysis in micellar systems and in newly developed immobilized microemulsions will have great impact on the economic competitiveness of biotransformation processes at an industrial level.

References

1. P. Wang, *Appl. Biochem. Biotechnol.*, 2009, **152**, 343–352.
2. O. Ghisalaba, H.-P. Meyer and R. Wohlgemuth, *Encyclopedia of Industrial Biotechnology: Bioprocess, Bioseparation, and Cell Technology*, John Wiley & Sons, Inc., 2010.
3. L. Pera, M. Mario, D. Baigori and G. R. Castro, *Indian J. Biotechnol.*, 2003, **2**, 356–361.
4. S. Kermasha, D. Ndeye and B. Bisakowski, *J. Mol. Catal. B: Enzym.*, 2001, **11**, 909–919.
5. L. E. Scriven, *Nature*, 1976, **263**, 123.
6. L. M. Prince, *Microemulsions: Theory and Practice*, Academic Press, NY, 1977.
7. M. Gratzel and K. Kalyanasundaram, *Kinetics and Catalysis in Microheterogeneous Systems*, Marcel Dekker, NY, 1991.
8. K. Martinek, A. V. Lavashov, N. Klayachlo, Y. L. Khmelnttski and I. V. Berezin, *Eur. J. Biochem.*, 1986, **155**, 453–468.
9. T. Dwars, E. Paetzold and G. Oehme, *Chem., Int. Ed.*, 2005, **44**, 7174–7199.
10. V. Papadimitriou, G. S. Theodore and A. Xenakis, *Langmuir*, 2007, **23**, 4.
11. S. K. Mehta and G. Kaur, *Microemulsions: Thermodynamic and Dynamic Properties, Thermodynamics*, 2011, ISBN: 978-953-307-544-0.
12. S. T. Moulik and A. K. Rakshit, *J. Surf. Sci. Technol.*, 2006, **22**, 159–189.
13. P. A. Winsor, *Trans. Faraday Soc.*, 1948, **44**, 376.
14. P. Fernades and J. Cabral, *Biocatalysis in Biphasic Systems: General*, Wiley-VCH VerlagGmbH &Co, 2008.
15. M. McCoy, *Chem. Eng. News*, 2001, **79**, 37–41.
16. E. Dickinson, *An Introduction to Food Colloids*, Oxford Science Publishers, 1992.
17. S. Friberg, K. Larsson and J. Sjoblom, *Food Emulsions*, Marcel Dekker, New York, 2004.
18. D. J. McClements, *Food Emulsions: Principles, Practice, and Techniques*, CRC Press, Boca Raton, 2005.
19. R. J. Hunter, *Foundations of Colloid Science*, Oxford University Press, 1986.
20. K. P. Plucknett, S. J. Pomfret, V. Normand, D. Ferdinando, C. Veerman, W. J. Frith and I. T. Norton, *J. Microsc.*, 2001, **201**, 279–290.
21. N. Loren, M. Langton and A. M. Hermansson, Confocal fluorescence microscopy for structure characterization, in *Understanding and Controlling the Microstructure of Complex Foods*, ed. D. J. McClements, Woodhead Publishing, Cambridge, UK, 2007.
22. R. Xu, *Microb. Cell Fact.*, 2002, **11**, 106.
23. D. J. Mcclements, *Critical Review of Techniques and Methodologies for Characterization of Emulsion Stability*, Massachusetts, Amherst, MA, 01003, USA, 2007.
24. N. Gomes, Y. Waché, J. A. Teixeira and I. Belo, *Biotechnol. Lett.*, 2011, **33**, 1601–1606.
25. P. J. Halling, *Enzyme Microb. Technol.*, 1984, **6**, 513–516.

26. A. C. Oliveira, M. F. Rosa, M. R. Aires-Barros and J. M. S. Cabral, *J. Mol. Catal. B: Enzym.*, 2001, **11**, 999–1005.
27. G. N. Kraai, J. G. M. Winkelmana, J. G. Vriesb and H. J. Heeresa, *Biochem. Eng. J.*, 2008, **41**, 87–94.
28. S. Tsai, G.-H. Wu and C.-L. Chiang, *Biotechnol. Bioeng.*, 1991, **38**, 761–766.
29. K. H. Kim, D. Y. Kwon and J. S. Rhee, *Lipids*, 1984, **19**, 975.
30. R. Dave and D. Madamwar, *Biocatal. Biotransform.*, 2010, **3**, 157–166.
31. A. C. Oliveira and M. F. Rosa, *J. Am. Oil Chem. Soc.*, 2006, **83**, 21–25.
32. P. Lozano, T. Diego, M. Vaultier and J. L. Iborral, in *Ionic Liquid Applications: Pharmaceuticals, Therapeutics, and Biotechnology*, American Chemical Society, Washington, DC, 2010, ACS Symposium Series.
33. P. Lozano, T. De Diego, D. Carrié, M. Vaultier and J. L. Iborra, *Chem. Commun.*, 2002, 692–693.
34. P. Lozano, T. De Diego, D. Carrié, M. Vaultier and J. L. Iborra, *Biotechnol. Prog.*, 2003, **19**, 380–382.
35. P. Lozano, T. De Diego, D. Carrié, M. Vaultier and J. L. Iborra, *Biotechnol. Prog.*, 2004, **20**, 661–669.
36. H. R. Hobbs and N. R. Thomas, *Chem. Rev.*, 2007, **107**, 2786–2820.
37. M. T. Reetz, W. Wiesenhofer, G. Francio and W. Leitner, *Adv. Synth. Catal.*, 2003, **345**, 1221–1228.
38. S. Garcia, N. M. T. Lourenco, D. Lousa, A. F. Sequeira, P. Mimoso, J. M. S. Cabral, C. A. M. Afonso and S. Barreiros, *Green Chem.*, 2004, **6**, 466–470.
39. P. Lozano, T. De Diego, D. Carrié, M. Vaultier and J. L. Iborra, *Biotechnol. Lett.*, 2006, **28**, 1559–1565.
40. L. E. S. Brink and J. Tramper, *Biotechnol. Bioeng.*, 1985, **27**, 1258–1269.
41. E. Antonini, G. Carrea and P. Cremonese, *Enzyme Microb. Technol.*, 1981, **3**, 291–296.
42. C. Mathur, R. Prakash, A. Ali, J. Kaur, S. S. Cameotra and N. T. Prakash, *Def. Sci. J.*, 2010, **60**, 251–254.
43. K. E. C. Smith, A. Rein, S. Trapp, P. Mayer and U. G. Karlson, *Environ. Sci. Technol.*, 2012, **46**, 4852–4860.
44. P. Bernat and J. Długonski, *Chemosphere*, 2006, **62**, 3–8.
45. P. Bernat and J. Długonski, *Int. Biodeterior. Biodegrad.*, 2012, **74**, 1–6.
46. P. C. Marques, F. Carvalho, C. C. C. R. Carvalho, J. M. S. Cabral and P. Fernandes, *Food Bioprod. Process.*, 2010, **8**, 12–20.
47. A. Beopoulos, J. Cescut, R. Haddouche, J.-L. Uribelarrea, C. Molina-Jouve and J.-M. Nicaud, *Prog. Lipid Res.*, 2009, **48**, 375–387.
48. P. F. F. Amaral, J. M. da Silva, M. Lehocky, A. M. V. Barros-Timmons, M. A. Z. Coelho, I. M. Marrucho and J. A. P. Coutinho, *Process Biochem.*, 2006, **41**, 1894–1898.
49. A. Braga, N. Gomes and I. Belo, *J. Am. Oil Chem. Soc.*, 2012, **89**, 1041–1047.
50. M. Aguedo, Y. Waché, F. Coste, F. Husson and J. M. Belin, *J. Mol. Catal. B: Enzym.*, 2004, **29**, 31–36.

51. D. Gómez-Díaz, N. Gomes, J. A. Teixeira and I. Belo, *Chem. Eng. J.*, 2009, **152**, 354–360.

52. C. Oldfield, R. B. Freedman and B. H. Robison, *Faraday Discuss.*, 2005, **129**, 247–263.

53. P. L. Luisi and L. J. Magid, *Crit. Rev. Biochem.*, 1986, **20**, 409.

54. K. Martinek, A. V. Levashov, N. Klyachko, Y. L. Khmelnitski and I. V. Berezin, *Eur. J. Biochem.*, 1986, **155**, 453.

55. P. L. Luisi, M. Giomini, M. P. Pileni and B. H. Robinson, *Biochim. Biophys. Acta*, 1988, **947**, 209.

56. N. Fragiskos, H. Stamatis, *Encycl. Ind. Biotechnol.*, DOI: 1002/9780470054581.eib528.

57. B. Orlich and R. Schomäcker, *Adv. Biochem. Eng./Biotechnol.*, 2002, **75**, 185–208.

58. D. Han and J. S. Rhee, *Biotechnol. Bioeng.*, 1986, **28**, 1250.

59. C. Laane, S. Boeren, K. Vos and C. Veeger, *Biotechnol. Bioeng.*, 1987, **30**, 81.

60. S. Y. Huang, H. L. Chang and M. Goto, *Enzyme Microb. Technol.*, 1998, **22**, 552.

61. S. H. Krishna, N. D. Srinivas, K. S. M. S. Raghavarao and N. G. Karanth, *Advances in Biochemical Engineering/Biotechnology*, ed. Th. Scheper, Springer-Verlag, Berlin Heidelberg, 2002.

62. M. Sagisaka, I. Shuho, H. Satosh, Y. Atsushi, M. Azmi, S. Cummings, S. E. Rogers, R. K. Heenan and J. Eastoe, *Langmuir*, 2011, **27**, 5772–5780.

63. J. O. Metzger, *Angew. Chem.*, 1998, **110**, 3145.

64. T. Dwars, E. Paetzold and G. Oehme, *Angew. Chem., Int. Ed.*, 2005, **44**, 7174–7199.

65. F. C. Marhuenda-Egea and M. J. Bonete, *Curr. Opin. Biotechnol.*, 2002, **13**, 385–389.

66. N. M. van Os, J. R. Haak and L. A. M. Rupert, *Physico-Chemical Properties of Selected Anionic, cationic, and Nonionic Surfactants*, Elsevier, Amsterdam, 1993.

67. A. M. Klibanov, *CHEMTECH*, 1986, **16**, 354.

68. R. Schomäcker, *J. Phys. Chem.*, 1991, **95**, 451.

69. F. N. Kolisis and H. Stamatis, *Reverse Micelles, Enzymes*, Wiley Encyclopedia of Industrial Biotechnology, 2010.

70. V. G. Rao, S. Mandal, S. Ghosh, C. Banerjee and N. Sarkar, *J. Phys. Chem. B*, 2013, **117**, 1480–1493.

71. J. Ranke, S. Stolte, R. Stormann, J. Arning and B. Jastorff, *Chem. Rev.*, 2007, **107**, 2183–2206.

72. A. Ballesteros, U. Bornscheuer, A. Capewell, D. Combes, J. S. Condoret, K. Koening, F. N. Kolisis, A. Marty, U. Menge, T. Scheper, H. Stamatis and A. Xenakis, *Biocatalysis*, 1995, **13**, 1–42.

73. C. M. L. Carvalho and J. M. S. Cabral, *Biochimie*, 2000, **82**, 1063–1085.

74. G. Haering and P. Luisi, *J. Phys. Chem.*, 1986, **90**, 5892–5898.

75. P. L. Luisi, R. Scartazzini, G. Haering and P. Schurtenberger, *Colloid Polym. Sci.*, 1990, **268**, 356–374.

76. C. Quellet, H. F. Eicke and W. Sager, *J. Phys. Chem.*, 1991, **14**, 5642–5655.
77. S. Bernardino, N. Estrela, V. Ochoa-Mendes, P. Fernades and L. Fonseca, *J. Sol-Gel Sci. Technol.*, 2011, **58**, 545–556.
78. R. Dave and D. Madamwar, *Process Biochem.*, 2008, **43**, 70–75.
79. M. Zoumpanioti, H. Stamatis and A. Xenakis, *Biotechnol. Adv.*, 2010, **28**, 395–406.
80. W. W. Zhang, N. Wang, Y. J. Zhou, T. He and X. Q. Yu, *J. Mol. Catal. B: Enzym.*, 2012, **78**, 65–71.
81. C. Blattner, M. Zoumpanioti, J. Kroner, G. Schmeer, A. Xenakis and W. Kunz, *J. Supercrit. Fluids*, 2005, **36**, 182–193.

Green Downstream Processing in the Production of Enzymes

P. F. F. AMARAL*[a] AND T. F. FERREIRA[a]

[a]Universidade Federal do Rio de Janeiro, Escola de Química, Department of Biochemical Engineering, Av. Athos da Silveira Ramos, 149, Bl E, Rio de Janeiro, 21941-909, Brazil
*E-mail: pamaral@eq.ufrj.br

8.1 Introduction

New enzyme technologies are emerging. In this context, an increased understanding of fundamental biology and bioinformatics is beginning to shape the discovery, development, purification, and application of biocatalysts to a much greater extent. This development creates new enzyme applications and increases the impact of enzyme technology in industry.[1]

Enzymes are very complex proteins, and their high specificity degree as catalysts is manifested only in their native state. The native conformation is obtained under specific conditions of temperature, pH and ionic strength. Hence, only mild and specific methods must be used for enzyme isolation.[2]

Different sectors of industry, like the food, feed, agriculture, paper, leather and textile industries, use enzyme technology because products, as well as raw materials, consist of biomolecules. These biomolecules can be produced, degraded or modified by enzymatic processes. Many enzymes are commercially available, and numerous industrial applications have been described.[1]

RSC Green Chemistry No. 45
White Biotechnology for Sustainable Chemistry
Edited by Maria Alice Z. Coelho and Bernardo D. Ribeiro
© The Royal Society of Chemistry 2016
Published by the Royal Society of Chemistry, www.rsc.org

The purity degree of commercial enzymes ranges from raw enzymes to highly purified forms, depending on the application.[2]

Normally, a bioprocess includes a fermentation section that occurs in a bioreactor, and a subsequent product recovery section.[3] Fermentation broths are complex mixtures containing cells, soluble extracellular products, intracellular products and unconverted substrate. The recovery and purification operations may require more equipment than all other parts of the process combined. The product recovery plant usually represents a major investment, with a substantial fraction of the total final product cost. Therefore, the recovery and purification processes must be well conceived and well projected.[4]

Downstream processing is a major cost factor (up to 50%) in bioprocesses, mainly because of rather dilute product streams. The success of enzyme technology strongly depends on the development of efficient and cost-effective bioseparation processes.[1]

The use of enzymes and microorganisms, which nature has developed, is undoubtedly an ideal choice toward "greening" chemical reactions. However, a real green catalyst is one produced by technologies that integrate reduced impact on the environment as a performance criterion in the design of all production steps, including the downstream ones.[5] Therefore, this chapter will focus on green enzyme separation techniques.

8.2 Initial Separation Steps for Enzyme Recovery

Each cell synthesizes a large number of different enzymes to maintain its metabolic reactions. The choice of procedure for enzyme purification depends on the enzyme's location. Isolation of intracellular enzymes involves cell disruption and separation of complex biological mixtures. Extracellular enzymes are generally released into the medium containing only a few other components.[2] Most industrially important microbial enzymes are extracellular enzymes secreted into the culture medium by the growing microorganism. They are readily isolated from the culture supernatant and are usually very stable. They have been purified on a vast scale for many years.[6]

The first step in enzyme downstream processing is the separation of cells from the supernatant containing the desired enzyme. This operation is difficult because of the small size of cells, especially bacteria. Continuous filtration and centrifugation are widely used in industry. Decantation can also be used for yeast, but the difference between the density of bacterial cells and the components of the fermentation broth makes this technique unsuitable for this microorganism.

Filtration separates solids from a liquid by forcing the liquid through a solid support or filter medium. The filtration rate is a function of the filter area, pressure, viscosity and resistance offered by the filter cake and medium. However, the small size and deformability of microorganisms make filtration of fermentation broth more complicated.[4] Filtration equipment varies widely, from conventional plate-and-frame filter presses to rotary vacuum filters. Small fermentation batches can be handled in a plate-and-frame filter, which gradually accumulates biomass, then is opened and cleared of filter cake.[3]

In industry, normally a continuous filter is used. The cumulative filtrate volume increases linearly with time. The thickness of the formed filter cake increases and the concomitant resistance decreases during the filtration process. Moreover, the compressibility of the biological material makes the process more difficult. The filtration rate and the resistance offered by the filter cake depend on the pressure applied.[2]

Centrifugation utilizes the difference between the density of solids and the surrounding fluid. When a suspension is at rest, the heavier solids settle to the bottom of the solution due to gravity, a process called sedimentation. When the settling is accelerated by a centrifugal field, the process is called centrifugation.[4] Centrifugation is widely used to remove cells from fermentation broths in industry, especially for yeast separation. Filamentous fungi are easily separated by centrifugation. Nevertheless, yeast and bacteria may require prior flocculation with polyelectrolytes or conventional agents. This method is efficient and cheap and allows the centrifugation to be performed using more simple equipment.[7]

The solid concentrate produced by centrifugation differs from that produced by filtration. Centrifugation can produce a paste or only a more concentrated suspension. Filtration produces a relatively dry cake, which is a major advantage. However, many biological feeds which can be centrifuged cannot be effectively filtered. Hence, centrifugation is often an attractive alternative.[4]

8.3 Concentration Steps in Enzyme Downstream Processing

The enzyme concentration in the fermentation broth is often very low, and the volume of material to be processed is generally very large. Thus, to make purification economic it is necessary to concentrate the volume of starting material.[2]

The concentration methods used include precipitation, which will be briefly discussed, and a more advanced method, ultrafiltration.

8.3.1 Precipitation

Enzymes are very complex protein molecules possessing ionizable and hydrophobic groups which interact with the solvent. Thus, by changing their environmental conditions, the proteins can agglomerate, and precipitate.[2] Precipitation techniques are quick and efficient, and are employed to remove impurities and increase the specific activity of the enzyme of interest.[7] The precipitation methods available include precipitation with salts, precipitation with organic solvents, precipitation with polymers and precipitation at the isoelectric point.

For the lipase separation process, a precipitation method is usually used as a first step. For example, Borkar *et al.*[8] purified a crude extract of lipase from *Pseudomonas aeruginosa* SRT 9 20 times by precipitation with 30% (w/v) ammonium sulfate. After this step, two chromatographic steps were performed in order to obtain a solution which was 98 times more pure than the crude extract. However, there are cases where only the precipitation step is sufficient to obtain a good purification degree, as for the separation of lipase from *Cunninghamella verticillata*.[9]

For other enzymes, such as α-amylases and proteases, ammonium sulfate is also used to precipitate this enzyme from cell-free crude extract.[10,11]

8.3.2 Membrane Separation

Membrane separation appears to be a simple and straightforward process where the membrane structure (*i.e.* its pore size distribution) and thickness determine the rejection characteristics and the resistance to flow, respectively.[12]

The membrane separation technique not only allows high flow rates of operation, but also permits the integration of several stages of downstream processing, such as separation, purification and concentration of the byproduct, in one step. Therefore, an increase in the overall yield of the product, and thereby a reduction of the production costs, is expected.[13] Furthermore, the membrane separation technique has a lower cost than other methods of separation and purification, such as chromatography, besides being easily operated and simpler to scale up.[13]

Among all membrane processes (ultrafiltration, reverse osmosis, nanofiltration, electrodialysis, *etc.*) ultrafiltration is highlighted here. Its mild operation conditions (*e.g.*, low temperature, low pressure, no phase shift or chemical additives) promote lower denaturation, deactivation and/or degradation rates of highly labile products, which make it particularly preferred for biological macromolecule separation.[14]

8.3.2.1 Ultrafiltration

The preservation of thermolabile compounds, energy saving, high yields and high selectivity are some of the characteristics of this technique. Due to these features, recently, the number of papers studying the application of the ultrafiltration technique in the process of separation and purification of bio-products has grown considerably. Products with high added value such as pharmaceuticals, biosurfactants and proteins are examples of these.[14]

The pore size of ultrafiltration membranes ranges from about 5 to 100 nm and they retain molecules in the range of molecular weight of 10 kDa–1 MDa; they are ideally used to concentrate macromolecules such as proteins and polymer molecules. Ultrafiltration membranes can be made of polymers, ceramics and metallic materials. However, it is the polymeric membrane that is the workhorse of ultrafiltration applications. Many polymers can be used to make membranes, but most commercial membranes for bioseparation are either polysulfone- or polyethersulfone-based.[14]

Fickers *et al.*[15] used an ultrafiltration apparatus equipped with a 10 m^2 polysulfone membrane (cutoff of 10 kDa) after removing cells from the culture broth. The ultrafiltration step allowed a significant reduction of the working volume without any loss of enzyme activity and contributed to partial lipase purification. Milk powder and gum arabic were important additives to minimize unfolding of the enzyme by thermal denaturation during dehydration by spray-drying, allowing the recovery of the enzymatic activity.

Lysozyme from chicken egg white was separated by ultrafiltration using a 50 kDa polysulfone membrane pre-treated with myoglobin. The transmission of lysozyme was about 26% higher with the pre-treated membrane than with the native membrane. At a transmembrane pressure of 120 kPa, the purity was greater than 96%.[13]

Inulinase produced by *Kluyveromyces marxianus* NRRL Y-7571 by solid state fermentation of sugarcane bagasse was purified from a culture broth by ethanol precipitation and ultrafiltration. After precipitation with 55% (v/v) ethanol at a flow rate of 10 mL min^{-1}, followed by ultrafiltration with a 100 kDa membrane, the inulinase could be purified by a factor of 5.5-fold with 86.1% yield.[16]

8.4 Purification Technologies for Enzymes

In recent decades, new purification technologies, such as two-phase systems, membrane separation, immune purification and the use of functionalized resins, have been studied. Additionally, old methods, such as chromatographic techniques, have been reformulated for more integrated use. Immune purification, for example, is a technique by which the enzyme is separated from other substances in the extract for specific binding to the protein, so that the yield and purity are maximized in a single process step. This technique has the advantage of a secure, reproducible, quick process which is easily scalable. However, it is still not industrially viable due to the cost of the binder matrix.[17] In the case of membrane separation, although this technique has been studied as a single purification step,[18] it is now more like a concentration step in the purification of proteins. Therefore, we will discuss here the new chromatographic methods and two-phase systems.

8.4.1 Chromatography

On a large scale, the most widely used and studied technique is separation by chromatography, either by ion exchange, affinity, hydrophobic interactions or molecular weight. Among them, ion exchange and affinity chromatography are the most used methods in studies of enzyme purification.[10,19] For both, the separation depends on the difference between the interaction of the substances in the extract with the mobile phase and the stationary phase. The type of interaction of the components in the extract with these two phases is what distinguishes the two techniques.

Gupta *et al.*[10] described that the most common affinity adsorbents used for alkaline proteases are hydroxyapatite, immobilized *N*-benzoyloxycarbonyl phenylalanine agarose, immobilized casein glutamic acid, aprotinin-agarose, and casein-agarose. Although affinity chromatography is one of the most successful purification techniques, a major limitation is the high cost of enzyme supports and the labile nature of the affinity ligands, which decrease their use at process scale.

In the case of lipase purification, the most commonly used stationary phases are diethylaminoethyl (DEAE)-cellulose, *Q*-Sepharose and phenyl or octyl Sepharose. Koblitz and Pastore[20] achieved purification factors of 3.9 and 6.8 when purifying lipase from *Rhizopus* sp. by ion exchange chromatography (DEAE) and hydrophobic interaction (phenyl Sepharose), respectively. In many cases, as in the purification of PSL2 from *Yarrowia lipolytica*,[21] aiming at higher purity, these techniques are employed sequentially.

Gummadi and Panda[22] reviewed the purification methods used for microbial pectinases and described the use of DEAE cellulose for exo-polygalacturonase purification with a 209-fold increase in specific activity and a recovery of 8.6%. They also described the success of CM Sepharose ion exchange chromatography for pectinase purification.

Turki *et al.*[23] used anion exchange chromatography followed by gel filtration for downstream processing of the lipase produced by *Yarrowia lipolytica*. After cell removal by centrifugation, the culture supernatant was clarified using a 0.2 μm Minisart filter and desalted on a Sephadex column. The clarified and desalted fraction was then loaded onto a 1 mL HiTrap Q column equilibrated with 25 mM Tris–HCl at pH 7. The lipase was then further purified on a Sephacryl column using PBS pH 7 at a flow rate of 0.5 mL min^{-1}. This procedure resulted in an overall yield of 72% and a 3.5-fold increase in the specific lipase activity.

Although very efficient for high-resolution separation and analysis of proteins, these processes are traditionally carried out using packed beds, which have several major limitations. A radically different approach to overcome the limitations associated with packed beds is to use synthetic microporous or macroporous membranes as chromatographic media.[24] Membrane chromatography is implemented by grafting specific ligands onto the pore surface in membranes and then adsorbing target biomolecules on these ligands during convective flow through the membrane pores. A larger pore size in the membranes would allow much easier access of protein molecules to the binding sites on the pore walls, thus significantly reducing the pressure drop and processing time.[25]

Ion-exchange membrane materials for membrane chromatography can be produced either by modification of commercially available MF membranes or by embedding IEX-resins into a polymeric porous matrix. Anion-exchange membrane chromatography with mainly quaternary amino groups or DEAE groups as ligands has been used for the separation of serum proteins, microbial proteins and enzymes, membrane proteins, cytokines and nucleic acids.[25]

8.4.2 Biphasic Systems

Two-phase systems are based on the partition principles of proteins in a two-phase aqueous system. They are easy to scale-up, have fast mass transfer and operate at room temperature. Additionally, the use of low cost materials and their compatibility with most proteins, among other advantages, make this technique potentially promising for obtaining industrial enzymes.[26]

Among the different biphasic systems studied in lipase separation, the polyethylene glycol-phosphate system is the most used.[27] The partitioning of proteins in this type of system depends on the concentration and molecular weight of the polymer, the concentration and type of salt and the pH of the system. Ooi *et al.*[28] have studied different alcohol salt systems for the purification of lipase from *Burkholderia pseudomallei*. A system of 2-propanol, potassium phosphate and sodium chloride showed the highest yield (99.3%) and purification factor (13.5) among all the systems studied.

8.4.2.1 Extractive Bioconversion

Growth and activity of microbial cultures may be limited either by the availability of or by the accumulation of toxic materials, which act as growth inhibitors and are usually metabolic products. Therefore, by removing the toxic metabolites from the culture medium during bioconversion, it is expected that higher cell population and process productivity can be achieved. This approach has been used in lactic acid production using various separation techniques.[29] In the case of enzymes, aqueous two-phase systems are preferred.

Aqueous two-phase systems contain about 80–90% water and therefore can provide an excellent environment for cells, cell organelles and biologically active substances. Extractive fermentation or extractive bioconversion is an emerging technique that involves the use of aqueous two-phase system based *in situ* fermentation processes. The advantages of such a system include rapid mass transfer due to low-interfacial tension, rapid and selective separation, biocompatibility, separation at room temperature, and easy and reliable scale-up of bench scale results to production scale.[30]

In extractive bioconversion, cells are confined to one of the aqueous phases and the product is made to partition into the other phase by appropriate manipulation of the system. High molecular weight dextran and polyethylene glycol are the two polymers that have been most widely used due to their desirable physical properties and non-toxicity.[31] Low molecular weight PEGs like PEG 200 and 1500 have an inhibitory effect on cell growth.[31,32] Polymer/salt systems are usually preferred for large scale operation due to their relatively low cost and shorter separation time.[30,33] Several salts, such as potassium phosphate, sodium citrate and chlorate, can form two phase systems with polymers. However, high salt concentrations can inhibit cell growth.[30] Bacterial cells have been mostly used in extractive bioconversions. However, successful extractive fermentations using fungal cultures have also been reported.[30]

A study was performed to find out the maximum partitioning of *Bacillus licheniformis* alkaline phosphatase in different molecular weights of PEG with salts (magnesium sulfate, sodium sulfate, sodium citrate) and polymers (dextran 40, dextran T500). PEG 4000 and dextran T500 were the most suitable system based on a higher partition coefficient ($k = 5.23$).[34]

The extracellular lipase derived from *Burkholderia pseudomallei* was extracted during bioconversion in an aqueous two-phase system composed of 9.6% (w/w) polyethylene glycol (PEG) 8000 and 1.0% (w/w) dextran T500. In this integrated process, biomass was accumulated in the bottom phase, whereas the lipase was extracted to the top phase.[28] Extractive microbial fermentation for production of lipase by *Serratia marcescens* was carried out in a cloud point system, composed of a mixture of nonionic surfactants with a ratio of Triton X-114 to Triton X-45 of 4:1 in aqueous solution. The lipase partitioned into the surfactant-rich phase, whereas the cells and other hydrophilic proteins were retained in the dilute phase of the cloud point system. Thus, a concentration factor of 4.2-fold and a purification factor of 1.3-fold of the lipase were achieved.[35]

8.5 Conclusions

A real green catalyst is one produced by technologies that integrate reduced impact on the environment as a performance criterion for the design of all production steps, including the downstream ones. Novel processes and reformulated old ones are being proposed to reduce environmental impact and process costs. Ultrafiltration can be used as a single purification step, but it is now more like a concentration step in the purification of proteins. Membrane chromatography is being implemented by grafting specific ligands onto the pore surfaces of membranes and then adsorbing target biomolecules on them. Extractive fermentation is an emerging technique that involves the use of aqueous two-phase system based *in situ* fermentation processes with the advantage of rapid and selective separation and biocompatibility.

Acknowledgements

Priscilla Amaral is grateful to Conselho Nacional de Desenvolvimento Científico e Tecnológico (CNPq) for a research scholarship.

References

1. J. B. Beilen and Z. Li, *Curr. Opin. Biotechnol.*, 2002, **13**, 338.
2. W. Aehle, in *Enzymes in Industry*, ed. W. Aehle, Wiley-VCh Verlag, Weinheim, 2007, p. 48.

3. J. E. Bailey and D. F. Ollis, in *Biochemical Engineering Fundamentals*, ed. J. E. Bailey and D. F. Ollis, McGraw-Hill, N.Y., 1986, p. 726.

4. P. A. Belter, E. L. Cussler and W. S. Hu, in *Bioseparations: Downstream Processing for Biotechnology*, John Wiley & Sons, New York, 1988, p. 456.

5. P. P. Anastas, in *Green Separation Processes*, ed. C. A. M. Afonso and J. G. Crespo, Wiley-VCh Verlag, Weinheim, 2005, p. v.

6. C. J. Bruton, A. R. Thomson and C. R. Lowe, *Philos. Trans. R. Soc. London, Ser. B*, 1983, **300**, 249.

7. M. A. Coelho, A. M. Salgado and B. D. Ribeiro, in *Tecnologia enzimática*, ed. EPUB, Rio de janeiro, 2008, p. 7.

8. P. S. Borkar, R. G. Bodade, S. R. Rao and C. N. Khobragade, *Braz. J. Microbiol.*, 2009, **40**, 358.

9. T. S. Kumarevel, S. C. B. Gopinath, A. Hilda, N. Gautham and M. N. Ponnusamy, *World J. Microbiol. Biotechnol.*, 2005, **21**, 23.

10. R. Gupta, Q. K. Beg, S. Khan and B. Chauhan, *Appl. Microbiol. Biotechnol.*, 2002, **60**, 381.

11. R. Gupta, P. Gigras, H. Mohapatra, V. K. Goswami and B. Chauhan, *Process Biochem.*, 2003, **38**, 1599.

12. G. P. Agarwal, *Advanced Process Biotechnology*, ed. S. N. Mukhopadhyay, Anshan Limited, Kent, 2006, p.334.

13. R. Ghosh and Z. F. Cui, *J. Membr. Sci.*, 2000, **167**, 47.

14. Z. Cui, *China Particuol.*, 2005, 3(6), 343.

15. P. Fickers, M. Ongena, J. Destain, F. Weekers and P. Thonart, *Enzyme Microb. Technol.*, 2006, **38**, 756.

16. S. Golunski, V. Astolfi, N. Carniel, D. Oliveira, M. Luccio, M. A. Mazutti and H. Treichel, *Sep. Purif. Technol.*, 2011, **78**, 261.

17. K. Pauwels and P. V. Gelder, *Protein Expression Purif.*, 2008, **59**, 342.

18. H. Sztajer and M. Bryjak, *Bioprocess Biosyst. Eng.*, 1989, **4**, 257.

19. R. K. Saxena, A. Sheoran, B. Giri and S. Davidson, *J. Microbiol. Methods*, 2003, **52**, 1.

20. M. G. Koblitz and G. M. Pastore, *Cienc. Tecnol. Aliment.*, 2004, **24**, 287.

21. M. Yu, S. Qin and T. Tan, *Process Biochem.*, 2007, **42**, 384.

22. S. N. Gummadi and T. Panda, *Process Biochem.*, 2003, **38**, 987.

23. S. Turki, A. Ayed, N. Chalghoumi, F. Weekers, P. Thonart and H. Kallel, *Appl. Biochem. Biotechnol.*, 2010, **160**, 1371.

24. R. Ghosh, *J. Chromatogr. A*, 2002, **952**, 13.

25. A. Saxena, B. P. Tripathi, M. Kumar and V. K. Shahi, *Adv. Colloid Interface Sci.*, 2009, **145**, 1.

26. R. Gupta, S. Bradoo and R. K. Saxena, *Curr. Sci.*, 1999, 77, 520.

27. K. E. Nandini and N. K. Rastogi, *Food Bioprocess Technol.*, 2011, **4**, 295.

28. C. W. Ooi, S. L. Hii, S. M. M. Kamala, A. Ariff and T. C. Ling, *Process Biochem.*, 2011, **46**, 68.

29. P. K. R. Choudhury, in *Advanced Process Biotechnology*, ed. S. N. Mukhopadhyay, Anshan Limited, Kent, 2006, p.271.

30. R. M. Banik, A. Santhiagu, B. Kanari, C. Sabarinath and S. N. Upadhyay, *World J. Microbiol. Biotechnol.*, 2003, **19**, 337.

31. J. Sinha, P. K. Dey and T. Panda, *Appl. Microbiol. Biotechnol.*, 2000, **54**, 476.

32. R. Kuboi, H. Umakoshi and I. Komasawa, *Biotechnol. Prog.*, 1995, **11**, 202.
33. A. Kaul, R. A. M. Pereira, J. A. Asenjo and J. C. Merchuk, *Biotechnol. Bio-eng.*, 1995, **48**, 246.
34. S. K. Pandey and R. M. Banik, *Bioresour. Technol.*, 2011, **102**, 4226.
35. T. Pan, Z. Wang, J. H. Xu, Z. Wu and H. Qi, *Appl. Microbiol. Biotechnol.*, 2010, **85**, 1789.

Lipases in Enantioselective Syntheses: Evolution of Technology and Recent Applications

DENISE MARIA GUIMARÃES FREIRE[a], ANGELO AMARO THEODORO DA SILVA[b], EVELIN DE ANDRADE MANOEL[c], RODRIGO VOLCAN ALMEIDA[a], AND ALESSANDRO BOLIS COSTA SIMAS*[b]

[a]Departamento de Bioquímica, Programa de Pós-Graduação em Bioquímica, Universidade Federal do Rio de Janeiro, Rio de Janeiro, Brazil; [b]Instituto de Pesquisas de Produtos Naturais Walter Mors (IPPN), Universidade Federal do Rio de Janeiro, Rio de Janeiro, Brazil; [c]Departamento de Biotecnologia Farmacêutica, Faculdade de Farmácia, Universidade Federal do Rio de Janeiro, Ilha do Fundão, Rio de Janeiro, RJ, Brazil
*E-mail: abcsimas@nppn.ufrj.br

9.1 Introduction

Stereochemical control over natural, synthetic and semi-synthetic compounds remains a challenge to organic chemists. Along with the fast-paced development of diverse enantioselective technologies (*e.g.*, chiral metallocatalysis, chiral organocatalysis), biocatalysis has emerged as a very sustainable and

RSC Green Chemistry No. 45
White Biotechnology for Sustainable Chemistry
Edited by Maria Alice Z. Coelho and Bernardo D. Ribeiro

Published by the Royal Society of Chemistry, www.rsc.org

Table 9.1   Biocatalyzed *versus* conventional reactions.

	Conventional reactions	Biocatalysis
Advantages	Relatively cheaper	More sustainable
	Ubiquitous reactions	High chemo-, regio- and stereoselectivities
	Formation of complex skeletons	Less toxic reagents
		More convenient purification
		Efficiency improvement by directed evolution
		Broad structural tolerance
Disadvantages	Many protection/deprotection steps	Higher cost
	Toxic effluents	Instability of some enzymes
	Undesired by-products	Require cofactors sometimes
	Poor substrate selectivity	

advantageous process. Thus, the synthetic use of biocatalysts, defined by IUPAC (International Union of Pure and Applied Chemistry) as "*...an enzyme or enzyme complex consisting of, or derived from, an organism or cell culture (in cell-free or whole-cell forms) that catalyses metabolic reactions in living organisms and/or substrate conversions in various chemical reactions*"[1] has enabled efficient and practical methodologies which comply with green chemistry principles.[2]

The tenets of sustainability for chemical processes include the use of bio-degradable catalysts and renewable materials (in raw state, mostly) high economy (conciseness, atom economy, high selectivity), mild conditions (regarding temperature, pressure and pH). In many instances, biocatalytic technologies provide competitive solutions to such demands.

The high cost of biocatalysis is regarded as the main disadvantage of such technology. However, with advances in protein engineering and immobilization technologies, enabling recycling and higher enzyme activity, a better cost–benefit relationship has resulted.[3–5]

A summary of the differences between the conventional techniques and biocatalysis is shown below (see Table 9.1).[6–8]

An outstanding example of the advantage of an enzyme-catalyzed sequence over an uncatalyzed one is the industrial hydrolysis of penicillin G (**1**) by a penicillin-acylase, yielding 6-aminopenicillinic acid (6-APA) (**2**). The latter substance is employed as a starting material for semi-synthetic penicillins (Scheme 9.1).[6,9] The biocatalytic alternative stands out due to its high practicality and economy.

9.2   Lipase-Catalyzed Enantioselective Syntheses

The enantioselective syntheses of chiral substances may exploit different strategies, relying on the use of racemates, prochiral or chiral substrates (Table 9.2).[8] From a green chemistry perspective, the ones enabling higher atom economy are preferred, that is, methods leading to a single antipode of the product in the most direct manner with low generation of waste.

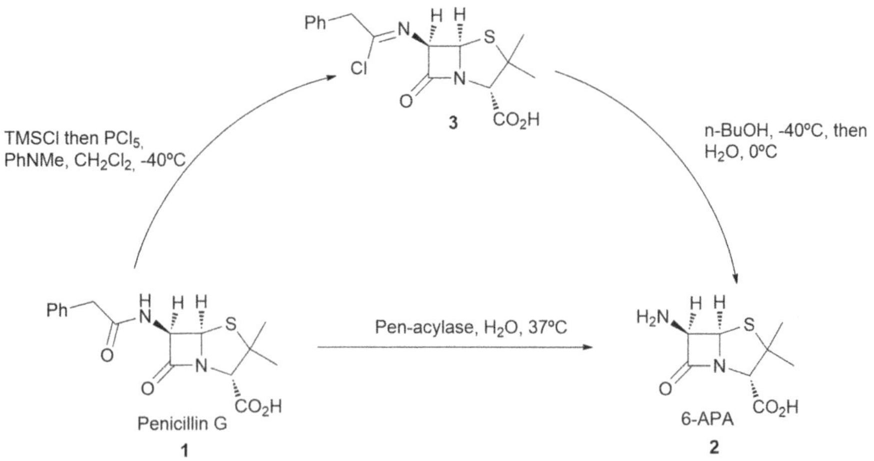

Scheme 9.1 Deacylation of penicillin G.

Table 9.2 Synthetic methods towards enantiopure compounds.

Starting material	Reaction type	Products
Racemate	Kinetic resolution (KR)	Both enantiomers
Racemate	Deracemization (dynamic kinetic resolution or cyclic deracemization)	Either enantiomer
Meso compound	Desymmetrization	Either enantiomer
Chiral substrate	Asymmetric induction	Single enantiomer
Latent symmetric substrate	Enantiodivergent synthesis	Both enantiomers

The known biocatalysts fall into the following major groups of classification: oxidoreductases, hydrolases, transferases, lyases, isomerases and ligases.[10] Among them, lipases (serine hydrolases) have been more widely employed due to their wide applicability, including in industry (Figure 9.1).[11]

Lipases work at the lipid/water interface, which naturally makes them efficient catalysts towards water-insoluble substrates in more hydrophobic solvents. An advantage of lipases (and esterases) is the fact that these biocatalysts do not require cofactors.[7]

Activated esters are the most suitable acyl donors (Figure 9.2) for lipase-catalyzed enantioselective reactions, since they avoid the reversibility of a process.[12–14] Vinyl acetate (**8**) is the most common acylating reagent used in these reactions, and may be employed both as solvent and acylating agent. Non-activated or slightly activated acyl donors are used with amines as substrates. New acyl donors for lipase-catalyzed reactions, such as diethyl malonate (**11**), continue to appear in the literature.[15]

Lipases may engage in enantioselective synthesis *via* processes of classical kinetic resolution (KR), desymmetrization and deracemization (Table 9.2 and Figure 9.3).[7,16] Naturally, due to the nature of the reactions they catalyze, lipases do not mediate the construction of stereocenters.

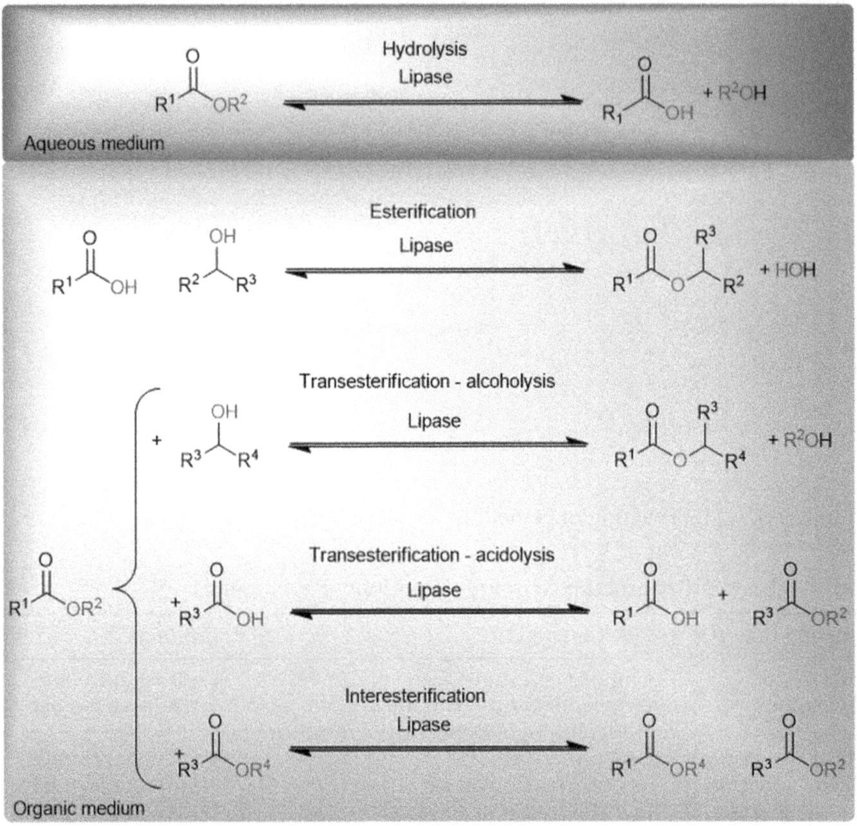

Figure 9.1   Types of lipase-catalyzed transformations (also applicable to amines and amino acids) (adapted).[11]

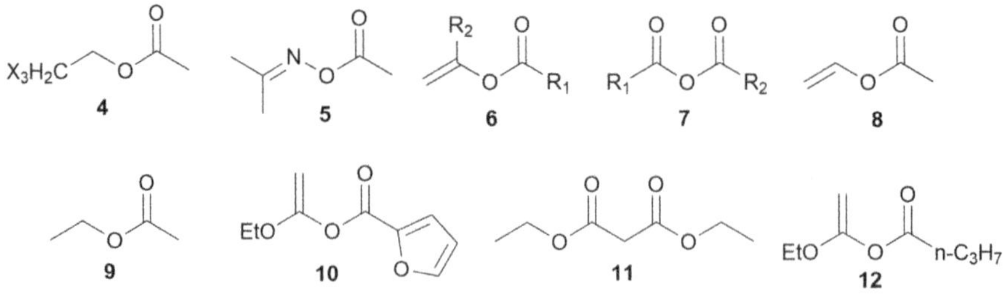

Figure 9.2   Some examples of acyl donors.

9.2.1   Classical Kinetic Resolution

In lipase-mediated kinetic resolution, a racemic substrate undergoes an enzymatic reaction wherein chiral discrimination of enantiomers takes place. A major drawback of this strategy is the maximum theoretical yield of 50% in the enantioselective transformation.

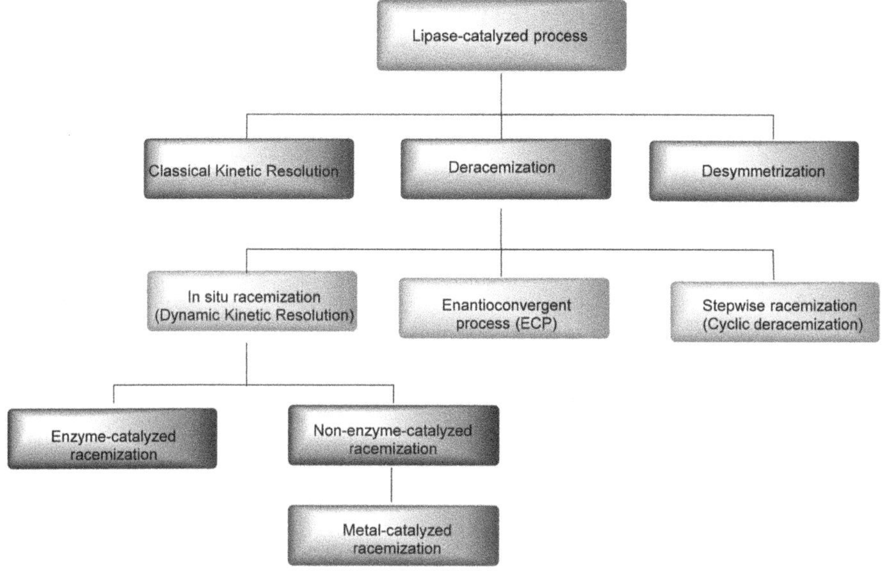

Figure 9.3 Different enantioselective lipase-catalyzed reactions.

Kinetic resolution was employed in the synthesis of (*S*)-monastrol (**13**), a promising anticancer agent, reported by Blasco *et al.*[17] The (*S*)-enantiomer is 15 times more active than the (*R*)-enantiomer. Thus, the racemic precursor was resolved by *Candida antarctica* lipase B (CaLB) to give enantioenriched ((*S*)-antipode **15**) remaining starting material. This compound was subjected to regioselective hydrolysis by *Candida rugosa* lipases, affording (*S*)-monastrol in high yield (Scheme 9.2).[18]

9.2.2 Deracemization Processes

A solution to the problem of 50% maximum yield is the transformation of a racemate into a non-racemate, a process referred to as deracemization.[19] In such cases, the yield of the reaction could in principle reach 100% of a single enantiomer. The racemization itself may be of chemical or enzymatic (racemases/isomerases) nature.[7]

9.2.2.1 Cyclic (Stepwise) Deracemization

By such methodology, the unreacted enantiomer is separated from a mixture also containing the product of the biocatalyzed-kinetic resolution and subjected to racemization *via* a non-selective (achiral) catalyst. The generated racemate is subjected again to KR and the cycle is repeated a few times. Such racemization commonly uses an oxidation–reduction sequence.[20] The overall selectivity relies on the selectivity of a single catalyst and therefore the process lacks selectivity amplification. The concentrations of the substrate **S** and the enantiomeric products $\mathbf{P}_R$ and $\mathbf{P}_S$ are dependent on the selectivity of the system, as well as on the relative rates of the forward and reverse reactions (Figure 9.4).

Deracemization reactions have been successfully applied to the kinetic resolutions of secondary alcohols, acids, amines, amino acids and their derivatives. Thus, a new manufacturing process for pregabalin ((S)-3-(aminomethyl)-5-methylhexanoic acid, **17**) synthesis (Scheme 9.3) relies on a cyclic deracemization protocol through a deprotonation–protonation (with NaOEt in ethanol) sequence.[21,22] This drug is a lipophilic GABA (γ-aminobutyric acid) analogue that was developed for the treatment of several central nervous system disorders. Moreover, it became the first medication approved by the U.S. Food and Drug Administration specifically for the treatment of fibromyalgia.

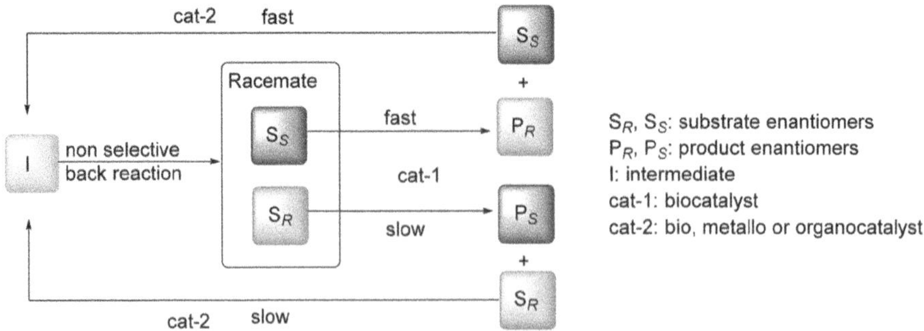

Scheme 9.2   Synthesis of (S)-monastrol (**13**).

Figure 9.4   Generic kinetic resolution involving racemization (in cyclic deracemizations or dynamic resolutions). *R*-enantiopreference is exemplified (adapted).[7]

Enzymatic resolution by *Thermomyces lanuginosus* (Lipolase) proved to be more cost-effective and environmentally harmless (higher yield and 80% reduction of waste) compared to a first-generation manufacturing process.

9.2.2.2 Enantioconvergent Processes (ECPs)

When a racemic mixture is transformed to yield a single stereoisomer by inversion of the configuration of one enantiomer and retention of that of the other, this process is referred to as an enantioconvergent process.[16] Sulcatol (**21**), the male-produced aggregation pheromone of ambrosia beetles, was subjected to a highly enantioselective kinetic resolution by *Candida antarctica* lipase B (Scheme 9.4).[23] The product mixture containing the remaining (*S*)-enantiomer [(*S*)-**21**], and the (*R*)-acylated derivative **22** was subjected to a microwave-irradiated (180 °C) Mitsunobu inversion with acetic acid in THF. This led to enantioconvergence to the (*R*)-acetate derivative **23** with high conversion (>99%) and ee (>98%). Finally, (*R*)-sulcatol [(*R*)-**21**] was formed by deacylation *via* reduction. A similar strategy was applied for the enantioconvergent synthesis of the antipode, (*S*)-sulcatol [(*S*)-**21**], *via* the enantioselective enzymatic hydrolysis of racemic sulcatol acetate.

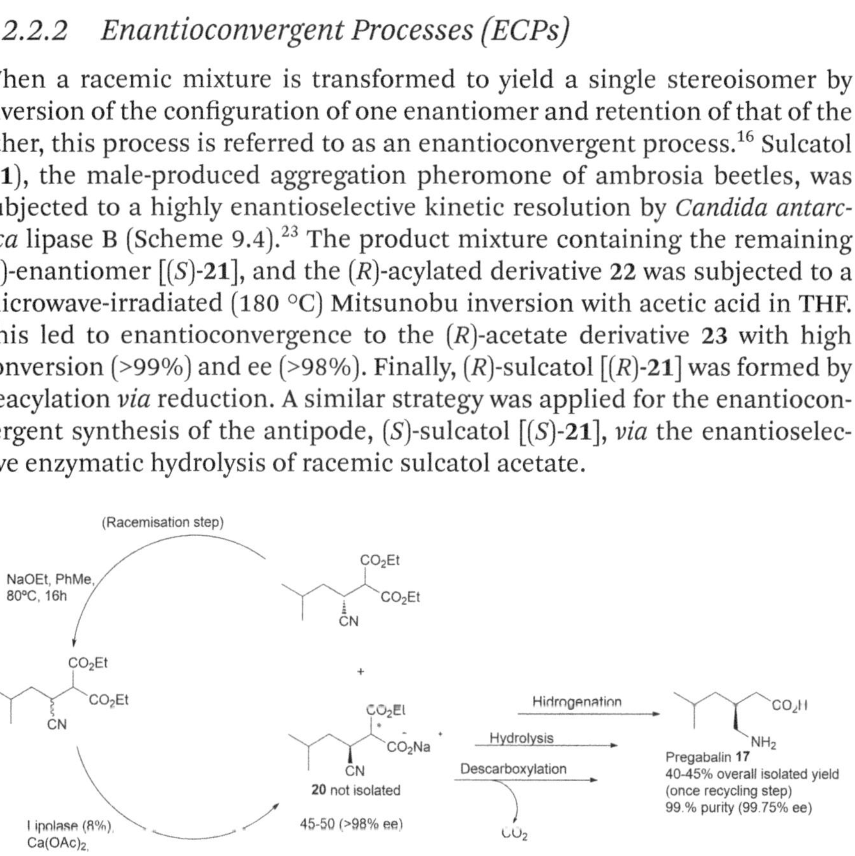

Scheme 9.3 Pregabalin synthesis using cyclic deracemization (adapted).[21]

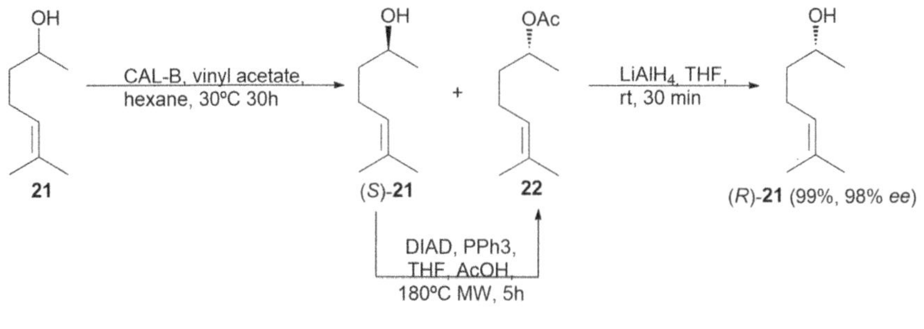

Scheme 9.4 Enantioconvergent resolution of racemic sulcatol *via* catalysis by *C. antarctica* lipase B and Mitsunobu reaction (adapted).[23]

A similar strategy was applied to the enantioselective synthesis of an antagonist of the calcium sensing receptor *via* kinetic resolution of a chlorohydrin by PS-D lipase.[24]

9.2.2.3 Dynamic Kinetic Resolution (DKR)

In biocatalytic dynamic kinetic resolution, along with the kinetic resolution itself, there occurs the *in situ* racemization of the slow reacting enantiomer.[7,25] Thus, with a theoretical yield of 100%, in practice, it allows the enantioselective transformation of most of the racemic starting material into one enantiomeric form (see Figure 9.4).

Kamal *et al.* cited three aspects required for an efficient DKR process: (1) the kinetic resolution step has to be irreversible; (2) the E value has to be at least 30, preferably in the range 50–100, and (3) apart from the E value, K_{rac} has to be at least equal to K_R (rate of reaction of fast reacting enantiomer).[7,26]

DKR racemization may be performed by enzymatic or non-enzymatic (mainly transition metal-mediated) catalysts (Figure 9.3),[27] or it can proceed spontaneously.[25,28]

9.2.2.3.1 DKR *via* Enzymatic Racemization. Biocatalyzed racemization is regarded as an attractive option in DKR, since it is performed under mild conditions (at ambient temperature and atmospheric pressure), preventing problems such as isomerization, racemization, epimerization and rearrangement.[27] A one-pot enantioconvergent synthesis of (*R*)-mandelic acid ethyl ester [(*R*)-**25**] from racemic mandelic acid (**24**) was achieved by an aqueous/organic two-phase system (Scheme 9.5), wherein the mandelate racemase (in the aqueous phase) worked in conjunction with the lipase-catalyzed KR in the organic phase. (*R*)-Mandelic acid [(*R*)-**24**] is the key intermediate for the production of semi-synthetic cephalosporins and penicillins. It is also used as a chiral resolving agent and chiral synthon for the synthesis of anti-tumor and anti-obesity agents.[29,30]

9.2.2.3.2 DKR *via* Chemical Racemization by Deprotonation–Protonation. If the stereocenter in the racemic material is attached to an acidic proton, an *in situ* deprotonation–protonation cycle may be explored for DKR. The following example of the DKR of a naproxen thioester *via* lipase MY employs an enolization (involving a benzylic and α-carbonyl C–H bond) by acid for *in situ* racemization (Scheme 9.6).[31]

9.2.2.3.3 DKR *via* Chemical Racemization by Addition–Elimination. Cyanohydrins, hemiacetals, hemiaminals and hemithioacetals may engage in such DKR (Scheme 9.7).[32]

9.2.2.3.4 DKR *via* Nucleophilic Substitution. DKRs of α-haloesters *via* racemization by substitution with halides are known.[33,34] In these cases, it was preliminarily shown that the racemization is slower in the products (carboxylic acids).

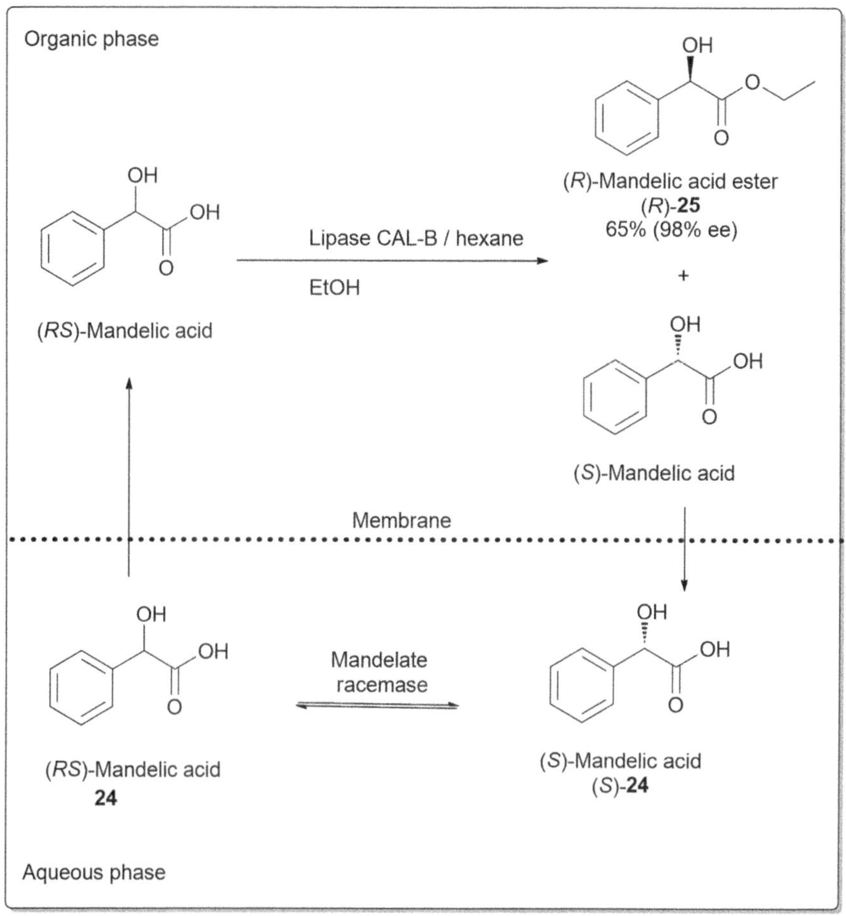

Scheme 9.5 DKR of mandelic acid *via* enzymatic racemization (adapted).[29]

Scheme 9.6 DKR of (*R,S*)-naproxen thioester.[7]

9.2.2.3.5 DKR *via* Metal Catalysis. A variety of metal catalysts for the racemization step have been reported (Table 9.3), but only a few complexes have proved to be compatible with the enzymatic reaction. The metal may interfere with the enzyme to give poor resolution, or the enzyme may slow down or inhibit the racemization by the metal catalyst.[28,30] However, the

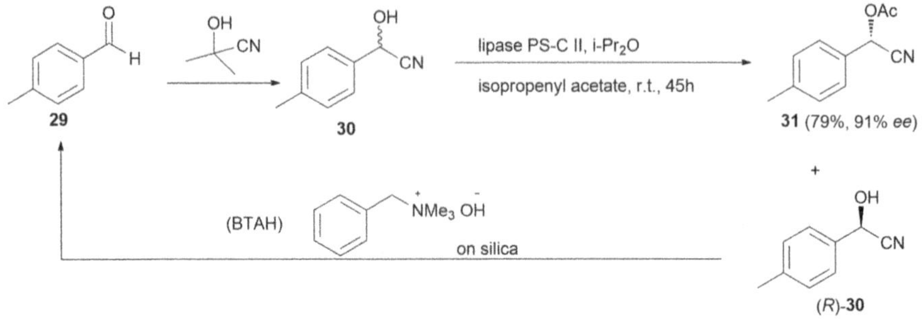

Scheme 9.7   Lipase-catalyzed DKR of aromatic cyanohydrin (adapted).[32]

Table 9.3   Examples of metallocomplex catalysts employed in the racemization step of DKR.

Metal	Examples
Ruthenium	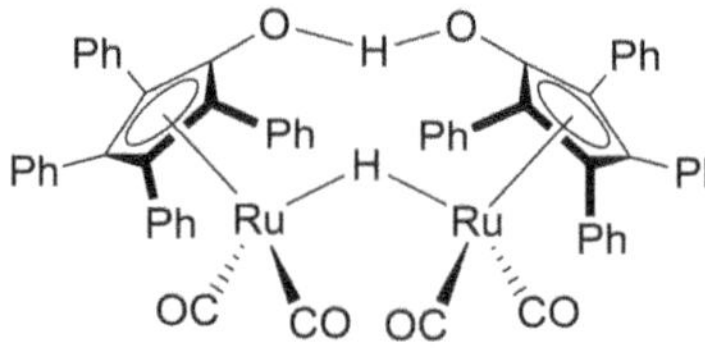 Shvo's catalyst 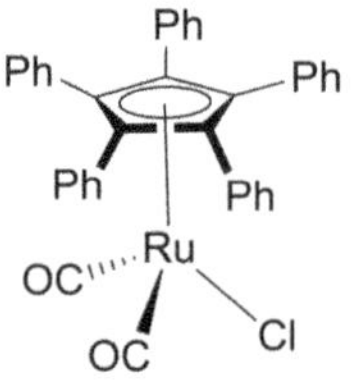 Bäckvall's Catalyst
Iridium	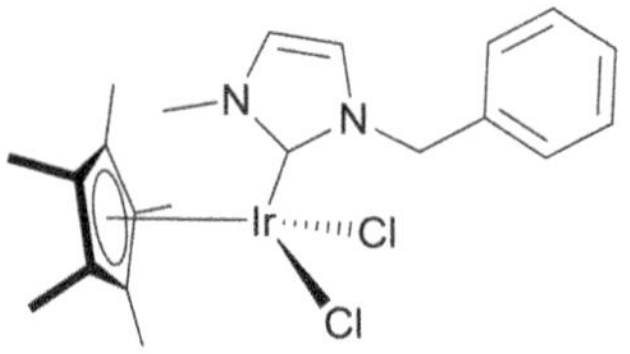
Palladium	$Pd(PPh_3)_4$
Vanadium	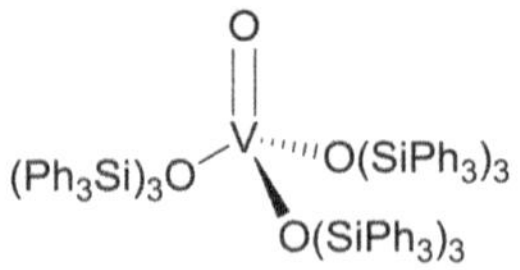 $(Ph_3Si)_3O$—V—$O(SiPh_3)_3$ $O(SiPh_3)_3$
Rhodium	$Rh_2(OAc)_4$
Aluminium	$AlMe_3$

use of bulky groups on the metal catalyst may prevent such unwanted interactions.[35]

Most metal-catalyzed racemizations in lipase-catalyzed DKR involve hydrogen transfer (redox mechanisms) (Scheme 9.8(a)). Racemizations *via* metal-π-allyl complexes are also known (Scheme 9.8(b)).[7,35]

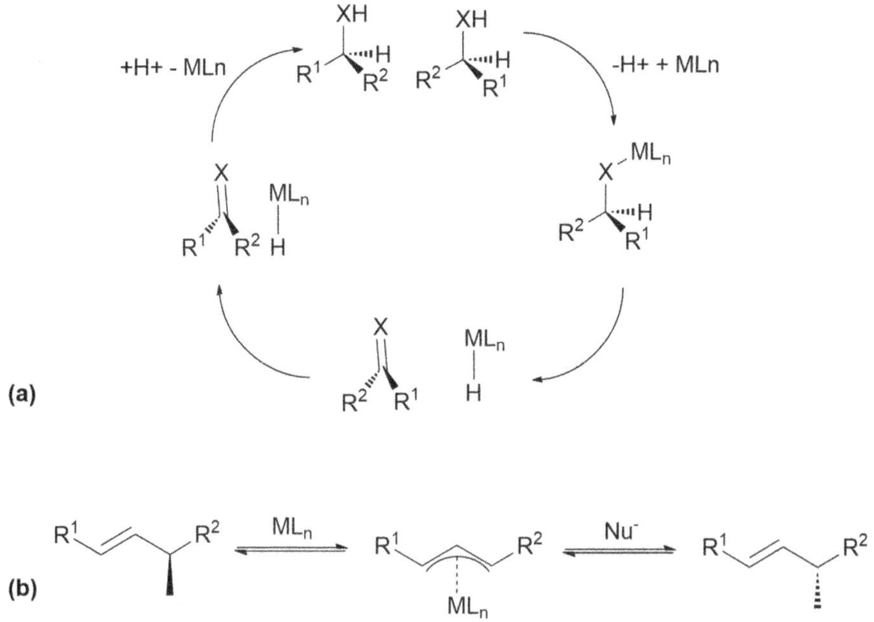

(a)

(b)

Scheme 9.8 Racemization by metal-mediated catalysis. (a) *Via* hydrogen transfer mechanisms. (b) *Via* metal-π-allyl complexes.[7,35]

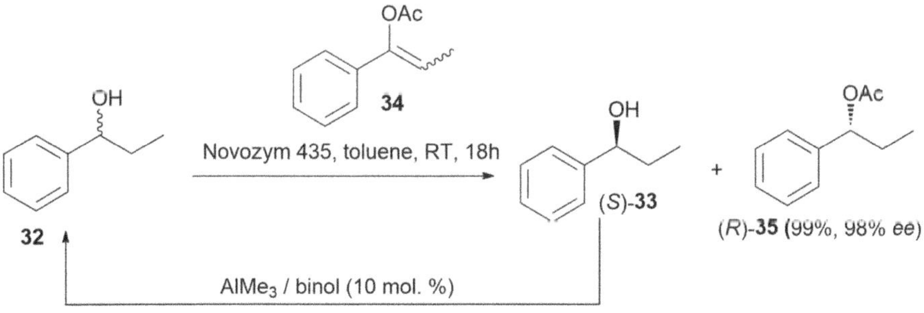

Scheme 9.9 DKR using AlMe3/binol ligand (adapted).[36]

The probable incompatibility of some catalysts (aluminium and vanadium complexes in Table 9.3) with lipases has not been a barrier for performing DKR jointly mediated by them (Schemes 9.9 and 9.10).[36,37]

Ruthenium complexes, such as the classical Shvo's[38] and Bäckvall's catalysts (and their derivatives), have been the most successful for DKR racemizations.[39]

Han *et al.* described the chemoenzymatic synthesis of rivastigmine (**39**) *via* DKR (Scheme 9.11).[40] Rivastigmine is an acetylcholinesterase inhibitor of the carbamate type. It is indicated for mild to moderate Alzheimer's disease, and it is also assumed to be effective in the treatment of dementia caused

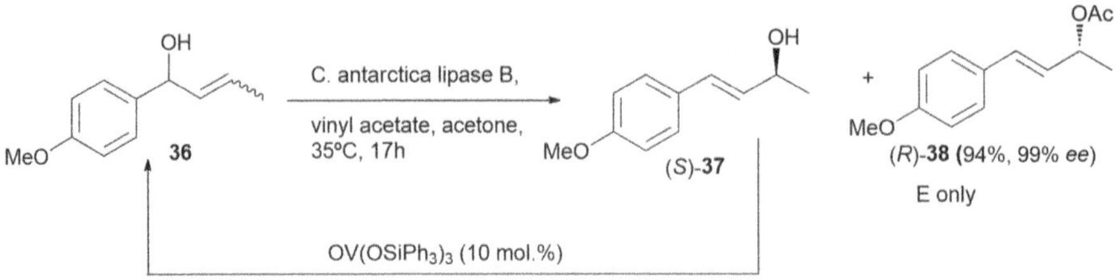

Scheme 9.10 DKR *via* vanadium-catalyzed racemization (adapted).[37]

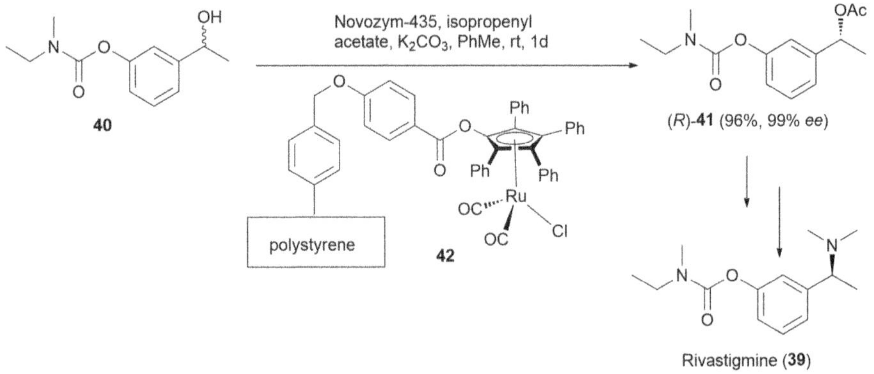

Scheme 9.11 Rivastigmine synthesis by DKR *via* ruthenium complex racemization.

by *e.g.* Parkinson's disease. In Han's work, they made use of a polymer-supported ruthenium catalyst **42**, since unsupported catalysts usually are soluble in the reaction medium, which precludes their recovery.

9.2.3 Enantioselective Desymmetrizations

The desymmetrization of prochiral compounds consists of a modification that eliminates one or more elements of symmetry of the substrate (Figure 9.5). The chirality is only established if the symmetry elements that prevent chirality are eliminated. Such a process proceeds enantioselectively whenever enantiotopic groups or faces of the substrate are efficiently differentiated by the chiral catalyst. In the case of lipases and other hydrolases as biocatalysts, desymmetrizations occur by the former means. *Meso* compounds, which bear two or more stereocenters, are also useful starting materials (*vide infra*). Naturally, in comparison to the classical resolutions, these processes are attractive due to their potential 100% yield.[41]

The synthesis of oseltamivir phosphate (Tamiflu®) (**43**) performed by Zutter *et al.* relies on a lipase-catalyzed desymmetrization.[42] Tamiflu® is the

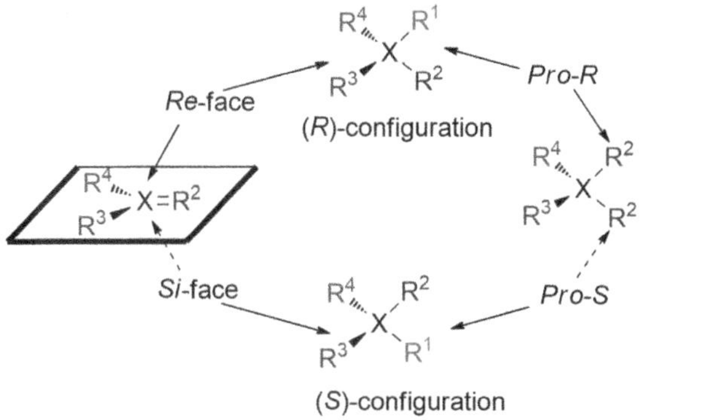

Figure 9.5 Chirality from prochiral compounds (adapted).[41]

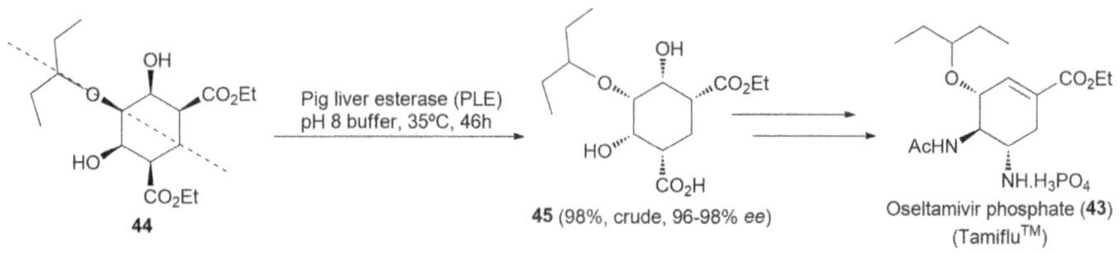

Scheme 9.12 Enantioselective enzymatic desymmetrization applied to oseltamivir synthesis (adapted).[42]

prodrug of the potent and selective inhibitor of influenza neuraminidases A and B, used to treat and prevent influenza infection viruses A and B. In this elegant chemoenzymatic synthesis, an all-*cis meso*-diester **44**, derived from trihydroxyisophthalic acid, underwent partial hydrolysis (*via* PLE) to the chiral monoacid in high ee and quantitatively (Scheme 9.12).[43]

9.3 Medium Engineering

Lipases' ability to catalyze transformations in dry organic solvents has propelled their widespread application in enantioselective syntheses. Besides allowing more hydrophobic compounds as substrates in these stereoselective reactions, the use of such media offers as advantages, ease of product recovery; higher stability of the involved substances and suppression of unwanted reactions that occur in aqueous media.

Lipases (as well as various other enzymes) display high stability in dry organic solvents.[44–46] In fact, although the presence of water molecules in these media is important, the stability of enzymes is prone to decrease as the water concentration is excessively raised in solvent–water mixtures due to denaturation processes.[47] Water plays a key role in both gradual degradative (hydrolysis, deamination, *etc.*) and reversible heat-induced unfolding processes related to a decrease in activity.[46] Studies have shown that enzymes retain their crystalline structure in dry organic media.[48] This is so because, although a drive to denature would be expected, the required pliability is lacking due to the lower conformational flexibility in organic media.

In organic solvents, lipases and other hydrolases may accept nucleophiles other than water (such as alcohols, amines, *etc.*), carrying out reactions that are not accessible in aqueous media for thermodynamic reasons.[49] Nevertheless, in organic solvents, enzymes are usually far less active than in water. It is worth noting that they nonetheless remain highly competent (showing similar kinetics to that in aqueous media) compared with the non-catalyzed processes in those media. Dry solvents may remove water molecules that are tightly bound to the protein structure. The essential layers of water molecules on the enzyme surface act as a lubricant, nurturing the conformational flexibility necessary for the catalytic process. Organic solvents do not share with water its ability to form multiple bonds and due to their lower dielectric constant do not alleviate the strong electrostatic forces within an enzyme structure making it more rigid.[50] As a matter of fact, early experimentation showed that, due to the lower flexibility in organic media, a lipase becomes inactive towards tertiary alcohols in transesterification with tributyrin.[44] Solutions to this problem have been devised (*vide infra*). Moreover, although lyophilized enzymes form suspensions in nearly all organic solvents, it appears that, provided that efficient stirring is in place, the mass transfer problem is not usually the main cause of the lower activity in these media.[51]

Most of the loss of enzyme activity caused by their use in dry organic solvents does not relate to the contact between the protein and the solvent itself. In the case of lyophilized enzymes, the lyophilization (gentle water evaporation method carried out under high vacuum and freezing temperatures) itself may affect the enzyme's structure *via* reversible denaturation. Actually, the fast dehydration process in the formation of enzyme powders may lead to deleterious conformational changes.[52] Such deleterious effects may be treated by the addition of small amounts of water or the increase of the thermodynamic water activity. Water mimicking by ethylene glycol or glycerol has an activating effect as well.[53,54]

The effect of water content (on catalytic activity) depends on the nature of the solvent, being more pronounced in hydrophobic ones, from which proteins strip water more easily.[50,55] Accordingly, water activity (a_w), rather than water content, is what really matters. In a report on the use of *Mucor miehei* lipase (Lipozyme) for the esterification of decanoic acid with dodecanol (solvent polarity ranging from hexane to 3-pentanone), it was shown that a better correlation occurred between water activity and enzyme activity (initial rates) than between water concentration and enzyme activity. Although activities varied among different solvents, the obtained curves were quite similar, with the maximum activity relating to an a_w close to 0.5.[56]

The problems with denaturation may also be remedied by the use of structure-preserving lyoprotectants, such as sugars and poly(ethyleneglycol), certain inorganic salts, substrate-resembling ligands, crown ethers, *etc.*[46]

Furthermore, pH is a very influential factor for enzyme activity in aqueous media. While it has no meaning for biocatalysis in organic media, it has been shown that the enzyme activity reflects the pH of the aqueous solution from which it was isolated. This effect is termed pH memory. Thus, subjecting an enzyme to an optimum pH prior to lyophilization would ensure higher activity.[49,57] Alternative protocols may have the same effect.[45]

The effects of medium engineering on the chemical selectivity of enzymes may parallel those of directed evolution technology, which demand extensive enzyme screening.[58] Thus, the change of solvent has proven to have a profound impact on the substrate specificity and (regio-, chemo- and stereo-) selectivity.

As for the specific issue of enantioselectivity, profound impacts have been reported by solvent change.[59,60] The structural basis for this appears to relate to the balance of hydrophobic forces in substrate recognition, which are less important in organic media. In one report, it was shown that the selected (*R*)-enantiomorph in a chymotrypsin-catalyzed transesterification of methyl 3-hydroxy-2-phenylpropionate is the least solvated in the active site.[61] In other words, desolvation energetics are of paramount importance. A seminal study on the esterification of 2-hydroxy acids with butanol *via Candida cylindracea lipase* (CCL) detailed the solvent contributions to the specific reactivities of the antipodes in the racemates.[62] For instance, it was shown that, in the kinetic resolutions of (±)-2-hydroxypropanoic acid (95% ee) and a less suitable substrate, (±)-2-hydroxy-3-methylbutanoic acid (17% ee) in toluene, the slow (*R*)-enantiomorphs in both cases reacted at a similar rate. Among the best solvents, toluene led to higher stereoselectivities than cyclohexane, irrespective of the higher conversion rates of the latter. More hydrophilic solvents, which led to lower enantioselectivities, had a deleterious effect on the reactivity of the faster (*S*)-enantiomorphs.

Other rationales have been proposed to explain the effects of dry organic solvents on the enantioselectivities of enzymes.[45] It is possible that organic solvents play different roles in the processes of enantiomorph discrimination or enantiospecificity.

The importance of medium engineering became even clearer after the first reports of the inversion of configuration in enzymatic kinetic resolutions by means of solvent change.[63] The same phenomenon was soon observed in kinetic resolutions by lipases.[64,65] In an extraordinary example, the desymmetrization of a prochiral 4-aryl-1,4-dihydro-2,6-dimethyl-3,5-pyridine dicarboxylate *via* hydrolysis by lipase AH (*Pseudomonas* sp.), which led to the (*S*)-monoacid (87%; 99% ee) when carried out in wet diisopropyl ether, became (*R*)-selective (88%; 89% ee) after the solvent was changed to wet cyclohexane.[66]

Since the seminal application of lipases in enantioselective synthesis, a massive number of reports have dealt with the use of such practical catalysts.[67–69] Currently, the use of such highly reputed biocatalysts remains widespread. With the continuous evolution of lipase technology, an ever growing molecular diversity of compatible substrates is observed. We herein discuss some recent applications of lipases for stereoselective synthesis of chiral compounds with a focus on the use and effects of solvents.

The hindered racemic *myo*-inositol derivative DL-**46**, a precursor of bioactive inositol phosphates, was successfully resolved *via* transesterification with vinyl acetate by three different lipases (Novozym 435, Amano PS-C and Amano PS-IM).[70] Novozym 435, as usual, performed better in more hydrophobic solvents, TBME being the best choice (hexane = 48%; ee_s = 88%) (Scheme 9.13). On the other hand, the use of EtOAc as solvent led to lower conversion and selectivity (29% conversion; E = 39). Interestingly, when the same transformation was mediated by PS-IM lipase, EtOAc (45% conversion; E > 200) performed better than TBME (34% conversion; E = 63) as solvent. In all cases, the acylations were highly regioselective. The biocatalyzed kinetic resolution *via* Novozym 435 has been recently optimized.[71]

The same group reported on the kinetic resolution of *myo*-inositol-derived diol DL-**48** (R = H) (Figure 9.6) *via* transesterification (vinyl acetate) by Novozym 435 with excellent results.[72,73] This compound is a relevant precursor of bioactive inositols. The resolution of racemate DL-**48** was enabled by regioselective acylation (vinyl acetate) to produce L-**49**. Despite the good results with vinyl acetate and isopropenyl acetate as solvents, TBME brought about the best results (48% conversion; E > 100), securing faster transformations (24 h). Such resolution in vinyl acetate was later optimized.[74]

Rhizomucor miehei lipase was employed in the kinetic resolution of cyanoaryl secondary alcohols **50** (R = H) (Figure 9.6), which is useful for the synthesis of biologically relevant 3-substituted-3,4-dihydroisocoumarins.[75] The biocatalyzed acylations with vinyl acetate worked better in TBME (8 h, c = 50%, E > 194) and, to a lesser extent, in toluene (23 h, c = 47% ee_s = 98%

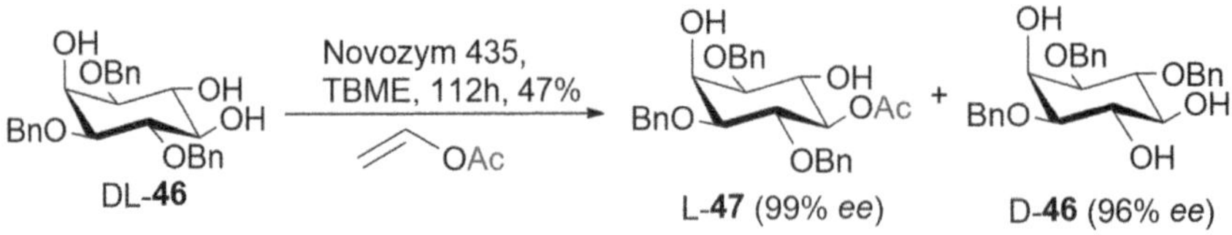

Scheme 9.13 Kinetic resolution of *myo*-inositol derivative DL-**46**.

DL-**48** (R=H) L-**49** (R=Ac)   (*RS*)-**50** (R=H), (R)-**51** (R=Ac)

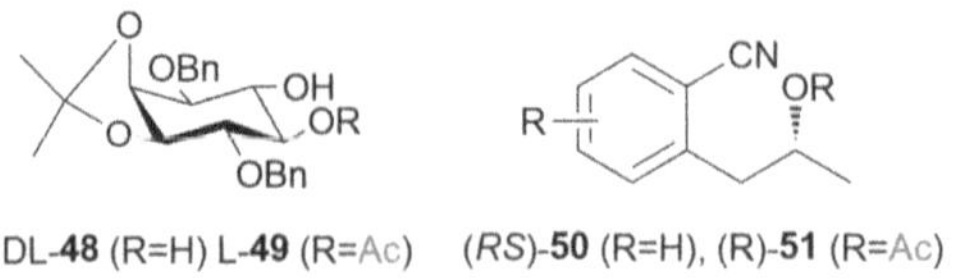

Figure 9.6 Recent examples of racemic *sec*-alcohols subjected to kinetic resolution by lipases.

$ee_s = 85\%$). Excellent biocatalytic performances were obtained in the kinetic resolution of some of the derivatives of **50**.

Kinetic resolution of racemic tetrahydroquinoline (R,S)-**52** was chosen as a means for the enantioselective synthesis of (R)-salsolinol, an endogenous catechol isoquinoline involved in the mechanism responsible for causing Parkinson's disease (Scheme 9.14).[76] In these biocatalyzed reactions (*via Candida antarctica* lipase A, CAL-A), the acylating agent screening (in toluene) pointed to 3-methoxyphenyl allyl carbonate as the acylating agent of choice. Solvent screening identified more hydrophobic solvents (toluene, diisopropyl ether, TBME) as the best media ($c \geq 49\%$; $E > 200$). While Et_2O was also an effective solvent, leading to high conversion (49%) with lower stereoselectivities ($ee_p = 95\%$ $ee_s = 93\%$), THF led to lower conversion (40%) and stereoselectivity ($ee_p = 94\%$, $E = 95$). Isopropanol as solvent virtually disabled this transformation ($c = 8\%$), irrespective of the good enantioselectivity ($E = 26$). The produced carbamate (R)-**53** was employed in the synthesis of the molecular target.

Novozym 435 was effective in resolving racemic diol monoester (RS)-**54** *via* alcoholysis of the remote ester function (Figure 9.7).[77] Upon screening (in acetonitrile), isobutanol was selected as acyl acceptor. Good correlation of the difference of activation free energies ($\Delta\Delta G$) with $\log P$ among more hydrophilic or (separately) more hydrophobic solvents was observed, acetonitrile being the most effective. Concerning enzyme activity (initial rate), however, TBME and diisopropylether were found to be the best media. Such results suggested the use of mixed solvents. Thus, a $1:3$ TBME–acetonitrile mixture was selected as the best solvent as it led to a better compromise of rate and

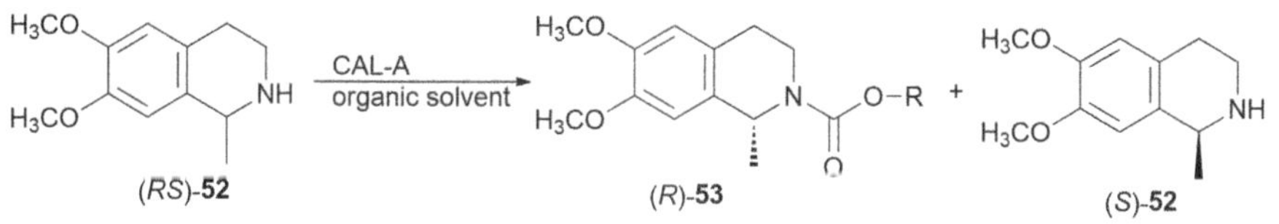

Scheme 9.14   Kinetic resolution of amine (RS)-**52**.

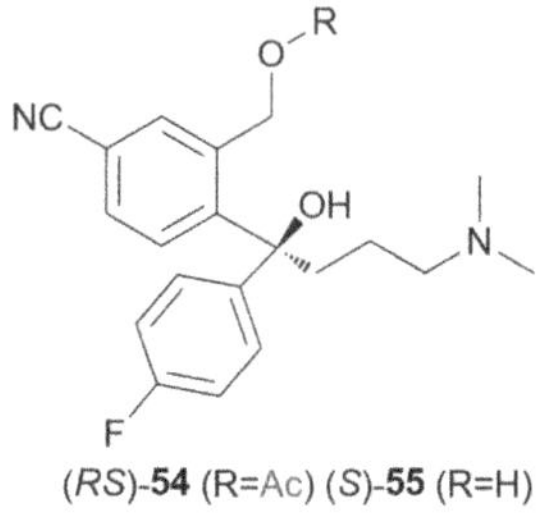

Figure 9.7   Ester (RS)-**54**, subjected to enantioselective alcoholysis at a remote function.

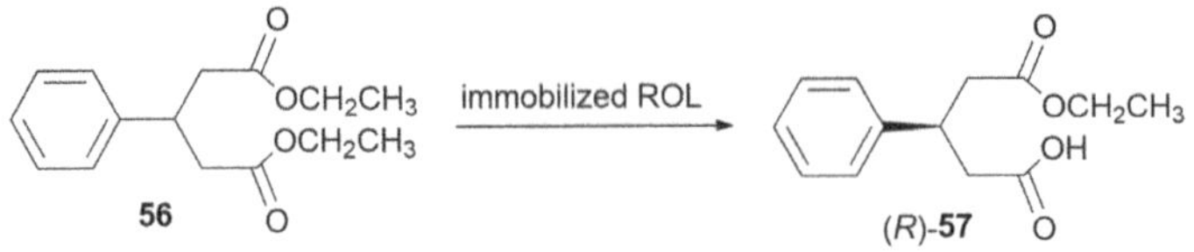

Scheme 9.15 Desymmetrization of prochiral substrate **56** by ROL.

E.[20] The remaining (S)-**55** is a precursor for citalopram, a highly selective inhibitor of serotonin (5-HT) reuptake and hence an efficient antidepressant.

3-Phenyl glutarate (**56**) is a model building block for a number of pharmaceutical substances. This prochiral substrate could be desymmetrized with newly immobilized *Rhizopus oryzae* lipase (ROL) (on Lewatit CNP 105) to produce monoester (R)-**57**[78] in a slow transformation, however (Scheme 9.15). Experimentation showed that the use of organic co-solvents (in 20 mM aq. sodium phosphate at pH = 7) improved the enantioselectivity of the biocatalyzed hydrolysis. Thus, dioxane (20%) showed a better performance than DMSO and acetone. Higher concentrations of dioxane eroded the selectivity nonetheless. Under the optimized conditions (5 °C), high enantioselectivity was achieved (ee = 92%).

9.4 Immobilization of Lipases

9.4.1 Brief Background

Through immobilization – the binding of biocatalysts to insoluble supports and confinement to defined spaces with catalyst activity maintenance – enzyme reuse can be enabled under continuous conditions or not.[79–81] Such technology has long been known, dating back to the early 1900s with a seminal report on the immobilization of invertase on active carbon.[82] From the 1960s on, a number of studies paved the way for the development of the immobilization technologies currently available.[83] Such innovative momentum is explained in part by the recognition of their potential for industrial applications, which demand catalyst recovery and reuse. Other advantages of enzyme immobilization are higher enantioselectivity, catalytic activity and thermostability.[80,84–102]

Immobilization technology continues to evolve *via* combination with protein engineering (site-directed evolution, *etc.*), the use of ionic liquids and nanotechnology.[103–108]

9.4.2 Immobilization Protocols

In the choice of the support for immobilization, criteria such as the maximum activity of the new biocatalyst, operational stability, support cost and toxicity of the required reagents are taken into account. Moreover, the support should have large superficial area, resistance to mechanical stress and to microorganisms, and a low diffusional barrier to the transport of substrates and products. Supports may be organic (from natural sources, *e.g.*, cellulose or synthetic sources like polystyrene) or inorganic (*e.g.* silica gel). Today, several commercial immobilization materials are available, *e.g.* Sepabeads®, Spherezyme™, Eupergit®, *etc.* The enzyme-support combination will furnish a biocatalyst with specific chemical, biochemical, mechanical and kinetic properties.[106,109–112] In this context, pore and particle sizes relate to the support capacity for enzyme loading. Although non-porous supports show less problems in diffusion (desirable), their enzyme load capacity is small. Thus, due to their high superficial area, porous materials are preferred as biocatalyst supports. Besides the higher load capacity shown by them, immobilized enzymes are protected from the external medium inside the pores.[113] Aiming at a particular biocatalyst performance (*e.g.* stereoselectivity), the immobilization technique should be customized. In some cases, even a set of different preparation methods may be the best solution.[88,89,114–116] A variety of immobilization means are currently known: physical, ionic or metallic adsorption, covalent bonding or confinement (Figure 9.8).[84,113,117,118]

9.4.2.1 Physical Adsorption

Physical adsorption is the attachment of enzymes to the surface of support particles by weak forces, such as van der Waals interactions and hydrogen bonds (Figure 9.8, 1). This technique is straightforward, accessible and allows facile recycling of the support at the end of enzyme life, through simple procedures such as the use of detergent and treatment with urea or guanidine, pH variation, saline solution treatment, *etc.*[121–123]

This is the most employed immobilization mode for continuous reactors and stirred tanks.[124,125] It works well in lipase-catalyzed reactions in organic solvents. Moreover, it has been used in industrial lipase preparations such as for Novozym 435 (*Candida antarctica* lipase B, CaLB, from Novo), a recombinant CaLB expressed in *Aspergillus niger* and immobilized in acrylic resin.

This immobilization mode is known to result in higher catalytic activity, good thermal stability and reusability of lipases.[111] Better hydrophobic character of the support may lead to a higher degree of immobilization and higher catalytic activity of adsorbed lipases, as shown in a report on the use of poly(vinyl acetate–acrylamide) microspheres with varying monomeric ratios.[126]

The adsorption of lipases on hydrophobic supports has been recognized as a factor for enantioselectivity improvement. In this case, the lipase may be fixed in an open conformation, with lid shifting that allows the substrate access to the active site. Other regions of the protein, besides the lid, may interact with the large hydrophobic surface of the support.[123,127]

Different reports have shown that lipase enantioselectivities may vary widely with different supports and immobilization procedures. As a matter of fact, lipase enantiospecificity may be reversed for the same reaction medium.[128]

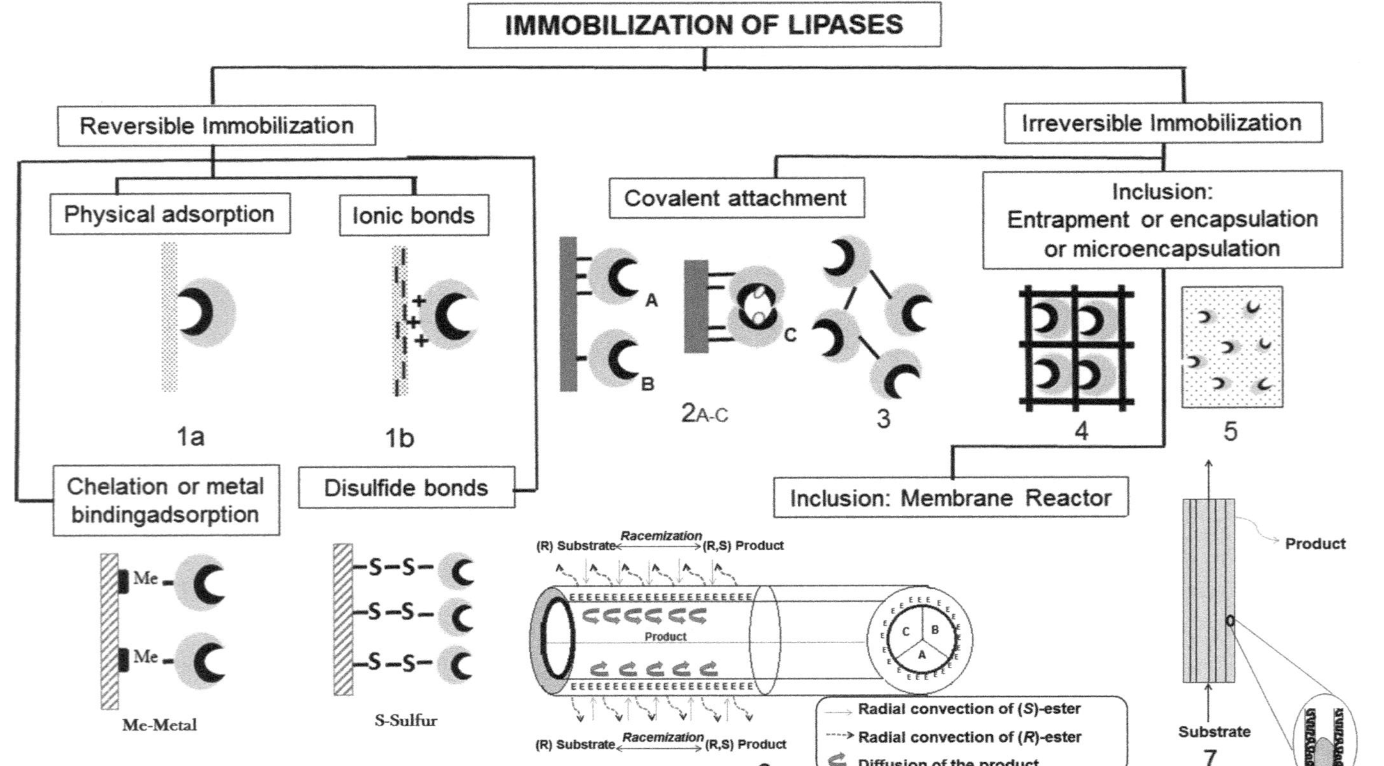

Figure 9.8 General methods of immobilization for biocatalysts: (1)-physical adsorption. (2)-Covalent attachment: A—multipoint-monomeric derivative, B—unipoint-monomeric derivative, C—multipoint-bi-molecular derivative (the filled in circle represents the active site). 3—Cross-linking. 4—Entrapment on matrix of a polymer. 5—Encapsulation on gels. 6 and 7: Membrane reactor; 6—enzyme immobilized on the external sponge layer of the membrane (adapted);[119] 7—the membrane contains the immobilized biocatalysts (adapted).[120]

In the enantioselective hydrolysis of (*R,S*)-2-hydroxy-4-phenylbutyric acid ethyl ester (HPBEt) by lipase from *Burkholderia cepacia* (formerly *Pseudomonas fluorescens*), higher activity (1.2 U mg^{-1}), in a highly enantioselective reaction, was achieved with octyl-Sepharose as support.[103] The authors supposed that the hydrophobic adsorption of active centers was a key factor for such a performance. Other studies have shown the impact of different supports on the same transformation catalyzed by lipase B from *Candida antarctica* (CaLB), lipase from *Thermomyces lanuginose* and lipase from *Bacillus thermocatenulatus*.[103,129,130]

9.4.2.2 Covalent Attachment

This immobilization mode has been intensively studied as it has the advantage of avoiding enzyme desorption from the support. It has a high impact on the enzyme conformation, which may result in an important rigidification of its three-dimensional structure.[84,117] Immobilization by covalent attachment explores the reactive functions of the polymer (support). Thus, the coupling techniques may involve: (A) activation of the support; (B) modification of the polymer backbone to generate a reactive function. In both cases, electrophilic groups are chosen in order to allow interaction with nucleophilic groups in the side chains of the following amino acids in the biocatalyst structure: lysine (ε-amino group), cysteine (thiol group), and aspartic and glutamic acids (carboxylic group). There is also an enzyme-cross-linking alternative, which is carried out by using bifunctional reagents. Glutaraldehyde is a common reagent for this purpose (since the 1960s) as it bridges amino groups (NH$_2$) on the surfaces of engaged protein molecules.[113]

The chemical modification of proteins, namely chemoselective transformations in regions close to the hydrophobic lid, may be regarded as an alternative to genetic engineering. A site-directed chemical modification of the free cysteine Cys64 residue of immobilized *Geobacillus thermocatenulatus* lipase (BTL2) (on CNBr-activated agarose or glyoxyl–agarose), using tailor-made polymers, enhanced its catalytic activity towards the hydrolysis of esters.[106] The Cys64 residue, located near the hydrophobic lid of the immobilized BTL2, was modified by reaction with PDP-activated aminated dextrans or PDP-activated carboxylated PEGs to form a disulfide linkage to these polymers. In the case of 2-*O*-butyroyl-2-phenylacetic acid as substrate, the immobilized glyoxyl–BTL2 modified with PEG1500CO$_2$H showed a 5-fold increase in hydrolytic activity compared to the non-modified immobilized biocatalyst. Inhibition experiments suggested that the site-directed modification of immobilized BTL2 caused lid opening in some of the prepared biocatalysts.

The use of bimolecular aggregates is another immobilization strategy and these are formed when biocatalyst molecules (bound by non-covalent interactions) are covalently attached (uni- or multipoint connection) as homodimers to the support (Figure 9.8, 2-C).[103,131,132]

Immobilization *via* multipoint attachment of bimolecular aggregates of *Burkholderia cepacia* lipase on glyoxyl supports has led to a biocatalyst with good performance. In the hydrolysis of (*R,S*)-HPBEt (*vide supra*), this immobilized lipase displayed higher stereoselectivity than those achieved with a one-point-attached derivative (*E* = 7) and with the multipoint-attached monomeric derivative (*E* = 20).[103] The same trend was observed concerning its stability in water and organic solvents. A bimolecular aggregate of the same lipase was mounted by binding a second monomer to the immobilized (and inactivated) enzyme (on glyoxyl–agarose in a multi-point manner) *via* hydrophobic interactions. This very stable catalyst displayed high stereoselectivity.[130]

Immobilization *via* Cross-Linking Enzyme Aggregates (CLEAs) or Cross-Linking Enzyme Crystals (CLECs) (Figure 9.8, 3) has gained a reputation as a practical and economical technology.[10,113,133] Such biocatalysts are commercially available from different companies. This technology has been successfully applied to lipase catalysis.[134–138]

9.5 Tailor-Made Lipases: Improving the Enantioselectivity

In catalysis studies, a great challenge for the chemist is catalyst customization for a specific reaction (*e.g.* high specificity, stability, low cost, *etc.*). In this way, we can say that with respect to enzyme catalysis we are approaching this stage of development. Some have argued that the "conventional paradigm" in the development of enzymatic processes has been overcome. The "conventional paradigm" refers to the exploration of reaction condition parameters (*e.g.* temperature, pH, organic solvents, *etc.*) which are to some extent compromised by the intrinsic limitations of enzymes. On the other hand, in the "ideal biocatalyst paradigm", the biocatalyst is built and produced towards the optimized process and, thus, its economy (Figure 9.9).[139,140]

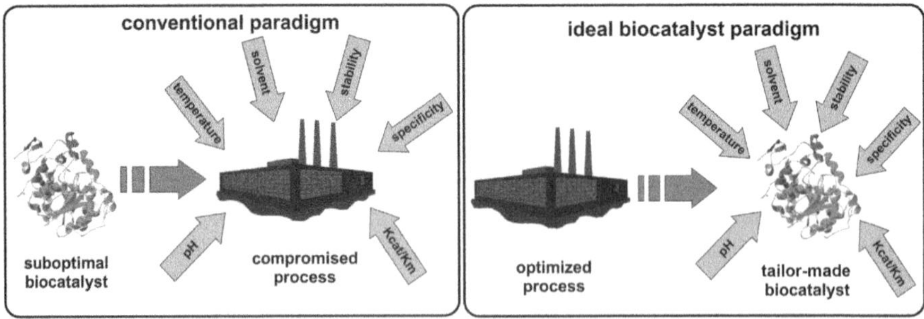

Figure 9.9   In the "conventional paradigm" the characteristics of the enzyme limit the process. On the other hand, in the "ideal biocatalyst paradigm", protein engineering enables the construction of novel biocatalysts for the optimized process.

Despite the availability of other useful alternative techniques for the preparation of efficient enzymes (*e.g.* immobilization, metagenomics, *etc.*), the protein engineering approach has gained prominence as a robust solution. It involves diverse methodologies for the design/manipulation of a target gene, leading to an optimized protein. As for the choice of the precise methodology towards this end, it depends on the knowledge of the target enzyme and the estimated cost.

The techniques in protein engineering are situated between two different approaches: rational design and directed evolution. The rational design of mutations by computational tools requires knowledge of the biocatalyst structure and the catalysis mechanism. Designed mutations are obtained using, in general, site-directed mutagenesis. On the other hand, in the directed evolution approach, such preliminary knowledge is not required. In this case, random mutations are made in the target gene, generating a library of mutants from which variants with the desired characteristics are selected (*e.g.* thermostability, enantioselectivity, *etc.*). The directed evolution approach is more demanding on laboratory infrastructure as it involves high-throughput screening techniques. It should be emphasized that in both strategies, after the evaluation of the first selected mutants, new rounds of mutations may be necessary for improvement of the engineered protein.[5,141,142] A number of other protein-engineering techniques for improved biocatalysts based on hybrid approaches have been developed, such as shuffling, circular permutation, iterative saturation mutagenesis (ISM), combinatorial active-site saturation testing (CASTing), cassette mutagenesis, restricted libraries, structure-guided consensus, *etc.*[5,141,143]

Table 9.4 lists selected reports between 2007 and 2013 using these techniques in the production of lipases and esterases with improved enantioselectivity. Apparently, there is a trend of decreased interest in traditional error prone PCR in the literature, as the screening efforts of large libraries are labor-intensive and time consuming. In fact, excellent results have been obtained using semi-rational (hybrid) methodologies.

An invaluable comparison of different methodologies has been made by Reetz's group regarding *Pseudomonas aeruginosa* lipase (PAL).[144,145] After the first application of directed evolution to a lipase for improved enantioselectivity,[146] different alternative techniques have emerged (epPCR, DNA-shuffling, combinatorial multiple cassette mutagenesis, cassette mutagenesis).[144] Thus, marked improvements in the enantioselectivity of PAL towards 2-methyldecanoic acid p-nitrophenyl ester were attained. From an $E = 1.1$ for wild-type PAL, the stereoselectivity evolved by means of epPCR ($E = 11.3$) and further on by combinatorial multiple cassette mutagenesis ($E = 51$).[144,146] Recently, Reetz *et al.* employed the ISM methodology to obtain a high-performance-engineered PAL with E superior to 200 ($E = 594$).[145]

In conclusion, the combination of modern protein-engineering techniques with traditional approaches (*e.g.* immobilization, process engineering, *etc.*) appears to bring lipase (and similar hydrolases) catalysis close to the "ideal biocatalyst paradigm".

Table 9.4   Studies on protein engineering approaches to improve enantioselectivity (from 2007 to 2013).

Enzyme	Protein engineering approach	Substrate(s)	Optimization^{a,b}	References
Esterase from **Bacillus subtilis** **(BS2)**	CASTing	1,1,1-Trifluoro-2-phenyl-but-3-yn-1-yl acetate	$E_{wt}(R) = 42$ $E_{ep}(S) = 64$	147
Esterase from **Pseudomonas** **fluorescens**	CASTing	3-Phenyl butyric acid p-nitrophenyl ester	$E_{wt}(R) = 3.2$ $E_{ep}(R) = 80$	148
Lipase A from **Candida** **antarctica** **(CAL-A)**	CASTing	2-Methylheptanoic acid p-nitrophenyl ester – and other derivatives	$E_{wt}(S) = 5.1$ $E_{ep}(S) = 52$	149
		2-Phenyl propanoic acid p-nitrophenyl ester	$E_{wt}(S) = 20$ $E_{ep}(R) = 276$	150
		2-(4-Isobutylphenyl) propionic acid p-nitrophenyl ester	$E_{wt}(S) = 3.4$ $E_{ep}(S) = 100$	151
Lipase A from **Pseudomonas** **aeruginosa** **(PAL)**	ISM	2-Methyldecanoic acid p-nitrophenyl ester – and other derivatives	$E_{wt}(S) = 1.1$ $E_{ep}(S) = 594$	145
Lipase B from **Candida** **antarctica** **(CAL-B)**	Circular permutation	2-(3-Fluoro-4-phenyl-phenyl)propionic acid and others	$E_{wt}(R) = 25$ $E_{ep}(R) = 40$	152
Lipase from **Yarrowia lipolytica** **(Lip2p)**	Rational design	2-Bromo-phenylacetic acid ethyl ester and 2-bromo-o-tolylacetic acid ethyl ester	$E_{wt}(S) = 5.5$ $E_{ep}(S) = 59$ and $E_{wt}(S) = 27$ $E_{ep}(S) = 111$	153
		2-Bromo-phenylacetic acid octyl ester	$E_{wt}(S) = 4.6$; ee = 42.3% $E_{ep}(R) > 200$; ee > 99%	154

^{a}The parameters shown in this table were estimated using different methods, so only comparisons inside the same reference are possible.
^{b}E_{wt} = enantiomeric ratio for wild type enzyme; E_{ep} = enantiomeric ratio for engineered protein – the letter in parenthesis means the enantiomer preference.

9.6   Reactor Configuration

The efficiency of large-scale enantioselective syntheses mediated by lipases depends on the configuration and operation mode of bioreactors. Such factors and the biocatalyst form (lyophilized or immobilized) altogether directly relate to the enzyme stability and reuse, and hence to the process economy. In the case of free (lyophilized) biocatalysts, ultrafiltration or centrifugation units should be coupled to the system. Nevertheless, the use of immobilized biocatalysts offers more process options.[155]

Most of the literature reports on small-scale-lipase-catalyzed reactions have made use of stirred tanks (BSTR) or shaken flasks operating in batch mode,[133] as in the kinetic resolution of a racemic solketal ester derivative *via* hydrolysis by Amano AK lipase (*Pseudomonas* sp., lyophilized)[156] or the kinetic resolution of racemic 1,3,6-tri-*O*-benzyl-*myo*-inositol *via* acylation by *Candida antarctica* lipase B with vinyl acetate.[70] The latter biocatalyzed transformation, after optimization, was carried out in gram-scale[71] but an investigation of the performance of this kinetic resolution in large scale, including catalyst reuse, was not undertaken. Physical destruction of the biocatalyst in such reactors by shear stress has been reported.

A number of reports have shown the advantages of packed-bed reactors operated under continuous flow over their batch counterparts in biocatalytic transformations. Good results have been obtained in the production of biodiesel and bioactive compounds.[157–170] In general, good conversions and increased enzyme stabilities in the presence of solvents or without solvents, resulted. Although the use of solvents makes costs higher, they may be important to avoid low bed pressure at industrial scale.

The effect of temperature on the kinetic resolution of racemic amines by *Candida antartica* lipase B under continuous-flow conditions (packed-bed reactors) was studied.[171] It depended on the precise combination of substrates and the type of immobilization. Higher mobility substrates and biocatalysts led to higher enantioselectivities at higher temperatures.

In general, kinetic resolutions under continuous-flow conditions display similar conversions and selectivities to those run in batch reactors. However, productivities in the former case are usually much higher. In the enantioselective acylation of different racemic alcohols (1-phenylethanol, 1-cyclohexylethanol and 1-phenylpropan-2-ol) by lyophilized or immobilized lipases, productivities were higher in continuous-flow reactors, while enantioselectivities were similar to the ones in batch reactors.[172] In another report, the kinetic resolutions of secondary alcohols by immobilized *Candida antartica* lipase B (Novozym 435) and sol–gel immobilized *Pseudomonas fluorescens* lipase in both shaken flasks and continuous-flow reactors were compared. The resolutions under both conditions were efficient but the productivities in the latter ones were always superior.[173]

Flurbiprofen was kinetically resolved by esterification with butanol (*Candida antartica* lipase B Novozym 435) in a continuous-flow reactor with high conversion and optical purity (>98%). The productivities were up to ten times higher than those in the batch reactor.[174] (*R*)-Flurbiprofen is a non-steroidal anti-inflammatory drug which is useful in therapy against metastasis in prostate cancer.

The use of a packed-bed reactor for the kinetic resolution of DL-1,3,6-tri-*O*-benzyl-*myo*-inositol by immobilized *Candida antartica* (Novozym 435) *via* acetylation made it 531 times more productive than the corresponding biocatalyzed transformation in batch mode (Figure 9.10 and see Scheme 9.13). The acylated product L-(–)-**47** was generated with high conversion and selectivity (ee > 99%, *E* > 100). The mentioned compound is a relevant precursor of bioactive *myo*-inositol derivatives.[170]

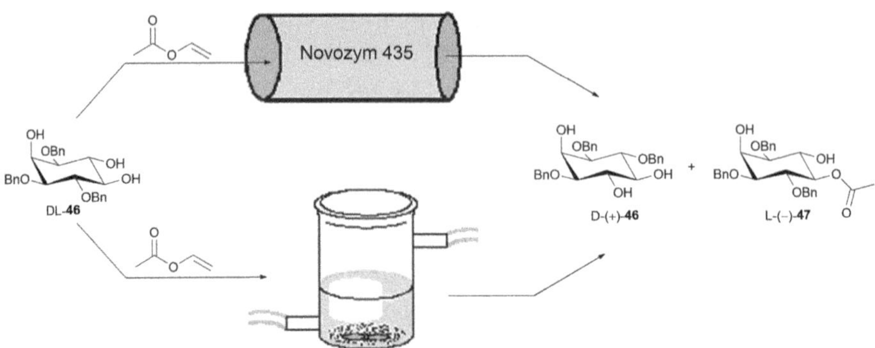

Figure 9.10 Kinetic resolution of DL-1,3,6-tri-*O*-benzyl-*myo*-inositol in different reactors.

Membrane bioreactors, wherein simultaneous separation of products and biocatalyst recovery occur, are relevant alternatives for lipase-catalyzed enantioselective synthesis. The membrane choice depends on the biocatalyst, which may be free or immobilized on the membrane.

A membrane reactor was employed in the 48 h dynamic kinetic resolution of racemic mandelic acid *via* esterification (see Scheme 9.5 in the introduction). The bioenzymatic process was carried out by means of a hollow fiber contactor, which separated the organic phase containing *Candida antartica* lipase B from the aqueous one containing mandelate racemase.[29] The ethylene dichloride phase was recirculated at the external side, while the latter one was contained inside the fiber into which the non-reacting (*S*)-antipode migrated.

Not many studies have dealt with lipase-catalyzed enantioselective synthesis on industrial scale. A kinetic resolution by immobilized *Candida antartica* lipase B (Novozym 435) leading to (*S*)-γ-fluoroleucine ethyl ester in a packed-bed reactor was carried out at 20–150 kg scales.[175] In this report, batch, fed-batch, plug flow and packed-bed reactors were compared *via* kinetic models. The continuous-flow process in the latter reactor type had the best performance as it minimized biocatalyst deactivation by shear stress and allowed the use of lower enzyme–substrate ratios (1:20, instead of 1:1 or 1:4, as employed in the batch and fed-batch reactor, respectively). Higher conversion (90%) and ee (86%) were also achieved.

In summary, the reactor configuration is an integral part of the development of lipase-catalyzed enantioselective synthesis. Along with reaction conditions and catalyst-type, it is a crucial factor in achieving economic and sustainable chemical processes.

9.7 The Use of Ionic Liquids

Researchers have shown that biocatalyst activity, enantioselectivity, thermal stability, and reusability may all be improved in ionic liquids (ILs).[176–181]

Room temperature ILs are liquids composed of cations and anions, usually organic salts, which present distinct properties favoring efficient catalysis. Such properties may be customized by the use of different cationic and anionic groups in a particular structure. Furthermore, reactions and processes which employ ILs are considered to be potentially environmentally friendly, and have attracted much interest from the academic and industrial research communities.[182–185]

As to the distinctive IL properties,[186,187] compared to the usual organic solvents, they show low vapour pressure, high ionic conductivity, high thermal stability, good dissolution capacity for various solutes and good physicochemical properties,[188,189] besides being non-flammable and displaying low toxicity.[185] The non-volatile property of ILs allows product removal by distillation and reuse of the catalytic agent.[190]

ILs are used in enzymatic reaction as a single solvent, as co-solvent in aqueous phase or in a two-phase system together with other solvents.[182,191] Moreover, ILs have been used in chemical reactions (transesterification, esterification, acylation, hydrolysis, alcoholysis, oxidation and polymerization) in biocatalytic processes, even with whole cell systems and for extraction of organic compounds. Uses as immobilization supports in mixtures with other non-classical solvents such as supercritical CO_2, or in biphasic systems for higher enzyme stability, have been reported.[192–194] In fact, ILs may be substitutes for organic solvents on toxicity and enzyme stability grounds.

Among the different enzymes used with ILs, lipases stand out as the most frequently assayed in such media,[192] displaying their usual tolerance to unnatural conditions.[194] In addition, the hydrophilic ionic head of ILs can have an important role in stabilizing the tertiary structure of the protein (unlike pure hydrophobic solvents like hexane).[191]

A growing number of publications on the use of ILs and lipases in organic synthesis can be seen in the literature (Figure 9.11).

The first reported enzyme-catalyzed reaction in ILs involved the thermolysin-catalyzed synthesis of Z-aspartame in [bmim][PF$_6$].[195] As for the use of lipases in ILs, Lau and co-workers[194] pioneered such a combination with a transesterification reaction (ethyl butanoate/butan-1-ol) catalyzed by lipase B from *Candida antarctica*, free and immobilized (Novozym 435) in [bmim][BF$_4$] and [bmim][PF$_6$]. A high yield (81%) was obtained when both ILs were utilized (4 h reaction time at 40 °C). In all cases the reaction rates were similar for all of the reactions investigated: alcoholysis, ammoniolysis, and perhydrolysis.

The esterases from *Bacillus subtilis* and *Bacillus stearothermophilus* catalyzed the transesterification of 1-phenylethanol in ILs and organic solvents in a comparative study on stability, activity and enantioselectivity. These enzymes, when immobilized on Celite, exhibited higher stability in ILs when compared to organic solvents (TBME, hexane and vinyl acetate).[196]

Kim and co-workers showed that ILs significantly enhanced the enantioselectivity of lipases in the transesterification of different pharmaceutical compounds compared to organic solvents. CaLB and PCL showed higher

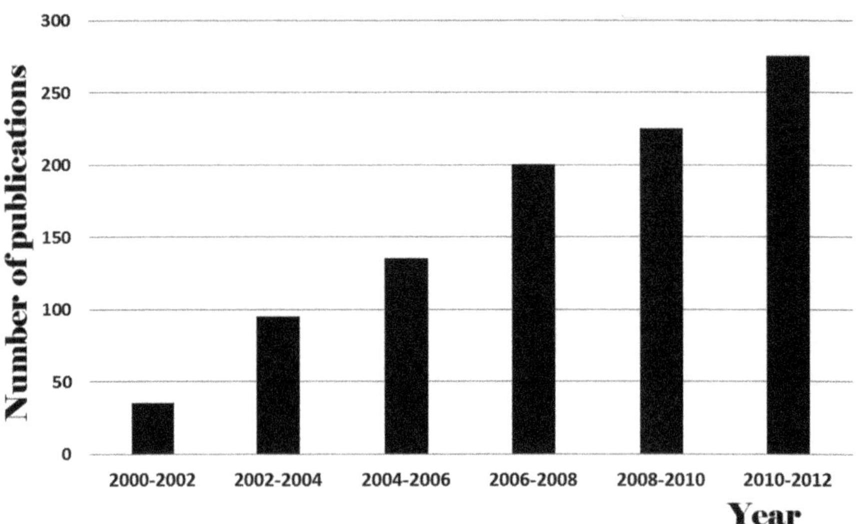

Figure 9.11   Number of publications containing 'ILs' and 'lipase' in the topic found on ISI Web of Science in the period 2000–2012.

enantioselectivity, especially with ILs bearing the [PF$_6^-$] counterion (Scheme 9.16).[197] Reusability and excellent transesterification activity of the immobilized CaLB and native *Pseudomonas cepacia* lipase (PCL) in [bmim][PF$_6$] and [emim][BF$_4$] proved the efficiency of the hydrophobic ILs.

The performance of lipase from *Pseudomonas aeruginosa* in the regioselective transesterification of *rac*-CDPP (a key intermediate of the *S*-lubeluzole drug) with vinyl butyrate (1:30 mM) in ILs was investigated by Singh *et al.* (Scheme 9.17). ILs were mixed with organic solvents in different proportions. The use of a 1:1 hexane–[bmim][PF$_6$] mixture led to the highest conversion (48.75%, 99.6% ee) after 6 h. The lipase showed high thermostability (at 60 °C) in [bmim][PF$_6$]. A higher temperature enabled a slight increase in conversion rate and ee (>99%).[191] The feasibility of enzyme recycling was demonstrated by repeated use ten times without a decrease in activity or selectivity.

The stability of CaLB in [emim][Tf$_2$N] (1-ethyl-3-methyl-imidazoliumbis(trifluoromethylsulfonyl)imide) and [btma][Tf$_2$N] (butyl-trimethylammonium bis(trifluoromethylsulfonyl)imide) was studied by Lozano *et al.* According to the authors, the stabilization of both enzymes by ILs may be associated with structural changes of the protein, observed by both fluorescence and circular dichroism spectroscopic studies.[181]

Roberts *et al.* reported the use of *Candida antarctica* lipase B (Novozym 435) for the resolution of 2,3,4,5-tetrahydro-4-methyl-3-oxo-1*H*-1,4-benzodiazepine-2-acetic acid methyl ester, a key lotrafiban intermediate. The (*S*)-enantiomer is a potent non-peptidic glycoprotein antagonist that inhibits platelet aggregation. The resolution was carried out in six ILs to afford (2*S*)-2,3,4,5-tetrahydro-4-methyl-3-oxo-1*H*-1,4-benzodiazepine-2-acetic acid,

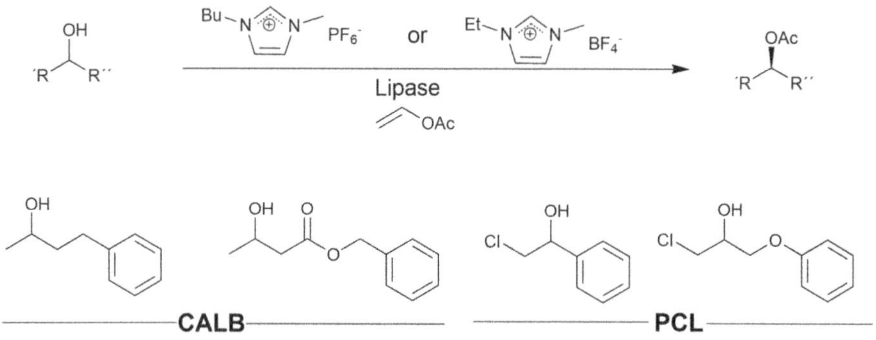

Scheme 9.16 Lipase-catalyzed transesterification (adapted).[197]

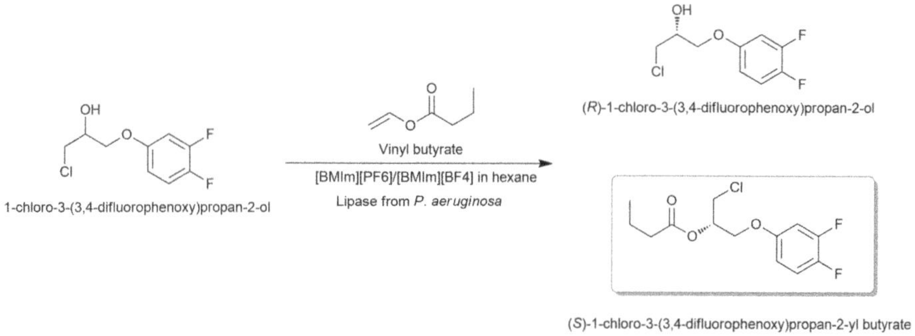

Scheme 9.17 Lipase-catalyzed enantioselective transesterification of *rac*-CDPP in IL–hexane (adapted).[191]

leaving the unreacted (*R*)-2,3,4,5-tetrahydro-4-methyl-3-oxo-1*H*-1,4-benzo-diazepine-2-acetic acid methyl ester and the (*R*)-2,3,4,5-tetrahydro-4-methyl-3-oxo-1*H*-1,4-benzodiazepine-2-acetic acid methyl ester in optically pure form.[198] The results were compared with the industrial process operated in *tert*-butanol.[199] The use of [bmim][PF$_6$] led to a two-fold increase in initial rate when compared with the industrial process.

Comparison between reactions performed with free and immobilized enzymes showed consistency with the results obtained in organic media, where suspensions of "free" lipases are generally less effective than immobilized lipases.[194,200] Lee and Kim observed this behavior in enzyme-catalyzed transesterifications wherein ILs significantly increased the catalytic activity and selectivity.[100,201]

ILs may function as substrate carriers in enantioselective lipase-catalyzed transformations, besides being the medium. In a seminal study by Naik *et al.*, (±)-ibuprofen, a non-steroidal anti-inflammatory drug, was anchored cova-lently to an IL for the purpose of kinetic resolution by different lipases. The experiment in [bmim][PF$_6$] as reaction medium performed slightly less effi-ciently (80% yield, 80% ee, while in DMSO, 87% yield, 86% ee were secured) (Scheme 9.18).[202]

Scheme 9.18   Lipase-catalyzed enantioselective hydrolysis of IL anchored ibuprofen ester (adapted).[202]

Notwithstanding the good performances shown in many cases,[192,203–206] ILs may damage the biocatalyst or reduce its productivity.[207] Zhou *et al.* studied different solvents/ILs, and observed degradation of the acrylic resin support of Novozym 435 in [bmim][BF$_4$] used in the kinetic resolution of a secondary alcohol. On the other hand, the authors observed that a monoether-functional IL (1-(3-ethoxypropyl)-2,3-dimethylimidazolium bis(trifluoromethylsulfonyl)imide) increases the enantioselectivity when compared to organic solvents. Furthermore, recycling was successfully carried out.[208]

References

1. B. Nagel, H. Dellweg and L. M. Gierasch, *Pure Appl. Chem.*, 1992, **64**, 143.
2. R. A. Sheldon, *Chem. Soc. Rev.*, 2012, **41**, 1437.
3. Díaz-Rodríguez and B. G. Davis, *Curr. Opin. Chem. Biol.*, 2011, **15**, 211.
4. E. M. Brustad and F. H. Arnold, *Curr. Opin. Chem. Biol.*, 2011, **15**, 201.
5. S. Bommarius, J. K. Blum and M. J. Abrahamson, *Curr. Opin. Chem. Biol.*, 2011, **15**, 194.
6. P. J. Dunn, *Chem. Soc. Rev.*, 2012, **41**, 1452.
7. M. Ahmed, T. Kelly and A. Ghanem, *Tetrahedron*, 2012, **68**, 6781.
8. T. Hudlicky and J. W. Reed, *Chem. Soc. Rev.*, 2009, **38**, 3117.
9. R. A. Sheldon, *Green Chem.*, 2007, **9**, 1273.
10. S. E. Milner and A. R. Maguire, *ARKIVOC*, 2012, 321.
11. A. Ghanem, *Tetrahedron*, 2007, **63**, 1721.
12. M. Paravidino and U. Hanefeld, *Green Chem.*, 2011, **13**, 2651.
13. V. Gotor-Fernández, R. Brieva and V. Gotor, *J. Mol. Catal. B: Enzym.*, 2006, **40**, 111.
14. Y. F. Wang, J. J. Lalonde, M. Momongan, D. E. Bergbreiter and C. H. Wong, *J. Am. Chem. Soc.*, 1988, **110**, 7200.
15. S. Simon, S. Oßwald, J. Roos and H. Gröger, *Z. Naturforsch. B*, 2012, **67**, 1123.
16. J. Steinreiber, K. Faber and H. Griengl, *Chem.–Eur. J.*, 2008, **14**, 8060.
17. M. A. Blasco, S. Thumann, J. Wittmann, A. Giannis and H. Gröger, *Bioorg. Med. Chem. Lett.*, 2010, **20**, 4679.
18. C. M. Clouthier and J. N. Pelletier, *Chem. Soc. Rev.*, 2012, **41**, 1585.
19. K. Faber, *Chem.–Eur. J.*, 2001, **7**, 5004.

20. L. Fransson and C. Moberg, *ChemCatChem*, 2010, **2**, 1523.
21. C. A. Martinez, S. Hu, Y. Dumond, J. Tao, P. Kelleher and L. Tully, *Org. Process Res. Dev.*, 2008, **12**, 392.
22. R. N. Patel, *ACS Catal.*, 2011, **1**, 1056.
23. A. Steinreiber, A. Stadler, S. F. Mayer, K. Faber and C. O. Kappe, *Tetrahedron Lett.*, 2001, **42**, 6283.
24. A. Kamal and G. Chouhan, *Tetrahedron: Asymmetry*, 2005, **16**, 2784.
25. R. S. Ward, *Tetrahedron: Asymmetry*, 1995, **6**, 1475.
26. A. Kamal, M. A. Azhar, T. Krishnaji, M. S. Malik and S. Azeeza, *Coord. Chem. Rev.*, 2008, **252**, 569.
27. H. Pellissier, *Tetrahedron*, 2011, **67**, 3769.
28. B. Martín-Matute and J.-E. Bäckvall, *Curr. Opin. Chem. Biol.*, 2007, **11**, 226.
29. W. J. Choi, K. Y. Lee, S. H. Kang and S. B. Lee, *Sep. Purif. Technol.*, 2007, **53**, 178.
30. A. H. Kamaruddin, M. H. Uzir, H. Y. Aboul-Enein and H. N. A. Halim, *Chirality*, 2009, **21**, 449.
31. H.-Y. Lin and S.-W. Tsai, *J. Mol. Catal. B: Enzym.*, 2003, **24–25**, 111.
32. T. Sakai, K. Wang and T. Ema, *Tetrahedron*, 2008, **64**, 2178.
33. M. M. Jones and J. M. J. Williams, *Chem. Commun.*, 1998, 2519.
34. L. Haughton and J. M. J. Williams, *Synthesis*, 2001, **6**, 0943.
35. C. L. Pollock, K. J. Fox, S. D. Lacroix, J. McDonagh, P. C. Marr, A. M. Nethercott, A. Pennycook, S. Qian, L. Robinson, G. C. Saunders and A. C. Marr, *Dalton Trans.*, 2012, **41**, 13423.
36. A. Berkessel, M. L. Sebastian-Ibarz and T. N. Müller, *Angew. Chem., Int. Ed.*, 2006, **45**, 6567.
37. S. Akai, R. Hanada, N. Fujiwara, Y. Kita and M. Egi, *Org. Lett.*, 2010, **12**, 4900.
38. M. C. Warner, C. P. Casey and J.-E. Backvall, in *Bifunctional Molecular Catalysis*, ed. T. Ikariya and M. Shibasaki, Springer-Verlag Berlin, Berlin, 2011, vol. 37, p. 85.
39. P. Hoyos, V. Pace and A. R. Alcántara, *Adv. Synth. Catal.*, 2012, **354**, 2585.
40. K. Han, C. Kim, J. Park and M.-J. Kim, *J. Org. Chem.*, 2010, **75**, 3105.
41. E. García-Urdiales, I. Alfonso and V. Gotor, *Chem. Rev.*, 2011, **111**, 110.
42. U. Zutter, H. Iding, P. Spurr and B. Wirz, *J. Org. Chem.*, 2008, **73**, 4895.
43. E. Barbayianni and G. Kokotos, *ChemCatChem*, 2012, **4**, 592.
44. A. Zaks and A. Klibanov, *Science*, 1984, **224**, 1249.
45. G. Carrea and S. Riva, *Angew. Chem., Int. Ed.*, 2000, **39**, 2226.
46. A. M. Klibanov, *Nature*, 2001, **409**, 241.
47. K. Griebenow and A. M. Klibanov, *J. Am. Chem. Soc.*, 1996, **118**, 11695.
48. P. A. Fitzpatrick, *Proc. Natl. Acad. Sci. U. S. A.*, 1993, **90**, 8653.
49. A. Zaks and A. M. Klibanov, *Proc. Natl. Acad. Sci. U. S. A.*, 1985, **82**, 3192.
50. A. Zaks and A. M. Klibanov, *J. Biol. Chem.*, 1988, **263**, 8017.
51. A. M. Klibanov, *Trends Biotechnol.*, 1997, **15**, 97.
52. K. Griebenow, *Proc. Natl. Acad. Sci. U. S. A.*, 1995, **92**, 10969.
53. Ö. Almarsson and A. M. Klibanov, *Biotechnol. Bioeng.*, 2000, **49**, 87.

54. G. Bell, P. J. Halling, B. D. Moore, J. Partridge and D. G. Rees, *Trends Biotechnol.*, 1995, **13**, 468.

55. C. Laane, S. Boeren, K. Vos and C. Veeger, *Biotechnol. Bioeng.*, 1987, **30**, 81.

56. R. H. Valivety, P. J. Halling and A. R. Macrae, *Biochim. Biophys. Acta, Protein Struct. Mol. Enzymol.*, 1992, **1118**, 218.

57. K. Xu and A. M. Klibanov, *J. Am. Chem. Soc.*, 1996, **118**, 9815.

58. C. R. Wescott and A. M. Klibanov, *Biochim. Biophys. Acta, Protein Struct. Mol. Enzymol.*, 1994, **1206**, 1.

59. P. A. Fitzpatrick and A. M. Klibanov, *J. Am. Chem. Soc.*, 1991, **113**, 3166.

60. H. Kitaguchi, P. A. Fitzpatrick, J. E. Huber and A. M. Klibanov, *J. Am. Chem. Soc.*, 1989, **111**, 3094.

61. C. R. Wescott, H. Noritomi and A. M. Klibanov, *J. Am. Chem. Soc.*, 1996, **118**, 10365.

62. S. Parida and J. S. Dordick, *J. Am. Chem. Soc.*, 1991, **113**, 2253.

63. S. Tawaki and A. M. Klibanov, *J. Am. Chem. Soc.*, 1992, **114**, 1882.

64. S.-H. Wu, F.-Y. Chu and K.-T. Wang, *Bioorg. Med. Chem. Lett.*, 1991, **1**, 339.

65. S. Ueji, R. Fujino, N. Ōkubo, T. Miyazawa, S. Kurita, M. Kitadani and A. Muromatsu, *Biotechnol. Lett.*, 1992, **14**, 163.

66. Y. Hirose, K. Kariya, I. Sasaki, Y. Kurono, H. Ebiike and K. Achiwa, *Tetrahedron Lett.*, 1992, **33**, 7157.

67. G. Kirchner, M. P. Scollar and A. M. Klibanov, *J. Am. Chem. Soc.*, 1985, **107**, 7072.

68. M. T. Reetz, *Curr. Opin. Chem. Biol.*, 2002, **6**, 145.

69. A. Ghanem and H. Y. Aboul-Enein, *Tetrahedron: Asymmetry*, 2004, **15**, 3331.

70. E. A. Manoel, K. C. Pais, A. G. Cunha, M. A. Z. Coelho, D. M. G. Freire and A. B. C. Simas, *Tetrahedron: Asymmetry*, 2012, **23**, 47.

71. E. A. Manoel, K. C. Pais, A. G. Cunha, A. B. C. Simas, M. A. Z. Coelho and D. M. G. Freire, *Org. Process Res. Dev.*, 2012, **16**, 1378.

72. A. G. Cunha, A. A. T. da Silva, A. J. R. da Silva, L. W. Tinoco, R. V. Almeida, R. B. de Alencastro, A. B. C. Simas and D. M. G. Freire, *Tetrahedron: Asymmetry*, 2010, **21**, 2899.

73. A. B. C. Simas, A. A. T. da Silva, A. G. Cunha, R. S. Assumpção, L. V. B. Hoelz, B. C. Neves, T. C. Galvão, R. V. Almeida, M. G. Albuquerque, D. M. G. Freire and R. B. de Alencastro, *J. Mol. Catal. B: Enzym.*, 2011, **70**, 32.

74. A. G. Cunha, A. A. T. da Silva, M. G. Godoy, R. V. Almeida, A. B. C. Simas and D. M. G. Freire, *J. Chem. Technol. Biotechnol.*, 2013, **88**, 205.

75. J. Mangas-Sánchez, E. Busto, V. Gotor-Fernández and V. Gotor, *Catal. Sci. Technol.*, 2012, **2**, 1590.

76. W. Ding, M. Li, R. Dai and Y. Deng, *Tetrahedron: Asymmetry*, 2012, **23**, 1376.

77. S. Wang, J. Wu, G. Xu and L. Yang, *Bioprocess Biosyst. Eng.*, 2012, **35**, 1043.

78. Z. Cabrera and J. M. Palomo, *Tetrahedron: Asymmetry*, 2011, **22**, 2080.

79. H. F. de Castro, G. M. Zanin, F. F. de Moraes and P. Sá-Pereira, Imobilização de enzimas e sua estabilização, in *Enzimas em biotecnologia: produção, aplicações e mercado*, ed. E. P. S. Bon, M. A. Ferrara and M. L. Corvo, Interciência, Rio de Janeiro, 2008, ch. 6, pp. 123–151.

80. J. M. Guisán, Immobilization of Enzymes as the 21st Century Begins, in *Immobilization of Enzymes and Cells*, ed. J. M. Guisan, Humana Press, Totowa, NJ, 2006, vol. 22, pp. 1–13.

81. A. Freeman and M. D. Lilly, *Enzyme Microb. Technol.*, 1998, **23**, 335.

82. J. M. Nelson and E. G. Griffin, *J. Am. Chem. Soc.*, 1916, **38**, 1109.

83. T. Tosa, T. Mori, N. Fuse and I. Chibata, *Enzymologia*, 1966, **31**, 214.

84. A. Mendes, P. C. Oliveira and H. F. de Castro, *J. Mol. Catal. B: Enzym.*, 2012, **78**, 119.

85. Y. Xu, G. Zhou, C. Wu, T. Li and H. Song, *Solid State Sci.*, 2011, **13**, 867.

86. X. Wang, G. Zhou, H. Zhang, S. Du, Y. Xu and C. Wang, *J. Non-Cryst. Solids*, 2011, **357**, 3027.

87. C. Calgaroto, R. P. Scherer, S. Calgaroto, J. V. Oliveira, D. de Oliveira and S. B. C. Pergher, *Appl. Catal., A*, 2011, **394**, 101.

88. K. Hernandez and R. Fernandez-Lafuente, *Enzyme Microb. Technol.*, 2011, **48**, 107.

89. R. C. Rodrigues, Á. Berenguer-Murcia and R. Fernandez-Lafuente, *Adv. Synth. Catal.*, 2011, **353**, 2216.

90. M. Hartmann and D. Jung, *J. Mater. Chem.*, 2010, **20**, 844.

91. S. Lu, J. He and X. Guo, *AIChE J.*, 2010, **56**, 506.

92. B. Zou, Y. Hu, D. Yu, J. Xia, S. Tang, W. Liu and H. Huang, *Biochem. Eng. J.*, 2010, **53**, 150.

93. R. Rufino, F. C. Biaggio, J. C. Santos and H. F. de Castro, *Int. J. Biol. Macromol.*, 2010, **47**, 5.

94. F. He, C.-F. Wang, T. Jiang, B. Han and R.-X. Zhuo, *Biomacromolecules*, 2010, **11**, 3028.

95. V. Caballero, F. M. Bautista, J. M. Campelo, D. Luna, J. M. Marinas, A. A. Romero, J. M. Hidalgo, R. Luque, A. Macario and G. Giordano, *Process Biochem.*, 2009, **44**, 334.

96. L. Lei, Y. Bai, Y. Li, L. Yi, Y. Yang and C. Xia, *J. Magn. Magn. Mater.*, 2009, **321**, 252.

97. J. M. Palomo, R. L. Segura, G. Fernandez-Lorente, R. Fernandez-Lafuente and J. M. Guisán, *Enzyme Microb. Technol.*, 2007, **40**, 704.

98. D. Koszelewski, A. Redzej and R. Ostaszewski, *J. Mol. Catal. B: Enzym.*, 2007, **47**, 51.

99. A. V. Paula, D. Urioste, J. C. Santos and H. F. de Castro, *J. Chem. Technol. Biotechnol.*, 2007, **82**, 281.

100. Y.-X. Bai, Y.-F. Li, Y. Yang and L.-X. Yi, *Process Biochem.*, 2006, **41**, 770.

101. P. D. Desai, A. M. Dave and S. Devi, *Food Chem.*, 2006, **95**, 193.

102. K. Bagi and L. M. Simon, *Biotechnol. Tech.*, 1999, **13**, 309.

103. L. N. de Lima, C. C. Aragon, C. Mateo, J. M. Palomo, R. L. C. Giordano, P. W. Tardioli, J. M. Guisan and G. Fernandez-Lorente, *Process Biochem.*, 2013, **48**, 118.

104. J.-K. Lee, *Bull. Korean Chem. Soc.*, 2012, **33**, 3458.

105. M. Bučko, D. Mislovičová, J. Nahálka, A. Vikartovská, J. Šefčovičová, J. Katrlík, J. Tkáč, P. Gemeiner, I. Lacík, V. Štefuca, M. Polakovič, M. Rosenberg, M. Rebroš, D. Šmogrovičová and J. Švitel, *Chem. Pap.*, 2012, **66**, 983.

106. C. A. Godoy, B. de las Rivas, M. Filice, G. Fernández-Lorente, J. M. Guisan and J. M. Palomo, *Process Biochem.*, 2010, **45**, 534.

107. P. Wang, *Appl. Biochem. Biotechnol.*, 2008, **152**, 343.

108. W. Xu, M. Yan, L. Xu, L. Ding and P. Ouyang, *Enzyme Microb. Technol.*, 2009, **44**, 77.

109. F. X. Malcata, H. R. Reyes, H. S. Garcia, C. G. Hill and C. H. Amundson, *J. Am. Oil Chem. Soc.*, 1990, **67**, 890.

110. M. Nasratun, H. A. Said, A. Noraziah and A. N. Abd Alla, *Am. J. Appl. Sci.*, 2009, **6**, 1653.

111. M.-M. Zheng, Y. Lu, F.-H. Huang, L. Wang, P.-M. Guo, Y.-Q. Feng and Q.-C. Deng, *J. Agric. Food Chem.*, 2013, **61**, 231.

112. U. Bornscheuer, O. Reif, R. Lausch, R. Freitag, T. Scheper, F. Kolisis and U. Menge, *Biochim. Biophys. Acta, Gen. Subj.*, 1994, **1201**, 55.

113. B. B. Brena and F. Batista-Vieira, Immobilization of enzymes, in *Immobilization of Enzymes and Cells*, ed. J. M. Guisan, Humana Press, Totowa, NJ, 2006, vol. 22, p. 15.

114. C. Garcia-Galan, Á. Berenguer-Murcia, R. Fernandez-Lafuente and R. C. Rodrigues, *Adv. Synth. Catal.*, 2011, **353**, 2885.

115. R. A. Sheldon, *Appl. Microbiol. Biotechnol.*, 2011, **92**, 467.

116. D. A. Cowan and R. Fernandez-Lafuente, *Enzyme Microb. Technol.*, 2011, **49**, 326.

117. N. Miletić, A. Nastasović and K. Loos, *Bioresour. Technol.*, 2012, **115**, 126.

118. B. Krajewska, *J. Mol. Catal. B: Enzym.*, 2009, **59**, 22.

119. S. Y. Lau, M. H. Uzir, A. H. Kamaruddin and S. Bhatia, *J. Membr. Sci.*, 2010, **357**, 109.

120. E. Nagy, *Basic equations of the mass transport through a membrane layer*, Elsevier, Amsterdam, Boston, 1st edn, 2012, ch. 9, p. 213.

121. J. M. Palomo, G. Fernández-Lorente, C. Mateo, R. Fernández-Lafuente and J. M. Guisan, *Tetrahedron: Asymmetry*, 2002, **13**, 2375.

122. J. M. Palomo, G. Muñoz, G. Fernández-Lorente, C. Mateo, R. Fernández-Lafuente and J. M. Guisán, *J. Mol. Catal. B: Enzym.*, 2002, **19–20**, 279.

123. A. M. Brzozowski, U. Derewenda, Z. S. Derewenda, G. G. Dodson, D. M. Lawson, J. P. Turkenburg, F. Bjorkling, B. Huge-Jensen, S. A. Patkar and L. Thim, *Nature*, 1991, **351**, 491.

124. R. L. Souza, W. C. Resende, C. E. Barão, G. M. Zanin, H. F. de Castro, O. A. A. Santos, A. T. Fricks, R. T. Figueiredo, Á. S. Lima and C. M. F. Soares, *J. Mol. Catal. B: Enzym.*, 2012, **84**, 152.

125. M. L. Foresti and M. L. Ferreira, *Colloid Surf., A*, 2007, **294**, 147.

126. D.-H. Zhang, L.-X. Yuwen, C. Li and Y.-Q. Li, *Bioresour. Technol.*, 2012, **124**, 233.

127. C. Mateo, J. M. Palomo, G. Fernandez-Lorente, J. M. Guisan and R. Fernandez-Lafuente, *Enzyme Microb. Technol.*, 2007, **40**, 1451.

128. J. M. Palomo, R. L. Segura, G. Fernández-Lorente, M. Pernas, M. L. Rua, J. M. Guisán and R. Fernández-Lafuente, *Biotechnol. Prog.*, 2008, **20**, 630.

129. G. Fernandez-Lorente, Z. Cabrera, C. Godoy, R. Fernandez-Lafuente, J. M. Palomo and J. M. Guisan, *Process Biochem.*, 2008, **43**, 1061.

130. J. M. Palomo, C. Ortiz, G. Fernández-Lorente, M. Fuentes, J. M. Guisán and R. Fernández-Lafuente, *Enzyme Microb. Technol.*, 2005, **36**, 447.

131. J. M. Palomo, M. Fuentes, G. Fernández-Lorente, C. Mateo, J. M. Guisan and R. Fernández-Lafuente, *Biomacromolecules*, 2003, **4**, 1.

132. G. Fernández-Lorente, J. M. Palomo, M. Fuentes, C. Mateo, J. M. Guisan and R. Fernández-Lafuente, *Biotechnol. Bioeng.*, 2003, **82**, 232.

133. U. T. Bornscheuer and R. J. Kazlauskas, *Hydrolases in organic synthesis: regio- and stereoselective biotransformations*, Wiley-VCH, Weinheim, 2nd edn, 2006, ch. 3, pp. 25–42.

134. J. Forde, A. Vakurov, T. D. Gibson, P. Millner, M. Whelehan, I. W. Marison and C. Ó'Fágáin, *J. Mol. Catal. B: Enzym.*, 2010, **66**, 203.

135. E. Caballero, L. Wilson and G. Aroca, *New Biotechnol.*, 2009, **25**, S138.

136. P. Hara, U. Hanefeld and L. T. Kanerva, *J. Mol. Catal. B: Enzym.*, 2008, **50**, 80.

137. Z. J. Dijkstra, R. Merchant and J. T. F. Keurentjes, *J. Supercrit. Fluids*, 2007, **41**, 102.

138. L. Wilson, G. Fernández-Lorente, R. Fernández-Lafuente, A. Illanes, J. M. Guisán and J. M. Palomo, *Enzyme Microb. Technol.*, 2006, **39**, 750.

139. S. G. Burton, D. A. Cowan and J. M. Woodley, *Nat. Biotechnol.*, 2002, **20**, 37.

140. S. Luetz, L. Giver and J. Lalonde, *Biotechnol. Bioeng.*, 2008, **101**, 647.

141. D. Böttcher and U. T. Bornscheuer, *Curr. Opin. Microbiol.*, 2010, **13**, 274.

142. R. Kourist, H. Brundiek and U. T. Bornscheuer, *Eur. J. Lipid Sci. Technol.*, 2010, **112**, 64.

143. G. A. Strohmeier, H. Pichler, O. May and M. Gruber-Khadjawi, *Chem. Rev.*, 2011, **111**, 4141.

144. M. T. Reetz, *Proc. Natl. Acad. Sci. U. S. A.*, 2004, **101**, 5716.

145. M. T. Reetz, S. Prasad, J. D. Carballeira, Y. Gumulya and M. Bocola, *J. Am. Chem. Soc.*, 2010, **132**, 9144.

146. M. T. Reetz, A. Zonta, K. Schimossek, K.-E. Jaeger and K. Liebeton, *Angew. Chem., Int. Ed.*, 1997, **36**, 2830.

147. S. Bartsch, R. Kourist and U. T. Bornscheuer, *Angew. Chem., Int. Ed.*, 2008, **47**, 1508.

148. H. Jochens and U. T. Bornscheuer, *ChemBioChem*, 2010, **11**, 1861.

149. A. G. Sandström, K. Engstrom, J. Nyhlen, A. Kasrayan and J.-E. Backvall, *Protein Eng., Des. Sel.*, 2009, **22**, 413.

150. K. Engström, J. Nyhlén, A. G. Sandström and J.-E. Bäckvall, *J. Am. Chem. Soc.*, 2010, **132**, 7038.

151. A. G. Sandström, Y. Wikmark, K. Engstrom, J. Nyhlen and J.-E. Backvall, *Proc. Natl. Acad. Sci. U. S. A.*, 2011, **109**, 78.

152. Z. Qian, C. J. Fields and S. Lutz, *ChemBioChem*, 2007, **8**, 1989.

153. M. Cancino, P. Bauchart, G. Sandoval, J.-M. Nicaud, I. André, V. Dossat and A. Marty, *Tetrahedron: Asymmetry*, 2008, **19**, 1608.

154. E. Cambon, R. Piamtongkam, F. Bordes, S. Duquesne, I. André and A. Marty, *Biotechnol. Bioeng.*, 2010, **106**, 852.

155. D. M. Guimarães Freire, J. S. de Sousa and E. d'Avila Cavalcanti-Oliveira, in *Biofuels*, Elsevier, 2011, p. 315.

156. A. C. O. Machado, A. A. T. da Silva, C. P. Borges, A. B. C. Simas and D. M. G. Freire, *J. Mol. Catal. B: Enzym.*, 2011, **69**, 42.

157. M. M. Hassan, M. Atiqullah, S. A. Beg and M. H. M. Chowdhury, *Chem. Eng. J. and Biochem. Eng. J.*, 1995, **58**, 275.

158. M. M. Hassan, M. Atiqullah, S. A. Beg and M. H. M. Chowdhury, *J. Chem. Technol. Biotechnol.*, 1996, **66**, 41.

159. C. Y. Kee, M. Hassan and K. B. Ramachandran, *Artif. Cells, Blood Substitutes, Biotechnol.*, 1999, **27**, 393.

160. N. N. Rao, S. Lütz, K. Würges and D. Minör, *Org. Process Res. Dev.*, 2009, **13**, 607.

161. K. Nie, F. Xie, F. Wang and T. Tan, *J. Mol. Catal. B: Enzym.*, 2006, **43**, 142.

162. Y.-H. Chen, Y.-H. Huang, R.-H. Lin and N.-C. Shang, *Bioresour. Technol.*, 2010, **101**, 668.

163. J. H. Lee, S. B. Kim, C. Park, B. Tae, S. O. Han and S. W. Kim, *Appl. Biochem. Biotechnol.*, 2009, **161**, 365.

164. K. Tongboriboon, B. Cheirsilp and A. H-Kittikun, *J. Mol. Catal. B: Enzym.*, 2010, **67**, 52.

165. J. Brask, M. L. Damstrup, P. M. Nielsen, H. C. Holm, J. Maes and W. Greyt, *Appl. Biochem. Biotechnol.*, 2010, **163**, 918.

166. K. Kawakami, Y. Oda and R. Takahashi, *Biotechnol. Biofuels*, 2011, **4**, 42.

167. I. I. Junior, F. K. Sutili, S. G. F. Leite, L. S. de, M. Miranda, I. C. R. Leal and R. O. M. A. de Souza, *J. Mol. Catal. B: Enzym.*, 2011, **72**, 313.

168. I. I. Junior, M. C. Flores, F. K. Sutili, S. G. F. Leite, L. S. de, M. Miranda, I. C. R. Leal and R. O. M. A. de Souza, *Org. Process Res. Dev.*, 2012, **16**, 1098.

169. I. Itabaiana, L. S. de Mariz e Miranda and R. O. M. A. de Souza, *J. Mol. Catal. B: Enzym.*, 2013, **85–86**, 1.

170. E. A. Manoel, K. C. Pais, M. C. Flores, L. S. de M. e Miranda, M. A. Zarur Coelho, A. B. C. Simas, D. M. G. Freire and R. O. M. A. de Souza, *J. Mol. Catal. B: Enzym.*, 2013, **87**, 139.

171. Z. Boros, P. Falus, M. Márkus, D. Weiser, M. Oláh, G. Hornyánszky, J. Nagy and L. Poppe, *J. Mol. Catal. B: Enzym.*, 2013, **85–86**, 119.

172. C. Csajági, G. Szatzker, E. Rita Tőke, L. Ürge, F. Darvas and L. Poppe, *Tetrahedron: Asymmetry*, 2008, **19**, 237.

173. A. Tomin, G. Hornyánszky, K. Kupai, Z. Dorkó, L. Ürge, F. Darvas and L. Poppe, *Process Biochem.*, 2010, **45**, 859.

174. L. Tamborini, D. Romano, A. Pinto, A. Bertolani, F. Molinari and P. Conti, *J. Mol. Catal. B: Enzym.*, 2012, **84**, 78.

175. M. D. Truppo, D. J. Pollard, J. C. Moore and P. N. Devine, *Chem. Eng. Sci.*, 2008, **63**, 122.

176. U. Kragl, M. Eckstein and N. Kaftzik, *Curr. Opin. Biotechnol.*, 2002, **13**, 565.

177. R. A. Sheldon, R. Lau, M. J. Sorgedrager, F. Van Rantwijk and K. R. Seddon, *Green Chem.*, 2002, **4**, 147.

178. S. Park and R. Kazlauskas, *Curr. Opin. Biotechnol.*, 2003, **14**, 432.

179. Z. yang and W. Pan, *Enzyme Microb. Technol.*, 2005, **37**, 19.

180. Y. H. Moon, S. M. Lee, S. H. Ha and Y. Koo, *Korean J. Chem. Eng.*, 2006, **23**, 247.

181. P. Lozano, *Green Chem.*, 2010, **12**, 555.

182. U. Kragl, M. Eckstein and N. Kaftzik, Biocatalytic Reactions in Ionic Liquids, *Ionic Liquids in Synthesis*, ed. P. Wasserscheid and T. Welton, Wiley-VCH, Weinheim, 2002, p. 336, reprinted in 2003.

183. J. S. Wilkes, *J. Mol. Catal. A: Chem.*, 2004, **214**, 11.

184. S. A. Forsyth, J. M. Pringle and D. R. Mac-Farlane, *Aust. J. Chem.*, 2004, **57**, 113.

185. J. L. Anthony, J. F. Brennecke, J. D. Holbrey, E. J. Maginn, R. A. Mantz, R. D. Rogers, P. C. Trulove, A. E. Visser and T. Welton, Physicochemical Properties of Ionic Liquids, in *Ionic Liquids in Synthesis*, ed. P. Wasserscheid and T. Welton, Wiley-VCH, Weinheim, Germany, 2002, pp. 41–118, reprinted in 2003.

186. G. A. Baker, S. Baker, S. Pandey and F. V. Bright, *Analyst*, 2005, **130**, 800.

187. M. Naushada, Z. A. ALOthmana, A. B. Khanb and M. Ali, *Macromolecules*, 2012, **51**, 555.

188. J.-C. Plaquevent, J. Levillain, F. Guillen, C. Malhiac and A.-C. Gaumont, *Chem. Rev.*, 2008, **108**, 5035.

189. T. Welton, *Chem. Rev.*, 1999, **99**, 2071.

190. S. H. Schöfer, N. Kaftzik, P. Wasserscheid and U. Kragl, *Chem. Commun.*, 2001, 425.

191. M. Singh, R. S. Singh and U. C. Banerjee, *J. Mol. Catal. B: Enzym.*, 2009, **56**, 294.

192. M. Sureshkumar and C.-K. Lee, *J. Mol. Catal. B: Enzym.*, 2009, **60**, 1.

193. H. Monhemi and M. R. Housaindokht, *J. Supercrit. Fluids*, 2012, **72**, 161.

194. R. M. Lau, F. van-Rantwijk, K. R. Seddon and R. A. Sheldon, *Org. Lett.*, 2000, **2**, 4189.

195. M. Erbeldinger, A. J. Mesiano and A. Russell, *J. Biotecnol. Prog.*, 2000, **16**, 1129.

196. M. Persson and U. T. Bornscheuer, *J. Mol. Catal. B: Enzym.*, 2003, **22**, 21.

197. K. W. Kim, B. Song, M. Y. Choi and M. J. Kim, *Org. Lett.*, 2001, **3**, 1507.

198. N. J. Roberts, A. Seago, J. S. Carey, R. Freer, C. Preston and G. J. Lye, *Green Chem.*, 2004, **6**, 475.

199. R. J. Atkins, A. Banks, R. K. Bellingham, G. F. Breen, J. S. Carey, S. K. Etridge, J. F. Hates, N. Hussain, D. O. Morgan, P. Oxley, S. C. Passey, T. C. Walsgrove and A. S. Wells, *Org. Process Res. Dev.*, 2003, **7**, 663.

200. F. Van Rantwijk, A. C. Kock-van Dalen and R. A. Sheldon, Stability and Stabilisation of Enzymes, in *Progress in Biotechnology*, ed. A. Ballasteros,

F. J. Plou, P. Iborra and Halling, Elsevier Science, New York, 1998, vol. 15, pp. 447–452.

201. J. K. Lee and M. J. Kim, *J. Mol. Cat. B: Enzym.*, 2011, **68**, 275.
202. P. U. Naik, S. J. Nara, J. R. Harjani and M. M. Salunkhe, *J. Mol. Catal. B: Enzym.*, 2007, **44**, 93.
203. S. J. Nara, S. S. Mohile, J. R. Harjani, P. U. Naik and M. M. Salunkhe, *J. Mol. Catal. B: Enzym.*, 2004, **28**, 39.
204. J. Y. Xin, Y. J. Zhao, G. L. Zhao, Y. Zheng, X. S. Ma, C. G. Xia and S. B. Li, *Biocatal. Biotransform.*, 2005, **23**, 353.
205. S. H. Lee, T. T. N. Doan, S. H. Ha and Y. M. Koo, *J. Mol. Catal. B: Enzym.*, 2007, **45**, 57.
206. S. H. Lee, S. H. Ha, N. M. Hiep, W. J. Chang and Y. M. Koo, *J. Biotechnol.*, 2008, **133**, 486.
207. J. A. Berberich, J. L. Kaar and A. J. Russell, *Biotechnol. Prog.*, 2003, **19**, 1029.
208. H. Zhou, J. Chen, L. Ye, H. Lin and Y. Yuan, *Bioresour. Technol.*, 2011, **102**, 5562.

Redox Biotechnological Processes Applied to Fine Chemicals

J. AUGUSTO R. RODRIGUES*[a], PAULO J. S. MORAN*[a], BRUNA Z. COSTA[a], AND ANITA J. MARSAIOLI*[a]

[a]University of Campinas, Chemistry Institute, Department of Organic Chemistry, 13083-970 Campinas, SP, Brazil
*E-mail: jaugusto@iqm.unicamp.br, moran@iqm.unicamp.br, anita@iqm.unicamp.br

10.1 Introduction

In the scenario of environmentally friendly redox processes, biocatalysis plays an important role by employing enzymes and whole cells as catalysts to perform chemical reactions that are considered green, *i.e.*, free from toxic solvents or reagents and generate easy-to-dispose waste.[1,2]

It is well known that enzymes are remarkable catalysts, capable of accepting a wide variety of complex molecules as substrates and of catalyzing reactions with high enantio- and regioselectivity. Therefore, biocatalysts can be applied to redox processes of multifunctional molecules, avoiding the time consuming protection and deprotection steps usually necessary in asymmetric organic synthesis.[3]

RSC Green Chemistry No. 45
White Biotechnology for Sustainable Chemistry
Edited by Maria Alice Z. Coelho and Bernardo D. Ribeiro
© The Royal Society of Chemistry 2016
Published by the Royal Society of Chemistry, www.rsc.org

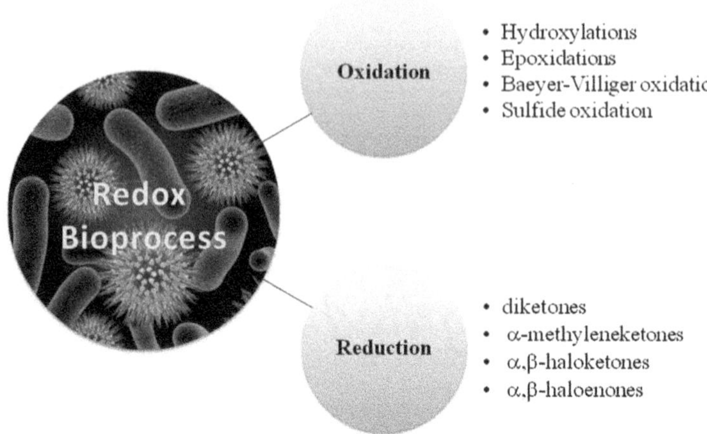

Figure 10.1   Redox bioprocesses discussed herein.

However, application of redox biocatalysts requires screening suitable enzymes or whole cells to obtain the desired product in good yields and with high enantiomeric excess. Whole cell processes are low cost and give autonomous regeneration of cofactors, however side reactions, and low tolerance for substrates and organic solvents may not favor these methods.[4] Nowadays, state-of-the-art redox processes use engineered whole cells which express only the desired enzyme plus a cofactor or isolated enzymes associated with the cofactor regenerating processes, thus giving high specificity, high productivity and cost effective products.

In the present chapter, we discuss different redox reactions with enzymes from bacteria, yeast and fungi (whole cells and isolated) that were selected for their efficient regio- and enantioselective conversions. We will therefore summarize some of the best redox processes from the Biocatalytic Group at the Chemistry Institute/UNICAMP (Figure 10.1).

10.2   Redox Enzymes

The redox enzymes are known as oxidoreductases and, according to their mechanism of action, are classified into four major groups: oxygenases, peroxidases, oxidases and dehydrogenases (Figure 10.2). These enzymes represent the second major class with applications in biocatalysis (25%).

Oxygenases, oxidases and peroxidases catalyze irreversible oxidation reactions due to their high reaction enthalpies mainly attributed to the exothermic reduction of O_2 or hydrogen peroxide. Oxidases use O_2 as electron acceptor, while oxygenases incorporate one or both oxygen atoms into their substrate. Moreover, dehydrogenases catalyze reversible reactions that can be used in both oxidative and reductive bioprocesses.[5]

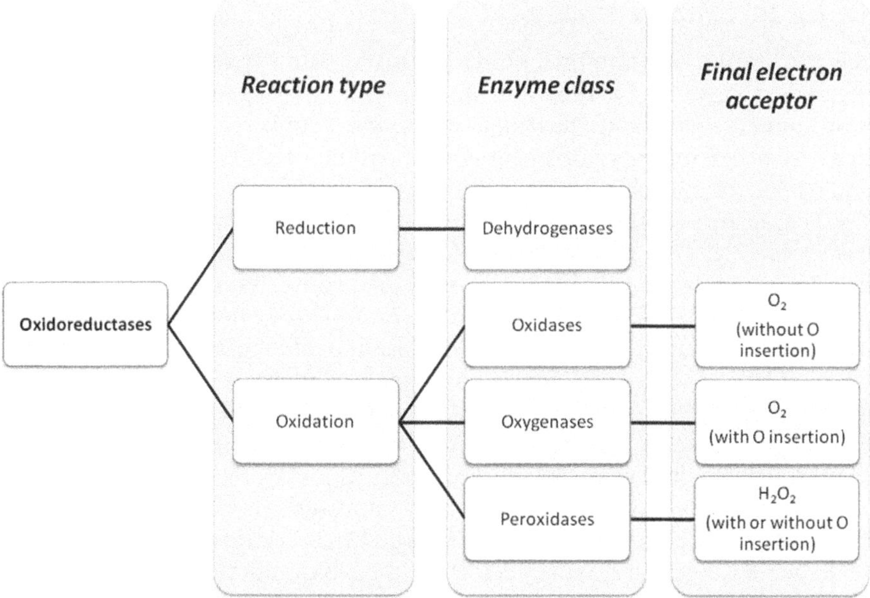

Figure 10.2 Mechanism-based oxidoreductase classification.

Typically, these electron transfer reactions involve additional organic molecules or transition metals, known as cofactors or coenzymes, to assist enzymatic catalysis. Exceptions include: peroxidases, which couple the reduction of hydrogen peroxide to water with two electron transfers and oxidation of the substrate; oxidases, which couple the reduction of O_2 to hydrogen peroxide or water by two- or four-electron transfers and oxidation of the substrate, and dioxygenases, which couple the reduction and incorporation of O_2 involving a four-electron transfer and the oxidation of an activated carbon scaffold.

This chapter focuses on redox biocatalysis applied to fine chemistry with no cofactor addition or regeneration systems.

10.3   Oxidation Reactions

10.3.1   Hydroxylation

Enantioselective microbial hydroxylation represents a green, simple and direct way to prepare chiral alcohols from cycloalkanes, carboxylic acids, amines and alcohols using whole cells or isolated enzymes.[6] However, on a preparative scale, there are almost no available methods for efficient hydrocarbon oxidation.[7] Problems with substrate acceptance, undesired side reactions, selectivity and hydroxylation site prediction hamper their general synthetic application. However, benzylic and allylic moieties undergo chemo and/or bio-oxidation.

10.3.1.1 Benzylic Moiety

Selection of the best strains for indan **1** and tetralin **2** hydroxylation involved the screening of 15 bacteria and fungi.[8] Parameter optimization for maximum biomass (in shaker flasks), and improved conversion yields and selectivity involved microorganism growth in different cultivation media at 25 and 30 °C, followed by substrate addition in buffer solution (pH 5.0 to 8.0). The best results for the production of **3** and **4** were achieved with *Mortierella isabellina* CCT 3498, *Mortierella ramanniana* CCT 4428 and *Beauveria bassiana* CCT 3161 (from ATCC 7159). *M. isabellina* converted 78% of **1** into (1*R*)-**3** (64% yield, 86% ee) in a 2-day-incubation, and 52% of **2** into (1*R*)-**4** (38% yield, 92% ee) in a 4-day-incubation. Over-oxidation of alcohols **3** and **4** during the reaction or after extended periods produced 1-indanone (**5**) and 1-tetralone (**6**), respectively (Scheme 10.1). The fungus was cultivated in potato/dextrose/carrot broth media and preparative scale (1 mmol) reactions were carried out in pH 6.0 potassium phosphate buffer at 30 °C.

The biohydroxylation described above represents a simple enantioselective method to prepare (*R*)-1-indanol (**3**) and (*R*)-1-tetralol (**4**) on a preparative scale, which is an alternative to low yielding bioreduction of indanone (**5**) and tetralone (**6**).

10.3.1.2 Allylic Moiety

Cis-jasmone (**7**) and derivatives are key intermediates in the production of prostaglandins (PGs) or prostaglandin-type compounds.[9] Currently, the synthesis and isolation of PG derivatives has attracted attention in various areas of knowledge, especially in medical and biological applications, due to their anti-inflammatory and anti-tumor activities.[10]

High-throughput screening with fluorescent probes selected several strains expressing monooxygenases. The best strain, *Trichosporon cutaneum* CCT 1903, was used to oxidize **7** in a 48 h process to yield 4-hydroxyjasmone (**8**), 7,8-epoxyjasmone (**9**) and 7,8-dihydroxyjasmone (**10**) (Scheme 10.2).

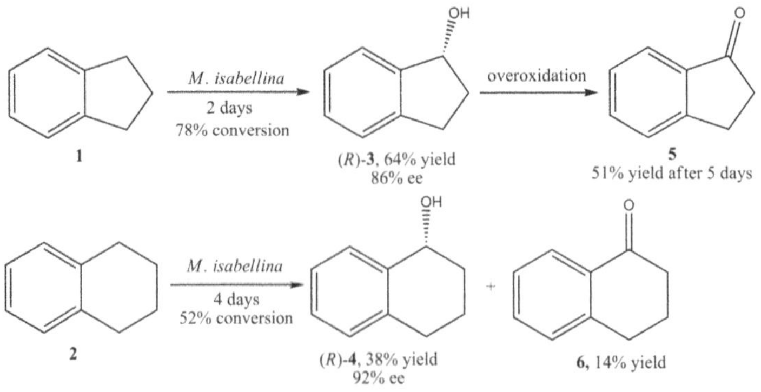

Scheme 10.1 Biotransformation of indan **1** and tetralin **2**.

Mosher's methodology[11] was applied to determine the absolute configuration of the (*S*)-4-hydroxyjasmone (**8**) asymmetric center. The enantiomeric excess was 86% (chiral GC), therefore hydroxylation of *cis*-jasmone (**7**) was regio- and stereoselective.[12]

10.3.2 Epoxidation

Epoxides are key intermediates directly associated with the biological activity of many synthetic drugs and natural products. They are highly appreciated in the pharmaceutical industry for their versatility, as they react with nucleophiles, electrophiles, acids and bases. These intermediates are commonly used in the syntheses of β-blockers, anti-obesity and anti-HIV drugs, among others.[13] Consequently there is a high demand for methods to access epoxides or vicinal diols in high enantiomeric excess by applying either chemo-[14] or biocatalytic[15] methods.

Trichosporum cutaneum CCT 1903 whole cells produced 7,8-epoxyjasmone (**9**) (Scheme 10.2) by enantioselective epoxidation of *cis*-jasmone (**7**). However, a major drawback is the simultaneous production of the alcohol (**8**) and diol (**10**), requiring process optimization to yield pure products.

Enzymatic epoxidations of carbon–carbon double bonds are stereospecific and the product stereochemistry is related to the *Z*- or *E*-alkene configuration. Notwithstanding this knowledge, the *cis*-stereochemistry of the 7,8-epoxyjasmone (**9**) was confirmed by ^{1}H NMR spectroscopy based on the presence of a double doublet at δ 3.01 (H-7) and a triple doublet at δ 2.84 (H-8) with a coupling constant of 4.5 Hz.[16] Comparison with a synthetic standard further confirmed the *cis* stereochemistry of the epoxide ring. Thus, *T. cutaneum* monooxygenase stereospecifically oxidizes *cis*-jasmone to *cis*-epoxyjasmone (over 99% de).

The absolute configuration of *cis*-(7*S*,8*R*)-epoxyjasmone (**9**, 92% ee) was assessed by ^{1}H NMR analyses of **12**. This derivative was obtained by a regioselective ring opening of **9** with trimethyl *ortho*-acetate and derivatizing the resulting methoxy-alcohol (**11**) with Mosher's acid and 1,3-dicyclohexylcarbodiimide (Scheme 10.3).[17]

The odoriferous properties of *cis*-jasmone and derivatives are of high value to the flavor and fragrance industries owing to their floral and woody notes.[18] Thus, new *cis*-jasmone derivatives like alcohols and epoxides are appreciated.

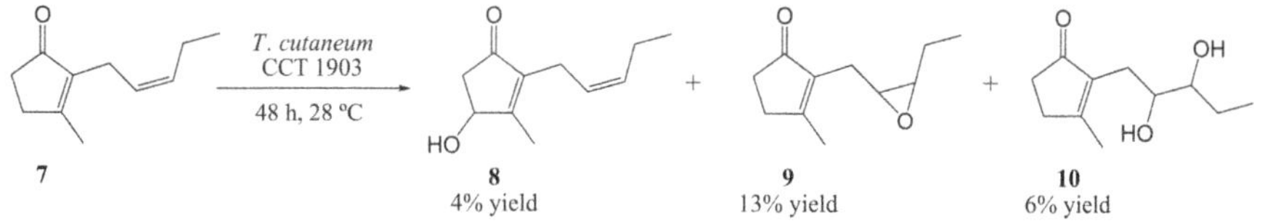

Scheme 10.2 *Cis*-jasmone (**7**) biotransformation products **8**, **9** and **10** using *Trichosporum cutaneum* CCT 1903 whole cells.

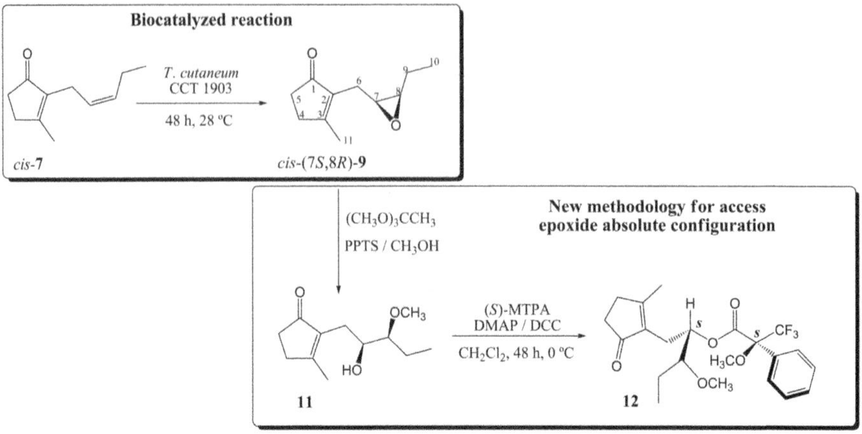

Scheme 10.3 diagram showing: "Biocatalyzed reaction" box with *cis*-7 converted by *T. cutaneum* CCT 1903, 48 h, 28 °C to *cis*-(7*S*,8*R*)-9. Below, "New methodology for access epoxide absolute configuration" box with reagents (CH₃O)₃CCH₃, PPTS / CH₃OH giving compound 11, then (*S*)-MTPA, DMAP / DCC, CH₂Cl₂, 48 h, 0 °C giving compound 12.

Scheme 10.3 *Cis*-jasmone epoxidation by *T. cutaneum* CCT 1903 whole cells and the new methodology to access the epoxide absolute configuration.

10.3.3   Baeyer–Villiger Oxidation

Baeyer–Villiger reactions involve carbon–carbon bond cleavage (adjacent to a carbonyl group) and the insertion of an oxygen atom. This is an important tool to prepare esters and lactones, which are essential building blocks for the synthesis of biologically active compounds.[19]

Chemically, the Baeyer–Villiger reaction is carried out using organic peracids, like *meta*-chloroperbenzoic acid (*m*-CPBA), as oxygen donor reagents (Scheme 10.4), which has limited industrial application.

Biocatalytic processes using isolated enzymes or whole cells are green alternatives to this type of reaction. Among the ketone oxidizing enzymes, Baeyer–Villiger monooxygenases (BVMO) play an important role. These enzymes require cofactors, and whole cell systems are the best choice either with wild or engineered organisms.

Whole cell screening of BVMO-producing microorganisms based on fluorogenic probes (**13–16**, Scheme 10.5) was optimized in our Biocatalysis Group in 2004 and was applied to Brazilian Culture Collection strains and to microbiota from human skin,[20] Brazilian petroleum[21] and copper mine drainage.[22]

The best BVMO-producing microorganisms were reinvestigated using milliliter assays with different ketones. Our best results were oxidation of the cyclic ketones **22** and **23** (Scheme 10.6) with fungi from the Culture Collection of André Tosello Tropical Foundation (Table 10.1).[23]

Baeyer–Villiger oxidation of 3-hexyl-cyclobutanone (**22**) was preferred when compared to 4-methyl-cyclohexanone (**23**), which is expected, as the five-member ring product (**24**) is more stable than the seven-member ring lactone (**26**) (Scheme 10.6). Cyclobutanone-MO was expressed by most evaluated fungi except by *Aspergillus oryzae* CCT 0975 and *Geotrichum candidum*

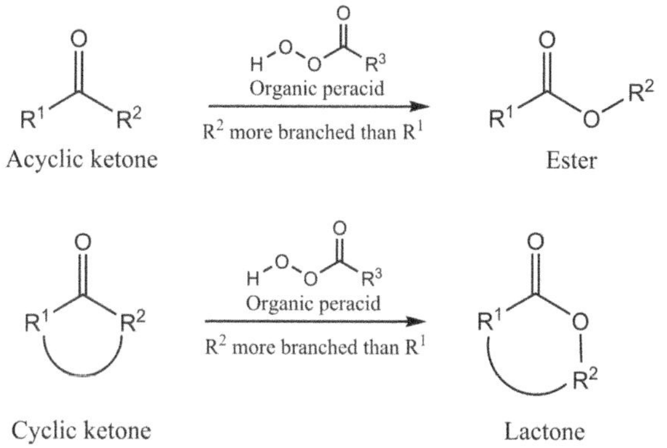

Scheme 10.4 General Baeyer–Villiger reaction using organic peracids.

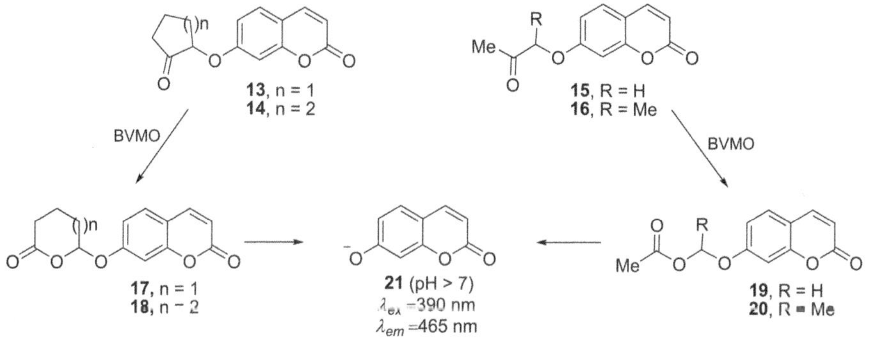

Scheme 10.5 BVMO screening assay based on fluorogenic probes (**13–16**) derived from umbelliferone (**21**).

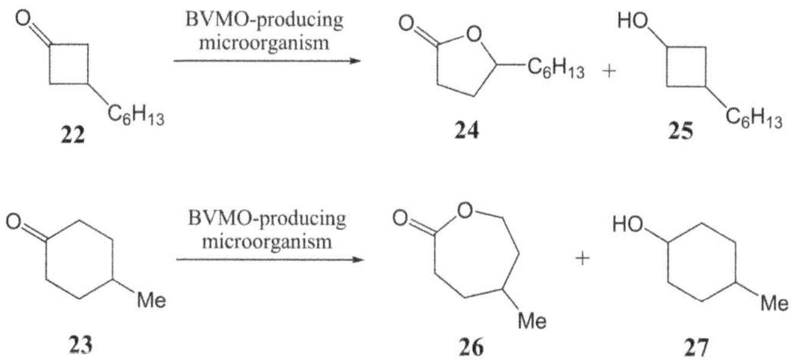

Scheme 10.6 Cyclic ketones used as substrates in testing BVMO-producing microorganisms from the Culture Collection of André Tosello Tropical Foundation.

Table 10.1 Biocatalysis reactions with ketones **22** and **23** and their products.[a]

Microorganisms	Products			
	24	**25**	**26**	**27**
Aspergillus niger CCT 5559	80	—	1	23
Aspergillus oryzae CCT 0975	—	—	54	23
Cunninghamella echinulata CCT 4259	80	—	—	81
Curvularia eragrostidis CCT 5634	20	10	2	75
Curvularia lunata CCT 5628	44	26	—	51
Curvularia lunata CCT 5629	80	—	3	94
Curvularia pallescens CCT 5654	54	—	1	97
Geotrichum candidum CCT 1205	—	—	100	—

[a]Reaction conditions: ketones **22–23** (approximately 20 mg each) and wet fungal biomass (2.0 g) were mixed in Erlenmeyer flasks (125 mL) containing 30 mL of phosphate buffer solution pH 7.0. The mixture was stirred on a rotary shaker (28 °C, 140 rpm) and monitored by GC-MS.

CCT 1205, which only expressed cyclohexanone-MO. All other evaluated fungi reduced **23** to **27**. Ketone reduction usually competes with Baeyer–Villiger oxidation (Table 10.1).

10.3.4 Sulfide Oxidation

In recent years, enantioselective oxidation of prochiral sulfides by biocatalytic methods has drawn considerable interest, with the goal of producing chiral sulfoxides, which have broad applicability in chemistry and biochemistry. They occur in many natural products (precursors of aromas and flavors),[24] compounds with biological activity (potential drugs) and metabolites,[25] and are widely used as chiral auxiliaries and asymmetric starting materials in synthesis.[26]

Enzymatic oxidation of sulfide can be performed by several enzymes, *e.g.*, cyclohexanone monooxygenase, chloroperoxidase, cytochrome P450, naphthalene and toluene dioxygenase, among others.[27] All of these enzymes require cofactors and sulfide bio-oxidations are usually performed with whole cell systems.

In this context, we performed the oxidation of ethyl phenyl sulfide (**28**, Scheme 10.7) using several different microorganisms.[28]

Phanerochaete chrysosporium CCT 1999, *Emericella nidulans* CCT 3119 and *Aspergillus terreus* CCT 3320 (isolated from the Atlantic Rain Forest, Brazil) were among the best microorganisms for the oxidation of **28** (Table 10.2).

The stereochemical outcomes of the reactions with *E. nidulans* CCT 3119 and *A. terreus* CCT 3320 were unexpected, with low initial and high final enantiomeric excess. These results were rationalized as a two-step process with moderate enantiotopic differentiation in the first step, followed by an efficient kinetic resolution of the resulting sulfoxides (Scheme 10.8). The efficiency of the kinetic resolution was confirmed by using a racemic mixture of (±)-sulfoxide **29**. The mixture was converted into sulfone **30** (50%), leaving enantiomerically pure (*S*)-**29**.

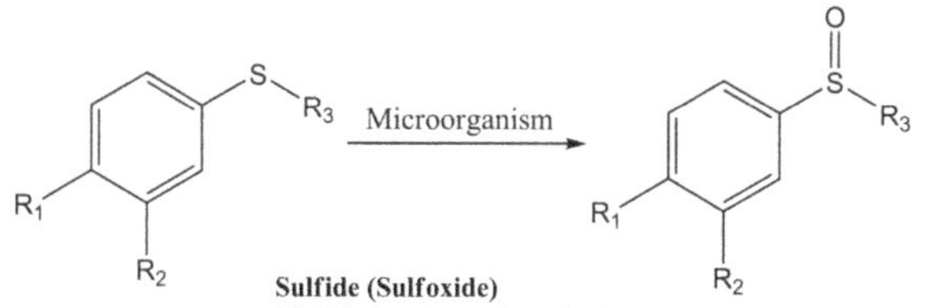

Sulfide (Sulfoxide)
28 (29): R_1=R_2=H, R_3=Ethyl
31 (36): R_1=R_2=H, R_3=Methyl
32 (37): R_1=Methyl, R_2=H, R_3=Ethyl
33 (38): R_1=Methyl, R_2=H, R_3=Methyl
34 (39): R_1=H, R_2=Cl, R_3=Ethyl
35 (40): R_1=H, R_2=Cl, R_3=Methyl

Scheme 10.7 Sulfides used in biotransformations with microorganisms.

Table 10.2 Activity monitoring of sulfide **28** oxidation with *A. terreus* CCT 3320, *E. nidulans* CCT 3119 and *P. chrysosporium* CCT 1999.[a]

	Microorganisms					
Parameters	*A. terreus*		*E. nidulans*		*P. chrysosporium*	
t (h)	48	96	48	96	48	96
Conversion (%)	10	100	75	100	74	100
Sulfoxide yield (%)	10	76	60	43	74	70
Sulfone yield (%)	—	24	15	57	—	30
ee[b] (*S*)-**1a**	26	>99	11	86	44	45

[a]Reaction conditions: the fungus was grown at 28 °C in malt extract (2 g L^{-1}, 72 h); cells were harvested by filtration, washed with phosphate buffer (pH 7.0, 0.1 mol L^{-1}) and added to reaction flasks (0.5 g wet weight) containing phosphate buffer (pH 7.0, 0.1 mol L^{-1}, 25 mL) and substrate **28** (20 μL). The mixture was shaken at 28 °C and the reaction was monitored by chiral GC.
[b]Enantiomeric excess.

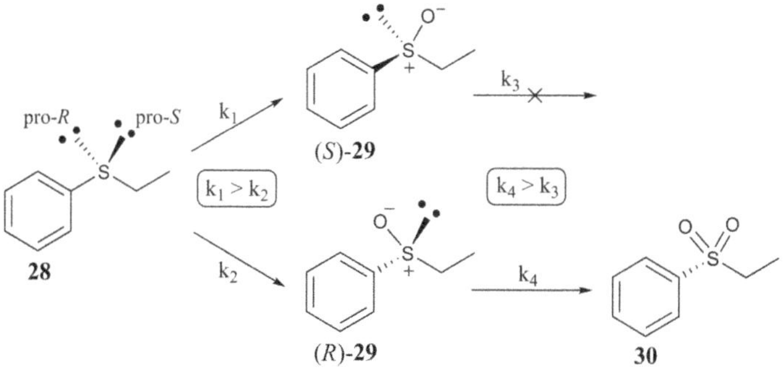

Scheme 10.8 Kinetic path proposed for the oxidation of sulfide **28** using *A. terreus* CCT 3320.

Table 10.3 Activity monitoring of sulfide **31–35** oxidation by *A. terreus* CCT 3320 over 96 h.[a]

Parameters (%)	Sulfides				
	31	**32**	**33**	**34**	**35**
Conversion	96	98	92	100	82
Sulfoxide yield	24	66	24	70	71
Sulfone yield	72	32	68	30	15
ee[b]	>95	>95	>98	>98	17

[a]Reaction conditions: the fungus was grown at 28 °C in malt extract (2 g L^{-1}, 72 h), cells were harvested by filtration, washed with phosphate buffer (pH 7.0, 0.1 mol L^{-1}) and added to reaction flasks (0.5 g wet weight) containing phosphate buffer (pH 7.0, 0.1 mol L^{-1}, 25 mL) and substrate **31–35** (20 µL). The mixture was shaken at 28 °C and the reaction was monitored by chiral GC.
[b]Enantiomeric excess.

The oxidation activity of *A. terreus* CCT 3320 was monitored using sulfides **31–35** (Scheme 10.7). The results in Table 10.3 clearly indicate that it is appropriate for the sulfide oxidation of those substrates and recommended for the production of compounds **36, 37, 38** and **39** with high enantiomeric excesses,[29] but strict control of the sulfone production is required.

In order to improve the biocatalytic process, *A. terreus* CCT 3320 cells were immobilized on two supports, chrysotile and cellulose–TiO$_2$. The immobilized cells showed similar biocatalytic behavior in the conversion rate and in the sulfoxide enantiomeric excess. Scanning electron micrographs (SEM) show that the cells are intertwined with the fibers of both supports, allowing fast separation from the reaction media and easy biocatalyst reuse. Supported cells stored for at least 3 months showed no loss of activity.[30]

10.3.5   Lipase-Mediated Oxidation

As previously described, biocatalysis has expanded rapidly in recent decades and one of the new frontiers deals with the discovery and development of enzymes with broad specificity, known as promiscuous.[31] Although often undervalued, this absence or low catalytic specificity plays a key role in the evolutionary aspects of enzymes[32] and indicates that active sites are not as rigid as afore-mentioned. Lipases are good examples of promiscuous performances, catalyzing several organic reactions in aqueous and organic media.[33]

Focusing on redox reactions, lipases oxidize ketones and alkenes through *in situ* generation of peracids by perhydrolysis of carboxylic acids in the presence of aqueous H$_2$O$_2$ (Figure 10.3).[34]

The application of lipases to the perhydrolysis of carboxylic acids in the presence of H$_2$O$_2$ was first reported in 1990 by Björkling and co-workers.[35] They also proposed the elegant use of *in situ* generated peracids for the oxidation of alkenes, ketones and sulfides.[36] Nowadays, this versatility is reported as promiscuous enzymatic activity and is mainly applied to Baeyer–Villiger reactions and epoxidations of alkenes.[37]

The mechanism of this catalytic process is not yet fully understood, but the most accepted proposal says that the enzymatic reaction is responsible for peracid formation only, while the oxidation reaction occurs outside the active site.[38]

Recently, we have evaluated nine commercial lipases for the oxidation of three cyclic ketones: 2-pentylcyclopentanone (**41**), 2-heptylcyclopentanone (**42**) and *cis*-jasmone (**43**) (Scheme 10.9, Table 10.4).

The results in Table 10.4 show that lipases from different sources exhibit distinct activities for conversion of the three evaluated cyclic ketones. In addition, it was observed that the lipase from *Candida antarctica* was best for

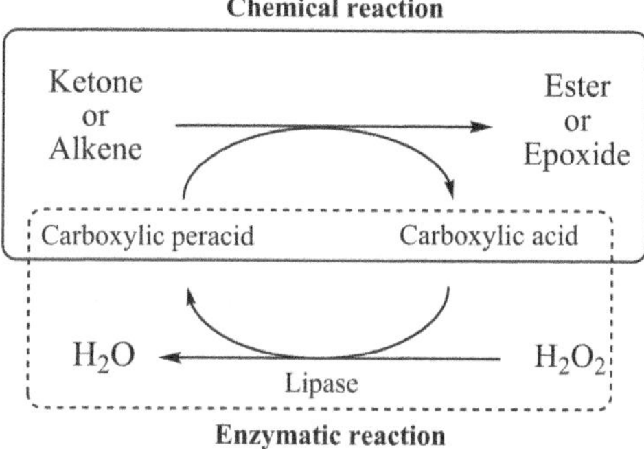

Figure 10.3 Ketone and alkene oxidations by *in situ* generation of peracids catalyzed by lipases.

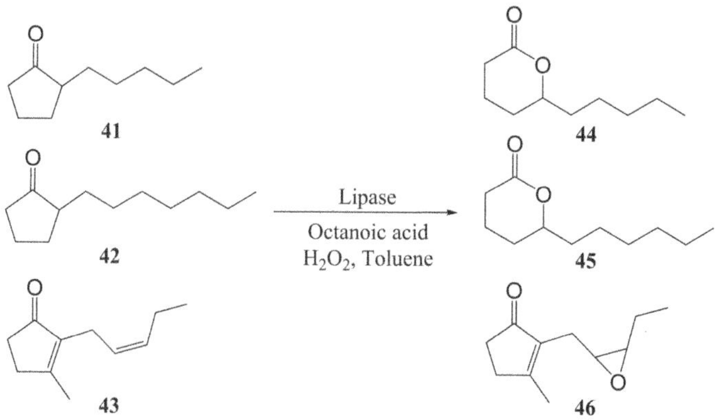

Scheme 10.9 Oxidation reactions catalyzed by lipases through *in situ* generation of peracids.

all oxidations. Participation of the enzyme was confirmed by the absence of products in the control reaction (absence of enzyme).

Oxidation of cyclic ketones **41** and **42** led to formation of the respective lactones **44** and **45**. We observed the same regioselectivity of the chemical reaction where the oxygen atom is inserted into the more substituted portion of the substrate. However, the oxidation product of ketone **43** was epoxide **46**, since epoxidation of the double bond is favored over oxidation of the cyclic ketone (Scheme 10.9).

Despite proven involvement of enzymes, chiral GC-FID analysis showed that these reactions are not enantioselective. This type of reaction is classified as chemo-enzymatic, since lipases only participate in perhydrolysis of the carboxylic acid, while the oxidation step is spontaneously carried out by free peracid.

Thus, considering that lipases exhibit a unique tertiary structure that exposes the catalytic active site only in the presence of a lipid phase, such as an emulsion or a micellar solution (a phenomenon known as interfacial activation),[39] we used several suitable techniques of ^{1}H-NMR to study this promiscuous reaction at the molecular level. From these analyses, we discovered and proved that reverse micelles are involved and arise as nano-heterogeneous systems where these lipase-mediated reactions may be conducted. The polar cores of these aggregates are nanometer-sized and able to host water and macromolecules, protecting the latter from direct contact with the organic layer; therefore, retaining their biological activity. Due to these features, there is growing interest in reverse micelles for future applications in the biotechnological industry, such as enzymatic catalysis in organic media, where substrates are hydrophobic and low water contents are desired.[40]

Table 10.4 Conversions (%) determined by GC-MS for oxidation of cyclic ketones **41–43** catalyzed by lipases.[a]

	Ketones		
Lipases	**41**	**42**	**43**
Aspergillus sp.	8	17	23
Candida antarctica	52	>99	73
Candida cylindracea	32	13	26
Mucor miehei	19	2	33
Pseudomonas cepacia	34	40	19
Pseudomonas fluorescens	4	7	3
Rhizopus arrhizus	14	21	13
Rhizopus niveus	7	6	5
Porcine pancreas	4	25	4

[a]Reaction conditions: all reactions were performed in a biphasic system in the presence of ketones (10 μL), octanoic acid (100 μL), 30% H_2O_2 v/v (1 mL) and lipases (5 mg) in toluene (10 mL). The reactions were maintained under stirring at 200 rpm and 40 °C. Aliquots were removed every 24 hours, washed with $NaHCO_3$ saturated solution, dried over anhydrous $MgSO_4$, derivatized with diazomethane and analyzed by GC-MS.

10.4 Reduction Reactions

10.4.1 Reduction of Diketones

It is well known that the reduction of 1-phenylpropane-1,2-dione **47** mediated by baker's yeast gives (1R,2S)-1-phenylpropanediol **50** in excellent yield and ee,[41] through a mechanism involving consecutive reactions where both C=O bonds are reduced.[42] As a consequence, two interesting reaction intermediates were detected (Scheme 10.10).

There are different biocatalytic approaches for the synthesis of enantiomerically enriched intermediate (R)-**49**, known as **L-PAC**, that are of particular value for the pharmaceutical industry and as fine chemicals because of their utility as building blocks for the production of larger molecules.[43] One of these approaches is a whole cell redox process catalyzed by several microorganisms, including the bio-oxidation of 1,2-diols. For this approach, a protocol of bio-oxidation of (1R,2S)-**50** mediated by whole cells of *S. cerevisiae* was developed, giving (S)-**48** in high ee in aerobic conditions, and also a protocol of kinetic resolution of (±)-*anti*-**50** was developed to obtain (1S,2R)-**50** in high ee using resting cells of microorganisms, among then *Geotrichum candidum* (Scheme 10.11).[44]

Recently, we reported a biocatalytic synthesis of (1S,2R)-1,2-indandiol (**53**) from 1,2-indanone (**51**) with high stereoselectivity, using resting whole cells of the yeast *Trichosporon cutaneum* (Scheme 10.12).[45] Later, we returned

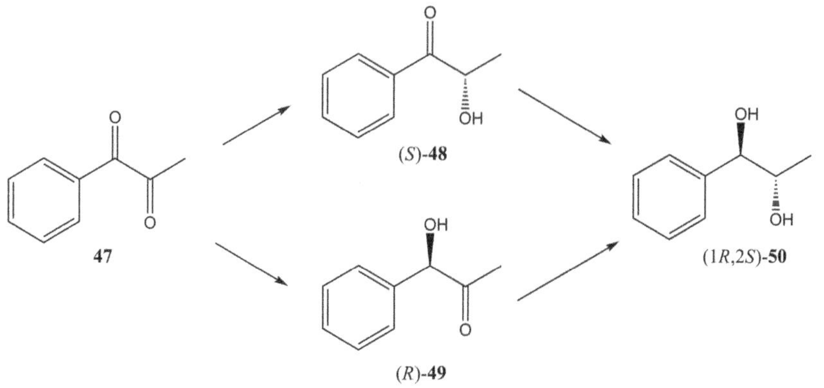

Scheme 10.10 Bioreduction of 1-phenylpropane-1,2-dione **47**.

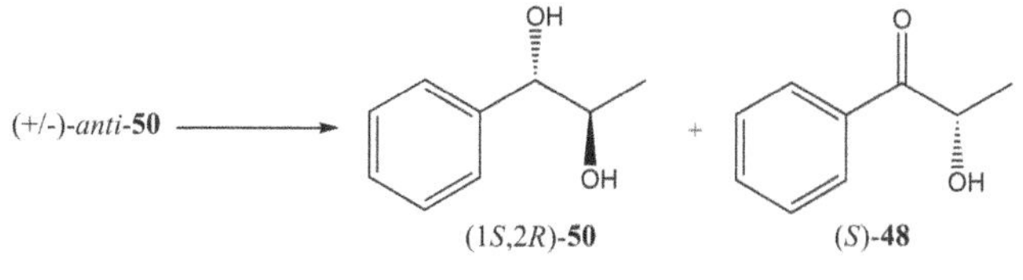

Scheme 10.11 Kinetic resolution of 1-phenylpropanediol **50**.

to this reaction to better understand what was happening during the biotransformation. Monitoring the reaction profile by chiral GC analysis disclosed that the biocatalytic reduction of 1,2-indandione (**51**) was not a simple one-step stereoselective reduction. In fact, it was observed that the (±)-2-hydroxy-1-indanone (**52**) was being formed as a racemate. Therefore, the results indicate that the reactive enantiomer is being depleted by stereoselective reduction to give **53** and the equilibrium (*R*)-**52**/(*S*)-**52** concentrations are constantly re-adjusted until complete conversion of the substrate occurs. In the proposed mechanism, the unreactive enantiomer (*S*)-**52** is isomerized to (*R*)-**52** in a concomitant step. As a result, the clean dynamic kinetic resolution (DKR) of racemic 2-hydroxy-1-indanone (**52**) afforded enantiomerically pure (1*S*,2*R*)-1,2-indandiol (**53**) in high yield (90%) and with an excellent ee (>99%).[13]

The first step to understand the biotransformation of (±)-**52** by *T. cutaneum* was to find the origin of the stereoinversion of the starting material observed in the proposed DKR process. That study showed no isomerization of the pure substrates (*R*)-**52** or (*S*)-**52**, which indicates that the DKR observed in the biotransformation of dione **51** by *T. cutaneum* is enzymatic. We observed that (*R*)-**52** was biotransformed by *T. cutaneum* in 22 hours in a clean fashion (no by-products like indanone **51** were detected) giving (1*S*,2*R*)-diol **53** as the sole product in both high yield and high enantiomeric excess (90% and 99%, respectively), as shown in Scheme 10.13. It is important to stress that (*R*)-**52** was not isomerized to (*S*)-**52** by the yeast under the experimental conditions.

On the other hand, the biotransformation of (*S*)-**52** to (1*S*,2*R*)-**53** took place through a different route to that observed for the (*R*)-**52** enantiomer. We have shown that the substrate (*S*)-**52** is isomerized to form enantiomer (*R*)-**52** through the intermediate formation of dione **51** by oxidation, followed by further reduction of the dione by *T. cutaneum* for give (1*S*,2*R*)-diol **53**. Therefore, the biotransformation of (*S*)-**52** requires the existence

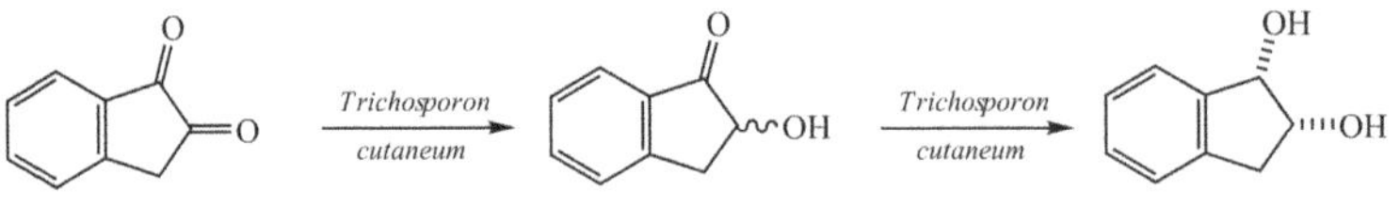

51　　　　benzoin **52**　　　　(1*S*,2*R*)-**53**

Scheme 10.12　　The benzoin **52** is proposed as an intermediate for bioreduction of 1,2-indanone **51** to give (1*S*,2*R*)-1,2-indandiol **53**.

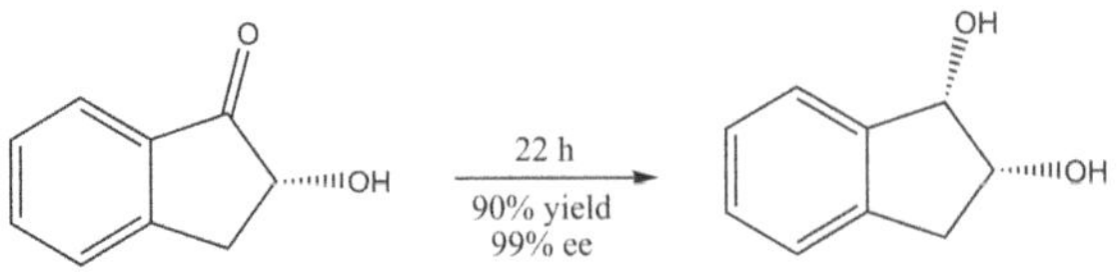

(*R*)-**52**　　　　(1*S*,2*R*)-**53**

Scheme 10.13　　Benzoin (*R*)-**52** was biotransformed by *T. cutaneum* in 22 hours in a clean fashion giving (1*S*,2*R*)-diol **53** as the sole product.

an equilibrium-controlled reduction–oxidation sequence, rather than the action of a single enzyme (like a racemase, see Scheme 10.14). Faber and co-workers observed a similar reduction–oxidation of (R)-2-hydroxy-1-indanone (**52**) by *Lactobacillus paracasei*.[46] The complete conversion of (S)-**52** by *T. cutaneum* was accomplished only after 42 hours but, in spite of the longer reaction time, the product (1S,2R)-diol **53** was also isolated in both high yield and ee (85% and 99%, respectively). Scheme 10.14 depicts a proposed general mechanism for the biotransformation.

To obtain a single enantiomer from a racemic mixture of enantiomers, one can resolve it either by conventional separation techniques or by using an existing difference in reactivity (kinetic resolution).[47] However, a major limitation of these techniques is that the maximum product yield is only 50%. The unwanted enantiomer must be separated, racemized and resubjected to resolution in order to increase this yield. These disadvantages can be overcome by employing DKR. DKR has recently become not only an alternative to traditional kinetic resolution, but also a new procedure for asymmetric synthesis.[16] It is one of the most useful and reliable methods to prepare a single chiral compound bearing two or more stereocenters starting from a racemate, with a theoretical yield of 100%. A number of successful examples of DKR methods have been reviewed recently.[48] Some of them involve solely conventional chemical methods,[49] whereas others combine chemical and biocatalytic steps.[50] In the latter case, enzymes or whole cells contribute either to racemization[51] or to kinetic resolution.[48]

We have studied the diastereo- and enantioselective preparation of (1S,2R)-1,2-dihydroxy-1,2,3,4-tetrahydronaphthalene (**55**) in high enantiomeric excess (>99% ee) and excellent chemical yield (83%) by the asymmetric reduction of racemic 2-hydroxy-1-tetralone (**54**) mediated by fresh resting cells of the yeast *Trichosporon cutaneum* through a DKR process.[52] Racemic 2-hydroxy-1-tetralone (**54**) was added to a slurry of resting cells of *T. cutaneum* (in distilled water). After incubation on a shaker (170 rpm) at

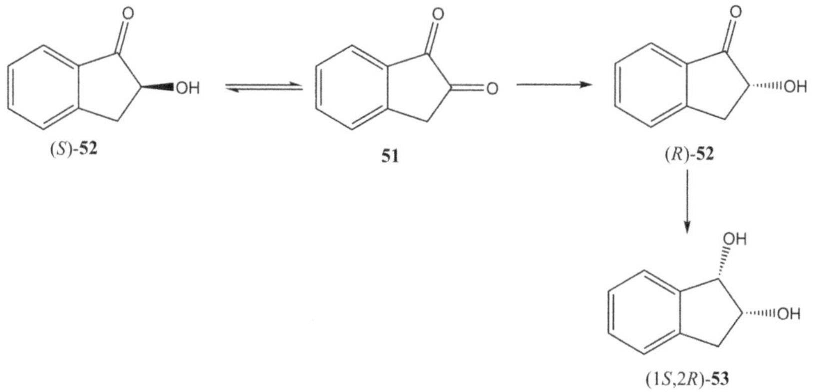

Scheme 10.14 Proposed general mechanism for the biotransformation of (S)-**52** by resting cells of *T. cutaneum*.

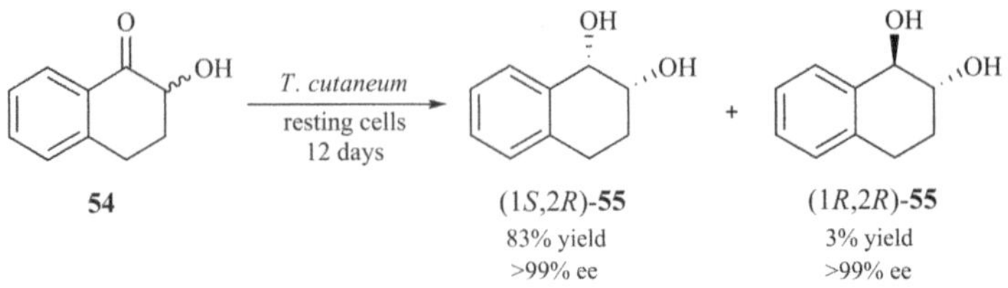

Scheme 10.15 Biotransformation of racemic **54** by *T. cutaneum* after 12 days of incubation.

30 °C for 12 days, (±)-**54** was converted into a mixture of (1*S*,2*R*)-*cis*-**55** and (1*R*,2*R*)-*trans*-1,2-dihydroxy-1,2,3,4-tetrahydronaphthalene (**55**). The major product, *cis*-(1*S*,2*R*)-diol **50**, was obtained in 83% isolated yield and with high enantiomeric excess (>99%), whilst the minor product, *trans*-(1*R*,2*R*)-diol **55**, was obtained in 3% isolated yield and >99% ee (Scheme 10.15). Over the course of the reaction, which was monitored by GC-MS analysis with a chiral column, the enantiomeric excess of (*S*)-**54** increased over the first 68 hours. After the first 68 hours of the reaction, (*R*)-**54** was almost entirely depleted and selectively converted into the *cis*-diol **55**. The remaining (*S*)-**54** was slowly converted into (*R*)-**54** until the end of the reaction. It is worth mentioning that after 143 hours of reaction, GC analysis indicated the formation of a racemic mixture of **55** and thereafter, the enantiomeric excess of (*S*)-**54** decreased. The formation of the products was followed by chiral HPLC analysis, which gave a high enantiomeric excess for the major *cis*-(1*S*,2*R*)-diol **55** over the course of the reaction. After 12 days of incubation, *cis*-(1*S*,2*R*)-diol **55** was isolated in 83% yield and >99% ee (Scheme 10.15). Then, (*R*)-**54** was recovered in 6% yield with 96% ee.

Experiments performed with chiral substrates show that deracemization of compound **50** is not a simple process. The formation of *cis*-(1*S*,2*R*)-diol **55** seems to require both the intervention of an isomerase to convert (*S*)-ketone **54** into (*R*)-ketone **54** and an equilibrium-controlled oxidation–reduction sequence to convert the *cis*-(1*R*,2*S*)-diol **55** back to the (*S*)-ketone **54** and keep the balance between the diols and hydroxyketones (Scheme 10.16). Concerning the former transformations (from the point of view of the substrates) a typical DKR could be considered. A closer look at the latter transformations (from the point of view of the products) shows that the product diol *cis*-(1*R*,2*S*)-diol **55** is ultimately converted to the *cis*-(1*S*,2*R*)-diol **55** through a stereoinversion-like process.

To summarize the topic, a biocatalytic process to deracemize (±)-2-hydroxy-1-tetralone **54** efficiently has been successfully devised using the versatile, easily-cultivated, non-conventional yeast, *Trichosporon cutaneum* CCT 1903, to prepare the (1*S*,2*R*)-diol **55** in enantiomerically pure form (>99% ee) and in high yield (83%). In order to understand the mechanism of the complete dynamic kinetic resolution of the substrate, a rationale based on a comprehensive study with chiral substrates was proposed. In this mechanism, the intervention of isomerase and the oxidation–reduction equilibrium are key

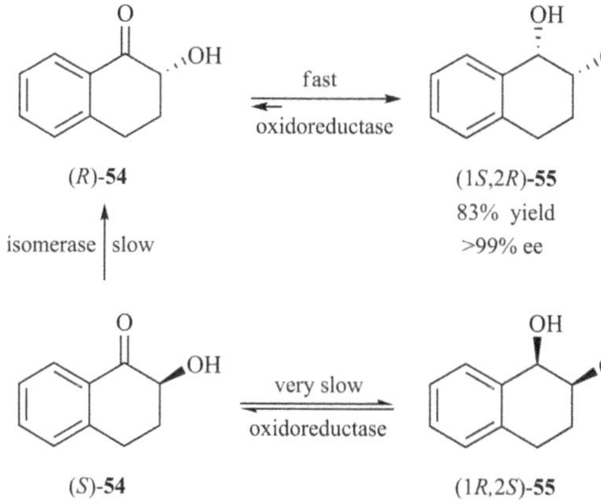

Scheme 10.16 Reductive deracemization of (±)-2-hydroxy-1-tetralone **54** to give (1*S*,2*R*)-diol **55**.

features to be highlighted. Moreover, the enzymatic DKR of (±)-**54** is unprecedented in the literature and represents a promising environmentally friendly method to obtain enantiopure (1*S*,2*R*)-diol **55** on larger scales.

10.4.2   Reduction of α-Methyleneketones

Unsaturated carbonyl compounds are prochiral substrates that can provide densely functionalized molecules with two or more consecutive stereogenic centers. Asymmetric reductive products such as optically pure α-hydroxy-β-methyl-γ-hydroxy esters are important building blocks for the synthesis of a variety of bioactive molecules[53] used as pharmaceutical intermediates[54] and are found in natural products.[55] In our continuing effort towards methodological studies to synthesize such useful intermediates, we developed two novel chemoenzymatic routes for the preparation of enantio- and diastereomerically pure α-hydroxy-β-methyl-γ-hydroxy esters. In this account, we present only one procedure of the two that we have developed and which involves biocatalysis in the key steps.[56] The relevant reactions in the synthesis include classical organic reactions such as Claisen condensation, a Wasserman chain homologation, a Mannich-type olefination with maintenance of the stereo-integrity of the involved intermediates and a Pd–C catalyzed hydrogenation to obtain a diastereo- and enantioselective product by the reduction of the appropriate enol formed in the keto–enol equilibrium (Scheme 10.17).

The bioreduction of **56** was performed accordingly to the Fadnavis protocol and the desired product **57** was achieved in 80% yield and 98% ee. A Mannich type α-methylenation was applied to *R*-**57**, yielding *R*-**58** in 80% and 99% ee with stereochemical integrity preserved. The asymmetric biocatalytic reduction of **58** should provide two chiral centers. Although the stereoselectivity achieved for this kind of reduction is excellent, the chemoselectivity of whole cell bioreductions with respect to C=C *versus* C=O bond reduction is

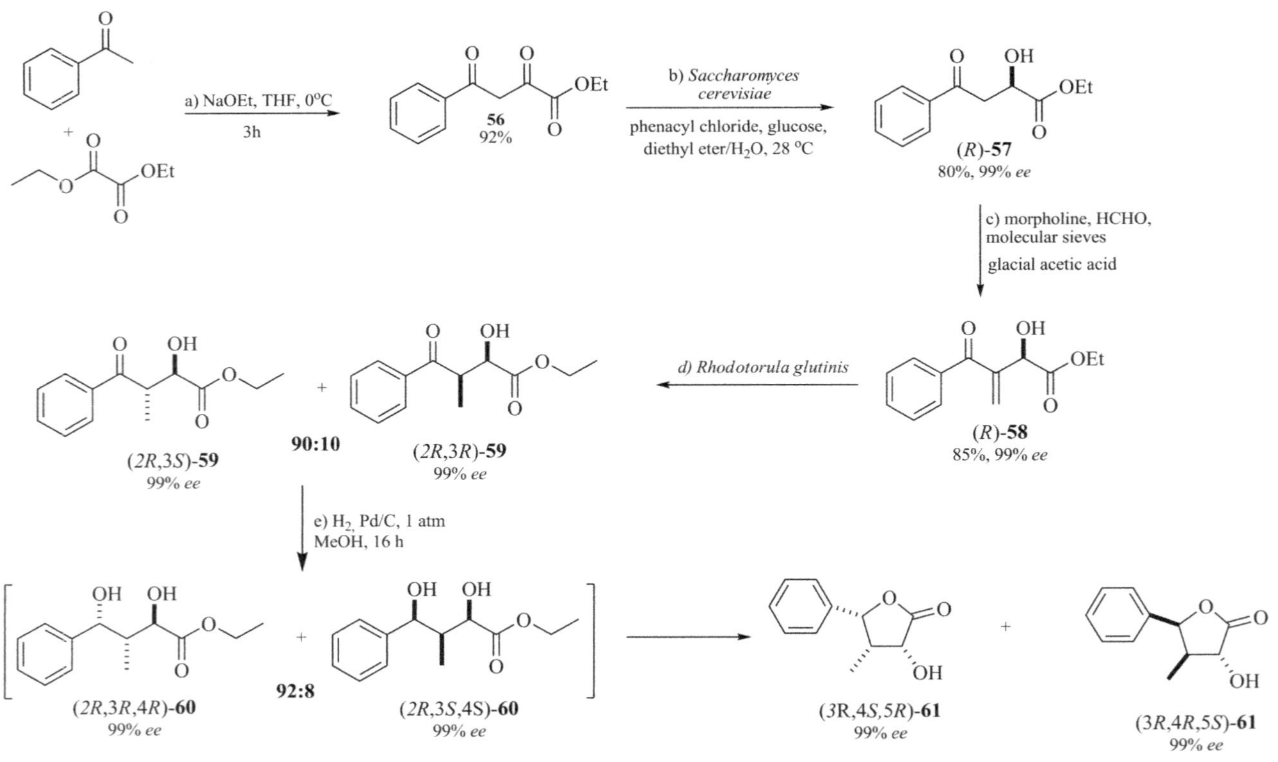

Scheme 10.17 Synthetic route developed for preparing lactones (3*R*,4*S*,5*R*)-**61** and (3*R*,4*R*,5*S*)-**61**.

Table 10.5 Microbial reduction of (*R*)-**58** using different yeasts.

		59	
Yeast[a]	Yield (%)	*anti*:*syn*[c] (ee%)	Time (h)
Saccharomyces cerevisiae[b]	—	—	48
Candida parapsilosis	73	77:23 (99)	24
Trichosporum cutaneum	75	74:26 (97)	30
Rhodotorula glutinis	76	85:15 (99)	24

[a]Growing cells, YM medium, *R*-**58** 0.4 mmol (100 mg) in ethanol (1 mL) orbital stirring, 28 °C.
[b]Commercial lyophilized yeast.
[c]Percentage of enantiomeric excess of each diastereoisomer; enantiomeric and diastereoisomeric excess were determined by HPLC.

often poor due to the presence of competing alcohol dehydrogenases, since enoate reductases and the alcohol dehydrogenases depend on the same nicotinamide cofactor.[57] Thus, some yeast strains were evaluated for the bioreduction of **58** in order to find the most reactive (Table 10.5).

In all cases, the bioreduction of *R*-**58** afforded exclusively **59** in good yield, with diastereoselectivity and excellent enantioselectivity. It is well known that in α,β-unsaturated ketone systems, the double bond is reduced preferably.[58] Only in rare cases, when α,β-unsaturated ketone resonance is destabilized by electron withdrawing groups bonded to double bonds, is the C=O reduced first, which justifies the absence of **62**. The **58** γ-carbonyl is not electrophilic enough to propitiate a NADH/NADPH hydride attack, thus formation of **60** is also not observed. The bioreduction of *R*-**58** was carried out on a multigram scale by the yeast *Rhodotorula glutinis* giving **59** in 90:10 diastereoisomeric ratio, favoring the *anti* isomer.

10.4.3 Reduction of α-Haloketones and α-Haloenones

Since the early 1990s, we have used the reduction of α-haloacetophenones mediated by baker's yeast to obtain halohydrins in high ee that can be used as chiral building blocks for the synthesis of pharmaceutical products like α-, β- and γ-adrenergic drugs,[59] following the route shown in Scheme 10.18. A large number of papers have appeared reporting the enantiomeric reduction of haloacetophenones by whole cells and isolated enzymes.[60] Recently, dehydrogenases in the form of whole cells for production of chiral styrene oxide have been used on a pilot-plant scale.[61] Generally the *R*/*S* configuration of the obtained halohydrin may be controlled by choosing an appropriate microorganism. For example, while *Geotrichum candidum* gives (*S*)-halohydrins, *Rhodotorula glutinis* gives (*R*)-haloydrins.[28]

A drawback of the bioreductions of α-haloketones is the reductive dehalogenation that may occur in some cases through a radical mechanism,[62] or through a mechanism involving glutathione,[63] generally observed with α-halo-α-ketoesters.[64] To overcome the reductive dehalogenation through a radical mechanism, it is necessary to add a small quantity of radical scavenger like *m*-dinitrobenzene (DNB) to the reaction medium (Scheme 10.19).[65]

Interestingly, the bioreductions of **67** and **68** mediated by *Saccharomyces cerevisiae* do not show the dehalogenation that is observed when *Pichia stipitis* is used, giving **72** and **73** in good de and ee.[66] In addition, the de and ee of **72** and **73** improved when the bioreductions were performed in a biphasic water–ionic liquid (IL) system. In this biphasic system, the substrate and products are mostly in the IL and the cells are dispersed in the water phase. In the case of hydrophobic substrates and products, the IL acts as a substrate reservoir, delivering the substrate to the aqueous phase and withdrawing the product from it. This implies a decrease in the concentrations of substrate and product in the aqueous phase, preventing cellular inhibition by them,[67] and favoring the enzyme with the lowest K_M, which must have a great effect on consecutive reactions (Figure 10.4). Therefore, the IL acts as an absorbing resin in an *in situ* extractive biocatalysis, controlling the substrate concentration of hydrophobic substrates and improving the enantio-chemoselectivity.[68]

In recent work, we described the asymmetric bioreduction of ethyl 3-halo-2-oxo-4-phenylbutanoate with several microorganisms, in particular with *S. cerevisiae*, which we optimized by changing the matrix of immobilization and the concentration of substrate, and by feeding with glucose as

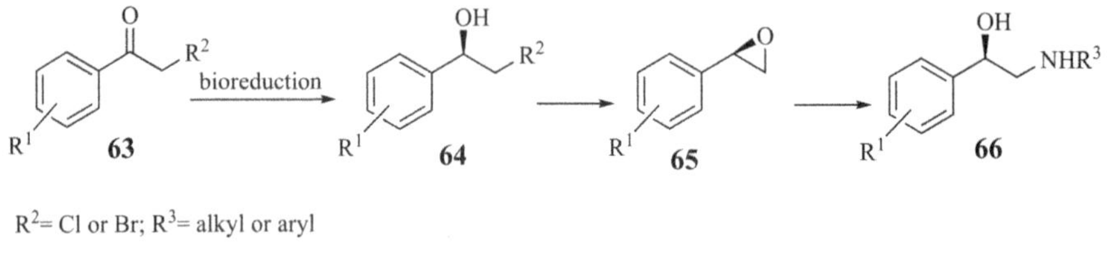

R^2= Cl or Br; R^3= alkyl or aryl

Scheme 10.18   Bioreduction of α-haloacetophenones.

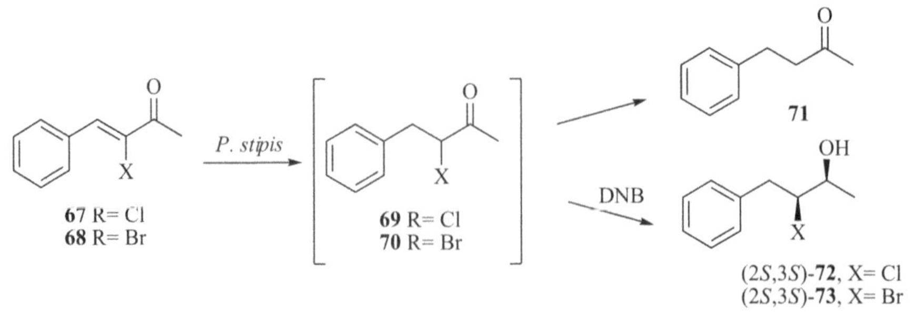

Scheme 10.19   Reduction of **67–68** with *Pichia stipitis*.

electron donor.[69] The entrapment of *S. cerevisiae* with double gel layers was fundamental for achieving high enantio- and diastereoselectivity.

Different strains of yeast and bacteria for microbiological reduction of ethyl 3-chloro-2-oxo-4-phenylbutanoate (**74**) and ethyl 3-bromo-2-oxo-4-phenyl-butanoate (**75**) were selected according to literature results (Scheme 10.20). Bioreduction was carried out using resting cells under non-fermenting conditions, *i.e.*, suspended in water without the addition of sugar. All microorganisms reduced **75** with a varied range of chemical and optical yields. A higher diastereomeric excess was observed at pH 4.0 than at pH 7.0, since **75** was selectively reduced to the ester (2*S*,3*S*)-**77**, while the substrates underwent a dynamic kinetic resolution due to epimerization occurring at C-2 *via* an enol intermediate under the effect of pH.[70] High chemical yields and reasonable diastereomeric excess were obtained with *S. cerevisiae* which was chosen for further studies to improve the preliminary results. The initial approach was to use free cells of *S. cerevisiae*, without the addition of glucose, and then to compare with immobilized cells. At low pH (4.0), the rate of enolization is fast and the *syn*-alcohol is formed as the major product [*syn*:*anti* 69 (83% ee): 31 (95% ee) in 87% yield]. At pH 7.0, when enolization becomes slower, enantioselectivity prevails, but optically pure diastereomers are formed in lower ratios [*syn*:*anti* 64 (78% ee):36 (94% ee) in 80% yield].

In an attempt to decrease the availability of the substrate, cells were immobilized in alginate beads, creating a barrier at the border of the cellular

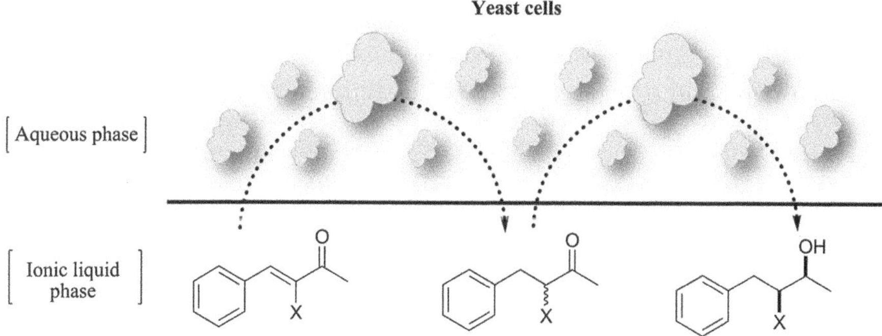

Figure 10.4	The consecutive bioreduction reactions of **64** and **65** performed in a biphasic water–ionic liquid system.

Scheme 10.20	Reduction of ethyl 3-halo-2-oxo-4-phenylbutanoate **74**–**75** mediated by baker's yeast.

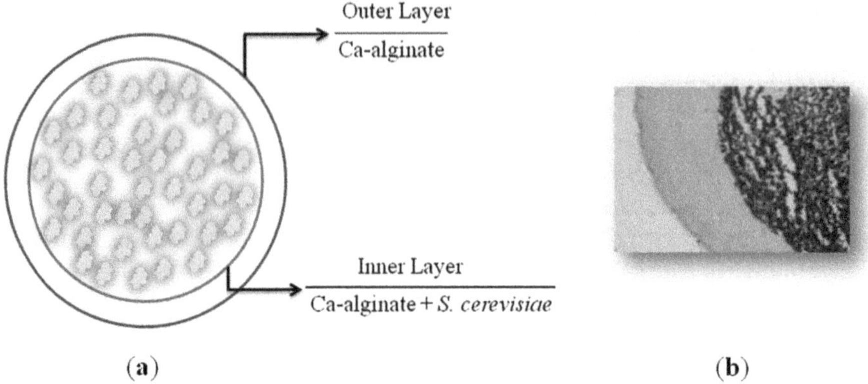

Figure 10.5   (a) Schematic diagram of a cross section of double layer Ca-alginate fibers and (b) microphotography of cross sections of fibers.

membrane.[71] During reduction, the concentration levels of these substrates inside the alginate beads were lower than in the solution, due to slow diffusive transport. The lower concentration of substrate in the beads helps the stereoselectivity, since only the enzyme with lowest K_M (Michaelis constant) is able to react with high V_{max}. It is well known that baker's yeast possesses several alcohol dehydrogenases with varying substrate selectivities and even opposing enantioselectivities.[59]

With immobilized cells, the selectivities of these reactions were better than those observed for free cells. The gradient of the concentration of substrate in the reaction medium is different from the gradient of concentration inside the calcium alginate beads, where the substrate is reduced by *S. cerevisiae*, due to slow diffusive transport.[59] In order to improve the selectivity further, a new procedure was established by decreasing even more the concentration of substrate available for the biocatalyst, by introducing a second barrier of alginate. Cells were immobilized with double gel layers, which also prevent cells from leaking out of the gel beads (see Figure 10.5). The conditions for bioreduction of **75** were optimized and the enantioselectivity was improved for both *syn* and *anti* enantiomers. Also, the *syn*:*anti* ratio increased to 85:15. With double gel layers, another barrier was created and the gradients of substrate concentration in the alginate beads affected the overall reduction performance (see Table 10.6). Bioreduction of **74** by *S. cerevisiae* entrapped in alginate beads with double gel layers gave ethyl (2*S*, 3*S*)-3-chloro-hydroxybutanoate (**76**) (85% ee) and its epimer, the (2*S*, 3*R*)-derivative **78** (>99% ee; ratio **76**:**78** = 30:70) in 85% yield.

The excellent enantioselectivity that was obtained with the entrapment of whole cell yeast in calcium alginate double layers prompted us to extend the methodology to other systems. The asymmetric reduction of ethyl benzoylformate **80** to ethyl (*R*)-mandelate **81** in a continuous process was carried out in a packed bed reactor with *Saccharomyces cerevisiae* immobilized on calcium alginate fibers with double gel layers (Scheme 10.21).[73]

Table 10.6 Comparison between the bioreduction of **74** and **75** mediated by *S. cerevisiae* immobilized on calcium alginate and literature data.[72]

			ee (%)	
Substrate	Yield (%)	*syn* : *anti*	*syn*	*anti*
Lit.[a,72]	50	34 : 66	43	95
74[b]	85	30 : 70	85	>99
75[b]	90	85 : 15	96	>99

[a]Reaction conditions: (a) 42.5 g of *S. cerevisiae* suspended in citrate buffer (300 mL, pH 7.0), 720 mg of **74** or **75** with the addition of 54.5 g of glucose.
[b]Reaction conditions: (b) 5.0 g of *S. cerevisiae* entrapped in calcium alginate beads with double gel layers on a large scale were suspended in citrate–phosphate buffer (60 mL, pH 4.0) containing glucose (8 g), 2.2 mmol of **74** or **75**. Every 6 h, more glucose (1.6 g) was added.

Scheme 10.21 Reduction of ethyl benzoylformate (**80**) to ethyl (*R*)-mandelate (**81**).

Continuous bioreduction was technically feasible and reproducible, and the conditions of the reactor were optimized by changing the concentration of **80** and the feed flow. The optimized concentration and feed flow were 1.1 mmol L^{-1} and 102 mL h^{-1}, respectively. High chemical yield (82%) and extent of reduction (92–97%) were obtained. The volumetric productivity was 0.12 mmol L^{-1} h^{-1}. The enantiomeric excess of (*R*)-**81** remained high (92%) over eight days, indicating the robustness of *S. cerevisiae* as a biocatalyst with the conditions used in the process. The reduction of **80** mediated by *S. cerevisiae* immobilized on alginate fibers with double gel layers showed stability, high chemical yield and enantiomeric excess. The reduction rate (0.03 mmol g^{-1} h^{-1}) obtained in this process was better than that obtained in our earlier approach, where the cells were adsorbed on chrysotile fibers.[74]

10.5 Conclusions

We have herein shown applications of yeasts, bacteria and fungi to bioprocesses to give oxidized sulfides, ketones, benzylic and allylic methylenes, reduced ketones and α-diketones with high yield and selectivity in accordance to the principles of green chemistry, white biotechnology and environmental sustainability. In some examples, processes involving the careful selection of the reaction medium and microorganism immobilization competed with isolated enzymes that are much more expensive and which also require specific enzymes to recycle the cofactors.

Acknowledgements

The authors thank the Brazilian science foundations FAPESP, CAPES, CNPq and FUNCAMP-UNICAMP for their financial support. We also thank Prof. Carol Collins for helpful suggestions on style and grammar.

References

1. P. Anastas and N. Eghbali, *Chem. Soc. Rev.*, 2010, **39**, 301.
2. P. T. Anastas and M. M. Kirchhoff, *Acc. Chem. Res.*, 2002, **35**, 686.
3. K. M. Koeller and C.-H. Wong, *Nature*, 2001, **409**, 232.
4. D. Gamenara, G. A. Seoane, P. Saenz-Méndez and P. D. de María, *Redox Biocatalysis: Fundamentals and Applications*, John Wiley & Sons, Hoboken, 2012, pp. 1–64.
5. L. M. Blank, B. E. Ebert, K. Buehler and B. Bühler, *Antioxid. Redox Signaling*, 2010, **13**(3), 349.
6. (a) A. Raadt and H. Griengl, *Curr. Opin. Biotechnol.*, 2002, **13**, 537; (b) Y. Chen, F. Lie and Z. Li, *Adv. Synth. Catal.*, 2009, **351**, 2107; (c) M. Kluge, R. Ullrich, K. Scheibner and M. Hofrichter, *Green Chem.*, 2012, **14**, 440; (d) S. Yadav, R. S. S. Yadav, S. Sudava and K. D. S. Yadav, *Catal. Commun.*, 2011, **12**, 781; (e) R. Kuriata-Adamusiak, D. Strub and S. Lochynski, *Appl. Microbiol. Biotechnol.*, 2012, **95**, 1427; (f) D. Kim, K. Y. Choi, M. Yoo, J. N. Choi, C. H. Lee, G. J. Zylstra, B. S. Kang and E. Kim, *Appl. Microbiol. Biotechnol.*, 2010, **86**, 1841.
7. (a) W. J. H. van Berkel, N. M. Kamerbeek and M. W. Fraaije, *J. Biotechnol.*, 2006, **124**, 670; (b) G. Haufe, D. Wolker and R. Frölich, *J. Org. Chem.*, 2002, **67**, 3022.
8. R. P. Limberger, C. V. Ursini, P. J. S. Moran and J. A. R. Rodrigues, *J. Mol. Catal. B: Enzym.*, 2007, **46**, 37.
9. (a) M. Krischke, C. Loeffler and M. J. Mueller, *Phytochemistry*, 2003, **62**, 351; (b) I. Vallikivi, L. Fransson, K. Hult, I. Järving, P. Tõnis, N. Samel, V. Tõugu, L. Villo and O. Parve, *J. Mol. Catal. B: Enzym.*, 2005, **35**, 62.
10. T. Rezanka and V. M. Dembitsky, *Eur. J. Org. Chem.*, 2003, **3**, 309.
11. (a) J. A. Dale and H. S. Mosher, *J. Am. Chem. Soc.*, 1973, **95**, 512; (b) R. Riguera, E. Quinoa and J. M. Seco, *Chem. Rev.*, 2004, **104**, 17.
12. L. Pinheiro and A. J. Marsaioli, *J. Mol. Cat. B: Enzym.*, 2007, **48**, 78.
13. (a) V. K. Aggarwal, D. M. Badine and V. A. Moorthie, *Aziridines and Epoxides in Organic Synthesis*, ed. A. K. Yudin, Wiley-VCH, Weinheim, 2006, pp. 1–35; (b) D. Yang, *Acc. Chem. Res.*, 2004, **37**, 497; (c) S. Grüschow and D. H. Sherman, *Aziridines and Epoxides in Organic Synthesis*, ed. A. K. Yudin, Wiley-VCH, Weinheim, 2006, pp. 349–398; (d) M. E. Jung, J. A. Berliner, L. Koroniak, B. G. Gugiu and A. D. Watson, *Org. Lett.*, 2008, **10**, 4207.
14. (a) O. A. Wong and Y. Shi, *Chem. Rev.*, 2008, **108**, 3958; (b) G. Sello, T. Fumagalli and F. Orsini, *Curr. Org. Synth.*, 2006, **3**, 457; (c) X.-T. Zhou, H.-B. Ji, J.-C. Xu, L.-X. Pei, L.-F. Wang and X.-D. Yao, *Tetrahedron Lett.*, 2007, **48**, 2691; (d) V. Schurig and F. Betschinger, *Chem. Rev.*, 1992, **92**, 873;

(e) T. Mukaiyama, T. Yamada, T. Nagata and K. Imagawa, *Chem. Lett.*, 1993, **2**, 327; (f) M. G. Finn and K. B. Sharpless, *J. Am. Chem. Soc.*, 1991, **113**, 113.

15. (a) T. Kubo, M. W. Peters, P. Meinhold and F. H. Arnold, *Chem.–Eur. J.*, 2006, **12**, 1216; (b) E. Y. Lee and M. L. Shuler, *Biotechnol. Bioeng.*, 2007, **98**, 318; (c) L. C. Nolan and K. E. O'Connor, *Biotechnol. Lett.*, 2008, **30**, 1878.

16. R. M. Silverstein, G. C. Bassler and T. C. Morrill, *Spectrometric Identification of Organic Compounds*, John Wiley & Sons, New York, 1994, p. 235.

17. L. Pinheiro, L. G. Oliveira and A. J. Marsaioli, *J. Mol. Catal. B: Enzym.*, 2009, **60**, 133.

18. (a) A. Pawelczyk and L. Zaprutko, *Eur. J. Med. Chem.*, 2006, **41**, 586; (b) T. J. A. Bruce, M. C. Matthes, K. Chamberlain, C. M. Woodcock, A. Mohib, B. Webster, L. E. Smart, M. A. Birkett, J. A. Pickett and J. A. Napier, *Proc. Natl. Acad. Sci. U. S. A.*, 2008, **105**, 4553; (c) L. Yeruva, J. A. Elegbede and S. W. Carper, *Anti-Cancer Drugs*, 2008, **19**, 766.

19. S. M. Roberts and P. W. H. Wan, *J. Mol. Catal. B: Enzym.*, 1998, **4**, 111.

20. C. P. da Silva, PhD Thesis, UNICAMP, 2012.

21. G. F. da Cruz, C. F. F. Angolini, L. G. Oliveira, P. F. Lopes, S. P. Vasconcellos, E. Crespim, V. M. Oliveira, E. V. Santos Neto and A. J. Marsaioli, *Appl. Microbiol. Technol.*, 2010, **87**, 319.

22. B. Z. da Costa, MSc Dissertation, UNICAMP, 2011.

23. R. Sicard, L. S. Chen, A. J. Marsaioli and J.-L. Reymond, *Adv. Synth. Catal.*, 2005, **347**, 1041.

24. J. Kusterer and M. Keusgen, *J. Agric. Food Chem.*, 2010, **58**, 1129.

25. R. Bentley, *Chem. Soc. Rev.*, 2005, **34**, 609.

26. H. Pellisier, *Tetrahedron*, 2006, **62**, 5559.

27. H. L. Holland, *Nat. Prod. Rep.*, 2001, **18**, 171.

28. A. L. M. Porto, PhD Thesis, UNICAMP, 2002.

29. R. B. Borges, A. Laverde Jr., A. L. M. Porto and A. J. Marsaioli, *Spectroscopy*, 2000, **14**, 203.

30. A. L. M. Porto, F. Cassiola, S. L. P. Dias, I. Joekes, Y. Gushiken, J. A. R. Rodrigues, P. J. S. Moran, G. P. Manfio and A. J. Marsaioli, *J. Mol. Catal. B: Enzym.*, 2002, **19–20**, 327.

31. K. Hult and P. Berglund, *Trends Biotechnol.*, 2007, **25**, 231.

32. O. Khersonsky, C. Roodveldt and D. S. Tawfik, *Curr. Opin. Chem. Biol.*, 2006, **10**, 498.

33. (a) K. Li, T. He, C. Li, X.-W. Feng, N. Wang and X.-Q. Yu, *Green Chem.*, 2009, **11**, 777; (b) U. K. Sharma, N. Sharma, R. Kumar, R. Kumar and A. K. Sinha, *Org. Lett.*, 2009, **11**, 4846; (c) F.-W. Lou, B. K. Liu, Q. Wu, D.-S. Lv and X.-F. Lin, *Adv. Synth. Catal.*, 2008, **350**, 1959; (d) C. Li, X.-W. Feng, N. Wang, Y.-J. Zhou and X.-Q. Yu, *Green Chem.*, 2008, **10**, 616; (e) M. T. Reetz, R. Mondière and J. D. Carballeira, *Tetrahedron Lett.*, 2007, **48**, 1679; (f) M. Moreira and M. G. Nascimento, *Catal. Commun.*, 2007, **8**, 2043; (g) M. Svedendahl, K. Hult and P. Berglund, *J. Am. Chem. Soc.*, 2005, **127**, 17988; (h) O. Torre, I. Afonso and V. Gotor, *Chem. Commun.*, 2004, 1724.

34. (a) G. J. ten Brink, I. W. C. Arends and R. A. Sheldon, *Chem. Rev.*, 2004, **104**, 4105; (b) M. Y. Rios, E. Salazar and H. F. Olivo, *J. Mol. Catal. B: Enzym.*, 2008, **54**, 61.

35. F. Björkling, S. E. Godtfredsen and O. Kirk, *J. Chem. Soc., Chem. Commun.*, 1990, 1301.

36. F. Björkling, H. Frykman, S. E. Godfredsen and O. Kirk, *Tetrahedron*, 1992, **48**, 4587.

37. (a) S. C. Lemoult, P. F. Richardson and S. M. Roberts, *J. Chem. Soc. Perkin Trans.*, *1*, 1995, 89; (b) B. K. Pchelka, M. Gelo-Pujic and E. Guibé-Jampel, *J. Chem. Soc. Perkin Trans. 1*, 1998, 2625; (c) M. Y. Rios, E. Salazar and H. F. Olivo, *Green Chem.*, 2007, **9**, 459; (d) M. Y. Rios, E. Salazar and H. F. Olivo, *J. Mol. Catal. B: Enzym.*, 2008, **54**, 61.

38. P. Carlqvist, R. Eklund, K. Hult and T. Brinck, *J. Mol. Model.*, 2003, **9**, 164.

39. (a) R. D. Schmid and R. Verger, *Angew. Chem., Int. Ed.*, 1998, **37**, 1608; (b) M. Kapoor and M. N. Kupta, *Process Biochem.*, 2012, **47**, 555.

40. (a) K. Tonova, Z. Lazarova, N. Nemestothy, L. Gubicza and K. Belafi-Bako, *Chem. Ind. Chem. Eng. Q.*, 2006, **12**(3), 175; (b) K. Naoe, C. Takeuchi, M. Kawagoe, K. Nagayama and M. Imai, *J. Chromatogr.*, 2007, **850**, 277.

41. (a) M. Takeshita and T. Sato, *Chem. Pharm. Bull.*, 1989, **37**, 1085; (b) E. C. S. Brenelli, P. J. S. Moran and J. A. R. Rodrigues, *Synth. Commun.*, 1990, **20**, 261.

42. (a) K. Nakamura, S. Kondo, Y. Kawai, K. Hida, K. Kitano and A. Ohno, *Tetrahedron: Asymmetry*, 1996, **7**, 409; (b) E. Lourenço, J. A. R. Rodrigues and P. J. S. Moran, *J. Mol. Catal. B: Enzym.*, 2004, **29**, 37.

43. H. Hoyos, J.-V. Sinisterra, F. Molinari, A. R. Alcántara and P. D. De Maria, *Acc. Chem. Res.*, 2010, **43**, 288.

44. R. S. Martins, D. S. Zampieri, J. A. R. Rodrigues, P. S. Carvalho and P. J. S. Moran, *ChemCatChem*, 2011, **3**, 1469.

45. T. Cazetta, I. Lunardi, G. J. A. Conceição, P. J. S. Moran and J. A. R. Rodrigues, *Tetrahedron: Asymmetry*, 2007, **18**, 2030.

46. (a) B. M. Nestl, W. Kroutil and K. Faber, *Adv. Synth. Catal.*, 2006, **348**, 873; (b) S. M. Glueck, M. Pirker, B. M. Nestl, B. T. Ueberbacher, B. Larissegger-Schnell, K. Csar, B. Hauer, R. Stuermer, W. Kroutil and K. Faber, *J. Org. Chem.*, 2005, **70**, 4028.

47. (a) F. F. Huerta, A. B. E. Minidis and J. E. Bäckvall, *Chem. Soc. Rev.*, 2001, **30**, 321; (b) O. Pamies and J. E. Bäckvall, *Chem. Rev.*, 2003, **103**, 3247; (c) H. Stecher and K. Faber, *Synthesis*, 1996, 1; (d) K. Faber, *Chem.–Eur. J.*, 2001, **7**, 5005.

48. (a) H. Pellissier, *Tetrahedron*, 2003, **59**, 8291; (b) R. S. Ward, *Tetrahedron: Asymmetry*, 1995, **6**, 1475; (c) S. Caddick and K. Jenkins, *Chem. Soc. Rev.*, 1996, 447.

49. R. Noyori and T. Ohkuma, *Angew. Chem., Int. Ed.*, 2001, **40**, 40.

50. (a) O. Pamies and J.-E. Bäkvall, *Trends Biotechnol.*, 2004, **22**, 130; (b) N. J. Turner, *Curr. Opin. Chem. Biol.*, 2004, **8**, 114; (c) R. Azerad and D. Buison, *Curr. Opin. Biotechnol.*, 2000, **11**, 565; (d) M-J. Kim, Y. Chung,

Y. K. Choi, H. K. Lee, D. Kim and J. Park, *J. Am. Chem. Soc.*, 2003, **125**, 11494.

51. B. Schnell, K. Faber and W. Kroutil, *Adv. Synth. Catal.*, 2003, **345**, 653.

52. I. Lunardi, T. Cazetta, G. J. A. Conceição, P. J. S. Moran and J. A. R. Rodrigues, *Adv. Synth. Catal.*, 2007, **349**, 925.

53. (a) R. Matsubara, Y. Nakamura and S. Kobayashi, *Angew. Chem., Int. Ed.*, 2004, **43**, 3258; (b) T. Wakabayashi, K. Mori and S. Kobayashi, *J. Am. Chem. Soc.*, 2001, **123**, 1372, and references therein; (c) K.-S. Yeung and I. Paterson, *Chem. Rev.*, 2005, **105**, 4237; (d) B. Schetter and R. Mahrwald, *Angew. Chem., Int. Ed.*, 2006, **45**, 7506.

54. (a) J. M. Woodley, *Trends Biotechnol.*, 2008, **26**, 321; (b) N. Ran, L. Zhao, Z. Chen and J. Tao, *Green Chem.*, 2008, **10**, 361; (c) C. S. Stauffer, P. Bhaket, A. W. Fothergill, M. G. Rinaldi and A. Datta, *J. Org. Chem.*, 2007, **72**, 9991; (d) J. Li and D. Menche, *Synthesis*, 2009, 2293.

55. (a) U. Dahn, H. Hagenmaier, H. Höhne, W. A. König, G. Wolf and H. Zähner, *Arch. Microbiol.*, 1976, **107**, 143; (b) H. Hagenmaier, A. Keckeisen, H. Zanher and W. A. König, *Liebigs Ann. Chem.*, 1979, 1494; (c) J. Delzer, H. P. Fiedler, H. Muller, H. Zähner, R. Rathmann, K. Ernst and W. A. König, *J. Antibiot.*, 1984, **37**, 80; (d) K. Kobinata, M. Uramoto, M. Nishii, H. Kusakabe, G. Nakamura and K. Isono, *Agric. Biol. Chem.*, 1980, **44**, 1709; (e) H. P. Fiedler, R. Kurth, J. Langhärig, J. Delzer and H. Zähner, *J. Chem. Tech. Biotechnnol.*, 1982, **32**, 271; (f) B. Lauer, R. Russwurn and C. Bormann, *Eur. J. Biochem.*, 2000, **267**, 1698; (g) Y. Hayashi, T. Urushima, M. Shin and M. Shoji, *Tetrahedron*, 2005, **61**, 11393, and references cited therein; (h) D. O'Hagan, *The Polyketide Metabolites*, Ellis Horwood, Chichester, 1991; (i) D. O'Hagan, *Nat. Prod. Rep.*, 1995, **12**, 1; (j) D. Menche, *Nat. Prod. Rep.*, 2008, **25**, 905; (k) I. Paterson, *Pure Appl. Chem.*, 1992, **64**, 1821.

56. C. D. F. Milagre, H. M. S. Milagre, P. J. S. Moran and J. A. R. Rodrigues, *J. Org. Chem.*, 2010, **75**, 1410.

57. (a) E. P. Siqueira Filho, J. A. R. Rodrigues and P. J. S. Moran, *Tetrahedron: Asymmetry*, 2001, **12**, 847; (b) M. Hall, B. Hauer, R. Stuermer, W. Kroutil and K. Faber, *Tetrahedron: Asymmetry*, 2006, **17**, 3058; (c) M. Hall, C. Stueckler, W. Kroutil, P. Macheroux and K. Faber, *Angew. Chem., Int. Ed.*, 2007, **46**, 3934; (d) R. Stuermer, B. Hauer, M. Hall and K. Faber, *Curr. Opin. Chem. Biol.*, 2007, **11**, 203; (e) A. Fryszkowaska, K. Fischer, J. M. Gardiner and G. M. Stephens, *J. Org. Chem.*, 2008, **73**, 4295; (f) M. Hall, C. Stueckler, H. Ehammer, E. Pointner, G. Oberdorfer, K. Gruber, B. Hauer, R. Stuermer, W. Kroutil, P. Macheroux and K. Faber, *Adv. Synth. Catal.*, 2008, **350**, 411; (g) M. Hall, C. Stueckler, B. Hauer, R. Stuermer, T. Friedrich, M. Breuer, W. Kroutil, P. Macheroux and K. Faber, *Eur. J. Org. Chem.*, 2008, **73**, 1511.

58. K. Faber, *Biotransformations in Organic Chemistry*, Springer-Verlag, Berlin, 4th edn, 2000.

59. (a) M. Carvalho, M. T. Okamoto, P. J. S. Moran and J. A. R. Rodrigues, *Tetrahedron*, 1991, **47**, 2073; (b) E. C. S. Brenelli, M. Carvalho, M. T. Okubo,

M. Marques, P. J. S. Moran, J. A. R. Rodrigues and E. P. M. Sorrilha, *Indian J. Chem.*, 1992, **31B**, 821.

60. L. C. Fardelone, J. A. R. Rodrigues and P. J. S. Moran, *Enzyme Res.*, 2011, 976368.

61. M. Brewer, K. Ditrich, T. Haibicher, B. Hauer, M. Keeler, R. Sturmer and T. Zelinski, *Angew. Chem., Int. Ed.*, 2004, **43**, 788.

62. (a) D. D. Tanner, G. E. Diaz and A. Potter, *J. Org. Chem.*, 1985, **50**, 2149; (b) D. D. Tanner and H. K. Singh, *J. Org. Chem.*, 1986, **51**, 5182; (c) D. D. Tanner, H. K. Singh, A. Kharrat and A. R. Stein, *J. Org. Chem.*, 1987, **52**, 2142; (d) D. D. Tanner and A. R. Stein, *J. Org. Chem.*, 1988, **53**, 1642; (e) L. M. Aleixo, M. Carvalho, P. J. S. Moran and J. A. R. Rodrigues, *Biorg. Med. Chem. Lett.*, 1993, **3**, 1637; (f) J. Yuasa and S. Fukusumi, *J. Phys. Org. Chem.*, 2008, **21**, 886.

63. (a) M. Bertau, *Tetrahedron Lett.*, 2001, **42**, 1267; (b) M. Bertau, *Biocatal. Biotransform.*, 2002, **20**, 363.

64. (a) M. Hamdani, B. De Jeso, H. Deleuze, A. Saux and B. Maillard, *Tetrahedron: Asymmetry*, 1993, **4**, 1233; (b) G. Jorg and M. Bertau, *ChemBioChem*, 2004, **4**, 87; (c) O. Cabon, M. Larchevêque, D. Buisson and R. Azerad, *Tetrahedron Lett.*, 1992, **33**, 7337; (d) O. Cabon, D. Buisson, M. Larchevêque and R. Azerad, *Tetrahedron: Asymmetry*, 1995, **6**, 2199.

65. D. S. Zampieri, L. A. Zampieri, J. A. R. Rodrigues, B. R. S. de Paula and P. J. S. Moran, *J. Mol. Catal. B: Enzym.*, 2011, **72**, 289.

66. D. S. Zampieri, B. R. S. de Paula, L. A. Zampieri, J. A. Vale, J. A. R. Rodrigues and P. J. S. Moran, *J. Mol. Catal. B: Enzym.*, 2013, **61**, 85.

67. (a) F. van Rantwijk, R. M. Lau and R. A. Sheldon, *Trends Biotechnol.*, 2003, **21**, 131; (b) W. Hussain, D. J. Pollard, M. Truppo and G. J. Lye, *J. Mol. Catal. B: Enzym.*, 2008, **55**, 19; (c) M. Sureshkumar and C.-K. Lee, *J. Mol. Catal. B: Enzym.*, 2009, **60**, 1; (d) S. G. Cull, J. D. Holbrey, V. Vargas-Mora, K. R. Seddon and G. J. Lye, *Biotechnol. Bioeng.*, 2000, **69**, 227; (e) J. Howarth, P. James and J. F. Daí, *Tetrahedron Lett.*, 2001, **42**, 7517; (f) U. Kragl, M. Ecktein and N. Kaftzik, *Curr. Opin. Biotechnol.*, 2002, **13**, 565; (g) T. Matsuda, Y. Yamagischi, S. Koguchi, N. Iwai and T. Kitazume, *Tetrahedron Lett.*, 2006, **47**, 4619; (h) S. A. Gangu, L. R. Weatherley and A. M. Scurto, *Curr. Org. Chem.*, 2009, **13**, 1242.

68. (a) P. D' Arrigo, M. Lattanzio, G. P. Fantoni and S. Servi, *Tetrahedron: Asymmetry*, 1998, **9**, 4021; (b) G. J. A. Conceição, P. J. S. Moran and J. A. R. Rodrigues, *Tetrahedron: Asymmetry*, 2003, **14**, 43; (c) S. Serra, C. Fuganti and F. G. Gatti, *Eur. J. Org. Chem.*, 2008, 1031.

69. H. M. S. Milagre, C. D. F. Milagre, P. J. S. Moran, M. H. A. Santana and J. A. R. Rodrigues, *Org. Process Res. Dev.*, 2006, **10**, 611.

70. Y. Naoshima, Y. Munakata, T. Nishiyama, J. Maeda, M. Kamezawa, T. Haramaki and H. Tachibana, *World J. Microbiol. Biotechnol.*, 1991, **7**, 219.

71. E. M. Buque, I. Chin-Joe, A. J. J. Straathof, J. A. Jongejan and J. J. Heijnen, *Enzyme Microb. Technol.*, 2002, **31**, 656.

72. H. Tanaka, Y. Kaneko, H. Aoyagi, Y. Yamamoto and Y. Funukaga, *J. Ferment. Bioeng.*, 1996, **81**, 220.
73. H. M. S. Milagre, C. D. F. Milagre, P. J. S. Moran, M. H. A. Santana and J. A. R. Rodrigues, *Enzyme Microbiol. Biotechnol.*, 2005, **37**, 121.
74. R. Wendhausen Jr, P. J. S. Moran and J. A. R. Rodrigues, *J. Mol. Catal. B: Enzym.*, 1998, **5**, 69.

Production of Polymers by White Biotechnology

S. SHODA*[a], A. KOBAYASHI[a], AND S. KOBAYASHI[b]

[a]Tohoku University, Department of Biomolecular Engineering, Aoba, Aoba-ku, Sendai 980-8579, Japan; [b]Kyoto Institute of Technology, Center for Fiber and Textile Science, Matsugasaki, Sakyo-ku, Kyoto 606-8585, Japan
*E-mail: shoda@poly.che.tohoku.ac.jp

11.1 Introduction – Production of Polymers *via* Conventional Chemical Processes

How was the modern polymer industry established? Since ancient times, human beings have used polymeric materials for daily necessities with wool, silk, food, rubber, amber, paper and wood being familiar examples. The main components of these polymeric materials are polymers that consist of repeating structural units created through polymerization. Polymerization is the process of reacting monomer molecules together, in most cases, catalyzed by an appropriate catalyst.[1,2] In nature, we can see many examples of polymerizations for the construction of biopolymers. Cellulose, the most abundant organic compound on earth, is formed by the polycondensation of uridine diphosphate glucose (UDP-glucose) as the monomer, catalyzed by the cellulose synthase enzyme.[3,4] Natural rubber is formed by the polycondensation of isopentenyl pyrophosphate by the action of the rubber transferase enzyme, where the monomer adds to the pyrophosphate end of the growing

RSC Green Chemistry No. 45
White Biotechnology for Sustainable Chemistry
Edited by Maria Alice Z. Coelho and Bernardo D. Ribeiro

Published by the Royal Society of Chemistry, www.rsc.org

polymer chain.[5] In spite of the great utility of these polymeric materials, it was not until Hermann Staudinger determined the structure of cellulose in 1920 that the existence of polymers gained wide acceptance in the scientific community.[6]

Most polymer production before 1920 was achieved by modifying naturally occurring polymeric substances. In 1839, Charles Goodyear invented the process of vulcanization, which made natural rubber into a tough elastic material.[7] In 1870, John Wesley Hyatt made celluloid (cellulose nitrate), the first thermoplastic made from chemically modified cellulose.[8] Even under these circumstances, a chemist created a completely artificial polymer by reacting small molecules together. In 1907, Leo Bakeland developed Bakelite, a thermosetting phenol formaldehyde resin, as the first plastic made from phenol and formaldehyde.[9] Since the beginning of the 20th century, various kinds of polymers, such as polyethylene, polyvinylchloride, polystyrene, synthetic rubbers, polyesters and nylons, have been created through rapid advances in organic chemistry, resulting in the realization of the huge polymer industry today.

Since the beginning of the polymer industry in the first half of the 20th century, most starting monomers have been prepared from petroleum. Ethylene is produced by steam cracking in which saturated hydrocarbons are broken down into smaller hydrocarbons at 750–950 °C, introducing unsaturation. Vinyl chloride is produced by the hydrochlorination of acetylene or the dehydrochlorination of 1,2-dichloroethane. Styrene is produced by the catalytic dehydrogenation of ethylbenzene, which is prepared on a large scale by alkylation of benzene with ethylene. The important monomer, 1,3-butadiene, used in the production of synthetic rubbers, is produced as a byproduct of the steam cracking process. Monomers containing a carbon–carbon double bond are extremely important for addition polymerizations.

Monomers for condensation polymerizations have also been produced in the petrochemical industry. Ethylene glycol, an important precursor to polyesters like polyethylene terephthalate (PET), is produced by the hydration of ethylene oxide prepared by oxidizing ethylene. The other monomer for PET synthesis, terephthalic acid, is produced by air-oxidation of *p*-xylene through a *p*-toluic acid intermediate. Hexamethylenediamine, the monomer for the production of nylon 66, is presently produced by the hydrogenation of adiponitrile. The other monomer for nylon 66 synthesis, adipic acid, is produced by several methods, for example, the oxidation of a mixture of cyclohexanol and cyclohexanone called "KA oil" by nitric acid, the carbonylation of butadiene, and the oxidation of cyclohexene by using hydrogen peroxide. In these examples, almost all monomers are produced using rather strong acids, bases, oxidants, and reductants.

In addition to monomer synthesis, there is one more important element for polymer production, namely, the use of catalysts.[2] In fact, the advancement of the polymer industry is indebted to the discovery of new polymerization catalysts. Historically, polymerization catalysts utilized classical catalysts of acids (Brønsted acids, Lewis acids, and various cations), bases (Lewis bases and various anions), and radical generating compounds. In the middle of the 20th century, transition metals in Ziegler–Natta catalysts and later in metathesis catalysts began to be used, as well as rare-earth metals in these catalysts. These catalysts still have major roles in polymer synthesis. We have so far described the importance of the preparation of monomers, as well as the development of new polymerization catalysts. These two elements are indispensable for the production of general-purpose polymers, as well as functional polymers, in the future polymer industry.

Let us turn our attention to fermentation, another activity of human beings for the production of useful materials.[10] A wider definition of fermentation is the transformation of organic substances into other useful compounds by the action of enzymes that are produced by microorganisms such as molds, yeasts, and bacteria. Since ancient times, human beings have utilized fermentation for making bread, cheese, wine, *etc.*, without understanding the existence of the microorganisms involved. After several breakthroughs in the second half of the 19th century, including Louis Pasteur's discovery[11] that living microorganisms cause fermentation, in 1897 Eduard Buechner finally succeeded in extracting a juice yeast that ferments a sugar solution, affording alcohol and carbon dioxide.[12] Triggered by these discoveries, fermentation technology advanced by mutating the microorganisms by physical and chemical treatment.

Early studies on the microbial production of chemicals were mainly carried out by identifying appropriate microorganisms that can produce target products with high efficiency. The performance of these microorganisms has been improved by random mutagenesis and optimization of the fermentation conditions. The recent great progress in biology has broadened the spectrum of target products where even unnatural chemicals can be produced to satisfactory levels.[13,14] Nowadays, microorganisms have been utilized in industrial fermentations, leading to the production of useful compounds such as vitamins, antibiotics, citric acid and amino acids.

Based on the great progress of the two fields with different cultural backgrounds, namely organic polymer chemistry and fermentation chemistry, researchers of the 21st century have taken advantage of their accumulated knowledge and technologies for polymer synthesis. This chapter reviews the production of monomers, as well as polymers, based on the concept of "White Biotechnology". The phrase "polymer production by white biotechnology" is defined as "technologies for synthesis of monomers and polymers by using microorganisms and enzymes" where less energy is required and less waste is produced. The focus is laid on from the viewpoint of synthetic reactions, rather than from the viewpoint of the environmentally benign properties of the resulting polymers. Consequently, the evaluation of biodegradable polymers will not be discussed here. The contents of the present chapter are summarized in Figure 11.1, where the notations in parentheses correspond to the section numbers describing each synthetic process of monomers or polymers.

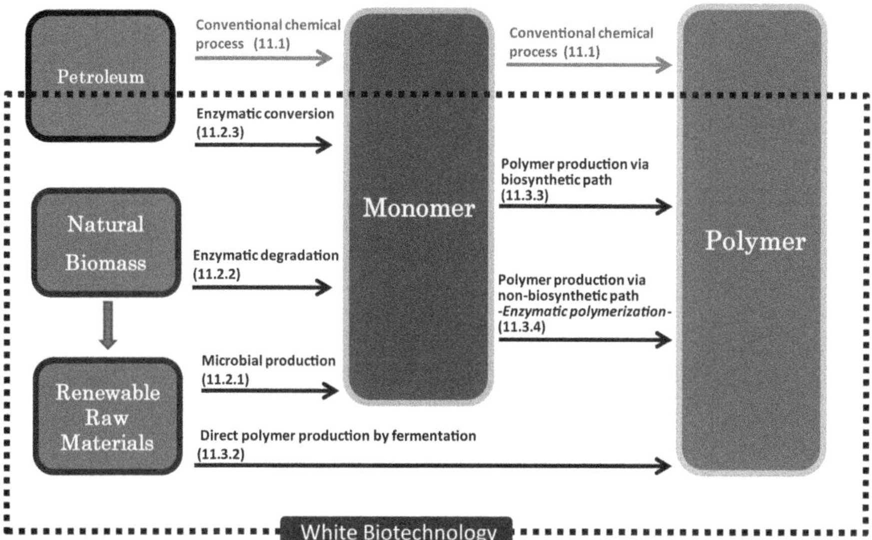

Figure 11.1 Production of monomers and polymers based on the concept of white biotechnology (notation in parentheses corresponds to section numbers).

11.2 Monomer Production by White Biotechnology

11.2.1 Microbial Production of Monomers

The great merit of using fermentation technology for the production of monomers is that valuable chemicals can be obtained from inexpensive feedstocks like sugars. Several important low molecular weight compounds employed for addition polymerizations and condensation polymerizations can be prepared *via* processes including microbial or enzymatic conversions (Table 11.1).

11.2.1.1 Ethylene

Ethylene can be prepared by the dehydration of ethanol. Most of our concern from the viewpoint of white biotechnology is, therefore, how to produce ethanol (bioethanol) by fermentation efficiently. Microbial ethanol is produced from corn, potato and sugar cane. Due to concern about increased food price, cellulosic ethanol has become more promising because cellulose fibers are a major component in plant cell walls and can be derived from plant waste.[15]

11.2.1.2 Propylene

Since propylene can be synthesized by metathesis between ethylene and 2-butene which is available from isobutanol *via* two steps, our concern is how to produce isobutanol by fermentation. Bioengineering technology has made

Table 11.1 Monomer production *via* microbial or enzymatic processes.

Monomer (or its precursor)	Bio-process	References
Monomer for addition polymerization		
Ethylene (bioethanol)	$(C_6H_{10}O_5)_n$ $\xrightarrow{\text{Fermentation}}$ CH_3CH_2OH $\xrightarrow{\text{Catalytic dehydration}}$ $CH_2{=}CH_2$	61
Propylene (bioisobutanol)	Cellulose $\xrightarrow{\text{C. cellulolyticum}}$ CH_3CHCH_2OH (CH_3) $\xrightarrow{\text{Catalytic dehydration}}$ $CH_3C{=}CH_2$ (CH_3)	16
	$\xrightarrow[\text{Rh-complex}]{\text{Isomerization}}$ $CH_3CH{=}CHCH_3$ $\xrightarrow[\text{Metathesis}]{CH_2{=}CH_2}$ $2\ CH_3CH{=}CH_2$	62 and 63
Isoprene	Glucose $\xrightarrow[\text{Agrobacterium, etc.}]{\text{Bacillus,}}$ (isoprene)	17
Acrylamide	$CH_2{=}CH{-}C{\equiv}N$ $\xrightarrow{\text{Nitrile hydratase}}$ $CH_2{=}CH{-}C({=}O){-}NH_2$	64
Acrylic acid	$CH_2{=}CH{-}C{\equiv}N$ $\xrightarrow{\text{Nitrilase}}$ $CH_2{=}CH{-}C({=}O){-}OH$	65
Monomers for condensation polymerization		
	$(C_6H_{10}O_5)_n$ $\xrightarrow{\text{Fermentation}}$ CH_3CH_2OH $\xrightarrow{\text{Catalytic dehydration}}$ $CH_2{=}CH_2$	66 and 67
Diols	$\xrightarrow[\text{thermal hydrolysis}]{\text{Oxidation followed by}}$ $HO{-}CH_2CH_2{-}OH$	
	(glycerol) $\xrightarrow[\text{or Lactobacillus}]{\text{Klebsiella, enterobacter, Clostridium,}}$ $HO{-}CH_2CH_2CH_2{-}OH$	22
	Biomass $\dashrightarrow$ (succinic acid) $\xrightarrow{\text{Catalytic Hydrogenation}}$ $HO{-}(CH_2)_4{-}OH$	68

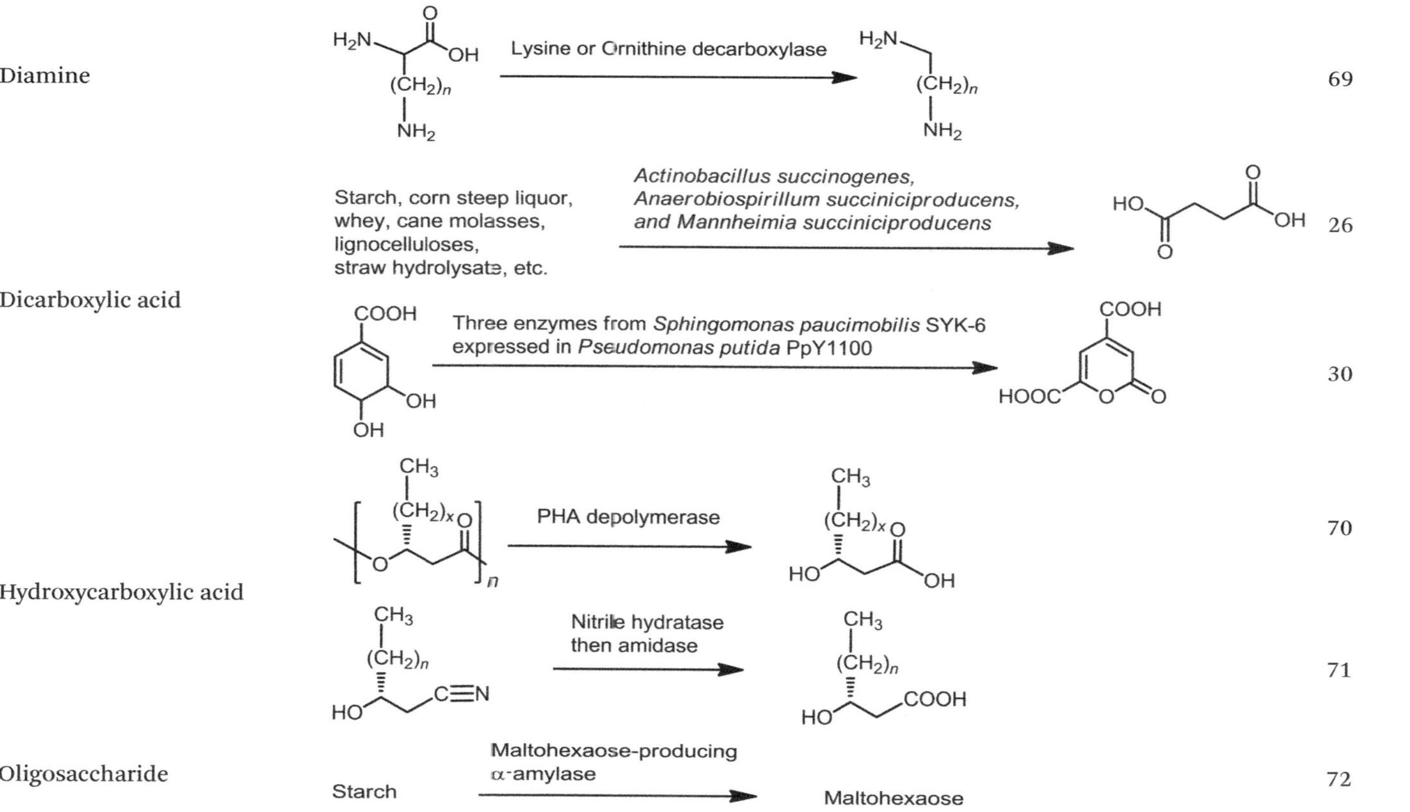
Diamine
Lysine or Ornithine decarboxylase
69
Dicarboxylic acid
Starch, corn steep liquor, whey, cane molasses, lignocelluloses, straw hydrolysate, etc.
Actinobacillus succinogenes, Anaerobiospirillum succiniciproducens, and Mannheimia succiniciproducens
26
Three enzymes from Sphingomonas paucimobilis SYK-6 expressed in Pseudomonas putida PpY1100
30
Hydroxycarboxylic acid
PHA depolymerase
70
Nitrile hydratase then amidase
71
Oligosaccharide
Starch
Maltohexaose-producing α-amylase
Maltohexaose
72

it possible to produce isobutanol by using microorganisms, such as an engineered strain of *Clostridium cellulolyticum* using cellulose as starting material.[16]

11.2.1.3 Isoprene

Isoprene is produced and released by many species of trees, such as oaks and eucalyptuses, into the atmosphere. Isoprene is an important five-carbon diene monomer for poly(isoprene) synthesis. Various bacterial species, both gram-negative and gram-positive, have been found to produce isoprene.[17,18]

11.2.1.4 Diols

DuPont Tate & Lyle BioProducts developed a new process for the production of 1,3-propanediol, a useful monomer in the synthesis of polytrimethylene terephthalate, starting from corn syrup by a genetically modified strain of *Escherichia coli.*[19] The process uses 40% less energy than the conventional process and reduces greenhouse gas emission by 20%. A sucrose-based process for the manufacture of 1,4-butanediol by an engineered microorganism is also expected to bring about substantial cost advantages and the reduction of greenhouse gas emissions. However, most 1,4-butanediol is still produced from chemical processes.

11.2.1.5 Diamines

Putrescine (1,4-diaminobutane) and cadaverine (1,5-diaminobutane) are attracting much interest due to their potential use in the polymer industry. Biotechnological production of putrescine has been developed by using an engineered strain of *Escherichia coli*. Cadaverine, 1,5-diaminopentane, has successfully been obtained from engineered *Corynebacterium* or *Escherichia coli.*[20]

11.2.1.6 Dicarboxylic Acid

Succinic acid, a linear four carbon dicarboxylic acid, can be used as a starting material for the production of 1,4-butanediol, tetrahydrofuran, γ-butyrolactone, 2-pyrrolidone, or N-methylpyrrolidone. Various bacteria such as *Actinobacillus succinogenes* have been reported to produce succinic acid industrially by using several feedstocks, for example, corn starch, lignocellulose, and straw hydrolysates.[21–25] Many companies are now pursuing succinic acid production aiming at its industrialization as a microbial production business.[26]

11.2.1.7 Hydroxycarboxylic Acid

Lactic acid (LA) is used as a monomer for the synthesis of poly(lactic acid) (PLA), which is utilized as a biodegradable polymer. The preparation of LA based on organic chemistry is not desirable because organic processes give a racemic mixture. For the PLA industry, L-lactic acid with high optical purity (98–99%) is required. Therefore, the most common process for LA production is *via* fermentation.[27] Two molecules of LA are produced from one molecule of glucose *via* the "glycolysis" pathway. Lactic acid can be dehydrated to give lactide, which can be subsequently used as a monomer for the synthesis of PLA.

11.2.2 Monomer Synthesis by Enzymatic Degradation of Naturally Occurring Polymers

In the previous section, we have seen several examples of monomer synthesis by fermentation, which is very suitable for the production of low molecular weight monomers with simple structures. To prepare more complex monomers with higher molecular weights, it is necessary to develop an alternative methodology for monomer production. The fragmentation of naturally occurring biopolymers is a promising candidate for this purpose. Typical chemical fragmentation processes are not suitable, because they require the use of strong acids or thermolysis, causing the decomposition of target monomers. Bacterial digestion is not suitable either for the production of complex monomers due to difficulty in controlling the fragmentation. Enzymatic degradation of naturally occurring polymers is the most promising method, being supported by neighborhood technologies, such as mechanical mills, to ensure degradation.

11.2.2.1 Oligosaccharides

Oligosaccharides, which are compounds that consist of a small number of monosaccharides connected through glycosidic bonds, are regarded as fragments of polysaccharides. To prepare oligosaccharide monomers, it is necessary to control the cleavage of the glycosidic bonds of polysaccharides to avoid overdegradation. There have been no general methods for the selective production of oligosaccharides with a definite degree of polymerization with high efficiency. The conventional method involves the partial hydrolysis of the glycosidic bonds of polysaccharides by the combined use of acids and glycosidase enzymes, followed by chromatographic separation, which gives an oligosaccharide in most cases with low efficiency. As exceptions, malto-oligosaccharides with definite DPs (3–6) can be produced efficiently by selective hydrolysis of starch catalyzed by an amylase from different origins, depending upon the required DPs.[28,29]

11.2.2.2 Lignin

Lignin is one of the most abundant natural carbon resources, which exists in trees at 15–36% by weight. Development of an efficient biotechnological process for the transformation of the highly networked structure of lignin into small molecules with well-defined structures is in demand. A new monomer for bio-based polymers, 2-pyrone-4,6-dicarboxylic acid (PDC), was produced from protocatechuate by using a soil microorganism, *Sphingomonas paucimobilis* SYK-6.[30]

11.2.3 Enzymatic Conversion of Vinyl Monomers

Enzymatic modification of petrochemically derived compounds is one of the most suitable techniques to synthesize refined monomers. One of the representative industrial processes is the conversion of acrylonitrile to acrylamide or acrylic acid using nitrile hydratase or nitrile aminohydrolase, respectively, as catalysts.[31] Glycerol can also be converted to 1,3-propanediol and dihydroxyacetone by biotechnological processes.[32]

11.3 Polymer Production by White Biotechnology

11.3.1 General Aspects

As a result of significant progress in biological chemistry, the mechanisms of various biosynthetic routes, including those of biopolymers, have been elucidated. These circumstances give us many opportunities to achieve polymerizations based on white biotechnology by using fermentation processes, as well as reactions catalyzed by various enzymes such as transferases, hydrolases, and oxidases. Depending on the reaction mode by which the polymerization proceeds, *via* a biosynthetic pathway or a non-biosynthetic pathway, these polymerizations can be classified into the following two categories.

1. Production of polymers *via* biosynthetic pathways *in vivo* or *in vitro, e.g.* polymerization of a phosphate monomer catalyzed by a polymerase enzyme in a living system or in the cell-free extract.
2. Production of polymers *in vitro via* non-biosynthetic pathways catalyzed by an isolated enzyme, *e.g.* dehydrative polycondensation between a diol and dicarboxylic acid catalyzed by a hydrolase enzyme in a test tube (enzymatic polymerization).

11.3.2 Polymer Production by Microorganisms (Table 11.2)

11.3.2.1 *Polyhydroxyalkanoates (PHAs)*

PHAs are synthesized by numerous microorganisms as energy reserve materials and completely degrade to water and carbon dioxide under aerobic conditions, and to methane under anaerobic conditions.[33–36] Polyhydroxybutyrate (PHB), a polyester isolated in 1925 by Maurice Lemoigne, is accumulated in many bacteria as a membrane enclosed inclusion at up to 80% of the dry cell weight.[37] In industrial production, PHAs are extracted and purified from bacteria by optimizing the conditions for microbial fermentation of sugar or glucose. In the 1980s, Imperial Chemical Industries developed

Table 11.2 Polymer production by microorganisms.

Polymer	Microorganisms/media conditions	References
Polyhydroxyalkanoate (PHA)		
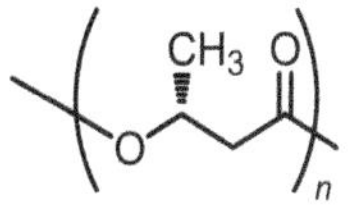	*Bacillus megaterium* *Bacillus subtilis*	37
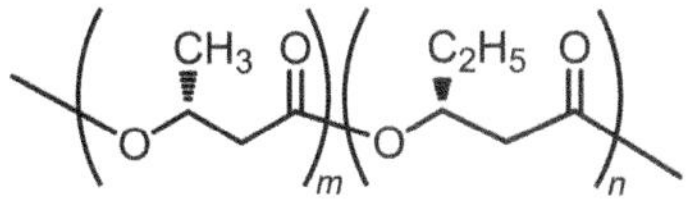	*Ralstonia eutropha*/propionic acid	73
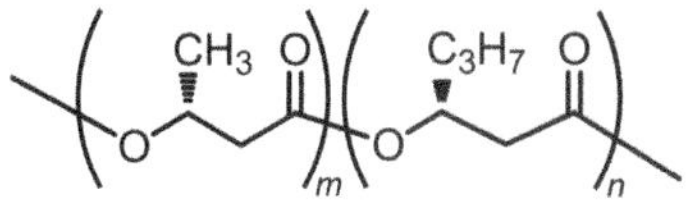	*A. latus*, *Aeromonas caviae*/1,5-pentandiol and olive oils	74
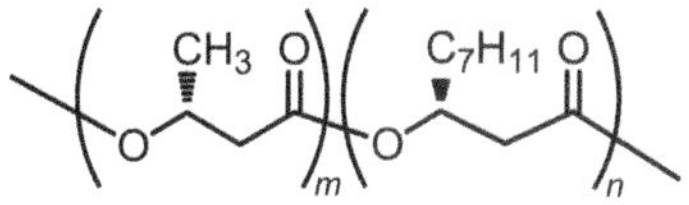	*Pseudomonas* sp., *R. eutropha*/sugar and 4-hydroxybutylic acid	75
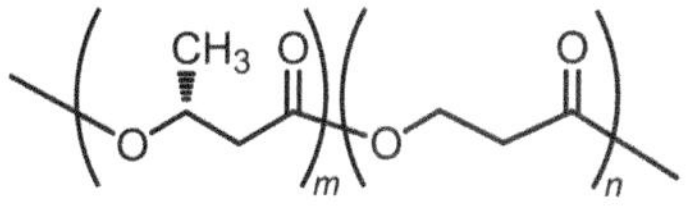	*R. eutropha*/pentanoic acid and 3-hydroxypropionic acid	76
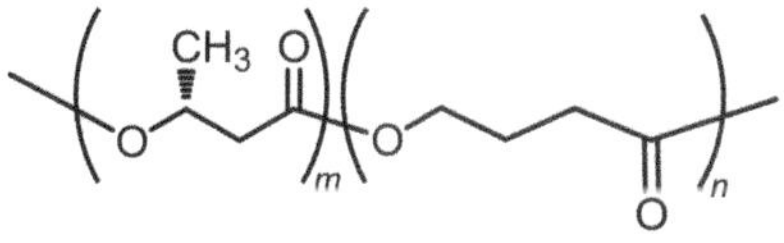	*A. lantas*, *Comamonas acidovorans*/γ-butyrolactone, 1,4-butanediol, 1,6-hexanediol	77 and 78
Poly(lactic acid) (PLA)		
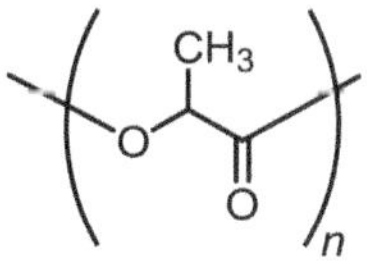	Metabolically engineered *Escherichia coli*/glucose	79
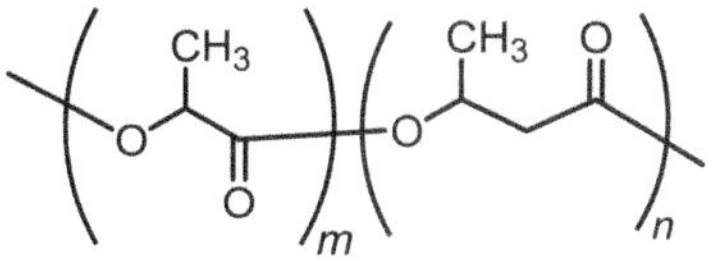	Recombinant strains of *Escherichia coli*	38
Polyamide		
Poly-ε-lysine(ε-PL)		
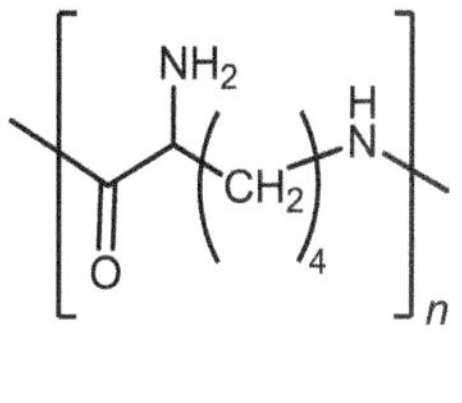	*Streptomyces albulus*	80

(*continued*)

Table 11.2   (*continued*)

Polymer	Microorganisms/media conditions	References
Poly-γ-glutamic acid (γ-PGA)	*Bacillus* strain	81
Cyanophycin	Recombinant strain of *Escherichia coli*	40
Polysaccharides		
Hyaluronic acid	*Bacillus subtilis*	82
	Lactococcus lactis	83
	Escherichia coli	84
Alginate	*Pseudomonas aeruginosa*	85
	Azotobacter vinelandii	86 and 87

poly(3-hydroxybutyrate-*co*-3-hydroxyvalerate) obtained *via* fermentation which was named "Biopol". Various PHA copolymers comprising other hydroxyalkanoic acids can be prepared by controlling the fermentation conditions.

11.3.2.2 *Poly(lactic acid) (PLA)*

PLA is chemically synthesized by a two-step process: fermentative production of lactic acid, followed by chemical polymerization. At present, industrial production of PLA has not been achieved by a perfect bioprocess. The first LA-polymerizing enzyme was discovered in 2008.[38] A new one-step method for the production of the PLA homopolymer and its copolymer, poly(3-hydroxybutyrate-*co*-lactate), P(3HB-*co*-LA), by direct fermentation of metabolically engineered *Escherichia coli* was reported. Using this engineered strain, the PLA homopolymer could be produced at up to 11 wt% from glucose. P(3HB-*co*-LA) copolymers containing 55–86 mol% lactate could be produced at up to 56 wt% from glucose and 3HB.[39]

11.3.2.3 Polyamides

Three different kinds of polymers of amino acids are known in nature: poly-γ-glutamic acid (γ-PGA), poly-ε-lysine (ε-PL) and cyanophycin. A major constituent of the Japanese food "Natto", γ-PGA, is formed by bacterial fermentation and has potential medical uses such as drug delivery. Poly-ε-lysine, a homopolymer of L-lysine, is used as a preservative in food. This polymer can be prepared by fermentation using *Streptomyces albulus* in commercial production. Cyanophycin, an amino acid polymer composed of an aspartic acid backbone and an arginine side chain, is produced by the use of *Escherichia coli on a* large scale.[40]

11.3.2.4 Bacterial Cellulose

Bacterial cellulose (BC) is synthesized by the acetic bacterium *Acetobacter xylinum*.[41] Bacterial cellulose has a wide range of current and future applications in the food, medical, and cosmetics industries. One of the most familiar bacterial celluloses, nata-de-coco, is now manufactured in a large quantity at the level of home industry in Southeast Asian countries such as the Philippines and Indonesia.

11.3.2.5 Hyaluronan (Hyaluronic Acid, HA)

Hyaluronan, a linear polysaccharide made of alternating *N*-acetyl-D-glucosamine and D-glucuronic acid, has significant structural, rheological, physiological, and biological functions, leading to a wide range of applications in the cosmetics and pharmaceutical industries.[42,43] HA has been successfully produced on an industrial scale with *Streptococcus* sp. as the main producer.

11.3.2.6 Alginates

Alginates are unbranched polysaccharides consisting of $1 \rightarrow 4$ linked β-D-mannuronic acid (M) and its C-5 epimer α-L-guluronic acid (G).[44] The natural copolymer is an important component of algae. While it is possible to obtain alginates from both algal and bacterial sources, commercially available alginates currently come only from algae. Industrial alginate production is approximately 30 000 metric tons annually, and is estimated to comprise less than 10% of biosynthesized alginate material.[45] The combination of chemical and biochemical techniques provides considerable potential for creating modified alginic acid derivatives with control over monosaccharide sequence, location and quantity of substituents.

11.3.3 Polymer Production *via* Biosynthetic Pathways *In vitro*

Several important biopolymers can be prepared in a test tube using biosynthetic reactions. The polymerase chain reaction (PCR) is an indispensable technology in biological research, including DNA cloning to amplify a piece of DNA.[46] *In vitro* β-glucan production has been achieved, catalyzed by digitonin-solubilized enzyme preparations from plasma membrane-enriched fractions of cotton fiber cells.[47]

11.3.4 Enzymatic Polymerization

The concept of "enzymatic polymerization" provides us with a new strategy for producing useful materials, including natural and unnatural (synthetic) polymers, giving more precise construction of well-defined structures than conventional chemical catalysis.[48–50] Enzymatic reactions take place under mild conditions and contribute promisingly to global sustainability, reducing the usage of energy, acids, bases, and toxic reagents.

11.3.4.1 Polyester Synthesis Catalyzed by Enzymes (Table 11.3)

11.3.4.1.1 Condensation Polymerization by Lipase[89–97]. Polyesters are polymers which contain ester moieties in their main chains. Organic ester moieties can be constructed by condensing a carboxylic acid with a hydroxyl compound such as an alcohol or phenol. To form a polyester, there are two possible modes of condensation, that is, self-polycondensation of a hydroxy acid and polycondensation between a dicarboxylic acid and a diol.

Lipases, enzymes for the hydrolysis of fat in nature, also catalyze esterification and transesterification, and can be employed for condensation

Table 11.3 Lipases for polyester synthesis.[88]

Origin	Name	Supplier
Microorganism		
Aspergillus niger	Lipase A	Amano Enzymes
Burkholderia cepacia	Lipase PS	Amano Enzymes
Candida antarctica	CALB	Sigma-Aldrich
	Novozym-435	Novozymes
	Lipase CA	Novozymes
Candida cylindracea	Lipase CC	Biocatalysts
Candida rugosa	Lipase CR	Sigma-Aldrich
Klebsiella oxytoca	Lipase K	Nagase Seikagaku
Mucor miehei	Lipase MM	Novozymes
Pseudomonas cepacia	Lipase PC	Amano Enzymes
Pseudomonas fluorescence	Lipase PF	Biocatalysts
Penicillium roqueforti	Lipase PR	Sigma-Aldrich
Rhizopus japonicus	Lipase RJ	Nagase Biochemicals
Others		
Hog liver	HLE (Esterase)	Sigma-Aldrich
Porcine pancreas	PPL	Sigma-Aldrich

polymerizations (Table 11.4). Lipases are highly stable, even in the presence of organic solvents and at higher temperatures. Various factors, such as temperature, reaction medium, and water concentration, affect the stability of lipases, which must be considered to industrialize an enzymatic process.

The key intermediate in lipase-catalyzed esterification is an active ester formed as a result of the participation of the hydroxy group on the serine residue in the enzyme catalytic site. The resulting intermediate is then attacked by an alcohol, affording an ester. Various carboxylic acid monomers, such as free acids, methyl esters, vinyl esters, and haloalkyl esters, have been employed as monomers for lipase-catalyzed polyester synthesis. The chain length of the monomer is also an important factor for the polymerization to occur. For example, in the polymerization of an α,ω-dicarboxylic acid and a glycol, the polymerization behavior was greatly affected by the methylene chain length of the monomers. The polyester was obtained in good yields from 1,10-decanediol, whereas no polymer formation was observed from 1,6-hexanediol, suggesting that the combination of monomers with appropriate hydrophobicity is needed for polymer production.

11.3.4.1.2 Ring-Opening Polymerization (ROP) of Lactones by Lipase[98-111]. Lactones, cyclic esters, are excellent starting substrates for the formation of the active ester by the participation of the serine residue in the catalytic site of the lipase catalyst. Since the ROP of ε-caprolactone (ε-CL, 7-membered) and δ-valerolactone (δ-VL, 6-membered) was discovered in 1993, lipase-catalyzed ROP has been extensively studied (Table 11.5).

11.3.4.2 Polysaccharide Synthesis Catalyzed by Enzymes

The use of an enzyme for the glycosylation process is considered to be one of the most promising methodologies for selective construction of a glycosidic linkage. The enzymes which have so far been utilized for glycosidic bond formation are glycosidases, glycosyl transferases, and phosphorylases. A glycosidic bond is formed as a result of the nucleophilic attack of a hydroxyl group of the acceptor on the anomeric center of the glycosyl–enzyme intermediate formed in the catalytic site of the enzyme.

11.3.4.2.1 Condensation Polymerization Catalyzed by Glycosidases[112-124]. Glycosyl fluorides, sugar derivatives whose anomeric hydroxyl group is replaced by a fluorine atom, are useful glycosyl monomers for enzymatic polymerization. For example, the *in vitro* synthesis of cellulose *via* a non-biosynthetic pathway has been achieved by enzymatic polymerization of β-D-cellobiosyl fluoride monomer catalyzed by cellulase from *Trichoderma viride*, an extracellular hydrolase of cellulose. Recently, hydrolytically inactive mutants of glycosidases (glycosynthase) have been developed to improve transglycosylation yields. An enzymatic polymerization by using one-step preparable glycosyl monomers having the 4,6-dimethoxy-1,3,5-triazin-2-yl

Table 11.4 Condensation polymerization by lipase.

Polymerization scheme	References
$HOOC-(CH_2)_x-COOH$ + $HO-(CH_2)_y-OH$ $\xrightarrow[\text{aqueous media}]{\textit{Aspergillus niger} \text{ Lipase}}$ $\left[\!\!-\overset{O}{\overset{\|}{C}}-(CH_2)_x-\overset{O}{\overset{\|}{C}}-O-(CH_2)_y-O-\!\!\right]_n$ $x = 4\text{-}6, 8, 10\text{-}12$; $y = 2,3$	89
$HOOC-(CH_2)_4-COOH$ + $HO-(CH_2)_4-OH$ $\xrightarrow[\text{Dialkyl ether}]{\substack{\text{Lipozyme IM-20}\\ \text{an immobilized lipase from } \textit{Mucor miehei}}}$ $\left[\!\!-\overset{O}{\overset{\|}{C}}-(CH_2)_4-\overset{O}{\overset{\|}{C}}-O-(CH_2)_4-O-\!\!\right]_n$	90
$HOOC-(CH_2)_x-COOH$ + $HO-(CH_2)_y-OH$ $\xrightarrow[\text{aqueous media}]{\textit{Lipase PC or CC}}$ $\left[\!\!-\overset{O}{\overset{\|}{C}}-(CH_2)_x-\overset{O}{\overset{\|}{C}}-O-(CH_2)_y-O-\!\!\right]_n$ $x = 4, 8, 12$; $y = 4, 8, 12$	91 and 92
$HOOC-(CH_2)_9-OH$ $\xrightarrow[\text{Benzene}]{\text{PEG-Lipase PF}}$ $\left[\!\!-\overset{O}{\overset{\|}{C}}-(CH_2)_9-O-\!\!\right]_n$	93
(ricinoleic acid) $\xrightarrow[\text{Hydrocarbon or Benzene}]{\text{Novozyme-435, PPL, CR, PS, PC}}$ (poly(ricinoleate))	94–96
(glucitol) + $HOOC-(CH_2)_4-COOH$ + $HO-(CH_2)_8-OH$ $\xrightarrow[\text{90 °C, vacuum}]{\text{Novozyme-435}}$ (copolyester)	97

Table 11.5 Ring opening polymerization of lactone and other cyclic monomers by lipases.

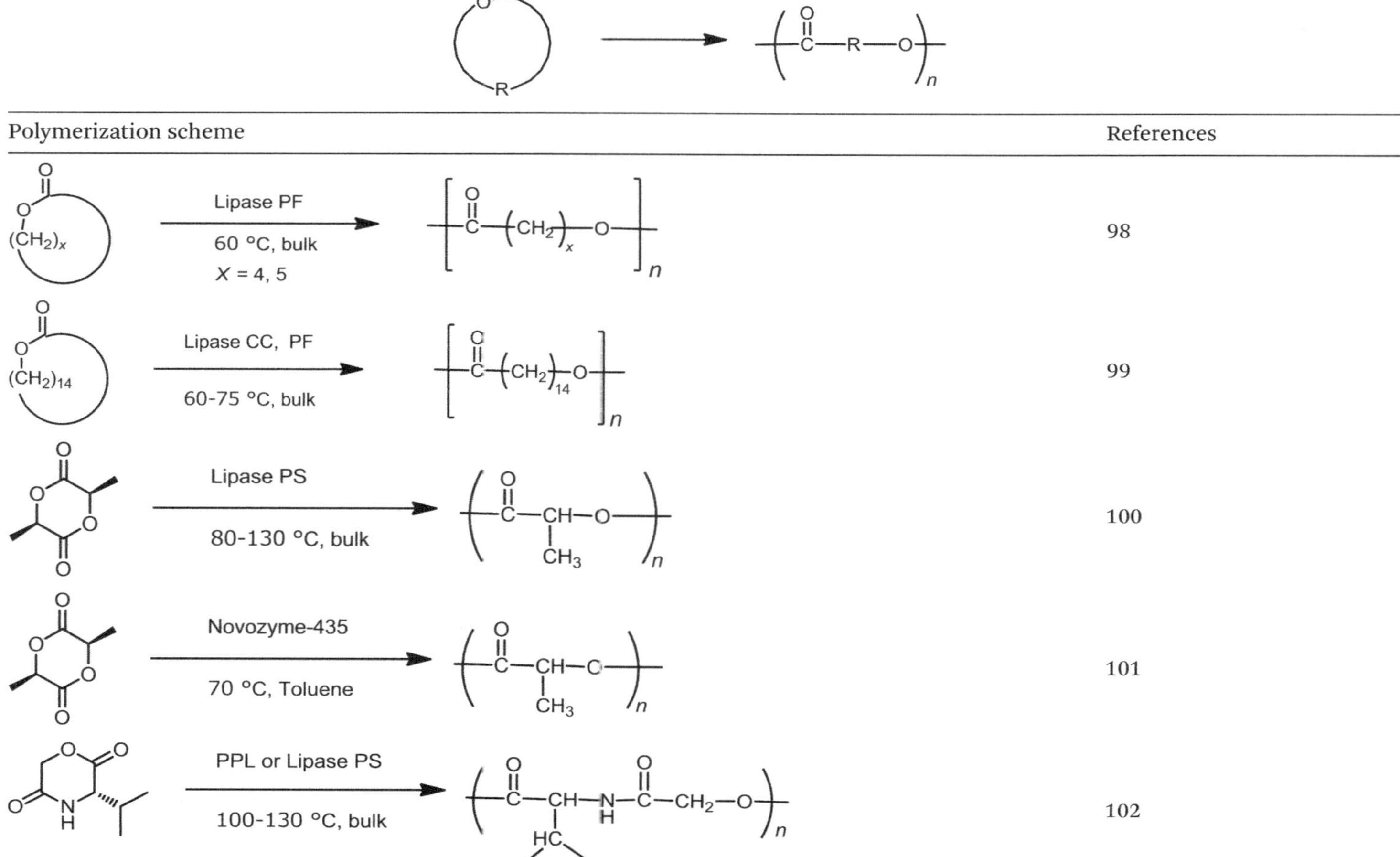

Polymerization scheme	References
Lipase PF, 60 °C, bulk, X = 4, 5	98
Lipase CC, PF, 60–75 °C, bulk	99
Lipase PS, 80–130 °C, bulk	100
Novozyme-435, 70 °C, Toluene	101
PPL or Lipase PS, 100–130 °C, bulk	102

(continued)

Table 11.5 (*continued*)

Polymerization scheme	References
PPL, 100–120 °C, bulk. R_1 = H, CH_3; R_2 = H, $CH(CH_3)_2$, $CH_2CH(CH_3)_2$, $CH(CH_3)CH_2CH_3$	103
PPL, 100 °C, bulk	104
Novozyme-435, 70 °C, solvent	105
Lipase CA, MM, PPL, PS, Novozyme-435, 60–100 °C, bulk	106–108
Novozyme-435, 90 °C, toluene	109
PPL, 40–100 °C, bulk. R = $CH(CH_3)_2$, $CH_2CH(CH_3)_2$	110 and 111

(DMT) group at the anomeric position has been demonstrated, where both the glycosyl monomer synthesis and the successive polymerization can be achieved in water without any protecting or deprotecting steps. In the nonaqueous medium of dimethylacetamide (DMAc)/LiCl, cellobiose was polycondensed by the action of a cellulase–surfactant complex (cellulase + dioleyl-*N*-D-glucona-L-glutamate), affording cellulose (DP > 100).

11.3.4.2.2 Addition Polymerization of Sugar Oxazolines Catalyzed by Glycosidases[125–129]. In the previous section, a review was made of enzymatic condensation polymerization using an activated glycosyl monomer having an appropriate leaving group at the anomeric center. The conformation of these glycosyl donors had to be very close to that of the product glycosides for the reaction to occur. The smooth acceptance of the glycosyl donor into the catalytic site of the enzyme was realized by the similarity in conformation between the product glycoside and the glycosyl donor. In this section, a new method for glycosylation by using a distorted glycosyl donor will be described, where the glycosyl donors are designed as a transition state analogue of enzymatic hydrolysis. Using sugar oxazoline as glycosyl donor, glycosidase-catalyzed addition polymerizations have been demonstrated, giving rise to various polysaccharides such as chitin and glycosaminoglycans (Table 11.6). It is to be noted that these addition reactions by the combined use of a transition state analogue substrate and *N*-acetylglucosaminidase proceed only in the direction of addition, while suppressing hydrolysis of the product in aqueous media.

11.3.4.2.3 Condensation Polymerization by Phosphorylases[130,131]. Amylose, α(1 → 4)glucan, can be synthesized by phosphorylase-catalyzed enzymatic polymerization of glucose 1-phosphate (G-1-P) monomer (Table 11.7). Phosphorylase is an enzyme that catalyzes the reversible phosphorolysis of α(1 → 4)glucan at the non-reducing end, giving G-1-P. Owing to the reversibility of the reaction, α(1 → 4)-glycosidic linkages can be formed by the phosphorylase-catalyzed chain-elongation using G-1-P as a glycosyl donor to give amylose. To initiate the polymerization, malto-oligosaccharides are required as a primer in the reaction medium.

Amylose is a well-known helical polysaccharide that forms inclusion complexes. It was found that inclusion complexes of amylose with synthetic polymers were formed when phosphorylase-catalyzed polymerization was performed in the presence of appropriate hydrophobic polymers, such as polyethers [poly(tetrahydrofuran) (PTHF), poly(oxetane) (POXT)] and polyesters [poly(ε-caprolactone) (PCL), poly(δ-valerolactone) (PVL)] (vine-twining polymerization).[51]

11.3.4.3 Polyaromatics Synthesis Catalyzed by Enzymes

In the past few decades, the enzymatic synthesis of polyaromatics has been extensively investigated (Table 11.8).[52] Several oxidoreductases (peroxidase, laccase, bilirubin oxidase, *etc.*) have been reported to catalyze the oxidative

Table 11.6 Glycanase-catalyzed synthesis of polysaccharides.

Polymerization scheme	References
Cellulase from *Trichoderma viride*, *Aspergillus niger, and Polyporus tulipiferae* — Acetonitrile/acetate buffer (pH 5.0)	112
Cellulase from *Trichoderma viride* — Acetonitrile/acetate buffer (pH 5.0)	113
α-Amylase from *Aspergillus oryzae, Bacillus subtilis, Bacillus licheniformis, porcine pancreas* — Methanol/phosphate buffer (pH 7.0)	114
Dextransucrase — buffer (pH 7.0)	115
Mutant cellulase EGI from *Trichoderma reesei* or *Humicola insoles* — Ammonium carbonate solution or phosphate buffer (pH 5.0)	116 and 117

118
119
120
121 and 122
Glycosynthase mutant of
Populus tremula x tremuloides xyloglucan
endo-transglycosylase
phosphate buffer (pH 5.0)
Cellulase from Trichoderma viride
Acetonitrile/acetate buffer (pH 5.0)
Glycosynthase mutant of
Populus tremula x tremuloides xyloglucan
endo-transglycosylase
phosphate buffer (pH 5.0)
2C$_{18}$Δ^{9}GE Lipid-coated Cellulase
From Trichoderma viride
8% LiCl/DMAc
2C$_{18}$Δ^{9}GE

(continued)

Table 11.6 (*continued*)

Polymerization scheme	References
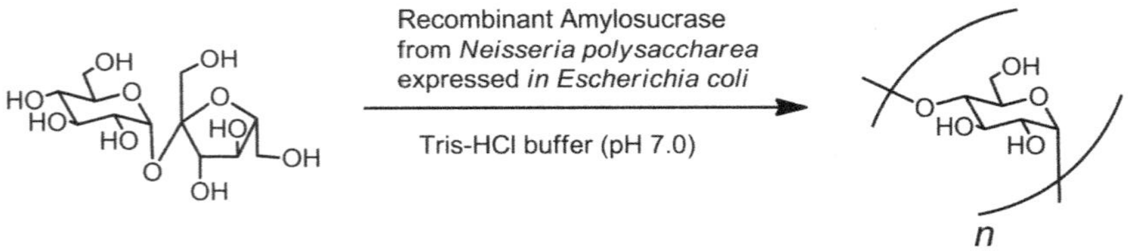 Recombinant Amylosucrase from *Neisseria polysaccharea* expressed *in Escherichia coli* → Tris-HCl buffer (pH 7.0)	123 and 124
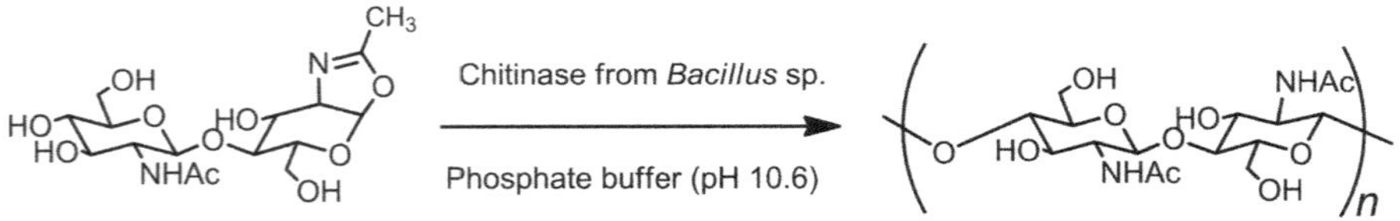 Chitinase from *Bacillus* sp. → Phosphate buffer (pH 10.6)	125 and 126
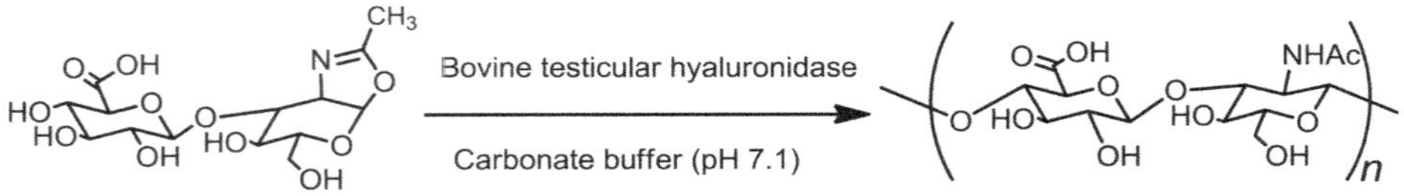 Bovine testicular hyaluronidase → Carbonate buffer (pH 7.1)	127
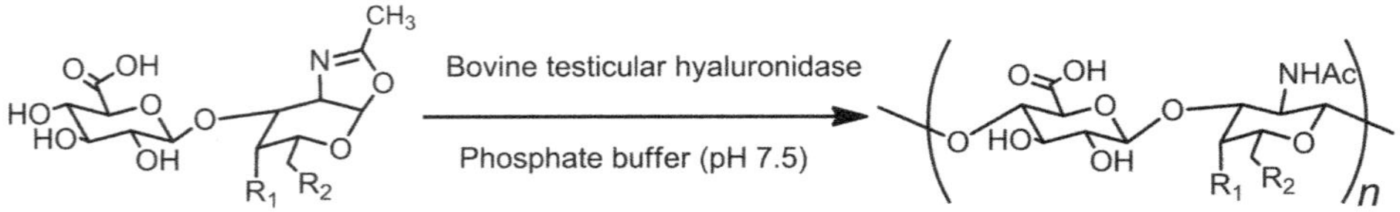 Bovine testicular hyaluronidase → Phosphate buffer (pH 7.5)	128 and 129

Chondroitin : $R_1=OH$, $R_2=OH$
Chondroitin 4-sulfate : $R_1=OSO_3^-$, $R_2=OH$
Chondroitin 6-sulfate : $R_1=OH$, $R_2=OSO_3^-$

Table 11.7   Phosphorylase-catalyzed synthesis of polysaccharides.

Polymerization scheme	References

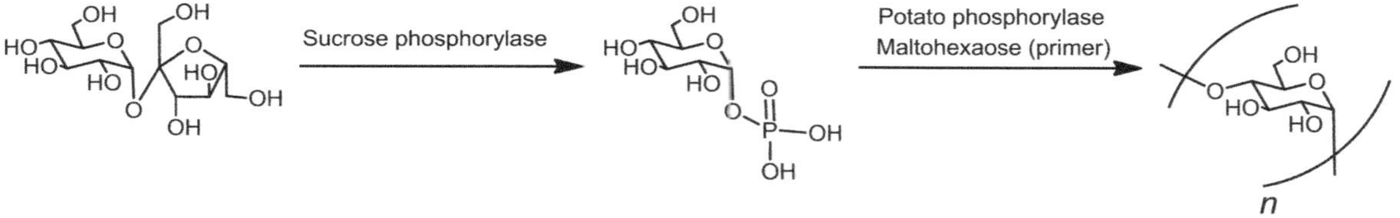

Sucrose phosphorylase → Potato phosphorylase Maltohexaose (primer) → 130

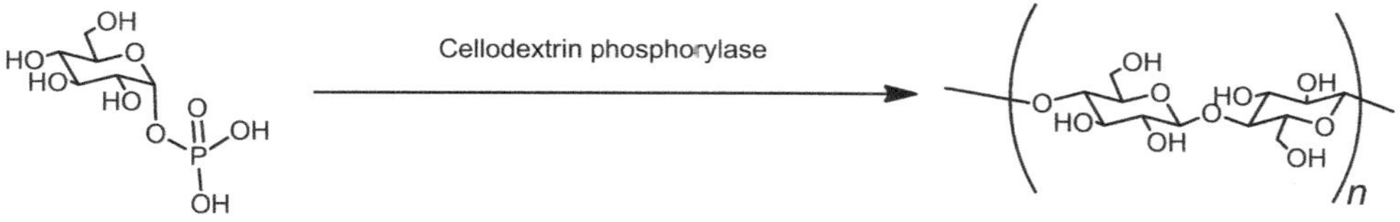

Cellodextrin phosphorylase → 131

Table 11.8   Enzymatic polymerization of phenol, polyphenol, and aniline derivatives.

Polymerization scheme	References

Phenol

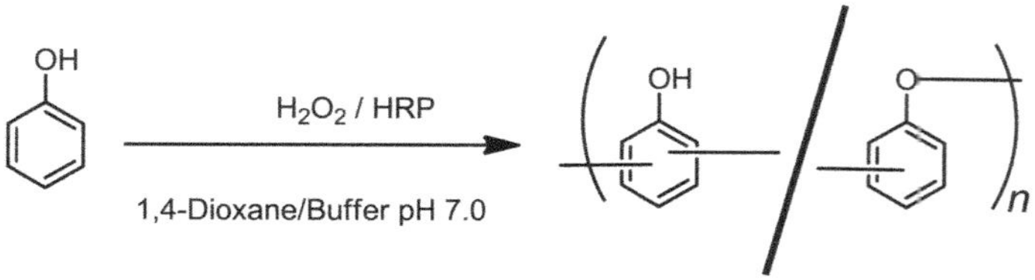

H_2O_2 / HRP, 1,4-Dioxane/Buffer pH 7.0 → 132–135

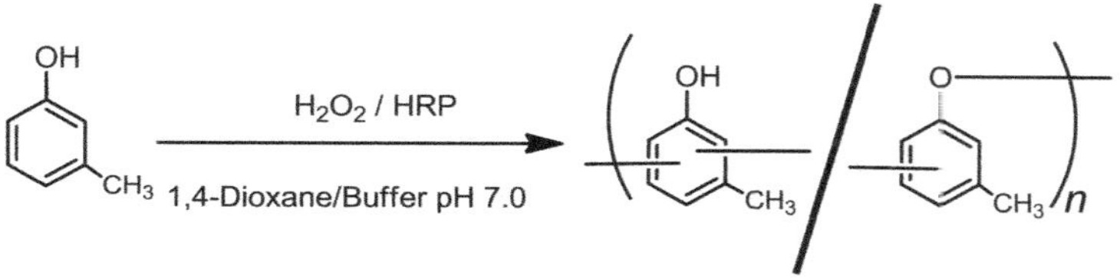

H_2O_2 / HRP, 1,4-Dioxane/Buffer pH 7.0 → 136

(*continued*)

Table 11.8 *(continued)*

Polymerization scheme	References
4-methylphenol $\xrightarrow[\text{1,4-Dioxane/Buffer pH 7.0}]{\text{H}_2\text{O}_2\,/\,\text{HRP}}$ poly(oxy-2-methyl-1,4-phenylene)	136
4-phenylphenol $\xrightarrow[\text{1,4-Dioxane/Buffer pH 7.0}]{\text{H}_2\text{O}_2\,/\,\text{HRP}}$ poly(oxy-2-phenyl-1,4-phenylene)	137
4-alkylphenol $\xrightarrow[\text{1,4-Dioxane/Buffer pH 7.0}]{\text{H}_2\text{O}_2\,/\,\text{HRP}}$ poly(oxy-2-alkyl-1,4-phenylene)	138

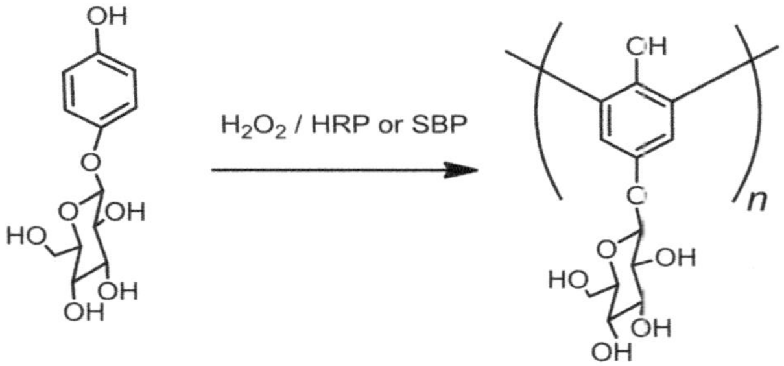OH HO OH O H2O2 / HRP or SBP n	139
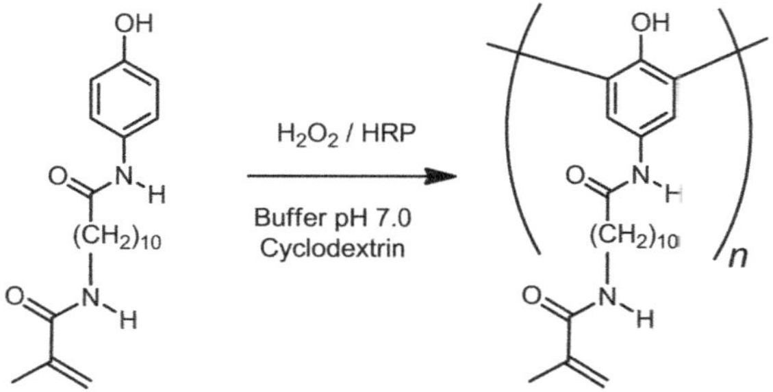OH O N H (CH2)10 H2O2 / HRP Buffer pH 7.0 Cyclodextrin n	140
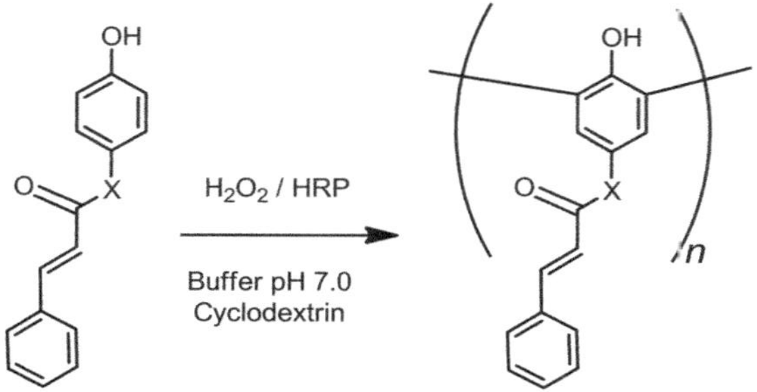OH O X H2O2 / HRP Buffer pH 7.0 Cyclodextrin n X = O or NH	141

(continued)

Table 11.8 (*continued*)

Polymerization scheme	References
	142
	143
	144

145
H$_2$O$_2$ / HRP or SBP
R =
Polyphenol
55
H$_2$O$_2$ / HRP or SBP
1,4-dioxane/acetate buffer pH 5.0 or phosphate buffer pH 7.0 reversed micelle
56
H$_2$O$_2$ / Trametes versicolor ATCC 200801 Laccase
Aceton/acetate buffer pH 5.0
146–149
H$_2$O$_2$ / Laccase
Aceton/acetate buffer pH 5.0

(continued)

Table 11.8 (*continued*)

Polymerization scheme	References
H_2O_2 / HRP — 1,4-Dioxane/Buffer pH 7.0	150
H_2O_2 / HRP — 1,4-Dioxane/Buffer pH 7.0 (R_1 = H or OH)	150
H_2O_2 / *Coprinus cinereus* Peroxidase	151

Aniline

NH2 H2O2 / HRP Sulphated Polystryrene H N n	152
NH2 COOH H2O2 / HRP HN COCH NH2 COOH n	153
NH2 SO3H NH2 H2O2 / HRP NH2 SO3H HN n	154
NH2 Glucose Penicillium vitale Glucose Oxidase HN n	155

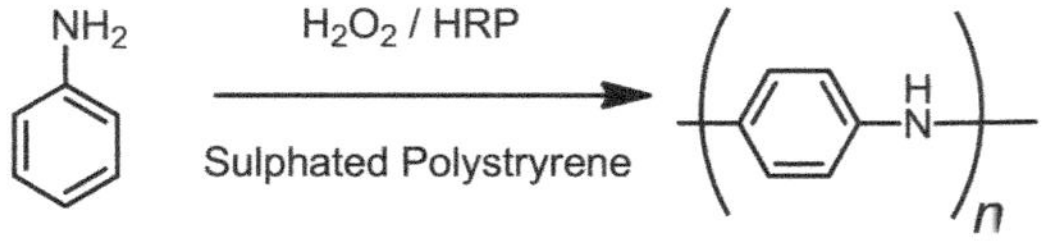

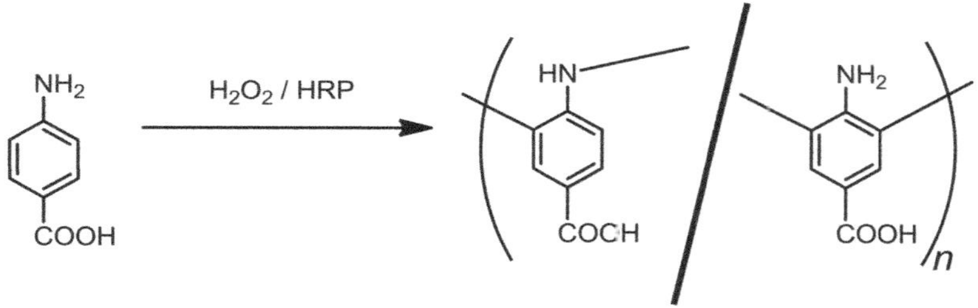

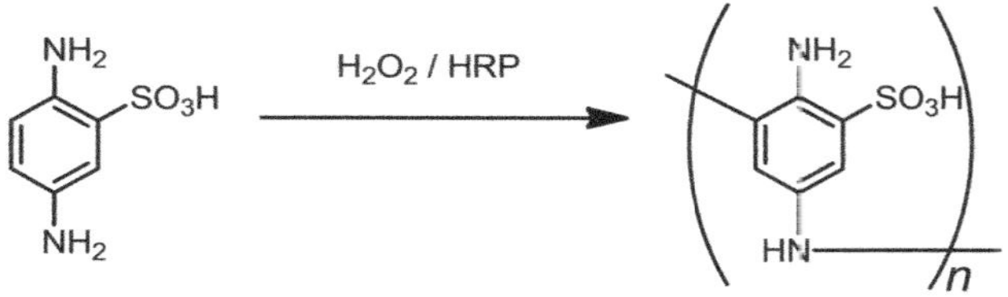

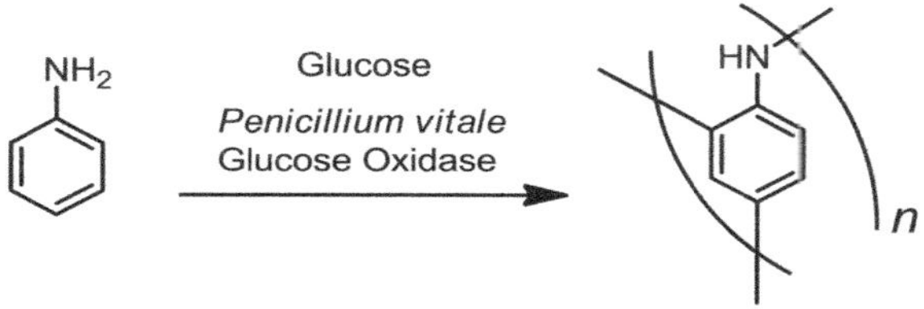

polymerization of phenol derivatives and, among these, peroxidase is most often used. For example, horseradish peroxidase (HRP), which has an Fe-containing porphyrin-type structure, is well known to catalyze the coupling of a number of phenol and aniline derivatives, using hydrogen peroxide as an oxidant.

11.3.4.3.1 Oxidative Polymerization of Phenolic Compounds[132–145].

The enzymes responsible for the oxidation–reduction process in maintaining the metabolism of living systems are called oxidoreductases, which normally require an oxidant or a reductant as reacting chemical species for substrates. Although there have been many publications on enzymatic oxidations and reductions in biochemistry and synthetic organic chemistry, it was only at the end of the 20th century that oxidoreductases were first used for polymerizations. Phenol and various phenol derivatives having a functional group have been polymerized by the action of oxidases.

11.3.4.3.2 Oxidative Polymerization of Polyphenolic Compounds[146–151].

Flavonoids, such as catechin and rutin, are called "polyphenols", and contain more than two phenolic OH groups in their aromatic ring. Polyphenols in red wine or green tea have received much attention because of their antioxidant action as anticancer agents.[53,54] The peroxidase catalyst induces the polymerization of catechol, which is considered to involve an unstable *o*-quinone intermediate to lead to poly(catechol).[55] Laccase also acted as a catalyst for catechol polymerization, which was conducted batch-wise in aqueous acetone to give a polymer with a molecular weight less than 1000.[56] Aromatic compounds were converted to catechol derivatives by the catalysis of toluene dioxygenase (TDO) and toluene *cis*-dihydrodiol dehydrogenase (TDD). The resulting catechol derivatives were polymerized, catalyzed by a peroxidase, giving rise to polymers with molecular weights of several thousand.[57] A biomimetic peroxidase catalyst, iron–porphyrin, was employed for catechol polymerization, and the structures of the isomeric dimer products were characterized in detail to discuss the coupling mechanism.[58]

11.3.4.3.3 Oxidative Polymerization of Aniline Derivatives[152–155].

Oxidative polymerization of aniline was conducted a century ago[59] to give polyaniline (PANI) as aniline black. In recent years, PANI has become one of the most popular conducting polymers because of its stability and its favorable electrical and optical properties. The well-known methods for the synthesis of PANI are either chemical or electrochemical oxidation polymerization of the aniline monomer. The reaction conditions are harsh with high temperatures, strong oxidants and highly toxic solvents being required. Enzymatic polymerization of aniline and its derivatives provides an alternative method using white biotechnology. These reactions are usually carried out at room temperature, in aqueous organic solvents at neutral pH. Reaction conditions were greatly improved and the purification process of the final products was simplified in comparison to conventional methods.[60]

11.4 Future Prospects

Even in ancient times, human beings had already started to use fermentations for producing various kinds of useful materials for daily essentials, without knowing about the existence of microorganisms, not to mention enzymes. After significant findings and technological developments in microbiology, enzymology, and biochemistry, we are now in the privileged position of having a constant large-scale supply of these biocatalysts that can be used in the production of useful compounds. Polymer chemistry has benefited from this progress by employing microorganisms or isolated enzymes for the production of monomers and polymers.

The introduction of the concept of white biotechnology has become essential to create new functional polymers with complicated structures and to minimize environmental stress according to environmental legislation. We described several examples of monomer synthesis and polymerization from the viewpoint of whether the process included a reaction catalyzed by microorganisms or enzymes. Further developments will be possible by modifying microorganisms on the basis of metabolism technology, by developing bio-machinery for selective production of certain bio-based polymers, and by introducing enzyme immobilization. Polymer synthesis *via* enzymatic polymer modification[49] will also be a vivid topic in white biotechnology, which has not been included in this chapter. It is expected that the production of monomers and polymers by white biotechnology will greatly contribute to the future innovation of polymeric materials that have been difficult to produce by conventional methodologies, including classical synthetic reactions that have high environmental stress.

References

1. D. J. Walton and J. P. Lorimer, *Polymers*, Oxford University Press, 2000.
2. *Catalysis in Precision Polymerization*, ed. S. Kobayashi, John Wiley & Sons, New York, 1997.
3. *Cellulose*, ed. R. A. Young and R. M. Rowell, John Wiley & Sons, New York, 1986.
4. *Biosynthesis and Biodegradation of Cellulose*, ed. C. H. Haigler and P. J. Weimer, Marcel Dekker, New York, 1991.
5. *Natural Rubber Biosynthesis and Physics-Chemical Studies on Plant Derived Latex*, ed. M. Elnashar, 2011.
6. R. Mülhaupt, *Angew. Chem., Int. Ed.*, 2004, **43**, 1054.
7. A. Y. Coran in *Science and Technology of Rubber*, ed. E. M. James, E. Burak and F. R. Eirich, Academic Press, Burlington, 3rd edn, 2005, p. 321.
8. *Essentials of Polymer Science and Engineering*, ed. P. C. Painter and M. M. Coleman, DEStech Publications, Inc., Lancaster, PA, 2008.
9. I. Amato, *Time*, March 29, 1999.
10. *Fermentation Microbiology and Biotechnology*, ed. E. El-Mansi, B. Charlie, H. Brian and D. Arnold, CRC Press, 2011.

11. *Studies on fermentation*, ed. L. Pasteur, Macmillan Publishers, 1897.
12. E. Buechner, *Ber. Dtsch. Chem. Ges.*, 1897, **30**, 117.
13. W. Soetaert and E. Vandamme, *Biotechnol. J.*, 2006, **1**, 756.
14. J. W. Lee, H. U. Kim, S. Choi, J. Yi and S. Y. Lee, *Curr. Opin. Biotechnol.*, 2011, **22**, 758.
15. National Renewable Energy Laboratory, *Research Advances: Cellulosic Ethanol*, 2007.
16. W. Higashide, Y. Li, Y. Yang and J. C. Liao, *Appl. Environ. Microbiol.*, 2011, **77**, 2727.
17. J. Kuzma, M. Nemecek-Marshall, W. Pollock and R. Fall, *Curr. Microbiol.*, 1995, **30**, 97.
18. J. Xue and B. K. Ahring, *Appl. Environ. Microbiol.*, 2011, **77**, 2399.
19. V. E. T. Maervoet, M. De Mey, J. Beauprez, S. De Maeseneire and W. K. Soetaert, *Org. Process Res. Dev.*, 2010, **15**, 189.
20. R. A. Benner, Jr., W. F. Staruszkiewicz and W. S. Otwell, *J. Food Prot.*, 2004, **67**, 124.
21. C. Kaneuchi, M. Seki and K. Komagata, *Appl. Environ. Microbiol.*, 1988, **54**, 3053.
22. M. V. Guettler, D. Rumler and M. K. Jain, *Int. J. Syst. Bacteriol.*, 1999, **49**, 207.
23. P. C. Lee, S. Y. Lee, S. H. Hong and H. N. Chang, *Bioprocess Biosyst. Eng.*, 2003, **26**, 63.
24. H.-W. Ryu, K.-H. Kang and J.-S. Yun, *Appl. Biochem. Biotechnol.*, 1999, **78**, 511.
25. H.-W. Ryu and Y.-J. Wee, *Appl. Biochem. Biotechnol.*, 2001, **91–93**, 525.
26. J. Xu and B.-H. Guo, in *Plastics from Bacteria*, ed. G. G.-Q. Chen, Springer, Berlin Heidelberg, 2010, vol. 14, p. 347.
27. K. J. Jem, J. Pol and S. Vos, in *Plastics from Bacteria*, ed. G. G.-Q. Chen, Springer, Berlin Heidelberg, 2010, vol. 14, p. 323.
28. M. Nakano, H. Chaen, T. Sugimoto and T. Miyake, US. Pat. 6,107,348, 1995.
29. H. Moon and G. Cho, *Biotechnol. Bioprocess Eng.*, 1997, **2**, 19.
30. Y. Otsuka, M. Nakamura, K. Shigehara, K. Sugimura, E. Masai, S. Ohara and Y. Katayama, *Appl. Microbiol. Biotechnol.*, 2006, **71**, 608.
31. M. Kobayashi, T. Nagasawa and H. Yamada, *Trends Biotechnol.*, 1992, **10**, 402.
32. C. E. Nakamura and G. M. Whited, *Curr. Opin. Biotechnol.*, 2003, **14**, 454.
33. *Microbial Polyesters*, ed. Y. Doi, VHC Publishers, New York, 1990.
34. K. Sudesh, H. Abe and Y. Doi, *Prog. Polym. Sci.*, 2000, **25**, 1503.
35. S. Khanna and A. K. Srivastava, *Process Biochem.*, 2005, **40**, 607.
36. S. Y. Lee, *Biotechnol. Bioeng.*, 1996, **49**, 1.
37. M. Lemoigne, *Bull. Soc. Chim. Biol.*, 1926, **8**, 770.
38. S. Taguchi, M. Yamada, K. I. Matsumoto, K. Tajima, Y. Satoh, M. Munekata, K. Ohno, K. Kohda, T. Shimamura, H. Kambe and S. Obata, *Proc. Natl. Acad. Sci. U. S. A.*, 2008, **105**, 17323.

39. Y. K. Jung, T. Y. Kim, S. J. Park and S. Y. Lee, *Biotechnol. Bioeng.*, 2010, **105**, 161.

40. K. M. Frey, F. B. Oppermann-Sanio, H. Schmidt and A. Steinbüchel, *Appl. Environ. Microbiol.*, 2002, **68**, 3377.

41. M. Iguchi, S. Yamanaka and A. Budhiono, *J. Mater. Sci.*, 2000, **35**, 261.

42. L. Liu, Y. Liu, J. Li, G. Du and J. Chen, *Microb. Cell Fact.*, 2011, **10**, 99.

43. B. Widner, R. Behr, S. Von Dollen, M. Tang, T. Heu, A. Sloma, D. Sternberg, P. L. Deangelis, P. H. Weigel and S. Brown, *Appl. Environ. Microbiol.*, 2005, **71**, 3747.

44. S. N. Pawar and K. J. Edgar, *Biomaterials*, 2012, **33**, 3279.

45. K. I. Draget, in *Handbook of Hydrocolloids*, ed. G. O. Phillips and P. A. Williams, Woodhead Publishing Ltd., 2nd edn, 2009, p. 924.

46. K. B. Mullis and F. A. Faloona, in *Methods in Enzymology*, ed. W. Ray, Academic Press, 1987, vol. 155, p. 335.

47. K. Okuda, L. Li, K. Kudlicka, S. Kuga and R. M. Brown, Jr, *Plant Physiol.*, 1993, **101**, 1131.

48. S. Kobayashi, S.-I. Shoda and H. Uyama, *Adv. Polym. Sci.*, 1995, **121**, 1.

49. S. Kobayashi and A. Makino, *Chem. Rev.*, 2009, **109**, 5288.

50. J.-I. Kadokawa and S. Kobayashi, *Curr. Opin. Chem. Biol.*, 2010, **14**, 145.

51. Y. Kaneko and J.-I. Kadokawa, *Chem. Rec.*, 2005, **5**, 36.

52. H. Uyama and S. Kobayashi, in *Enzyme-Catalyzed Synthesis of Polymers*, ed. S. Kobayashi, H. Ritter and D. Kaplan, Springer Berlin Heidelberg, 2006, vol. 194, p. 51.

53. J. Jankun, S. H. Selman, R. Swiercz and E. Skrzypczak-Jankun, *Nature*, 1997, **387**, 561.

54. A. Bordoni, S. Hrelia, C. Angeloni, E. Giordano, C. Guarnieri, C. M. Caldarera and P. L. Biagi, *J. Nutr. Biochem.*, 2002, **13**, 103.

55. S. Dubey, D. Singh and R. A. Misra, *Enzyme Microb. Technol.*, 1998, **23**, 432.

56. N. Aktaş, N. Şahiner, Ö. Kantoğlu, B. Salih and A. Tanyolaç, *J. Polym. Environ.*, 2003, **11**, 123.

57. G. Ward, R. E. Parales and C. G. Dosoretz, *Environ. Sci. Technol.*, 2004, **38**, 4753.

58. D. Šmejkalová, P. Conte and A. Piccolo, *Biomacromolecules*, 2007, **8**, 737.

59. R. Willstätter and S. Dorogi, *Ber. Dtsch. Chem. Ges.*, 1909, **42**, 4118.

60. P. Xu, A. Singh and D. Kaplan, in *Enzyme-Catalyzed Synthesis of Polymers*, ed. S. Kobayashi, H. Ritter and D. Kaplan, Springer Berlin Heidelberg, 2006, vol. 194, p. 69.

61. H. Huang, in *Plastics from Bacteria*, ed. G. G.-Q. Chen, Springer Berlin Heidelberg, 2010, vol. 14, p. 389.

62. X. Luo, D. Tang and M. Li, *J. Mol. Struct.*, 2005, **731**, 139.

63. X. Li, J. Guan, A. Zheng, D. Zhou, X. Han, W. Zhang and X. Bao, *J. Mol. Catal. A: Chem.*, 2010, **330**, 99.

64. H. Yamada and M. Kobayashi, *Biosci., Biotechnol., Biochem.*, 1996, **60**, 1391.

65. J.-S. Gong, Z.-M. Lu, H. Li, J.-S. Shi, Z.-M. Zhou and Z.-H. Xu, *Microb. Cell Fact.*, 2012, **11**, 142.

66. H. Köpnick, M. Schmidt, W. Brügging, J. Rüter and W. Kaminsky, in *Ullmann's Encyclopedia of Industrial Chemistry*, Wiley-VCH Verlag GmbH & Co. KGaA, 2000.

67. S. Rebsdat and D. Mayer, in *Ullmann's Encyclopedia of Industrial Chemistry*, Wiley-VCH Verlag GmbH & Co. KGaA, 2000.

68. D. Minh, M. Besson, C. Pinel, P. Fuertes and C. Petitjean, *Top. Catal.*, 2010, **53**, 1270.

69. J. Schneider and V. Wendisch, *Appl. Microbiol. Biotechnol.*, 2011, **91**, 17.

70. G.-Q. Chen and Q. Wu, *Appl. Microbiol. Biotechnol.*, 2005, **67**, 592.

71. E. C. Hann, A. E. Sigmund, S. K. Fager, F. B. Cooling, J. E. Gavagan, A. Ben-Bassat, S. Chauhan, M. S. Payne, S. M. Hennessey and R. DiCosimo, *Adv. Synth. Catal.*, 2003, **345**, 775.

72. M. Nakano, H. Chaen, T. Sugimoto and T. Miyake, EP670368A2, 1995.

73. M. Akiyama, Y. Taima and Y. Doi, *Appl. Microbiol. Biotechnol.*, 1992, **37**, 698.

74. M. Akiyama, T. Tsuge and Y. Doi, *Polym. Degrad. Stab.*, 2003, **80**, 183.

75. B. Füchtenbusch and A. Steinbüchel, *Appl. Microbiol. Biotechnol.*, 1999, **52**, 91.

76. E. Shimamura, M. Scandola and Y. Doi, *Macromolecules*, 1994, **27**, 4429.

77. M. Hiramitsu, N. Koyama and Y. Doi, *Biotechnol. Lett.*, 1993, **15**, 461.

78. Y. Saito and Y. Doi, *Int. J. Biol. Macromol.*, 1994, **16**, 99.

79. Y. K. Jung and S. Y. Lee, *J. Biotechnol.*, 2011, **151**, 94.

80. Y. Hamano, *Biosci., Biotechnol., Biochem.*, 2011, **75**, 1226.

81. I. Bajaj and R. Singhal, *Bioresour. Technol.*, 2011, **102**, 5551.

82. L. J. Chien and C. K. Lee, *Biotechnol. Prog.*, 2007, **23**, 1017.

83. L.-J. Chien and C.-K. Lee, *Appl. Microbiol. Biotechnol.*, 2007, **77**, 339.

84. Z. Mao, H.-D. Shin and R. Chen, *Appl. Microbiol. Biotechnol.*, 2009, **84**, 63.

85. W. Sabra, A. P. Zeng and W. D. Deckwer, *Appl. Microbiol. Biotechnol.*, 2001, **56**, 315.

86. A. Linker and R. S. Jones, *Nature*, 1964, **204**, 187.

87. P. A. J. Gorin and J. F. T. Spencer, *Can. J. Chem.*, 1966, **44**, 993.

88. R. Gupta, N. Gupta and P. Rathi, *Appl. Microbiol. Biotechnol.*, 2004, **64**, 763.

89. S. Okumura, M. Iwai and Y. Tominaga, *Agric. Biol. Chem.*, 1984, **48**, 2805.

90. F. Binns, S. M. Roberts, A. Taylor and C. F. Williams, *J. Chem. Soc., Perkin Trans. 1*, 1993, 899.

91. S. Kobayashi, H. Uyama, S. Suda and S. Namekawa, *Chem. Lett.*, 1997, 105.

92. S. Suda, H. Uyama and S. Kobayashi, *Proc. Jpn. Acad., Ser. B*, 1999, **75**, 201.

93. A. Ajima, T. Yoshimoto, K. Takahashi, Y. Tamaura, Y. Saito and Y. Inada, *Biotechnol. Lett.*, 1985, **7**, 303.

94. S. Matsumura and J. Takahashi, *Makromol. Chem., Rapid Commun.*, 1986, **7**, 369.

95. H. Ebata, K. Toshima and S. Matsumura, *Macromol. Biosci.*, 2007, **7**, 798.

96. H. Ebata, M. Yasuda, K. Toshima and S. Matsumura, *J. Oleo Sci.*, 2008, **57**, 315.

97. A. Kumar, A. S. Kulshrestha, W. Gao and R. A. Gross, *Macromolecules*, 2003, **36**, 8219.

98. H. Uyama and S. Kobayashi, *Chem. Lett.*, 1993, **22**, 1149.

99. H. Uyama, H. Kikuchi, K. Takeya and S. Kabayashi, *Acta Polym.*, 1996, **47**, 357.

100. S. Matsumura, K. Mabuchi and K. Toshima, *Macromol. Rapid Commun.*, 1997, **18**, 477.

101. M. Hans, H. Keul and M. Moeller, *Macromol. Biosci.*, 2009, **9**, 239.

102. Y. Feng, J. Knüfermann, D. Klee and H. Höcker, *Macromol. Rapid Commun.*, 1999, **20**, 88.

103. Y. Feng, D. Klee, H. Keul and H. Höcker, *Macromol. Chem. Phys.*, 2000, **201**, 2670.

104. Y. Feng, D. Klee and H. Höcker, *Macromol. Biosci.*, 2004, **4**, 587.

105. Z. Jiang, H. Azim, R. A. Gross, M. L. Focarete and M. Scandola, *Biomacromolecules*, 2007, **8**, 2262.

106. S. Kobayashi, H. Kikuchi and H. Uyama, *Macromol. Rapid Commun.*, 1997, **18**, 575.

107. S. Matsumura, K. Tsukada and K. Toshima, *Macromolecules*, 1997, **30**, 3122.

108. K. S. Bisht, Y. Y. Svirkin, L. A. Henderson, R. A. Gross, D. L. Kaplan and G. Swift, *Macromolecules*, 1997, **30**, 7735.

109. L. W. Schwab, R. Kroon, A. J. Schouten and K. Loos, *Macromol. Rapid Commun.*, 2008, **29**, 794.

110. J. Wen and R.-X. Zhuo, *Macromol. Rapid Commun.*, 1998, **19**, 641.

111. F. He, R. X. Zhuo, L. J. Liu, D. B. Jin, J. Feng and X. L. Wang, *React. Funct. Polym.*, 2001, **47**, 153.

112. S. Kobayashi, K. Kashiwa, T. Kawasaki and S. Shoda, *J. Am. Chem. Soc.*, 1991, **113**, 3079.

113. S. Kobayashi, X. Wen and S. Shoda, *Macromolecules*, 1996, **29**, 2698.

114. S. Kobayashi, J. Shimada, K. Kashiwa and S. Shoda, *Macromolecules*, 1992, **25**, 3237.

115. D. S. Genghof and E. J. Hehre, *Proc. Soc. Exp. Biol. Med.*, 1972, **140**, 1298.

116. L. F. Mackenzie, Q. Wang, R. A. J. Warren and S. G. Withers, *J. Am. Chem. Soc.*, 1998, **120**, 5583.

117. S. Fort, V. Boyer, L. Greffe, G. J. Davies, O. Moroz, L. Christiansen, M. Schülein, S. Cottaz and H. Driguez, *J. Am. Chem. Soc.*, 2000, **122**, 5429.

118. K. Piens, A.-M. Henriksson, F. Gullfot, M. Lopez, R. Faure, F. M. Ibatullin, T. T. Teeri, H. Driguez and H. Brumer, *Org. Biomol. Chem.*, 2007, **5**, 3971.

119. M. Noguchi, T. Tanaka, M. Ishihara, A. Kobayashi and S. Shoda, 2nd International Cellulose Conference, Tokyo, 2007.

120. T. Tanaka, M. Noguchi, M. Ishihara, A. Kobayashi and S. Shoda, *Macromol. Symp.*, 2010, **297**, 200.

121. S. Egusa, T. Kitaoka, M. Goto and H. Wariishi, *Angew. Chem., Int. Ed.*, 2007, **46**, 2063.

122. S. Egusa, M. Goto and T. Kitaoka, *Biomacromolecules*, 2012, **13**, 2716.

123. G. Potocki-Veronese, J.-L. Putaux, D. Dupeyre, C. Albenne, M. Remaud-Siméon, P. Monsan and A. Buleon, *Biomacromolecules*, 2005, **6**, 1000.

124. F. Grimaud, C. Lancelon-Pin, A. Rolland-Sabaté, X. Roussel, S. Laguerre, A. Viksø-Nielsen, J.-L. Putaux, S. Guilois, A. Buléon, C. D'Hulst and G. Potocki-Véronèse, *Biomacromolecules*, 2013, **14**, 438.

125. S. Kobayashi, T. Kiyosada and S. Shoda, *J. Am. Chem. Soc.*, 1996, **118**, 13113.

126. S. Kobayashi, T. Kiyosada and S. Shoda, *Tetrahedron Lett.*, 1997, **38**, 2111.

127. S. Kobayashi, H. Morii, R. Itoh, S. Kimura and M. Ohmae, *J. Am. Chem. Soc.*, 2001, **123**, 11825.

128. S. Kobayashi, S.-I. Fujikawa and M. Ohmae, *J. Am. Chem. Soc.*, 2003, **125**, 14357.

129. S.-I. Fujikawa, M. Ohmae and S. Kobayashi, *Biomacromolecules*, 2005, **6**, 2935.

130. H. Waldmann, D. Gygax, M. D. Bednarski, W. Randall Shangraw and G. M. Whitesides, *Carbohydr. Res.*, 1986, **157**, c4.

131. M. Hiraishi, K. Igarashi, S. Kimura, M. Wada, M. Kitaoka and M. Samejima, *Carbohydr. Res.*, 2009, **344**, 2468.

132. J. S. Dordick, M. A. Marletta and A. M. Klibanov, *Biotechnol. Bioeng.*, 1987, **30**, 31.

133. H. Uyama, H. Kurioka, J. Sugihara and S. Kobayashi, *Bull. Chem. Soc. Jpn.*, 1996, **69**, 189.

134. T. Oguchi, S.-I. Tawaki, H. Uyama and S. Kobayashi, *Macromol. Rapid Commun.*, 1999, **20**, 401.

135. T. Oguchi, S.-I. Tawaki, H. Uyama and S. Kobayashi, *Bull. Chem. Soc. Jpn.*, 2000, **73**, 1389.

136. H. Kurioka, H. Uyama and S. Kobayashi, *Polym. J.*, 1998, **30**, 526.

137. H. Uyama, H. Kurioka and S. Kobayashi, *Chem. Lett.*, 1995, **24**, 795.

138. H. Uyama, H. Kurioka, J. Sugihara, I. Komatsu and S. Kobayashi, *J. Polym. Sci., Part A: Polym. Chem.*, 1997, **35**, 1453.

139. P. Wang, B. D. Martin, S. Parida, D. G. Rethwisch and J. S. Dordick, *J. Am. Chem. Soc.*, 1995, **117**, 12885.

140. M. H. Reihmann and H. Ritter, *Macromol. Chem. Phys.*, 2000, **201**, 798.

141. M. H. Reihmann and H. Ritter, *Macromol. Chem. Phys.*, 2000, **201**, 1593.

142. Y. Pang, H. Ritter and M. Tabatabai, *Macromolecules*, 2003, **36**, 7090.

143. H. Tonami, H. Uyama, S. Kobayashi, T. Fujita, Y. Taguchi and K. Osada, *Biomacromolecules*, 2000, **1**, 149.

144. S. Kobayashi, R. Ikeda, H. Oyabu, H. Tanaka and H. Uyama, *Chem. Lett.*, 2000, **29**, 1214.

145. T. Tsujimoto, R. Ikeda, H. Uyama and S. Kobayashi, *Macromol. Chem. Phys.*, 2001, **202**, 3420.

146. H. Uyama and S. Kobayashi, in *Advances in Polymer Science*, Springer GmbH, 2006, vol. 194, p. 51.

147. H. Uyama, *Macromol. Biosci.*, 2007, **7**, 410.

148. M. Kurisawa, J. E. Chung, Y. J. Kim, H. Uyama and S. Kobayashi, *Biomacromolecules*, 2003, **4**, 469.
149. M. Kurisawa, J. E. Chung, H. Uyama and S. Kobayashi, *Macromol. Biosci.*, 2003, **3**, 758.
150. L. Mejias, M. H. Reihmann, S. Sepulveda-Boza and H. Ritter, *Macromol. Biosci.*, 2002, **2**, 24.
151. Y. H. Kim, E. S. An, S. Y. Park, J.-O. Lee, J. H. Kim and B. K. Song, *J. Mol. Catal. B: Enzym.*, 2007, **44**, 149.
152. L. A. Samuelson, A. Anagnostopoulos, K. S. Alva, J. Kumar and S. K. Tripathy, *Macromolecules*, 1998, **31**, 4376.
153. K. S. Alva, K. A. Marx, J. Kumar and S. K. Tripathy, *Macromol. Rapid Commun.*, 1996, **17**, 859.
154. K. S. Alva, J. Kumar, K. A. Marx and S. K. Tripathy, *Macromolecules*, 1997, **30**, 4024.
155. A. Kausaite, A. Ramanaviciene and A. Ramanavicius, *Polymer*, 2009, **50**, 1846.

Production of Aroma Compounds by White Biotechnology

JULIANO LEMOS BICAS[*,a,b], GUSTAVO MOLINA[a,c], FRANCISCO FÁBIO CAVALCANTE BARROS[a], AND GLÁUCIA MARIA PASTORE[a]

[a]Department of Food Science, University of Campinas, Cx. Postal 6121, Campinas, SP, 13083-862, Brazil; [b]Department of Chemistry, Biotechnology and Bioprocess Engineering, University of São João Del-Rei, Cx. Postal 131, Ouro Branco, MG, 36420-000, Brazil; [c]Institute of Science and Technology – Food Engineering, UFVJM, Diamantina, MG, 39100-000, Brazil
*E-mail: jlbicas@gmail.com

12.1 Introduction

The sensory evaluation of food products involves the five senses. Among them, vision (color, appearance) and olfaction (aroma, odor) are perhaps the most important because they are the first to be noticed and directly reflect the quality and acceptance of the analyzed material. Also, aroma is one of the main attributes that defines the specific flavor of each food.[1]

Aroma compounds are volatile molecules with low molecular weight (usually <400 Da) which are able to sensitize the receptor cells from our olfactory cavity. The contact between these cells and volatile molecules may be

RSC Green Chemistry No. 45
White Biotechnology for Sustainable Chemistry
Edited by Maria Alice Z. Coelho and Bernardo D. Ribeiro
© The Royal Society of Chemistry 2016
Published by the Royal Society of Chemistry, www.rsc.org

brought about by two routes: direct inhalation (orthonasal), which will give rise to odor perception, or *via* retronasal (internal mouth–olfactory cavity communication), which will result in aroma perception.[2]

In general, aroma compounds show detection thresholds on the order of parts per billion (ppb).[3] As a result, they are usually present as minor components (about 50 ppm), but depending in their quality and application, they may represent up to 50% of the cost of the final product (*e.g.* some beverages[4]).

With more than 6500 volatiles already identified, an uncountable number of combinations, and, therefore, aroma perceptions are possible. Products with complex aromas, such as wine, coffee, cocoa, beer, *etc.*, present more than 500 volatile compounds which, together, are responsible for the formation of each specific aroma.[4] In other cases, however, one single substance (character impact aroma) may be responsible for expressing the main sensory characteristic of that food, such as isoamyl acetate in bananas.[1] In this context, an efficient analytical technique to evaluate the aromatic profile of foodstuffs is gas chromatography–olfactometry, capable of separating each volatile compound which would then be analyzed by a trained taster and, subsequently, identified by mass spectrometry.[5]

Chemically, aroma compounds may belong to different chemical functional groups, including hydrocarbons, alcohols, aldehydes, ketones, acids, esters or lactones. Among these, there are many terpenes [$(C_5H_8)_n$], which are the main chemical class of natural products. Mono- and sesquiterpenes, in fact, are commonly found in high amounts in different essential oils (Table 12.1).

At the end of the 19th century and at the beginning of the 20th century, with the intense expansion of organic chemistry, chemical synthesis of aroma compounds faced strong development and wide application. However, environment and health concerns related to synthetic products have led to the development of biotechnological tools to produce natural aroma compounds in a sustainable manner. Therefore, this chapter will discuss the main concepts and achievements involved with the industrial production of aroma compounds by microbial bioprocesses.

Table 12.1 Main components of some essential oils (adapted from Margetts, 2005[6]).

Essential oils	Main components[a]
Caraway	*S*-(+)-Carvone, limonene, myrcene, α-phellandrene, α-pinene, β-pinene
Dill	*S*-(+)-Carvone, α-phellandrene, limonene
Eucalyptus	1,8-Cineol, α-pinene, *p*-cymene, limonene
Ginger	Zingiberene, α-curcumene, β-sesquiphellandrene, bisabolene, camphene, β-phellandrene, 1,8-cineol
Orange peel	Limonene, myrcene, linalool, citronellal, neral, geranial, valencene, α- and β-sinensal
Rose	Citronellol, geraniol, nerol, methyl eugenol, geranylacetate, eugenol, rose oxide
Mint	*R*-(−)-Carvone, limonene, myrcene, 1,8-cineol, dihydrocarveol

[a]The compounds presented are terpenes or terpene derivatives.

12.2 Methods for Producing Aroma Compounds

There are basically three methods to produce aroma compounds: direct extraction from nature, chemical transformations or biotechnological synthesis (Figure 12.1).

Traditionally, natural aromas have been recovered from plant materials, although many disadvantages are encountered: low concentrations of the product of interest, which increase the extraction and purification costs; dependency on seasonal, climatic and political features, and possible ecological problems involved with the extraction.[7]

As already mentioned, chemical synthesis underwent intense development after the second part of the 19th century. This technique resulted in satisfactory yields, good production rates and, consequently, resulted in low cost products. Nevertheless, chemical procedures may require harsh conditions (toxic catalysts, high pressure and temperature, *etc.*) and usually lack adequate regio- and enantioselectivity to the substrate, resulting in a mixture of products. Additionally, the compounds generated are labeled as "artificial" or "nature identical", decreasing their economic value.[7]

Alternatively, the biotechnological production of aroma compounds has emerged as an attractive alternative, since it occurs at mild conditions, presents high hydroxy- and enantio-selectivity, does not generate toxic waste, and the products obtained may be labeled as "natural".[7,8] Many studies describe the biotechnological production of aroma compounds (see section 12.4). However, the yields obtained so far are not sufficient to enable commercial application of most published processes.

The biotechnological production of aroma compounds may be performed in two basic ways: through *de novo* synthesis or by biotransformation (Figure 12.2), both being possible with genetically modified organisms. *De novo* synthesis refers to the production "from the new", *i.e.* the synthesis of substances from simple building block molecules (sugars, amino acids, nitrogen salts, minerals, *etc.*), which will be metabolized by organisms to form a different and complex structure. Biotransformations, in turn, are single reactions catalyzed enzymatically (as pure enzymes or within microbial cells). Therefore, the substrate is metabolized by the organism (usually a breakdown or an oxidation/reduction process) in a single (biotransformation) or

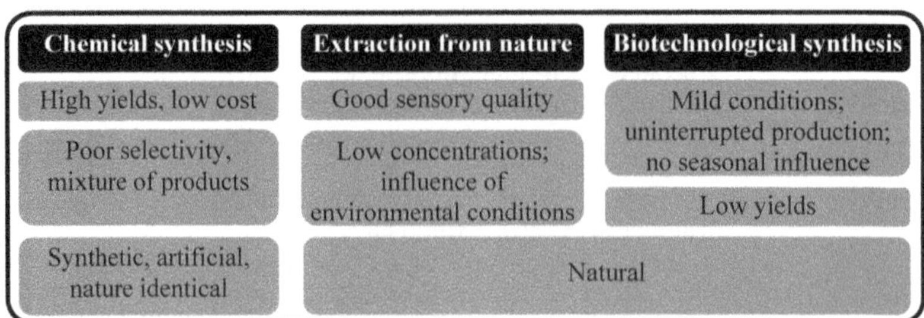

Figure 12.1   Some features related to each aroma-producing method.

a few (bioconversion) reactions to produce a structurally similar molecule.[1,7] Figure 12.3 indicates some examples of bioaromas of industrial interest.

The production of aroma compounds by *de novo* synthesis usually generates a mixture of products, whose maximal concentrations are commonly below 100 mg L^{-1}.[9] Therefore, biotransformations have higher potential for the production of bioaromas on a commercial scale. More precisely, the biotransformation of abundant and cheap terpene substrates (*e.g.* limonene and α-pinene), which implicates yields higher than 1 g L^{-1}, has attracted particular interest.[10] The features of the industrial biotechnology processes for the production of these bioaromas will be discussed in the following sections.

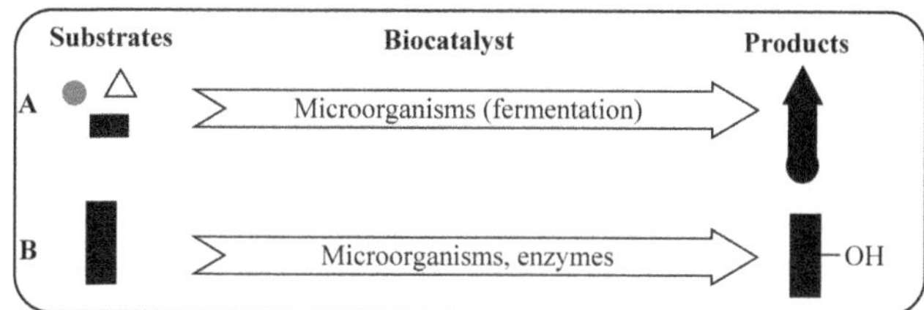

Figure 12.2   Schematic representation of aroma compound production by (A) *de novo* synthesis or (B) by biotransformation.

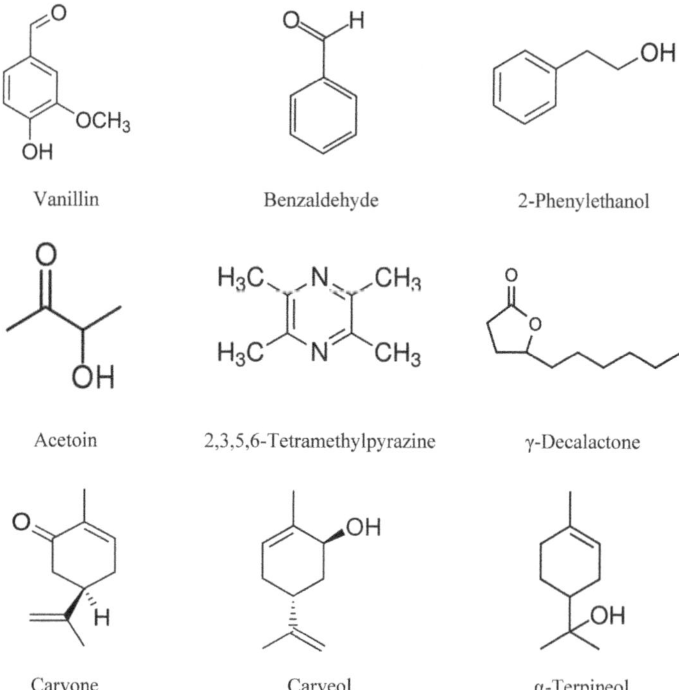

Figure 12.3   Examples of bioflavors obtained through biotechnological processes with industrial interest.

12.3   Why Use White Biotechnology to Produce Aroma Compounds?

With the advent of biotechnological processes, a new division was proposed to cover compounds obtained at industrial scale, which has become a commercial reality. The European Commission, for example, defines "White Biotechnology" as a modern version of biotechnology "based on the use of renewable resources and clean production and less pollution and less energy intensive processes in biological systems, such as whole cells or enzymes, used as reagents or catalysts". It is mainly these ecological and sustainability aspects that differentiate the novel term "White Biotechnology" from previous descriptions of biotechnology.[11]

Industrially, flavors and fragrances have wide application in the food, feed, cosmetics, chemical and pharmaceutical sectors and have a global market estimated at €15 billion, equally divided into flavor and fragrance sectors. This market is concentrated in developed countries: 36% for Europe, Africa and the Middle East region, 32% for North America, 26% for the Asia–Pacific region and 6% for South America.[12]

Despite the low yields obtained in most reported processes, the production of many bioaromas is, in some cases, economically feasible. Among the approximately 300 aroma compounds available for industry, about 100 are actually produced biotechnologically, while the rest are chemically synthesized. In Germany, most of the aromas used in beverages are of natural origin. Examples of commercial natural aroma chemicals with probable biotechnological origin can be found: ethyl butanoate, 2-heptanone, β-ionone, nootkatone, 1-octen-3-ol, 4-undecalactone and vanillin, among others.[4,13] Additionally, a number of recent patents or patent applications may be found for this subject[14–16] – particularly for those processes involving vanillin production by the biotransformation of ferulic acid[17–20] – which is a reflection of the development of advanced and (possibly) more viable procedures.

The main motivation for the microbial production of aroma compounds is the market value of the so-called "biotechnological aroma" compounds, which is commonly far above their synthetic counterparts, although usually lower than those extracted from nature. For example: synthetic vanillin has a price of ~US\$ 11 kg^{-1}, natural vanilla flavor extracted from fermented pods of *Vanilla* orchids costs US\$ 1200–4000 kg^{-1}, while "biotech vanillin" is sold for a price of ~US\$ 1000 kg^{-1}.[21,22] Equivalent data are also observed for other aroma compounds, such as γ-decalactone (synthetic = US\$ 150 kg^{-1}; natural = US\$ 6000 kg^{-1}; "biotech" = US\$ 300 kg^{-1}) and ethyl butyrate (synthetic = US\$ 4 kg^{-1}; natural = US\$ 5000 kg^{-1}; "biotech" = US\$ 180 kg^{-1}), for example.[23] In fact, microbial aromas are becoming more and more competitive, since their market price has decreased in the last few years. The most notable example is γ-decalactone. In the early 1980s, natural γ-decalactone was rare and expensive. With the adoption and further optimization of its microbial production process, the price has continuously decreased, from US\$ 12 000 kg^{-1} in 1986 to US\$ 500 kg^{-1} in 1997, and the market volume increased to several hundred tons per annum.[24,25] In this sense, the possibility of producing aroma compounds from agroindustrial residues and by-products is a promising possibility that may be considered to reduce costs.[26] Additionally, according to Berger (2009),[13] the biotechnological production of aroma compounds has been stimulated by increasing consumer preference for "natural"-labeled products and also by the vivid discussion of functional foods and healthy ingredients.

Given the above, research has revealed that many terpene compounds may present potential applications against certain types of cancers. The anticancer activity associated with monoterpene compounds results from: (i) blocking effects (initiation phase), characterized by the induction of phase I and phase II enzymes from the metabolism of xenobiotics, which is associated with the detoxification of the carcinogen; (ii) suppressing effects (promotion phase), characterized (a) by the inhibition of cell proliferation, induction of apoptosis or differentiation or (b) the inhibition of post-translational isoprenylation of cell growth-regulating proteins.[27]

Among these monoterpenes, one of the most significant examples is R-(+)-limonene, a major constituent (>90%) of orange peel oil. According to some studies on rodents, this compound seems to exert chemopreventive effects against chemically induced breast, skin, liver and lung cancers and other spontaneous tumors. Besides, it has been reported that when in conjunction with menthol, limonene inhibits the enzyme HMG-CoA reductase, the key enzyme in the biosynthesis of cholesterol.[27]

Carvone is a ketone monoterpenoid of great importance as a flavoring agent in the food industry. Besides its interesting aroma character (mint for R-(−) and caraway for the S-(+) isomer), carvone also presents anti-sprouting activity in potatoes, and antimicrobial and insect repellent properties, among others. Studies also indicate that carvone increases the expression of glutathione-S-transferase, a detoxifying enzyme, which may be related to its chemopreventative action.[28]

Perillyl alcohol, a monoterpenoid alcohol isolated from essential oils of lavender, mint, cherry, bergamot and other plants, has been extensively studied as an antitumor agent, including in clinical trials. In rodents, this compound presents chemopreventive activity against chemically induced cancers in the liver, colon, skin, lung and prostate. Additionally, it can prevent recurrent tumors or secondary tumors during chemotherapy and also has therapeutic potential in breast and pancreatic cancers.[27] Recently, clinical studies have also confirmed the potential of this terpene alcohol in humans. A team headed by researchers from Rio de Janeiro (Brazil) has already applied perillyl alcohol, *via* intranasal administration, in patients with malignant gliomas in the central nervous system. The results revealed that perillyl alcohol is a promising compound for the treatment of brain tumors, especially when the conventional methods have failed.[29,30] Another pilot study with eight patients has also indicated the potential of perillyl alcohol for treating pancreatic cancer.[31]

Only a few reports are available on the biological activities of other biotransformation products of limonene, such as α-terpineol, and with the promising results for perillyl alcohol, its counterpart monoterpene alcohol. Three examples were recently described, which demonstrate the need for more detailed tests: (i) Maróstica Jr *et al.* (2009)[32] concluded that some monoterpenes and the "extract" (mainly α-terpineol) resulting from the biotransformation of orange oil (>90% *R*-(+)-limonene) presented important antioxidant activity; (ii) Hassan *et al.* (2010)[33] evidenced the potential of α-terpineol to inhibit the growth of tumors *in vitro*, especially against small cell lung carcinoma; (iii) Bicas *et al.* (2011)[34] found results pointing in the same direction after they reported the antioxidant potential and the *in vitro* antiproliferative action of α-terpineol against certain tumor cell lines.

In this section will be described the main examples already reported for the production of aroma compounds through biotransformation, the main strategy compatible with industrial scale.

12.4 Examples of Aroma Compounds Produced Through White Biotechnology

12.4.1 Background and Overview: Processes, Advantages and Developments

As already mentioned, aroma biosynthesis can be based on *de novo* microbial processes (fermentation) and on bioconversions of natural precursors with microbial cells or enzymes (biotransformation). In general, microorganisms are capable of producing an amazingly broad array of flavor compounds by *de novo* synthesis. However, the production levels are very poor, and thus constitute a limit for industrial exploitation. For this reason, biotechnologists have focused on biotransformation or bioconversion processes that offer more economic advantages, while processes and products may be inserted into the white biotechnology definition.[35]

Therefore, many of the industrially relevant microbial aroma production processes follow this precursor approach, such as vanillin from ferulic acid or eugenol, 4-decanolide from ricinoleic acid and 2-phenylethanol from L-phenylalanine, reaching a production volume of one to several tons per year.[25] Despite the great industrial interest in this strategy, up to now, most studies dealing with biotransformations for aroma production are not feasible for industrial applications due to problems commonly encountered, such as low transformation rates, high production costs and other limiting features (*e.g.* low product concentrations and product yields, short biocatalyst lifetimes and long incubation times).[36]

To overcome these drawbacks that impact the process feasibility, several studies have been conducted to make the bioprocess more economically viable. Berger (2009)[13] recently discussed the latest advances in the bio-production of flavor compounds. For this author, progress is expected from the toolbox of genetic engineering, which is likely to help in identifying metabolic bottlenecks and in creating novel high-yielding strains. Bioengineering, in a complementary way, could provide promising technical options, such as improved substrate dosage, gas-phase or two-phase reactions and *in situ* product recovery.

Some other recent advances have been made in order to achieve a feasible yield and product concentration. An interesting example of these alternatives is the use of statistical tools for simultaneous evaluation of the different parameters of a process. Thus, these techniques may be associated with several improvements in processes and products, minimizing costs, and maximizing efficiency and productivity.[37,38]

Another important strategy to attempt to maximize the industrial production of aroma compounds is the use of biphasic systems (organic and aqueous phases), which can reduce the problems associated with the low solubility of substrate and product, while minimizing their toxic effects towards microorganisms.[39] This is in accordance with the approach provided by Berger (2008)[11] who discussed that a key parameter for both economic and environmental performance is biomass concentration, considering that the concentration of the effective catalyst governs the overall reaction rate.

The use of agro-industrial residues as substrates has also been extensively investigated for the production of aroma compounds by biotransformation processes. This strategy seems to be a growing trend to overcome the high costs involved in microbial transformations. The main examples include the use of cassava wastewater and coffee husk for microbial supports, and orange and turpentine oils as alternatives sources of limonene and pinenes, respectively.[7]

It is considered that space–time yields of commercial bioprocesses range from milligram to three-digit grams per liter and hour. In an economic approach, Berger (2008)[11] found an inverse double-log correlation of product concentration and sales price. In this sense, a market price of a bioaroma of 100 to 500 US\$ kg^{-1} would correspond to yields in the range of 100 mg L^{-1} to 1 g L^{-1}. Although data from the literature are patchy, many reported processes have beaten the one gram per liter threshold. Based on the above statements, it is important to highlight some examples in the bioprocess field that could associate the yield achieved with a good production process, reaching a concentration of production in this presented range, focusing on products with yields higher than one gram per liter.

12.4.2 Products Obtained—An Industrial Perspective

One of the most studied and developed examples of aroma bioprocesses corresponds to the production of vanillin. Several reports deal with vanillin production at concentrations ranging from 3.75 g L^{-1} after 150 h[40] to 7.56 g L^{-1} using *Bacillus fusiformis*.[41] However, the highest yields reported for biotechnologically-produced natural vanillin refer to patented processes comprising the bioconversion of ferulic acid into vanillin by strains of *Amycolatopsis* sp. or *Streptomyces setonii* in a 10 L bioreactor,[42] with final yields of 11.5 g L^{-1}[43] and 13.9 g L^{-1},[44] respectively. In fact, the microbial transformation of ferulic acid is recognized as the most attractive and promising alternative source of natural vanillin.[7] This compound can be considered the most widely used flavoring agent in food products, with an estimated market of 5000 tons per year for biovanillin, 12 000 tons per year for synthetic vanillin and 50 tons per year for natural vanilla extracts.[35] The recent developments of molecular tools and genetic engineering have opened up the possibility of overcoming the different draw-backs related to the bioconversion processes, leading to the development of recognized industrial processes to obtain bioflavors,[45] such as biovanillin.[35]

Another interesting example is the biotechnological production of benzalde-hyde, used as an ingredient in cherry and other natural fruit flavors. This is the second most important aroma molecule for the flavor and fragrance industries, after vanillin, and has a world market of approximately 7000 tons per year.[35] Bio-technological production of this compound achieved almost 1 g L^{-1} when the media was supplemented with L-phenylalanine using different basidiomycetes, such as *Trametes, Ischnoderma, Polyporus* and *Bjerkandera* species.[46,47]

2-Phenylethanol, with a world market of about 7000 tons per year, is also one of the most used fragrances in perfumes and cosmetics due to its rose-like odor.[48] This compound can be synthesized by normal microbial metab-olism (*de novo* synthesis), however, the final concentrations in the culture broth generally remain very low, insufficient for an economically viable bio-process.[25] Much effort has been directed towards obtaining 2-phenylethanol, by improving growth-associated formation based on Ehrlich bioconversion from L-phenylalanine.[49]

Short-chain aliphatic aldehydes, such as acetaldehyde, 2-methyl-1-pro-panal, 2-methylbutanal and 3-methylbutanal (isovaleraldehyde), impart fruity and roasted characters to flavor compositions. Thus, the production of such compounds by a bio-route has become an important alternative.[50] Natural acetaldehyde, for example, achieved a final concentration of 140.6 g L^{-1} in a cyclic-batch fermentation process,[51] recognizing *Pichia pastoris* as an interesting biocatalyst for aldehyde production from alcohols in general.[25]

Some other bioprocesses employ fatty acids as substrates for bio-oxidation routes and the products obtained, such as 2-pentanone (20 g L^{-1}), 2-hepta-none (75 g L^{-1}) and 2-nonanone (60 g L^{-1}), may vary depending on the start-ing material. Methylketone formation, for example, is an aerobic process which is strongly favored when fungal growth is restricted and which does not occur with long-chain fatty acids.[25]

Meanwhile, another interesting industrial process has reported the use of *Yarrowia lipolytica* for the production of γ-decalactone, yielding over 10 g L^{-1} by the bioconversion of ricinoleic acid (12-hydroxy-C18:1). In this case, cas-tor oil represents an abundant and low cost source of that starting material.[52]

Another important strategy for the biotechnological production of bio-aromas is the biotransformation of inexpensive terpene substrates. Many scientific reports are available on this subject and some reviews[8,53,54] are rec-ommended for more detailed information. Research and development of the oxyfunctionalization of limonene is still one of the most studied processes. A variety of conversion products, such as the perillic compounds, carveol and carvone, may be formed in significant amounts.

One of the most promising bio-oxidation routes for limonene is the pro-duction of α-terpineol. Concentrations ranging from 3.2 g L^{-1}[55] to 4 g L^{-1}[56] of α-terpineol have been reported for the biotransformation of terpene by fungi. From an industrial point of view, perhaps the most efficient alternative for the commercial production of "bio" *R*-α-terpineol was reported recently.

Employing a two-phase system using sunflower oil as organic phase, limonene could be transformed into α-terpineol by *Sphingobium* sp. at concentrations reaching 120–130 g L^{-1}.[57]

Another limonene-derived alcohol, carveol, can be efficiently produced using the bacteria *Cellulosimicrobium cellulans*,[58] at a concentration of approximately 2 g L^{-1}. This compound, which has a spearmint-like odor, could also be produced using *Aspergillus niger*[59] or *Rhodococcus opacus*[60] as biocatalysts. Following the same metabolic route, carvone can also be formed. This is a very important monoterpene ketone widely applied in food, beverages and oral hygiene products. One of the most notable examples of carvone production involves the biotransformation of limonene using *Rhodococcus opacus* as biocatalyst.[61]

α-Pinene is another abundantly available substrate, which has also been extensively studied in biotransformation processes. Verbenol, for example, has a fresh pine, ozone odor[62] that could be produced by performing the biotransformation of α-pinene using *Aspergillus niger* and *Penicillium digitatum*[63] or *Pseudomonas putida*.[59] Some other possible compounds derived from α-pinene are verbenone, obtained from *Aspergillus niger*[64] or *Hormonema* sp.,[65] and myrtenol, produced by *Bacillus pallidus*.[66] Despite the great diversity of compounds obtained from this substrate, their concentrations remains too low from an industrial standpoint.

The bio-oxidation of linalool may produce linalool oxide, an important ether aroma compound with a sweet woody and earthy-flower odor.[62] *Corynespora cassiicola* DSM 62485 was identified as a novel highly stereoselective linalool-transforming biocatalyst, showing the highest productivity reported so far, with a conversion yield close to 100% and a productivity of 120 mg L^{-1} per day linalool oxide, which opens up prospects for scale-up studies.[67]

Geraniol was used as a starting material for a bioconversion process under aerobic conditions, and at alkaline pH, this compound was enantioselectively oxidised to (*E*)-geranic acid (85%) and (+)-citronellic acid (15%) using commercial baker's yeast. Geranic acid reached a maximum concentration of 3.6 g L^{-1} after 48 h at small scale.[26]

12.4.3 Production of Aroma Compounds in Bioreactors

Although many biotransformation procedures are still performed in Erlenmeyer flasks, scale-up in bioreactors has already been considered by many researchers. This is the first step to obtain compounds on a larger scale or in satisfactory yields, which is essential for industrial processes.

The use of a bioreactor allows good air exchange and homogenization during cultivation, as well as monitoring of some essential parameters such as pH, dissolved oxygen, temperature and others. Also, many researchers have outlined the basic problems in biotransformation processes at laboratory scale (*e.g.* limitation by liquid–liquid transport phenomena in flasks, low yields achieved, *etc.*), which have allowed the use of bioreactors and a scale of production closer to industrial reality.[68]

In this sense, improvements in the process performance and yields were observed for acetoin (3-hydroxy-2-butanone) production, for example. This important flavor compound, present in dairy products and some fruits, was synthesized by a *Bacillus subtilis* mutant strain, reaching 43.8 and 46.9 g L^{-1} in a flask and a 10 L bioreactor, respectively, when the culture medium was composed of glucose, yeast extract, corn steep liquor, ammonium sulfate and manganese sulfate.[69]

The production of 2,3,5,6-tetramethylpyrazine reached 4.33 g L^{-1} after 64.6 h of cultivation in a 5 L bioreactor, using the optimized medium which contained 20% glucose, 5% soytone, 3% $(NH_4)_2HPO_4$, and vitamin supplements.[70] The production of this pyrazine could be important for industrial applications, due to its musty, fermented and coffee-like odor, and it can be applied in beverages, baked goods, meat products and frozen dairy products.[62]

Still on limonene bioconversion, *Pseudomonas putida* GS1 was able to convert limonene to perillic acid (up to 11 g L^{-1}) when the bacteria was cultivated in a fed-batch culture with non-limiting amounts of glycerol, ammonium, and limonene.[71] Considering the same conversion, an efficient integrated bioprocess was developed using a method for *in situ* product recovery (ISPR) to overcome the product inhibition, leading to a cumulative perillic acid concentration of 31 g L^{-1} after 7 days. This represents the highest perillic derivative product concentration achieved in a microbial monoterpene oxyfunctionalization so far.[72] The toxic effects of the substrate could also be minimized in a gas loop system bioreactor.[68]

The biotransformation of carveol into carvone in concentrations up to 150 g L^{-1} has been achieved in small scale column reactors, using the bacteria *Rhodococcus globerus*[60] or *R. erythropolis*[73] as biocatalysts.

Boontawan and Stuckey (2006)[74] used a membrane bioreactor for the biotransformation of α-pinene oxide to isonovalal by *Pseudomonas fluorescens* NCIMB 11671, using a 2 L conventional stirred tank bioreactor. The final concentration of isonovalal obtained was 108 g L^{-1} after 400 h of operation. The same biotransformation process was achieved by using permeabilized cells of *Pseudomonas rhodesiae* CIP 107491. In this case, 400 g L^{-1} isonovalal was produced after 2.5 h.[75]

Aspergillus niger IFO 8541 was found to be an efficient biocatalyst for the biotransformation of β-ionone into hydroxy and oxo derivatives. The biotransformation was carried out in fed-batch mode (sequential precursor addition). These conditions allowed recovery of about 2.5 g L^{-1} aroma compounds after 230 h cultivation with a molar yield close to 100%.[76]

12.5 Green Chemistry in the Production of Aroma Compounds

Green chemistry approaches have already been considered in processes for the production of aroma compounds.[8,77] Basically, these kinds of technology aim to reduce or eliminate the use or the production of hazardous materials during the conception, manufacturing and utilization of chemical products,[78–80] based on the concepts of reducing, recycling and reusing.

Considering these premises, one should always focus on the reduction of energy demand, the use of safe procedures and avoiding the use and production of hazardous chemicals during manufacturing and throughout the products' life cycle. Additionally, the final product must be non-toxic, safe and degradable, forming innocuous substances and a minimum amount of waste.[79,81]

As previously mentioned, there has been increasing interest in natural aroma compounds, especially in the food market, where flavor and aroma extracts free of solvents or toxic compounds are required. Thus, bioaromas have become a suitable choice due to their better acceptance when compared to their chemically synthesized counterparts.[82] Moreover, minimizing the use of water and solvents, the formation of wastewater, the consumption of fossil energy and the generation of hazardous substances are also important features to be considered.[77,83] In this sense, chemical synthesis and recovery of aromas often results in environmentally unfriendly procedures.[84] Thus, "green chemistry" techniques, typically those requiring less solvents and energy demand, can be used to change this scenario.[77] Among such strategies, we highlight the use of alternative solvents (pressurized liquids, ionic liquids), energy-efficient extraction methods (ultrasound and microwave), and waste minimization (use of "renewable resources" or "agroindustrial residues" and "biocatalyst recycling"), which will be cited in this section. The two first approaches have been applied to recover aroma compounds from traditional (vegetable) sources, but they also present good applicability in microbial bioaroma-producing processes.

12.5.1 Alternative Solvents

Solvent extraction is generally performed by immersing/mixing a flavoring source in a suitable solvent, mostly derived from petroleum. Subsequent separation by distillation is carried out at specific temperatures, which causes the selective evaporation of that solvent.

Traditional extraction techniques include: Soxhlet extraction, sonication, blending and solid–liquid extraction. Usually, these approaches are time-consuming and require large amounts of samples, sorbents and organic solvents, which increase costs and cause disposal problems, besides having possible negative impacts on human health.[83] Therefore, the use of organic solvents is now seen as an undesirable choice. This is certainly the case when it comes to the production of food-grade materials, especially those with "organic" quality.[18] Consequently, novel approaches have been envisaged.

12.5.1.1 Supercritical Fluids

There are a wide range of solvents used in supercritical fluid extraction (SFE).[85] Among them, CO_2 is the most used for the recovery of aroma chemicals. This is the preferred solvent since it is nonflammable, odorless, tasteless, inexpensive, readily available, nontoxic, chemically stable, solubilizes most volatiles, is eliminated (evaporated) by decompression at relatively low temperature and, thus, produces a solvent-free extract.[82,85,86] An additional advantage refers to a possible solvent recovery step which can be made to require less energy when compared to ordinary liquid extraction.[85]

SFE has been described as a technique that provides extracts with an aroma character closely related to the original sources,[85] due to the production of extracts free of off-notes, with more top notes and with higher concentration of aromatic compounds.[86] Therefore, this technique has been used for aroma extraction from different food products, such as strawberries, soy sauce and various spices, coffee and fermented products.[82,86]

12.5.1.2 *Pressurized Liquid Extractions (PLEs)*

Pressurized liquid extractions have been considered as a promising extraction technology to achieve green technology's demands.[83,87] The elevated pressure and temperature used in this technique enhance the solubility and diffusion rate of solvents, improving mass transfer and extraction yields, besides decreasing the extraction time and solvent consumption.[87–89] Furthermore, it is considered a relatively simple and low cost technique, which is also environmentally friendly (low solvent use) and amenable to automation.[87] Moreover, when combined with other methods, such as ultrasound-assisted extraction and solid phase extraction, lesser amounts of sample and solvents are needed, and better efficiency and selectivity are observed.[83] Therefore, it is a convenient, reliable and flexible technique, which is potentially useful for flavor isolation.[88]

On the other hand, it is important to note that although PLE equipment could provide protection for oxygen- and light-sensitive molecules, special caution should be taken with thermolabile compounds, as these could be degraded if a careful optimization of the extraction parameters is not conducted.[83]

To illustrate, PLE has been used for the extraction of essential oils from different types of herbs and medicinal plants, such as coriander (*Coriandrum sativum*), *Acorus tatarinowii* Schott, boldo (*Penumus boldus* M.), *Cyperus rotundus*,[83] turmeric (*Curcuma domestica*),[87] coffee (*Coffea arabica* L. cv. Catimor)[88] and lemongrass (*Cymbopogon citratus*).[89]

12.5.1.3 *Ionic Liquids (ILs)*

Ionic liquids have received increasing interest over the last few years. The attention to this technology, often indicated as "green", is due to their properties in liquid–liquid extraction processes: low vapor pressure (when compared with organic solvents), thermal and chemical stability, commercial availability and adjustable properties (*e.g.* polarity, hydrophobicity and miscibility) by appropriate ion modification.[90–92] Moreover, the possible solvent reusability reinforces their environmental appeal and future industrial applications.[93] However, little information about their toxicity has been found in the literature,[92] although this feature must be carefully considered for food/aroma applications.[91]

ILs can be applied in several fields of aroma production, such as essential oil deterpenation[91,94] and a wide number of applications in enzymatic transformations.[92,93] In biotransformations, ILs may be used in biphasic systems, replacing part or all of the organic solvent to enhance stability, selectivity, enantioselectivity, product yield and reaction rate in several types of enzymatic reactions, *e.g.* esterification, transcyanidation, oxidoreduction, hydrogenation, transglycosylation.[93]

12.5.2 Alternative Extraction Methods

12.5.2.1 Ultrasound Assisted Extraction (UAE)

Ultrasound extraction offers many advantages in terms of productivity, yield, selectivity and processing time. Other interesting features related to this technique are the reduced thermal degradation and the low chemical/physical hazards, and thus it is considered an environmentally friendly technology.[77]

Distilling is one of the most conventional methods for aroma extraction. This process has been improved by combining it with ultrasound or even through the development of new ultrasound-assisted extraction (UAE) methods, which improve the extraction yields.[77,95] Moreover, this is a versatile technique that can be used both on a small and large scale.[95]

The extraction of natural aromas by UAE has been reported: essential oils from aromatic plants (peppermint leaves, *Artemisia*, lavender, almond) or other vegetal matrices (garlic and citrus flowers); herbal extracts from fennel, hops, marigold, mint, ginger, *Salvia* sp.; aroma compounds from vanilla pods (vanillin), caraway seeds (carvone) and Greek saffron (safranal),[77,96] and honey extracts.[97] In addition, different aroma components from fermented products, such as oak-aged grape brandies[98] and wine,[99,100] have also been recovered through this technique.

Other studies have considered the use of ultrasound as an extraction technique for the winemaking industry,[99,100] particularly for aroma quantification by combining ultrasound and organic solvent extractions.[98,99] The advantages of using UAE are high reproducibility and the possibility of simultaneous extraction of several samples, which makes it an interesting alternative for the analysis of wine flavor components.[100]

12.5.2.2 Microwave Heating

In recent years, there has been increasing interest in the application of microwave heating for solvent extractions. Several classes of compounds, such as essential oils, aromas, pesticides, phenols, dioxins, and other organic compounds have been efficiently extracted from a variety of matrices (mainly soils, sediments, animal tissues, food and plant material).[101,102] The reported data have shown that microwave-assisted solvent extraction (MAE) is a viable alternative to conventional techniques. The main benefit of this technique is related to the reduction of extraction time.[101]

Solvent-free microwave extraction (SFME), in turn, is a technique which combines microwave heating and dry distillation at atmospheric pressure to isolate and concentrate, for example, essential oils in fresh plant materials.[103]

The main advantages of SFME are: shorter extraction times, substantial energy saving and reduced environmental impact (less CO_2 released into the atmosphere). All these advantages make SFME a good alternative for the extraction of essential oils from aromatic plants,[101] such as spices (ajowan, cumin and star anise), aromatic herbs (basil, garden mint, thyme and cardamom seeds) and others (oregano and orange essential oil).[101–104]

12.5.3 Alternative Substrates

One recurrent strategy in bioprocesses consists of the use of alternative substrates, particularly biomass,[26,105–109] to substitute conventional cultivation media. This proposal is based on the fact that such materials are usually inexpensive, underused and are readily available waste or by-products derived from (agro)industrial processing, but whose disposal could be harmful to the environment. Therefore, the use of such renewable materials – which may be provided by different sectors: agriculture (agricultural and waste products), silviculture (forest and waste products), industry (waste and leftover processes) and aquaculture (micro-algae and macro-algae) – represents an interesting choice to reduce manufacturing costs, minimize waste disposal and add value to underutilized by-products.

12.5.3.1 *Alternative Substrates and Culture Media for Bioaroma Production*

An ideal culture medium must meet the microbial needs and ensure the best conditions for the formation of products, besides permitting a simpler purification process.[26] Depending on the process, the value of a culture medium may represent 38 to 73% of the total production costs;[110] consequently, the use of cheaper raw materials is a key element to make an industrial-scale bioprocess viable and competitive.[107,110] In this sense, agro-industrial wastes have become an attractive alternative due to their low commercial value and to their composition, usually rich in carbohydrates, nitrogen and minerals, which can be easily absorbed by microbes.[26,84,111,112]

As suggested by Berger (2009),[13] the use of non-conventional media could also be a suitable choice in biotechnological processes for flavor production. Therefore, in recent years, solid state fermentation (SSF) has received much attention, since several agro-industrial residues (bagasse, straw, husks, pulp) are suitable for this kind of fermentation. Additionally, some studies on flavors and other substances have shown that the use of SSF may result in higher yields than submerged fermentation (SmF).[84,112] Thus, several examples of applying agro-industrial residues as a promising strategy for aroma production can be cited.[106,113–117] The main studies showing the application of residues for bioaroma production by "*de novo*" synthesis are presented in Table 12.2.

The use of agro-residues and by-products as substrates in bioprocesses has also been evaluated for the production of bioaromas by biotransformation. As already stated above, this strategy seems to be a rising trend to overcome the

Table 12.2   Agroindustrial residues in "*de novo*" aroma production.[26,84]

Substrate	Microorganism	Aroma	Additional information	References
Citric pulps	*Ceratocystis fimbriata*	Fruity	Solid state fermentation (SSF)	Rossi *et al.*, 2009[118]
Coffee pulp and husk	*Pachysolen tannophilus*	Fruity aroma, pineapple, banana	SSF	Soares *et al.*, 2000[119]
Sugarcane bagasse	*C. fimbriata*	Fruity aroma	SSF; supplemented with a synthetic medium containing glucose, leucine or valine	Christen *et al.*, 1997[120]
Cassava bagasse, apple pomace, amaranth and soybean	*C. fimbriata*	Fruity aroma	SSF	Bramorski *et al.*, 1998[121]
Cassava bagasse, apple pomace, amaranth and soybean	*Rhizopus oryzae*	Fruity aroma	SSF	Christen *et al.*, 2000[122]
Cassava bagasse, sugarcane bagasse, sunflower seeds and palm bran	*Kluyveromyces marxianus*	Fruity aroma	SSF, supplemented with 10% glucose	Medeiros *et al.*, 2000[123]
Cassava wastewater	*Geotrichum fragrans*	Fruity aroma	SmF	Damasceno *et al.*, 2003[124]

high manufacturing costs involved in microbial production of bioflavors.[26] The main examples of volatile synthesis through biotransformations by applying agro-industrial residues and by-products are presented in Table 12.3.

12.5.3.2   Rational Use of Biocatalysts

Even in optimized systems, in which a target compound is produced in large quantities, other molecules are still generated by the cell. In these cases, the complete exploitation of the potential of fermentations, by recovering these "other molecules" as well, has become a promising possibility. These coproduction approaches are an important strategy, since they allows simultaneous production of more than one (bio)product in the same bioprocess. Therefore, manufacturing costs may be diluted (many products recovered from a single process), besides the obvious environmental advantages (better use of raw materials and energy).

Table 12.3   Agroindustrial residues in biotransformation aroma production.[26]

Substrate	Microorganism	Precursor	Aroma/volatile compound	References
Cereal bran or sugar-beet pulp	*Pycnoporus cinnabarinus*	Ferulic acid	Vanillin	Thibaut *et al.*, 1998[125]
Rice bran	*P. cinnabarinus*	Ferulic acid	Vanillin	Zheng *et al.*, 2007[126]
Maize bran	*P. cinnabarinus* and *Aspergillus niger*	Ferulic acid	Vanillin	Lesage-Meessen *et al.*, 2002[127]
Wheat bran	Engineered *Escherichia coli* (strain JM109/pBB1), containing the genes for feruloyl-SCoA synthetase and feruloyl-SCoA hydratase/ aldolase from *Pseudomonas fluorescens* BF13	Ferulic acid	Vanillin	Di Gioia *et al.*, 2007[128]; Torres *et al.*, 2009[129]
Cassava wastewater	*Penicillium* sp.	Citronellol	*Cis-* and *trans-* rose oxides	Maróstica Jr. and Pastore, 2006[130]
Cassava wastewater	*Fusarium oxysporum*	Orange peel oil (*R*-(+)-limonene)	*R*-(+)-α-terpineol	Maróstica Jr. and Pastore, 2007[131]

In terms of bioaroma production, this strategy may be illustrated by a recent study. *Fusarium oxysporum* 152B is a fungal strain known to produce extracellular alkaline lipase in synthetic media containing inducing agents (*e.g.* vegetable oils)[132] and also has the ability to convert *R*-(+)-limonene to *R*-(+)-α-terpineol.[37] However, the first process results in biomass being discarded, while in the second, biomass is used and the fermentation supernatant is disposed. Bicas *et al.* (2010)[56] proposed an integrated process by using the biomass resulting from lipase production as a biocatalyst for the biotransformation of *R*-(+)-limonene to *R*-(+)-α-terpineol. The authors concluded that although the yield of *R*-(+)-α-terpineol was lower than that of the conventional process, this is an interesting concept from an industrial point of view.

12.6   Concluding Remarks

As already mentioned by Baines and Seal (2012),[133] in the 1960s the words "natural" and "food" were rarely used simultaneously, not even as a marketing strategy. This has drastically changed in the last 50 years. Today, consumer's preference for "natural" has been influenced in the market, and food laws world-wide have reflected the "natural/artificial" discrimination made by consumers. As an example, 90% of all beverage flavors in Europe, and 80% in the USA are "natural".[13] In parallel, concerns regarding environmental conservation have been raised. The fact is that there is currently an intense demand for natural and environmentally friendly (aroma) production processes. Therefore, as presented in this chapter, white biotechnology appears to be a promising alternative to meet these requirements. Besides the green technologies potentially applied in microbial processes for bioaroma production, optimization tools allied with a rational use of resources may be used to maximize yields and reduce costs. Thus, it is expected that the flavor and fragrance supply derived from bioprocesses, which is now less than 10%,[13] will certainly rise in future, guided by the recent trends and positive improvements in biotechnological processes.

References

1. G. Molina, J. L. Bicas, M. R. Maróstica Jr. and G. M. Pastore, *Biotecnologia de Alimentos*, ed. G. M. Pastore, J. L. Bicas and M. R. Maróstica Jr., Atheneu, São Paulo, 1st edn, 2013, ch. 11, pp. 273–317.
2. D. Linares, Etudes sur la voie de dégradation de l'α-pinène chez Pseudomonas rhodesiae en milieu biphasique liquide-liquide, PhD in Life and Health Sciences, Université Blaise Pascal, Clermont-Ferrand, France, 2008.
3. P. Fontanille, Biotransformation de α-pinene oxyde en cis-2-methyl-5-isopropylhexa-2,5-dienal (isonovalal) par Pseudomonas rhodesiae CIP 107491, PhD in Life and Health Sciences, Université Blaise Pascal, Clermont-Ferrand, France, 2002.
4. R. G. Berger, *Aroma Biotechnology*, Springer-Verlag, Berlin, 1st edn, 1995, p.240.
5. R. Wagner, Composição de voláteis e aroma de salames nacionais tipos Italiano e Milano, PhD in Food Science, University of Campinas, Campinas, Brazil, 2007.
6. J. Margetts, *Chemistry and Technology of Flavors and Fragrances*, ed. D. J. Rowe, Blackwell Publishing, Oxford, 1st edn, 2005, ch. 8, pp. 169–197.
7. J. L. Bicas, J. C. Silva, A. P. Dionísio and G. M. Pastore, *Ciênc. Tecnol. Aliment.*, 2010, **30**, 7.
8. J. L. Bicas, A. P. Dionísio and G. M. Pastore, *Chem. Rev.*, 2009, **109**, 4518.
9. Y. Gounaris, *Flavour Fragrance J.*, 2010, **25**, 367.
10. A. P. Dionísio, G. Molina, D. S. Carvalho, R. Santos, J. L. Bicas and G. M. Pastore, *Natural food additives, ingredients and flavourings*, ed. D. Baines and R. Seal, Woodhead Publishing Limited, Cambridge, 1st edn, 2012, ch. 11, pp. 231–259.
11. R. G. Berger, *Expression of Multidisciplinary Flavour Science*, ed. I. Blank, M. Wüst and C. Yeretzian, Switzerland, 2008, pp. 319–327.
12. M. Guentert, *Flavors and Fragrances: Chemistry, Bioprocessing and Sustainability*, ed. R. G. Berger, Springer, Berlin, 1st edn, 2007, ch. 1, pp. 1–13.

13. R. G. Berger, *Biotechnol. Lett.*, 2009, **31**, 1651.
14. L. Caputi and E. Aprea, *Recent Pat. Food, Nutr. Agric.*, 2011, **3**, 9.
15. M. Müller, K. Dirlam, H. H. Wenk, R. G. Berger, U. Krings and R. Kaspera, Method for the production off flavor-active terpenes, US Patent No. 7803606 B2, 2010.
16. H. Zorn, M. A. Fraatz, S. J. L. Reimer, M. Takenberg, U. Krings, R. G. Berger and S. Marx, Enzymatic synthesis of nootkatone, US Patent Application no. US 2012/0045806 A1, 2012.
17. P. S. J. Cheetham, M. L. Gradley and J. T. Sime, Flavor/aroma materials and their preparation, US Patent no. 6,844,019 B1, 2005.
18. S. Heald, S. Myers, T. Walfords, K. Robbins and C. Hill, Preparation of vanillin from microbial transformation media by extraction by means supercritical fluid or gases, US Patent 20110268858–A1, 2011.
19. A. A. Torres, M. T. Martinez, A. B. Mir and R. M. M. Ochoa, Process of production of vanillin with immobilized micro-organisms. US Patent Application no. US 2011/0065156 A1, 2011.
20. P. Xu, D. Hua, L. Song, C. Ma, Z. Zhang, Y. Du, H. Chen, L. Gan, Z. Wei and Y. Zeng, Streptomyces strain and the method for converting ferulic acid to vanillin by using the same, US Patent Application no. US 2009/0186399 A1, 2009.
21. J. Schrader, M. M. W. Etschmann, D. Sell, J.-M. Hilmer and J. Rabenhorst, *Biotechnol. Lett.*, 2004, **26**, 463.
22. P. Xu, D. Hua and C. Ma, *Trends Biotechnol.*, 2007, **25**, 571.
23. S. A. Dubal, Y. P. Tilkari, S. A. Momin and I. V. Borkar, *Adv. Biotechnol.*, 2008, **6**, 20.
24. I. L. Gatfield, *Flavor chemistry: 30 years of progress*, ed. R. Teranishi, E. L. Wick and I. Hornstein, Kluwer/Plenum, New York, 1st edn, 1999, ch. 19, pp. 211–227.
25. J. Schrader, *Flavors and Fragrances: Chemistry, Bioprocessing and Sustainability*, ed. R. G. Berger, Springer, Berlin, 1st edn, 2007, ch. 23, pp. 507–574.
26. J. L. Bicas, A. P. Dionísio, J. C. Silva, F. F. C. Barros and G. M. Pastore, *Industrial fermentations: Food process, nutrient sources and production strategies*, ed. J. Krause and O. Fleischer, Nova Publishers, New York, 1st edn, 2010b, ch. 9, pp. 275–296.
27. P. Crowell, *J. Nutr.*, 1999, **129**, 775S.
28. C. C. C. R. de Carvalho and M. M. R. da Fonseca, *Food Chem.*, 2006, **95**, 413.
29. C. O. Da Fonseca, G. Schwartsmann, J. Fischer, J. Nagel, D. Futuro, T. Quirico-Santos and C. R. Gattass, *Surg. Neurol.*, 2008, **70**, 259.
30. C. O. Da Fonseca, R. M. Teixeira, R. Ramina, G. Kovaleski, J. T. Silva, J. Nagel and T. Quirico-santos, *J. Cancer Ther.*, 2011, **2**, 16.
31. J. M. Matos, C. M. Schmidt, H. J. Thomas, O. W. Cummings, E. A. Wiebke, J. A. Madura, L. J. Patrick and P. L. Crowell, *J. Surg. Res.*, 2008, **147**, 194.
32. M. R. Maróstica Jr., T. A. A. R. Silva, G. C. Franchi, A. Nowill, G. M. Pastore and S. Hyslop, *Food Chem.*, 2009, **116**, 8.
33. S. B. Hassan, H. Gali-Muhtasib, H. Goransson and R. Larsson, *Anticancer Res.*, 2010, **30**, 1911.

34. J. L. Bicas, I. A. Neri-Numa, A. L. T. G. Ruiz, J. E. De Carvalho and G. M. Pastore, *Food Chem. Toxicol.*, 2011, **49**, 1610.

35. G. Feron and Y. Waché, *Food Biotechnology*, ed. K. Shetty, G. Paliyath, A. Pometto and R. Levin. CRC Press, Boca Raton, 2nd edn, 2006, ch. 1.16, pp. 407–441.

36. M. J. van der Werf, J. A. M. De Bont and D. J. Leak, *Adv. Biochem. Eng./Biotechnol.*, 1997, **55**, 147.

37. J. L. Bicas, F. F. C. Barros, R. Wagner, H. T. Godoy and G. M. Pastore, *J. Ind. Microbiol. Biotechnol.*, 2008a, **35**, 1061.

38. I. Rottava, P. F. Cortina, E. Martello, R. L. Cansian, G. Toniazzo, O. A. C. Antunes, E. G. Oestreicher, H. Treichel and D. de Oliveira, *Appl. Biochem. Biotechnol.*, 2011, **164**, 514.

39. J. L. Bicas, P. Fontanille, G. M. Pastore and C. Larroche, *J. Appl. Microbiol.*, 2008b, **105**, 1991.

40. D. Hua, C. Ma, S. Lin, L. Song, Z. Deng, Z. Maomy, Z. Zhang, B. Yu and P. Xu, *J. Biotechnol.*, 2007, **130**, 463.

41. L.-Q. Zhao, Z.-H. Sun, P. Zheng and J.-Y. He, *Process Biochem.*, 2006, **41**, 1673.

42. A. Daugsch and G. M. Pastore, *Quím. Nova*, 2005, **28**, 642.

43. J. Rabenhorst and R. Hopp, Process for the preparation of vanillin and suitable microorganisms. DE Patent. 19532317, 2000.

44. A. Muheim and K. Lerch, *Appl. Microbiol. Biotechnol.*, 1999, **51**, 456.

45. G. Molina, J. L. Bicas, E. Moraes, M. R. Maróstica Jr and G. M. Pastore, *Applications of Microbial Engineering*, ed. V. K. Gupta, M. Schmoll, M. Maki, M. Tuohy and M. A. Mazutti, CRC Press, Boca Raton, 1st edn, 2013b, ch. 5, pp. 122–157.

46. A. Lomascolo, M. Asther, D. Navarro, C. Antona, M. Delattrc and L. Lesage-Meessen, *Lett. Appl. Microbiol.*, 2001, **32**, 262.

47. R. G. Berger, A. Boker, M. Fischer and J. Taubert, *Flavor chemistry: 30 years of progress*, ed. R. Teranishi, E. L. Wick and I. Hornstein, Kluwer/Plenum, New York, 1st edn, 1999, ch. 20, pp. 229–238.

48. C. E. Fabre, P. J. Blanc and G. Goma, *Biotechnol. Prog.*, 1998, **14**, 270.

49. M. M. W. Etschmann, W. Bluemke, D. Sell and J. Schrader, *Appl. Microbiol. Biotechnol.*, 2002, **59**, 1.

50. H. Surburg and J. Panten, *Common Fragrance and Flavor Materials. Preparation, Properties and Uses*, Wiley-Vch Verlag, Weinheim, 5th edn, 2006, p. 330.

51. W. D. Murray, S. J. B. Duff and T. Beveridge, *Appl. Environ. Microbiol.*, 1990, **56**, 2378.

52. E. J. Vandamme and W. Soetaert, *J. Chem. Technol. Biotechnol.*, 2002, **77**, 1323.

53. U. Krings and R. G. Berger, *Appl. Microbiol. Biotechnol.*, 1998, **49**, 1.

54. W. A. Duetz, H. Bouwmeester, J. B. Van Beilen and B. Witholt, *Appl. Microbiol. Biotechnol.*, 2003, **61**, 269.

55. Q. Tan, D. F. Day and K. R. Cadwallader, *Process Biochem.*, 1998, **33**, 29.

56. J. L. Bicas, C. P. de Quadros, I. A. Neri-Numa and G. M. Pastore, *Food Chem.*, 2010c, **120**, 452.

57. J. L. Bicas, P. Fontanille, G. M. Pastore and C. Larroche, *Process Biochem.*, 2010d, **45**, 481–486.

58. Z. Wang, F. Lie, E. Lim, K. Li and Z. Li, *Adv. Synth. Catal.*, 2009, **351**, 1849.

59. M. S. Divyashree, J. George and R. Agrawal, *J. Food Sci. Technol.*, 2006, **43**, 73.

60. W. A. Duetz, A. H. M. Fjällman, S. Ren, C. Jourdat and B. Witholt, *Appl. Environ. Microbiol.*, 2001, **67**, 2829.

61. C. C. C. R. de Carvalho and M. M. R. da Fonseca, *Tetrahedron: Asymmetry*, 2003, **14**, 3925.

62. G. A. Burdock and G. Fenaroli, *Fenaroli's handbook of flavor ingredients*, CRC Press, Boca Raton, 6th edn, 2010, p. 2159.

63. S. C. V. Rao, R. Rao and R. Agrawal, *Biotechnol. Appl. Biochem.*, 2003, **37**, 145.

64. R. Agrawal and R. Joseph, *Appl. Microbiol. Biotechnol.*, 2000, **53**, 335.

65. M. S. van Dyk, E. van Rensburg and N. Molelek, *Biotechnol. Lett.*, 1998, **20**, 431.

66. N. Savithiry, D. Gage, W. Fu and P. Oriel, *Biodegradation*, 1998, **9**, 337.

67. M. A. Mirata, M. Wu, A. Mosandl and J. Schrader, *Food Chem.*, 2008, **56**, 3287.

68. M. Pescheck, M. A. Mirata, B. Brauer, U. Krings, R. G. Berger and J. Schrader, *J. Ind. Microbiol. Biotechnol.*, 2009, **36**, 827.

69. H. Xu, S. Jia and J. Liu, *Afr. J. Biotechnol.*, 2011, **10**, 779.

70. Z. Xiao, N. Xie, P. Liu, D. Hua and P. Xu, *Appl. Microbiol. Biotechnol.*, 2006, **73**, 512.

71. A. E. Mars, J. P. L. Gorissen, I. van den Beld and G. Eggink, *Appl. Microbiol. Biotechnol.*, 2001, **56**, 101.

72. M. A. Mirata, D. Heerd and J. Schrader, *Process Biochem.*, 2009, **44**, 764.

73. C. C. C. R. de Carvalho, A. Poretti and M. M. R. da Fonseca, *Appl. Microbiol. Biotechnol.*, 2005, **69**, 268.

74. A. Boontawan and D. C. Stuckey, *Appl. Microbiol. Biotechnol.*, 2006, **69**, 643.

75. P. Fontanille and C. Larroche, *Appl. Microbiol. Biotechnol.*, 2003, **60**, 534.

76. C. Larroche, C. Creuly and J. B. Gros, *Appl. Microbiol. Biotechnol.*, 1995, **43**, 222.

77. F. Chemat, Z.-e. Huma and M. K. Khan, *Ultrason. Sonochem.*, 2011, **18**, 813.

78. R. Hatti-Kaul, U. Törnvall, L. Gustafsson and P. Börjesson, *Trends Biotechnol.*, 2007, **25**, 119.

79. M. M. Kirchhoff, *Resour., Conserv. Recycl.*, 2005, **44**, 237.

80. P. Anastas and J. Warner, *Green Chemistry: theory and practice*, Oxford University Press, New York, 1998, p. 135.

81. F. Cherubini, *Energy Convers. Manage.*, 2010, **51**, 1412.

82. Z.-M. Lu, W. Xu, N.-H. Yu, T. Zhou, G.-Q. Li, J.-S. Shi and Z.-H. Xu, *Intern. J. Food Sci. Technol.*, 2011, **46**, 1508.

83. A. Mustafa and C. Turner, *Anal. Chim. Acta*, 2011, **703**, 8.

84. M. A. Longo and M. A. Sanromán, *Food Technol. Biotechnol.*, 2006, **44**, 335.

85. M. B. King and T. R. Bott, *Extraction of natural products using near-critical fluids*, Blackie Academic & Professional, Glasgow, 1st edn, 1993, p. 325.

86. E. Ramos, E. Valero, E. Ibáñez, G. Reglero and J. Tabera, *J. Agric. Food Chem.*, 1998, **46**, 4011.

87. A. H. Zaibunnisa, S. Norashikin, S. Mamot and H. Osman, *LWT–Food Sci. Technol.*, 2009, **42**, 233.

88. M.-W. Cheong, A.-A. Tan, S.-Q. Liu, P. Curran and B. Yu, *Talanta*, 2013, **115**, 300.

89. A. H. Nur Ain, A. H. Zaibunnisa, H. M. S. Zahrah and S. Norashikin, *Int. Food Res. J.*, 2013, **20**, 451.

90. F. Rantwijk and R. A. Sheldon, *Chem. Rev.*, 2007, **107**, 2757.

91. M. Francisco, S. Lago, A. Soto and A. Arce, *Fluid Phase Equilib.*, 2010, **296**, 149.

92. H. Olivier-Bourbigou, L. Magna and D. Morvan, *Appl. Catal., A*, 2010, **373**, 1.

93. A. Fehér, V. Illeová, I. Kelemen-Horváth, K. Bélafi-Bakó, M. Polakovič and L. Gubicza, *J. Mol. Catal. B: Enzym.*, 2008, **50**, 28.

94. A. Arce, A. Pobudkowska, O. Rodríguez and A. Soto, *Chem. Eng. J.*, 2007, **133**, 213.

95. M. Vinatoru, *Ultrason. Sonochem.*, 2001, **8**, 303.

96. S. R. Shirsath, S. H. Sonawane and P. R. Gogate, *Chem. Eng. Proc.*, 2012, **53**, 10.

97. C. E. Manyi-Loh, R. N. Ndip and A. M. Clarke, *Int. J. Mol. Sci.*, 2011, **72**, 9514.

98. I. Caldeira, R. Pereira, M. C. Clímaco, A. P. Belchior and R. B. Sousa, *Anal. Chim. Acta*, 2004, **513**, 125.

99. D. H. Vila, F. J. H. Mirab, R. B. Lucena and M. A. F. Recamalesa, *Talanta*, 1999, **50**, 413.

100. J. F. G. Martín and D.-W. Sun, *Trends Food Sci. Technol.*, 2013, **33**, 40.

101. M. E. Lucchesi, F. Chemat and J. Smadja, *J. Chromatogr. A*, 2004, **1043**, 323.

102. M. E. Lucchesi, J. Smadja, S. Bradshaw, W. Louw and F. Chemat, *J. Food Eng.*, 2007, **79**, 1079.

103. B. Bayramoglu, S. Sahin and G. Sumnu, *J. Food Eng.*, 2008, **88**, 535.

104. M. A. Ferhat, B. Y. Meklati, J. Smadja and F. Chemat, *J. Chromatogr. A*, 2006, **1112**, 121.

105. A. Pandey, C. R. Soccol, P. Nigam, V. T. Soccol, P. S. L. Vandenberghe and R. Mohan, *Bioresour. Technol.*, 2000, **74**, 81.

106. A. Pandey, C. R. Soccol, P. Nigam and V. T. Soccol, *Bioresour. Technol.*, 2000, **74**, 69.

107. R. S. Makkar and S. S. Cameotra, *Appl. Microbiol. Biotechnol.*, 2002, **58**, 428.

108. F. R. Timofiecsyk and U. Pawlowsky, *Bol. Cent. Pesqui. Process. Aliment.*, 2000, **18**, 221.

109. G. Laufenberg, B. Kunz and M. Nystroem, *Bioresour. Technol.*, 2003, **87**, 167.

110. P. F. Stanburry, A. Whitaker and S. J. Hall, *Principles of Fermentation technology*, Butterworth Heinemann, Oxford, 2nd edn, 1995, p. 357.
111. Z. Konsoula and M. Liakopoulou-Kyriakides, *Bioresour. Technol.*, 2007, **98**, 150.
112. M. C. Orzua, S. I. Mussatto, J. C. Contreras-Esquivel, R. Rodriguez, H. de la Garza, J. A. Teixeira and C. N. Aguilar, *Ind. Crops Prod.*, 2009, **30**, 24.
113. R. R. Singhania, A. K. Patel, C. R. Soccol and A. Pandey, *Biochem. Eng. J.*, 2009, **44**, 13.
114. A. Pandey, *Biochem. Eng. J.*, 2003, **13**, 81.
115. A. Pandey, C. R. Soccol and D. Mitchell, *Process Biochem.*, 2000, **35**, 1153.
116. Z. Wang and S.-T. Yang, *Bioresour. Technol.*, 2013, **137**, 116.
117. U. Hölker and J. Lenz, *Curr. Opin. Microbiol.*, 2005, **8**, 301.
118. S. C. Rossi, L. P. S. Vandenberghe, B. M. P. Pereira, F. D. Gago, J. A. Rizzolo, A. Pandey, C. R. Soccol and A. B. P. Medeiros, *Food Res. Intern.*, 2009, **42**, 484.
119. M. Soares, P. Christen, A. Pandey and C. R. Soccol, *Process Biochem.*, 2000, **35**, 857.
120. P. Christen, J. C. Meza and S. Revah, *Mycol. Res.*, 1997, **101**, 911.
121. A. Bramorski, C. R. Soccol, P. Christen and S. Revah, *Rev. Microbiol.*, 1998, **29**(3), DOI: 10.1590/S0001-37141998000300012.
122. P. Christen, A. Bramorski, S. Revah and C. R. Soccol, *Bioresour. Technol.*, 2000, **71**, 211.
123. A. B. P. Medeiros, A. Pandey, R. J. S. Freitas, P. Christen and C. R. Soccol, *Biochem. Eng. J.*, 2000, **6**, 33.
124. S. Damasceno, M. P. Cereda, G. M. Pastore and J. G. Oliveira, *Process Biochem.*, 2003, **39**, 411.
125. J. Thibault, V. Micard, C. Renard, M. Asther, M. Delattre, L. Lesage Meessen, C. Faulds, P. Kroon, G. Williamson, J. Duarte, J. C. Duarte, B. C. Ceccaldi, M. Tuohy, D. Couteau, S. Van Hulle and H.-P. Heldt Hansen, *LWT-Food Sci. Technol.*, 1998, **31**, 530.
126. L. Zheng, P. Zheng, Z. Sun, Y. Bai, J. Wang and X. Guo, *Bioresour. Technol.*, 2007, **98**, 1115.
127. L. Lesage-Meessen, A. Lomascolo, E. Bonnin, J. F. Thibault, A. Buleon, M. Roller, M. Asther, E. Record, B. C. Ceccaldi and M. Asther, *Appl. Biochem. Biotechnol.*, 2002, **102–103**, 141.
128. D. Di Gioia, L. Sciubba, L. Setti, F. Luziatelli, M. Ruzzi, D. Zanichelli and F. Fava, *Enzyme Microb. Technol.*, 2007, **41**, 498.
129. B. R. Torres, B. Aliakbarian, P. Torre, P. Perego, J. M. Domínguez, M. Zilli and A. Converti, *Enzyme Microb. Technol.*, 2009, **44**, 154.
130. M. R. Maróstica Jr. and G. M. Pastore, *Cienc. Tecnol. Aliment.*, 2006, **26**, 690.
131. M. R. Maróstica Jr. and G. M. Pastore, *Food Chem.*, 2007, **101**, 345.
132. C. P. Quadros, J. L. Bicas, I. A. Néri-Numa and G. M. Pastore, *Food Sci. Biotechnol.*, 2009, **18**, 1519.
133. D. Baines and R. Seal, *Natural food additives, ingredients and flavourings*, ed. D. Baines and R. Seal, Woodhead Publishing Limited, Cambridge, 1st edn, 2012, ch. 1, pp. 1–21.

Biotransformation Using Plant Cell Culture Systems and Tissues

BERNARDO DIAS RIBEIRO*[a], EVELIN ANDRADE MANOEL[a], CLAUDIA SIMÕES-GURGEL[b], AND NORMA ALBARELLO[b]

[a]Federal University of Rio de Janeiro, Brazil; [b]State University of Rio de Janeiro, Brazil
*E-mail: bernardo@eq.ufrj.br

13.1 Biotransformation and Green Chemistry

Biotransformations are chemical reactions that may be used to carry out specific conversions of complex substrates using plant, animal or microbial cells or purified enzymes as catalysts, and have great potential to generate novel products or to produce known products more efficiently. These processes explore the unique properties of biocatalysts, namely their stereo- and regiospecificity and their ability to carry out reactions at nonextreme pH values and temperatures. It is not necessary for the compounds to be natural intermediates, meaning that they can be of synthetic origin. Moreover, biotransformation is also gaining considerable attention as a step towards green chemistry by reducing the usage of hazardous chemicals. Plant cell cultures exhibit vast biochemical potential for the production of specific secondary metabolites. Plant bioconversion systems may be used alone to produce novel chemicals or in combination with organic synthesis. Multistep

RSC Green Chemistry No. 45
White Biotechnology for Sustainable Chemistry
Edited by Maria Alice Z. Coelho and Bernardo D. Ribeiro
© The Royal Society of Chemistry 2016
Published by the Royal Society of Chemistry, www.rsc.org

processes catalyzed by cell or organ cultures often generate intermediary metabolites which help to establish biosynthetic pathways.[1,2]

The biotransformation capabilities of microorganisms and their enzymes for the production of a wide variety of fine chemicals are well known. Microbial systems are advantageous in that biomass doubling times are short and hence, the production of biomass can be achieved quickly. In addition, methods for genetic manipulation of microbes are well established. Plant systems, on the other hand, produce a more limited range of enzymes and undifferentiated plant cells have doubling times larger than those of microbial cells In addition, the desired enzymes are often produced in minute quantities. Despite these drawbacks, the plant kingdom contains some unique enzymes, which produce a variety of chemicals. Cell cultures have a higher rate of metabolism than intact differentiated plants because the initiation of cell growth in culture leads to fast proliferation of cell mass and to a condensed biosynthetic cycle.[3] However, chemical synthesis of some of these compounds is extremely complex and costly. Hence, biotransformations using plant cells and isolated enzymes have immense potential for the production of pharmaceuticals. Plant enzyme biocatalysts may be applied to the production of completely new drugs and also may be used to modify existing drugs by improving their bioactivity spectrum, known as second generation phytopharmaceuticals, with improved pharmaceutical properties, such as lower toxicity, improved solubility and pharmacokinetics, and they are primarily natural product analogs.[2,4,5] Plant secondary metabolites are unique resources for pharmaceuticals, food additives, for example, providing original materials for use in other areas.[6]

For comparison between plant and microbial cell cultures, Zafar *et al.*[7] studied the biotransformation of ethynodiol diacetate, a semi-synthetic steroidal drug used as an oral contraceptive, with fungi *Cunninghamella elegans* and plant cell suspension cultures of basil (*Ocimum basilicum*) and neem (*Azadirachta indica*). While fungal biotransformation promoted mainly hydroxylation, plant cell biotransformation aided in hydrolysis and oxidation of the substrate. A similar report was made by Musharraf *et al.*[8] using (−)-ambrox, a highly fragrant constituent of ambergris, a metabolite of the sperm whale, as substrate, and a fungal cell culture of *Macrophomina phaseolina* and a plant cell suspension culture of *Peganum harmala* as biocatalysts. In this case, the substrate went through hydroxylation at C-3β, C-6β, and C-1α and regio-controlled keto formation at C-3 with fungal biotransformation, whereas plant cell biotransformation resulted in the production of hydroxylated products at C-2α and C-3β, besides lactone formation in the sclareolide skeleton.

13.2 Plant Cell Cultures

Plants have been present in our lives since times immemorial as sources of carbohydrates, proteins, and fats for food and are also used for shelter. They synthesize a wide variety of secondary metabolites, such as terpenoids, cardenolides, flavonoids, glucosinolates, and alkaloids, which are related to the interaction of plants with their environment and are used as drugs, flavours, pigments (food ingredients), and agrochemicals. The distribution and production of these compounds in plants are generally restricted to a few species, or even within a few varieties within a species, and at low content (<1%). Moreover, secondary metabolites often accumulate in specialized cells or organs, and their synthesis depends greatly on plant's physiological and developmental stages.[4,9,10] Thus, these limitations need to be overcome, in addition to other limitations such as:

* Some species containing high-value compounds are difficult to cultivate.
* Highly complex molecules are not economically feasible by chemical synthesis.
* Seasonal and geographical influences.
* Diseases (insects, microorganisms and viruses) in the cultivation of medicinal plants.
* Political and socio-economic situations.

Secondary metabolites have been used by humans to treat health disorders and illnesses. Plant cell culture systems have been developed and have become an alternative to extraction from the whole plant material. Tissue culture techniques play an important role in providing a valuable tool for the production of plant chemicals and have been extended to commercial use for biosynthesis of various high-value metabolites of importance to the pharmaceutical, food and chemical industries.[11] When compared with the intact plants, plant cell cultures often produce different quantities of secondary metabolites with different profiles.

Furthermore, the quantitative and qualitative features may change with time.[10] *In vitro* culture of ginseng (*Panax ginseng*) is a prime example of a plant cell culture system which achieves higher productivity, when compared to the whole plant. In these systems, the content of saponins reached 27%, while extraction from the whole plant resulted in 4.5%.[12]

Some metabolites have already been produced on large scale in plant cell and tissue cultures, such as naphthoquinone shikonin (from *Lithospermum erythrorhizon*) and alkaloid berberine (from *Coptis japonica*) produced by the Japanese Mitsui Petrochemical Industry Company, or ginsenosides from *Panax ginseng* produced by the Japanese Nitto Denko Company, and diterpenic alkaloid paclitaxel (*Taxus brevifolia*) manufactured by different companies such as ESC Genetic (USA), Phyton (USA), Phyton Biotech (Germany), and Samyang Genex (South Korea).[13–16]

By definition, plant cells, tissue and organ cultures are cultivated under sterile conditions, totally independent of geographical and climatic factors with a controlled physical environment and reduced space.[17] These techniques offer several technologies for producing metabolites.

The cultures are started by inoculation of small tissue fragments named explants on solid or in liquid nutrient medium, and can regenerate organs, embryos or whole plants or even develop cellular masses denominated calluses. These responses are based on the totipotency of the plant cells, which allows them to express the ability to initiate the formation of a new individual.[18,19]

For a long time, micropropagation was considered the main economic application of *in vitro* tissue culture. However, other applications of plant biotechnology, such as callus and organ cultures, cell suspension cultures and somatic embryogenesis, represent alternative continuous sources of metabolites.[2,11,20–22] An analytical application of plant cell culture would be their use in biosensors.[23]

Callus cultures represent *in vitro* systems resulting from cell division originating from tissues which lose their initial identity and undergo redifferentiation, performing specific and coordinated functions.[24] Calluses can be classified as being compact or friable. The latter are frequently used to turn solid cultures into liquid media under stirring, producing suspension systems.[25,26] Cell suspensions are the *in vitro* systems which are most frequently used for the production of various high-value metabolites of commercial interest. These cultures can be maintained in a dedifferentiated state with rather uncoordinated cell division over long periods of time. Cells in suspension proliferate faster than callus cells and they can be cultivated on a large scale using commercial bioreactors. These features make cell suspensions an efficient alternative source for the production of plant secondary metabolites and biotransformations.[22] Moreover, these cultures have some additional advantages:[10,26–28]

* Continuous supply of products with uniform quality and yield.
* Automated control of cell growth.
* Regulation of metabolic processes, decreasing cost price and increasing productivity.
* Production of novel compounds that are not normally found in the parent plant.
* Efficient downstream production.

Roots have received considerable attention as a potential production system for plant-derived chemicals due to their capacity for stable metabolite production. However, root systems of higher plants generally exhibit slower growth and are difficult to harvest. Hence, alternative methods, such as hairy root culture systems, have been developed. The ability of *Agrobacterium rhizogenes* to induce hairy roots in a range of host plants has led to studies on it as a source of root-derived pharmaceuticals. Hairy roots are induced by transfer of T-DNA from the plasmid of *A. rhizogenes* to host tissues, resulting in root formation by virtue of auxin synthesis of genes coded by bacterial DNA. The Ri plasmid of *A. rhizogenes* also induces the synthesis of opines such as agropine or mannopine. The interest in hairy roots is mainly due to their ability to grow fast without needing an external supply of auxins. Often, they do not need incubation under light and are fairly well stable in metabolite yield due to their genetic stability.[3,29] In recent years, hairy root cultures are gaining preferential advantage as biocatalysts over cell suspension cultures, mostly because of their genetic/biochemical stability, multienzyme biosynthetic potential similar to that of the parent plant and relatively low-cost culture requirements. The hairy root cultures of different plant systems represent rich repositories of enzymes as they mimic their respective parent plants. Consequently, the natures of biotransformation reactions differ depending upon the availability of the plant enzymes as biocatalysts, as well as on the structure and functional groups of the exogenous substrate.[5]

Some factors have been studied to evaluate the economic and industrial feasibility of plant cell cultures, especially related to costs minimization, optimization of the nutrient medium and physical parameters, permeabilization (dimethylsulfoxide, isopropanol or chitosan), the use of elicitors (xanthan gum, chitosan, jasmonate, UV) and precursors (amino acids, for example), the selection of highly productive cell lines by genetic and metabolic engineering, besides the use of bioreactors and *in situ* extraction (adsorbents and biphasic systems). These factors have aided in overcoming some difficulties as low yields, biochemical and/or genetic instability (somaclonal variation), metabolite accumulation in vacuoles and scale-up.[30] Using bioreactors for plant cell culture, there are some usual controls, such as temperature (23–29 °C), pH (5–6), low aeration (1–10 mmol O_2 L^{-1} h^{-1}) and stirring speed (<120 rpm), since plant cells are sensitive to shear stress and oxidation. Hence, alternative aeration systems have been tested with success, such as air-lift systems, bubble columns and dispersion. Then, in some cases, bioreactor configurations used for microorganisms cannot be adapted to plant cell cultures. For some metabolite types, like anthocyanins, light control is necessary (in cycles or constant) using photobioreactors. Other differences between plant cell cultures and microbial processes are the use of growth regulators, such as auxins and cytokinin, in the aeration, which utilizes O_2, CO_2 (0.1–5%) and small volumes (ppm) of ethylene, and the use of inoculum with high cell concentrations (50–100 g L^{-1}), decreasing the lag phase of the cultivate.[1,4,10,13,15,16,29,31–39]

13.3　Use of Plant Cell Cultures in Biotransformation

The biochemical potential of plant cell cultures to produce specific secondary metabolites, such as drugs, flavours, pigments and agrochemicals, is of considerable interest in connection with their biotechnological utilization. However, it has been reported that the formation and accumulation of some secondary metabolites do not normally occur in the cell cultures of higher plants. There is evidence that such cultures retain an ability to specifically transform exogenous substrates administered to cultured cells. Therefore, plant cell cultures are considered to be useful for transforming cheap and plentiful substances into rare and expensive substances by using the cell culture as a bioreactor. For a successful process, the culture must have the necessary enzymes, the substrate or precursor must not be toxic to the culture, the substrate must reach the cellular compartment of the cell, and the rate of product formation must be faster than its further metabolism; the solubility of precursors, the amount of enzyme activity present, the localization of enzymes, the presence of side reactions producing undesired byproducts and the presence of enzymes degrading the desired product also have to be considered.[2,10,29]

The advantages of using plant cell cultures as biocatalysts are the following:

* The cultured material is homogeneous.
* The experiments can be performed and reproduced the whole year round.
* The cultured cells can accumulate large amounts of the products wanted.
* The growth cycles are normally between 1 and 2 weeks, which facilities the experiments.
* The cultured cells can be grown up to an almost unlimited quantity of biological material.

Therefore, plant cell cultures have been used as a good tool (biocatalyst) for organic synthesis, in addition to the use of microorganisms, such as yeast, fungi, and bacteria. The reaction type, stereospecificity, enantioselectivity, and mechanism involved in the biotransformations of exogenous substrates by plant cell cultures are hydroxylation, oxidation, reduction, glycosylation, hydrolysis, esterification, demethylation, hydrogenation and isomerization reactions,[3,4,9,29,40,41] which can be seen in Table 13.1.

Many biotransformation reactions are related to plant defense and detoxification. Hydroxylation is realized by cytochrome P450 monoxygenase,[42] acting stereospecifically and enantiomerically at the C=C double bond in the allylic position of monoterpenoids.[2,9,40]

Enantioselective oxidation of alcohols into ketones, and the reverse, reduction of ketones and aldehydes into alcohols are useful for the preparation of chiral compounds, mainly in asymmetric synthesis. These oxido-reductions are governed by an NAD-dependent alcohol dehydrogenase;[9,40] however, in some plant cell cultures, the best reductions occurred with substrates bearing a carbonyl function conjugated with an aromatic or heteroaromatic ring, giving hydroxy compounds with *S*-chirality.[2,5] The reduction of enones uses different enzymes, such as enone reductases, acting at C=C double bonds, as a hydrogenation, but also requires a NADH or NADPH coenzyme. These enzymes are stereospecific in the reduction of unsaturation, which gives hydrogen in the β-position in relation to the carbonyl group.[2,9,40,42]

Glycosylation reactions are of special interest because they facilitate the conversion of water-insoluble compounds to water-soluble compounds, associated with a detoxification mechanism using glycosyltransferases, accumulating the exogenous compounds in glucosylated form, enhancing their bioavailability. Plant cell cultures play an important role in this regard since it is difficult to perform this reaction using microorganisms or by chemical synthesis. These cultures are capable of glucosylation of a variety

Table 13.1 Biotransformations by plant cell culture systems.

Types of chemical reaction	Substrates	Products	Plant species	Culture type	References
Cyclization	Germacrone-4,5-epoxide	Neoprocurcumenol	*Cichorium intybus*	Suspension	Piet *et al.*[43]
Epoxidation	Hyoscyamine	Scopolamine	*Nicotiana tabacum* (transgenic)	Suspension	Moyano *et al.*[44]
Epoxidation	Tabersonine	Lochnericine, lochnerinine	*Catharanthus roseus*	Suspension	Furuya *et al.*[45]
Hydroxylation	Dihydroartemisinic and artemisinic acids	3-α- and 3-β-Hydroxydihydroartemisinic acids, and 3-α- and 3-β-hydroxyartemisinic acids	*Catharanthus roseus*	Suspension	Zhu *et al.*[46]
Hydroxylation	6-*n*-Pentyl-2*H*-pyran-2-one	5-(2-Pyron-6-yl)pentan-5-ol	*Pinus radiata*	Suspension	Cooney *et al.*[47]
Hydroxylation	β-Pinene	*trans*-Pinocarveol, myrtenol, α-terpineol, pinocamphone and pinocarvone	*Picea abies*	Suspension	Lindmark-Henriksson *et al.*[48]
Hydroxylation	Limonene	Carveol and carvone	*Solanum aviculare* and *Dioscorea deltoidea*	Suspension	Vanek *et al.*[49]
Hydroxylation	Curdione	(2*S*)- and (2*R*)-2-Hydroxycurdione, (8*S*)-6-hydroxycurdione, (2*R*,8*S*)-8-hydro-2-hydroxycurdione, (1*S*,10*S*)- and (1*R*,10*R*)-1,10-epoxycurdione	*Lonicera japonica*	Suspension	Horiike *et al.*[50]
Hydroxylation	Geraniol, nerol, carvone	10-Hydroxygeraniol,10-hydroxycitronellol, 10-hydroxynerol, 4α-, 4β-, 5α- and 10-hydroxycarvone, 5α-hydroxydihydrocarvone, 5α- and 5β-hydroxyneodihydrocarveol	*Catharanthus roseus*	Suspension	Hamada *et al.*[51]
Hydroxylation	Ambrox	3-Oxoambrox, 3β-hydroxyambrox, 1α-hydroxyambrox, 3β,6β-dihydroxyambrox, 1α,6β-dihydroxyambrox, 1α,3β-dihydroxyambrox	*Actinidia deliciosa*	Suspension	Nasib *et al.*[52]

(continued)

Table 13.1 (*continued*)

Types of chemical reaction	Substrates	Products	Plant species	Culture type	References
Hydroxylation	Sinenxan A	9α-Hydroxy-2α,5α,10β,14β-tetra-acetoxy-4(20), 11-taxadiene, 9α,10β-dihydroxy-2α,5α,14β-triacetoxy-4(20), 11-taxadiene	*Ginkgo biloba*	Suspension	Dai *et al.*[53]
Hydroxylation	Cinobufagin	Desacetylcinobufotalin, 3-*epi*-desacetylcinobufagin	*Catharanthus roseus, Platycodon grandiflorum*	Suspension	Ye *et al.*[54]
Hydroxylation	Valdecoxib	4-(3-Phenyl-4-isoxazolyl) benzenesulfonamide, 4-[3-(4-hydroxyphenyl)-5-methyl-4-isoxazolyl] benzenesulfonamide, 4-[5-carboxy-3-phenylisoxazol-4-yl] benzenesulfonamide, 4-[5-hydroxymethyl-3- phenylisoxazol-4-yl] benzenesulfonamide	*Catharanthus roseus, Azadirachta indica, Capsicum annuum, Ervatamia heyneana, Nicotiana tabacum*	Suspension	Molmoori *et al.*[55]
Hydroxylation	Agroclavine, elymoclavine	Setoclavine, isosetoclavine	*Ajuga reptans, Armoracia rusticana, Atropa belladonna, Duboisia myoporoides, Euphorbia calyptrata, Papaver somniferum, Solanum aviculare*	Suspension	Ščigelová, *et al.*[56]
Hydroxylation	Withanolide A, withaferin A, withanone	Interconversion products	*Withania somnifera*	Suspension	Sabir *et al.*[57]
Hydrolysis and esterification	14-Deacetoxyl sinenxan A	10β-Hydroxy- and 10β-butyloxy-2α,5α-diacetoxy-4(20),11-taxadiene	*Ginkgo biloba*	Suspension	Zhan *et al.*[58]

Hydroxylation and oxidation–reduction	Bufotalin, telocinobufagin, gamabufotalin	3-*epi*-Bufotalin, 3-*epi*-desacetylbufotalin, 3-*epi*-bufotalin 3-*O*-β-D-glucoside, 1β- and 5β-hydroxybufotalin, 3-dehydroscillarenin, 3-dehydrobufalin, 3-*epi*-telocinobufagin, 3-*epi*-gamabufotalin, 3-dehydrogamabufotalin, 3-dehydro-Δ¹-gamabufotalin	*Saussurea involucrata*	Suspension	Zhang *et al.*[59]
Hydroxylation and oxidation	*n*-Hexadecane	Hexadecanol, palmitic acid	*Cinchona robusta, Dioscorea composita*	Suspension	Vega-Jarquin *et al.*[60]
Hydroxylation and reduction	Coronaridine, conopharyngine, vobasine, tabersonine	(19*S*)-Heyneanine, (19*S*)-hydroxyconopharyngine, tabernaemontanine, dregamine, voaphylline	*Tabernaemontana divaricata*	Suspension	Dagnino *et al.*[61]
Hydroxylation and reduction	Perillaldehyde, piperonal, myristicin aldehyde, hydrocinnamaldehyde, α-pinene, furfural, D-limonene	Perilla alcohol, piperonol, myristicin alcohol, hydrocinnamyl alcohol, verbenol, verbenone, furfuryl alcohol, limonene epoxide, *p*-2,8-menthadien-1-ol, carveol	*Peganum harmala*	Suspension	Zhu *et al.*[62]
Hydroxylation and reduction	(1*S*)-2-Carene, (1*S*)-3-carene	(1*S*)-2-Caren-4-one, (1*S*)-3-caren-5-one, (1*S*)-2-caren-4-one	*Picea abies*	Suspension	Dvorakova *et al.*[63]
Hydroxylation and reduction	α-Santonin	11β-Hydroxysantonin, 14-hydroxysantonin, 4,5-dihydro-α-santonin, 3α-hydroxy-4,5-dihydro-7α(*H*),6:11β(*H*)-eudesman-6:12-olide	*Catharanthus roseus, Ginkgo biloba, Platycodon grandiflorum, Taxus cuspidata, Phytolacca acinosa*	Suspension	Yang *et al.*[64]
Oxidation–reduction	3,4,5-Trimethoxy-benzaldehyde, 3,4,5-trimethoxyacetophenone	3,4,5-Trimethoxy benzoic acid, 3,4,5-trimethoxy benzyl alcohol, 1-(3,4,5-trimethoxyphenyl) ethanol	*Atropa belladonna*	Hairy root	Srivastava *et al.*[65]

(*continued*)

Table 13.1   (*continued*)

Types of chemical reaction	Substrates	Products	Plant species	Culture type	References
Prenylation	Chrysin (1), genistein (2), sophoricoside (2), diosmetin (2)	5,7-Dihydroxy-8-(3,3-dimethylallyl)flavone (1), 5,7,4′-trihydroxy-6-(3,3-dimethylallyl)isoflavone (2), 5,7,3′-trihydroxy-4′-methoxy-6-(3,3-dimethylallyl)flavone (2)	*Cudrania tricuspidata* (1), *Morus alba* (2)	Suspension	Yin *et al.*[66]
Reduction	Citronellal	Menthane-3,8-diol, citronellol, isopulegol	*Solanum aviculare*	Suspension	Vanek *et al.*[67]
Reduction	Acetophenone	(*S*)-Phenylethanol	*Daucus carota*	Hairy root	Caron *et al.*[68]
Reduction	Acetophenone and its derivatives	1-Phenylethanol and its derivatives	*Raphanus sativus*	Hairy roots	Orden *et al.*[69]
Reduction	Camphorquinones	2α-, 2β-, 3α- and 3β-Hydroxycamphor	*Nicotiana tabacum*, *Catharanthus roseus*	Suspension	Chai *et al.*[70]
Reduction	Artemisinin	Deoxyartemisinin	*Catharanthus roseus*, *Lavandula officinalis*	Suspension	Patel *et al.*[71]
Reduction	1-(5-Acetyl-2-hydroxyphenyl)-3-methylbut-2-en-1-one, 6-acetyl-2,2-dimethyl-2,3-dihydro-4*H*-chromen-4-one	6-(1(*S*)-Hydroxyethyl)-2,2-dimethyl-2,3-dihydro-4*H*-chromen-4-one	*Brassica napus*	Hairy roots	Orden *et al.*[72]
Combined	Isoliquiritigenin	Daidzein	*Genista tinctoria*	Shoot and hairy root	Łuczkiewicz and Kokotkiewicz[73]
Combined	β-Artemether	Tetrahydrofuran (THF)–acetate derivative	*Glycyrrhiza glabra*, *Lavandula officinalis*	Suspension	Patel *et al.*[74]
Glycosylation	7-Hydroxy-4-methylcoumarin, 7-hydroxy-4-phenylcoumarin, 5,7-dihydroxy-4-methylcoumarin, 7,8-dihydroxycoumarin	4-Methylcoumarin-7-*O*-β-D-glucopyranoside, 4-phenylcoumarin-7-*O*-β-D-glucopyranoside, 4-phenylcoumarin-7-*O*-β-D-glucopyranosyl (1 → 6)β-D-glucopyranoside, 7-hydroxy-4-methylcoumarin-5-*O*-β-D-glucopyranoside, 7-hydroxycoumarin-8-*O*-β-D-glucopyranoside	*Catharanthus roseus*	Suspension	Xue *et al.*[75]

Reaction	Substrate	Product	Species	Culture	Reference
Glycosylation	Menthol and geraniol	Menthol and geraniol glycosides	*Levisticum officinale*	Hairy root	Nunes et al.[76]
Glycosylation	Vanillin and vanillyl alcohol	Vanillyl alcohol-4-O-β-D-glucopyranoside	*Pharbitis nil*	Hairy root	Kanho et al.[77]
Glycosylation	Thymol, carvacrol, eugenol	Thymol-, carvacrol-, eugenol-β-gentibiosides	*Eucalyptus perriniana*	Suspension	Shimoda et al.[78]
Glycosylation	Daidzein	Daidzein 7-O-β-D-glucopyranoside	*Eucalyptus perriniana*	Suspension	Shimoda et al.[79]
Glycosylation	Naringin, naringenin	Naringenin 4',7-O-β-D-diglucopyranoside, naringenin 7-O-β-D-glucopyranoside, naringenin 5,7-O-β-D-diglucopyranoside	*Eucalyptus perriniana*	Suspension	Shimoda et al.[80]
Glycosylation	Paeonol	Paeonol 2-O-β-D-glucopyranoside, paeonol 2-O-β-D-xylopyranoside	*Panax ginseng*	Suspension and roots	Li et al.[81]
Glycosylation	Umbelliferone	Umbelliferone 7-O-β-D-glucopyranoside, umbelliferone 7-O-β-D-glucopyranosyl (1 → 6) β-D-glucopyranoside	*Panax ginseng*	Hairy roots	Li et al.[82]
Glycosylation	Cinnamyl alcohol and tyrosol	Rosin, hydrocinnamyl alcohol glycosides, salidroside	*Rhodiola rosea*	Compact callus aggregate	György et al.[83]
Glycosylation	Tocopherols	Tocopheryl 6-O-β-D-glucopyranoside, tocopheryl 6-O-(6-O-β-D-glucopyranosyl)-β-D-glucopyranoside, tocopheryl 6-O-(6-O-α-L-rhamnopyranosyl)-β-D-glucopyranoside	*Eucalyptus perriniana*	Suspension	Shimoda et al.[84]
Glycosylation	p-Hydroxybenzyl alcohol	Gastrodin	*Datura tatula*	Hairy roots	Peng et al.[85]
Glycosylation	Methanol, ethanol, 2-propanol	Methyl β-D-glucopyranoside, methyl β-D-ribo-hex-3-ulopyranoside, ethyl β-D-glucopyranoside, 2-propyl β-D-glucopyranoside	*Coleus forskohlii*	Hairy root	Li et al.[86]
Glycosylation	Hydroquinone	Arbutin	*Brugmansia candida*	Hairy roots	Casas et al.[87]
Glycosylation	Eugenol	2-Methoxy-4-(2-propenyl) phenyl-O-β-D-glucopyranoside	*Nicotiana tabacum*	Suspension	Yang et al.[88]

(continued)

Table 13.1   (*continued*)

Types of chemical reaction	Substrates	Products	Plant species	Culture type	References
Glycosylation	Capsaicin, 8-nordihydrocapsaicin	Capsaicin 4-O-β-D-glucopyranoside, capsaicin 4-O-(6-O-β-D-xylopyranosyl)-β-D-gluco-pyranoside, capsaicin 4-O-(6-O-α-L-arabino-pyranosyl)-β-D-glucopyranoside, 8-nordihydrocapsaicin 4-O-β-D-glucopyranoside, 8-nordi-hydrocapsaicin 4-O-(6-O-β-D-xylopyranosyl)-β-D-glucopyranoside, 8-nordihydrocapsaicin 4-O-(6-O-α-L-arabinopyranosyl)-β-D-glucopyranoside	*Catharanthus roseus*	Suspension	Shimoda *et al.*[89]
Glycosylation	p-, m- and o-Hydroxybenzoic acids	p- and m-Carboxyphenyl β-D-glycopyranoside, p- and m-hydroxybenzoic acid β-D-glycopyranosyl ester	*Panax ginseng*	Hairy root	Chen *et al.*[90]
Glycosylation	1,4-Benzenediol, 4-hydroxybenzaldehyde, 4-hydroxybenzyl alcohol, 4-hydroxybenzoic acid	Arbutin, gastrodin, 4-carboxyphenyl α-D-glucopyranoside	*Polygonum multiflorum*	Hairy root	Yan *et al.*[91]
Glycosylation and esterification	Digitoxigenin	Digitoxigenin palmitate, 3-epidigitoxigenin, digitoxigenin β-D-glucoside, digitoxigenin β-D-glucoside malonyl ester	*Panax ginseng*	Suspension	Kawaguchi *et al.*[92]
Glycosylation and oxidation	Betuligenol	4-(p-Hydroxyphenyl)-2-Butanone, betuloside	*Atropa belladonna*	Hairy roots	Srivastava *et al.*[93]

Hydroxylation and glycosylation	Cinnamic, *p*-coumaric, caffeic, and ferulic acids	Cinnamic acid β-D-glucopyranosyl ester, 4-*C*-β-D-glucopyranosylcoumaric acid, *p*-coumaric acid β-D-glucopyranosyl ester, 4-*O*-β-D-glucopyranosylcoumaric acid β-D-glucopyranosyl ester, 3- and 4-*O*-β-D-glucopyranosylcaffeic acid, 3- and 4-*O*-β-D-glucopyranosylcaffeic acid β-D-glucopyranosyl ester, 4-*O*-β-D-glucopyranosylferulic acid, 4-*O*-β-D-glucopyranosylferulic acid β-D-glucopyranosyl ester	*Eucalyptus perriniana*	Suspension	Katsuragi *et al.*[94]
Hydroxylation and glycosylation	Dehydroabietic acid (DHA)	DHA-18-*O*-glucoside (1), 17-hydroxy-DHA (2), DHA-17-*O*-glucoside (2)	*Nicotiana tabacum* (1), *Catharanthus roseus* (2)	Suspension and hairy roots	Häkkinen *et al.*[95]
Hydroxylation, reduction and glycosylation	21-*O*-Acetyl-deoxycorticosterone	Deoxycorticosterone, 3β,21-dihydroxy-5α-pregnan-20-one, 2β,21-dihydroxy-4-pregnen-3,20-dione, deoxycorticosterone 21-*O*-β-glucoside	*Digitalis lanata* (strain W.1.4)	Suspension	Pádua *et al.*[96]
Glycosylation, oxidation and reduction	Geraniol, borneol, menthol, thymol, farnesols	Nerol, geranial, neral, isoborneol, camphor, menthone, menthyl acetate, 1,8-cineole, carvacrol, limonene, α-pinene, farnesene and its glycosylated derivatives	*Achillea millefolium* L. ssp. *millefolium*	Suspension	Figueiredo *et al.*[97]
Glycosylation, oxidation and reduction	β-Ionone, dehydrovomifoliol, geraniol, linalool	3-Oxo-α-ionol, 4-oxo-β-ionol, 4-oxo-7,8-dihydro-β-ionol, 3-hydroxy-β-ionone, 3-hydroxy-7,8-dihydro-β-ionol, 3-oxo-7,8-dihydro-α-ionol, 2,6-dimethylocta-2,6-diene-1,8-diol, 6,7-dihydro-3-hydroxylinalool, 8-hydroxylinalol and its glycosylated derivatives	*Vitis vinifera* L. cv. Gamay Fréau	Suspension	Mathieu *et al.*[98]

(*continued*)

Table 13.1 (*continued*)

Types of chemical reaction	Substrates	Products	Plant species	Culture type	References
Esterification and oxidation	Menthol, geraniol	Menthyl acetate, linalool, α-terpineol, citronellol, neral, geranial, citronellyl, neryl and geranyl acetates	*Anethum graveolens*	Hairy root	Faria *et al.*[99]
Hydroxylation, reduction and glycosylation	4-(*p*-Hydroxyphenyl)-2-butanone (raspberry ketone), zingerone	4-[4-(β-D-Glucopyranosyloxy)phenyl]-2-butanone, 4-[(3S)-3-hydroxybutyl]phenyl-β-D-glucopyranoside, (2S)-4-(4-hydroxyphenyl)but-2-yl-β-D-glucopyranoside, 2-hydroxy-4-[(3S)-3-hydroxybutyl]phenyl-β-D-glucopyranoside, 2-hydroxy-5-[(3S)-3-hydroxybutyl]phenyl-β-D-glucopyranoside, (2S)-4-(4-hydroxy-3-methoxyphenyl)but-2-yl-β-D-glucopyranoside, (2S)-2-(β-D-glucopyranosyloxy)-4-[4-(β-D-glucopyranosyloxy)-3-methoxyphenyl]butane, (2S)-4-(4-hydroxy-3-methoxyphenyl)-2-butanol, 4-[4-(β-D-glucopyranosyloxy)-3-methoxyphenyl]-2-butanone, 4-[(3S)-3-hydroxybutyl]-2-methoxyphenyl-β-D-glucopyranoside	*Phytolacca americana*	Suspension	Shimoda *et al.*[100]
Hydroxylation and glycosylation	Digitoxin	Deacetyllanatoside C	*Digitalis lanata*	Suspension	Choi *et al.*[101]
Oxidation	*rac*-Phenylethanol and its derivatives	(*S*)-Acetophenone and (*S*)-derivatives	*Gardenia jasminoides*	Suspension	Magallanes-Noguera *et al.*[102]

of exogenously added compounds, *e.g.* phenols, phenylpropanoid acid and their analogs.[2,3,10,42] By definition, glycosylation is the enzymatic progression that links saccharides to produce glycans, either free or attached to proteins or lipids, involving the coupling of a sugar moiety to a glycosyl acceptor to form glycosides, whereas glucosylation is an enzymatic reaction that links glucose to produce glucosides.[5] Another type of glycosyl conjugation involves esterification between the carboxylic acid and sugar moiety, a little different from glycosylation, with ether formation by the hydroxyl group and sugar moiety.[9,40]

Both esterification and hydrolysis can be realized by intracellular lipases. Enantioselective hydrolysis of acetoxy groups is considered to be useful for the optical resolution of racemic acetates. The enantiomers with the *R*-configuration at the carbon atom bearing the acetoxy group are preferentially hydrolyzed. Another example reported is the hydrolysis of the hydroxy-imine group and an ether bond.[2,9,40]

13.3.1 Biotransformation Using Cell Immobilization

Immobilization is defined as a technique which confines a catalytically active enzyme or cells on a fixed support and prevents their entry into the liquid phase. Immobilized plant cells have been used for single and multistep bio-transformations of precursors to desired products, as well as for the *de novo* biosynthesis of secondary metabolites, and to utilize and recycle essential cofactors and co-enzymes. Isolated enzymes may be sensitive to denaturing conditions, including pH extremes, heat and specific organic solvents. Use of whole cell immobilized system may help overcome some stability problems. Immobilized plant cells have some additional advantages over freely suspended cells:[2–4,10,29,41]

- Cells become more resistant to shear damage by immobilization.
- Immobilized cells can be used repeatedly over a prolonged period (over six months), with high retention of cell viability.
- High concentrations of biomass are possible, thus giving high conversions of substrate.
- The method facilitates recovery of the cell mass and products.
- Sequential chemical treatments are possible.
- Occurrence of non-growth conditions allied with improved cell–cell adhesion and eventual aggregation under which the production of secondary metabolites may be improved.
- Immobilization may produce a microenvironment (nutrients gradients, lower water activity and O_2 tension) that resembles the organized tissue in the intact plant, causing differentiation and production of secondary metabolites.
- Minimization of fluid viscosity increase, which in cell suspension causes mixing and aeration problems.
- Reducing the risk of contamination.

Compared to immobilized systems, enzymes and whole cells have the advantage of a higher activity rate, while immobilized cells have distinct advantages: (a) they can carry out multienzyme operations; (b) by selecting highly biosynthetic cells, catalytic activity can be enhanced; (c) there is no need to provide cofactors since cells themselves produce them and (d) immobilized cells can be easily handled as compared to immobilized enzymes. However, some issues have to be discussed and improved, including: production is decoupled from cell growth; the initial biomass must be grown in suspension; the gel matrix introduces an additional diffusion barrier; secretion of the product into the extracellular medium is imperative, and extracellular degradation of the products can still occur.[10,29]

General methods for immobilization of plant cells are adsorption (*e.g.*, onto polystyrene or glass), covalent attachment (*e.g.*, by glutaraldehyde or carbodiimide coupling) and gel entrapment by ion exchange. Calcium alginate is the most widely used matrix due to its simplicity and relative lack of toxicity. Besides this, other gels such as agar, agarose, gelatin, carrageenan and polyacrylamide have also been used. Other alternative supports are polyurethane foam and hollow-fibre membranes.[2,4,9,29,40] The efficient entrapment of whole cells should allow for the subsequent porosity of the formed gel to be uniform and controllable (free exchange of substrates, products, co-factors, and gases is essential for efficient performance of the immobilized cells); the gel should retain good mechanical, chemical, and biological stability (it should not be easily degraded by enzymes, solvents, pressure changes, or shearing forces), and the gel should be composed of reasonably priced components.[10,22]

Vaněk *et al.*[103] studied the influence of different immobilization methods on the biotransformation of verbenol by cell cultures of *Solanum aviculare*. The immobilization techniques used included entrapment in alginate, pectate and carrageenan gels, in polyurethane foam, and on the surface of polyphenylene oxide activated by glutaraldehyde. Free cells acted as reduction biocatalysts, transforming 1*S*-*cis*-verbenol into 1*S*(−)-verbenone. This biotransformation with the same yield was maintained using polyurethane foam as support. However, with other types of immobilization, the yield of 1*S*-*trans*-verbenol increased, indicating direct isomerization of the substrate or a two-step biotransformation by reduction and oxidation.

Ramachandra Rao and Ravishankar[104] used freely suspended and immobilized cells of *Capsicum frutescens* Mill for the conversion of protocatechuic aldehyde and caffeic acids to vanillin and capsaicin. The increase in vanillin accumulation was well correlated with an increase in the specific activity of caffeic acid *O*-methyltransferase in the protocatechuic aldehyde and *S*-adenosyl-L-methionine-treated immobilized *C. frutescens* cell culture. Biotransformation of isoeugenol to vanillin flavour metabolites and capsaicin was more effective in immobilized rather than free cells and this activity was further enhanced by β-cyclodextrin and fungal elicitor.

13.3.2 β-Cyclodextrins in Biotransformation

Some precursors are either insoluble or very poorly soluble in the aqueous phase, resulting in very low bioconversion rates. A new approach to solve the problem of the bioconversion of water-insoluble precursors is to combine the advantages of apolar systems (higher solubility of the substrate) and of aqueous systems (maintenance of cell viability) by carrying out bioconversions in the presence of clathrating agents, such as cyclodextrins. Cyclodextrins are cyclic oligosaccharides, consisting of 7 glucose units linked by $(1 \rightarrow 4)$ glycosidic bonds, that are able to form inclusion complexes with a variety of apolar ligands. Cyclodextrins may be modified by substituting various functional compounds in the primary or secondary phase of the molecule. Chemically modified cyclodextrins are more water soluble than native cyclodextrins. Since tolerance of plant cell cultures towards organic phases is low, cyclodextrin-complexed precursors could be used to facilitate the bioconversion of water-insoluble precursors in a more compatible aqueous environment.[2,4,29] One example of the use of complexation with β-cyclodextrin was reported by Woerdenbag *et al.*,[105] where the phenolic steroid 17β-estradiol had its solubility increased from almost insoluble to 660 μM, and it could be *ortho*-hydroxylated into a catechol, mainly 4-hydroxyestradiol (40%), by a phenoloxidase preparation from suspended cells of *Mucuna pruriens*. Mushroom tyrosinase was even tested, and converted 17β-estradiol, as a complex with β-cyclodextrin, solely into 2-hydroxyestradiol, with a maximal yield of 30% after 6–8 h, while uncomplexed 17β-estradiol was not converted at all in any of these systems.

13.4 Use of Whole or Parts of Plants in Biotransformation

According to the above, whole cell, isolated enzymes or plant cell cultures can be used as biocatalysts in biotransformation. However, these sources are sometimes limited by their cost and large scale production. Furthermore, plant cell cultures, exhibit relatively low growth rates and the culture process can be too long.[106,107] In order to avoid the time consuming and care-requiring preparation of plant cell cultures, the use of whole plants has been considered and has received significant attention, mainly due to the wide biotechnological potential of enzymatic reactions[108–110] (Table 13.2). This technique still offers low cost of the process and high enantioselectivity. It does not require preliminary preparation, extraction, purification or transformation, promoting the preservation of the maximum catalytic activity of the enzymes, besides adhering effectively to green chemistry principles, once biodegradable waste can be generated.[107,110–112] Whole plants are also an excellent alternative to isolated enzymes, since the cofactor regeneration system is all within the cell, for example. On the other hand, whole plants can present low cellular conversion and productivity (this is due to the toxicity of these substrates towards

Table 13.2 Major examples of biotransformations with whole plants.

Types of chemical reaction	Substrates	Products	Species	Plant part	References
Reduction	2,4,6-Trinitrotoluene (TNT)	4-Amino-2,6-dinitrotoluene (4ADNT) and 2-amino-2,6-dinitrotoluene	*Cyperus esculentus* L.	Root, leaves, rhizomes and tubes	Palazzo *et al.*[148]
Hydrolysation	(±)-1-Phenylethyl acetate	L-Phenylethanol	*Solanum Tuberosum* L., cv. *Saturna* (potato)	Root	Mironowicz[149]
Reduction	2,4,6-Trinitrotoluene (TNT)	4-Amino-2,6-dinitrotoluene (4ADNT) and 2-amino-2,6-dinitrotoluene (2ADNT)	*Cyperus esculentus* L.	Root	Best *et al.*[123]
Reduction	(±)-2-Methylcyclohexanone	(1S,2R)- and (1S,2S)-2-Methylcyclohexanol	*Daucus carota* L. (fresh carrot)	Root	Baldassarre *et al.*[108]
Reduction	Acetophenone **1**	S-Alcohol	*Daucus carota* L. (carrot)	Root	Bruni *et al.*[109]
Hydrolysation (1) and reduction (2)	Enol acetate (1) 1-acetoxy-2-methylcyclohexene (2)	1-Acetoxy-2-methylcyclohexene (1), S-alcohol(2)	*Daucus carota* L. (carrot)	Root	Bruni *et al.*[109]
Hydrolysation	Bromo- and methoxy-substituted 1-phenylethanol acetates	L-Phenylethanol	*Apium graveolens* L. var. *rapaceum* (celeriac)	Root	Mironowicz[149]
Lactonization	γ,δ-Epoxy esters	δ-Hydroxy-γ-lactones; γ-hydroxy-δ-lactones	*Malus sylvestres* L. (Mill) (apple) and *Helianthus tuberosus* A. Gray (bulbs of Jerusalem artichoke)	Fruit and root	Olejniczak *et al.*[129]
Oxidation	(*RS*)-1-Phenylethanol derivatives	S-Alcohol (1,3,4), R-alcohol (2,5)	*Brassica rapa* L. (1) *Arctium lappa* L. (2) *Polymnia sonchifolia,* Poeppig and Endlicher (3) *Zingiber officinale* Roscoe (4) *Allium schoenoprasum* L. (5)	Root	Andrade *et al.*[117]

Reduction	Acetophenone, 4'-chloroacetophenone (1) ethyl 4-chloroacetoacetate (2)	*S*-Chiral alcohol (1a), *R*-chiral alcohol (1b), *R*-chiral alcohol (2a), *S*-chiral alcohol (2b)	*Cucumis sativus* L. (cucumber) (1a) *Malus pumila* auct. (apple) (1b) *Allium cepa* L. (onion) (2a) *Daucus carota* L. (carrot) (2b)	Fruit, root	Yang *et al.*[132]
Hydrolysation and reduction	Aldehydes, ketones, and esters), amides, or nitrobenzene	Alcohol or aldehydes	*Cocos nucifera* L.	Juice of the fruit	Fonseca *et al.*[111]
Hydrolysation	5-Acetoxy-4-methyl-3-heptanone	Isomers of 5-hydroxy-4-methyl-3-heptanone with higher occurrence: (1) preference (4*S*,5*R*) (2) preference (4*R*,5*R*) (3) preference (4*R*,5*S*)	(1) *Ipomoea batatas* (potato) (2) *Daucus carota* L. (carrot) (2) *Apium graveolens* L. (celery), (3) *Solanum melongena* L. (eggplant), (3) *Petroselinum crispum* (Mill.) Fuss (parsley)	Root	Bohman *et al.*[120]
Reduction	Acetophenone	Benzaldehyde	*Daucus carota* L. (carrot)	All vegetal in small thin pieces	Chang *et al.*[140]
Reduction	Acetophenone	Benzaldehyde	*Raphanus sativus* (radish)	All vegetal in small thin pieces	Chang *et al.*[140]
Reduction	*trans*-4-Phenylbut-3-en-2	(*S*)-*trans*-4-Phenylbut-3-en-2-ol	*Vigna unguiculata* (L.) Walp. (beans)	Seeds	Bizerra *et al.*[126]
Reduction	α-Hydroxy aromatic ketones	(*R*)-Aryl vicinal diols	*Daucus carota* L. (carrot)	Root	Liu *et al.*[146]
Hydrolysation	(±)-1-Phenylethyl acetate	(*R*)-Phenylethanol.	*Beta vulgaris* L.	Root	Vandenberghe *et al.*[107]
Reduction	*trans*-4-Phenylbut-3-en-2	(*S*)-*trans*-4-Phenylbut-3-en-2-ol	*Daucus carota* L. or *Apium graveolens* L. var. *rapaceum* (celeriac) *Beta vulgaris* L. subsp. *Vulgaris* (beetroot)	Root	Majewska and Kozłowska[110]

living organisms or to a dramatic modification of the cellular system). For this reason, the use of whole plants sometimes needs a large amount of biomass to obtain reasonable yields and requires adequate and careful growing operations and a more difficult large-scale setup.[108]

The term *whole vegetable*, termed *botanochemistry* by some authors,[107,113] refers to the use of plant tissue without relation to the cultivation process. In this case, it consists of the use of the vegetable itself, the intact plant (full plant, whole plant, whole vegetable) or part of the plant (skin and pieces) like leaves, roots (usually the most common part of the vegetable used), fruit, stamens, *etc.* Of all the species of plant, to date, *Daucus carota* L. (carrot) is the most common species used and it has proved be one of the most efficient plant parts reported by many authors, with high enantiomeric excess (>95%) in many reactions, mainly asymmetric reduction of prochiral ketones.[109,114–120] However the method of preparation of the plant can result in great impact on the selectivity and yield.[116,118,120] Each part can be designated as a biocatalyst for a specific reaction medium.

The use of the whole or parts of plants in biotransformations has been applied in different technological sectors for phytoremediation and biosensors.

13.4.1 Phytoremediation

Biotransformations using these organisms can be used to transform a compost in order to reduce the toxicity of the same.[121,122] In this case, the study of these plants in soil contaminated with cyanide can be cited.[122] Another case that we can comment on is that of TNT (2,4,6-trinitrotoluene). It is widely used in military explosives and its manufacture and handling can lead to contamination of the environment by improper disposal. For this reason, the remediation is sometime necessary. Best *et al.*[123] related the first work on phytoremediation study of aquatic and wetland plants which showed proven persistence of the explosives concentrations employed in the test, in groundwater with a chemical composition typical for explosive-contaminated aquifers.

13.4.2 Biosensors

The integration of a biological component within a transducer results in biosensors, excellent detectors of various analytes, including drugs, hormones, toxicants, neurotransmitters and amino acids. Besides animal cells, whole or parts of plants have been utilized as immobilized cells in this technology. Species like *Helianthus annuus* L. (sunflower) have been grown in chamber environments for experiments.[124,125] We can mention the construction of a plant tissue biosensor coupled with a sample flow injection system for the determination of glycolic acid. The feasibility of sunflower leaves in this system was checked. The authors showed the design of a bioelectrode of low cost, high sensitivity and with rapid response.[124]

Depending on the plant species, different reactions can be occur. Reactions like bioreduction of nitro compounds,[117,126] hydrolysis of esters,[107,127,128] oxidation reactions,[117] and enzymatic lactonization[129] are common in whole plant systems. We can cite:

13.4.3 Reduction

This is the most common reaction within whole plants. Although previous studies have reported the use of microbes as biocatalysts for most biotransformations, the literature also presents excellent results using whole plants as potential biocatalysts in different types of reactions[117,130–133] as asymmetric reduction reactions. In the last decade, various examples of the reduction of prochiral ketones to chiral alcohols have been reported using growing[134–137] or immobilized plant cell cultures.[138,139] Only recently, the possibility of directly using parts of plants as biocatalysts has been investigated. For example, the use of freshly cut carrot root in the reduction of 2-methyl- and 2-hydroxy-cyclohexanone.[108] In fact, there are a few results showing the asymmetric reduction reaction of pro-chiral aromatic ketones by whole plants.[109,140]

It is common to find good to excellent selectivities in the reduction of aromatic and aliphatic ketones, as well as β-ketoesters by whole plants. The results will depend on the conversions and the chemoselectivity of the substrate structure. The reduction of carbonyl, nitro and alkene groups has been related. The species *Vigna unguiculata* (L.) Walp. (Fabaceae) has been utilized as a biocatalyst. It is a legume, also called feijão corda or feijão caupi, of high interest in the Brazilian diet. For almost all reactions, the biocatalyst showed excellent stereo- and chemoselectivity for the production of different organic compounds of interest (ee > 99%).[126] Another example is the use of *Cocos nucifera* L. (coconut juice), a species of the Arecaceae family, which has been utilized as a biocatalyst for reduction and hydrolytic reaction processes of aldehydes and ketones.[141,142]

Bioreductions in organic solvents are described in the literature. Isooctane in the presence of comminuted roots of popular vegetables such as carrots (*Daucus carota* L.), celeriac (*Apium graveolens* L. var. *rapaceum*), and beetroot (*Beta vulgaris* L. subsp. *vulgaris*) proceeded in good to excellent yields with regio- and stereoselective reduction of *trans*-4-phenylbut-3-en-2-one. The highest yield was achieved with carrots (96%), and the yields were lower for celeriac (78%) and beetroot (71%). Roots in isooctane can be viewed as a further useful tool for organic chemists by virtue of their simplicity. With these results, whole plants in bioreduction reactions appear to be new possibilities for the reduction of selected α,β-unsaturated carbonyl compounds that are insoluble in water, as a critical step in a synthetic organic pathway, specifically avoiding the use of non-sustainable, hydride reducing agents.[110]

Whole plants have been widely applied in bioreductions of carbonyl groups[108,109,112,121,128,130,133,143,144] in the preparation of chiral alcohols and chiral building blocks, resulting in important products of interest in the pharmaceutical and (agro)chemical industry.[132,145–147] For the last one, we can mention the use of parts of plants that are not for human consumption. These can be used in biotransformations. They seem to be an agro-industrial alternative which uses these parts of the plant.

13.4.4 Hydrolysis

Species of the Salicaceae family have been widely used in hydrolysis in phytoremediation. *Populus deltoides* is a species that is widely used to hydrolyze 4-monochlorobiphenyl (CB3), mainly an airborne pollutant, producing hydroxylated metabolites (OH-CB3s).[121]

13.4.5 Oxidation

Oxidation by enzymatic systems of 15 different whole plants has been related by Andrade *et al.*[117] The authors used the alcohol oxidation of a racemic mixture to produce the corresponding ketones or the enantiomeric enrichment of the alcohol to evaluate the catalytic potential (Figure 13.1).

Some species of plants like *Allium schoenoprasum* L. and *Raphanus sativus* L. showed high conversion (54%) and enantiomeric excesses (98%) after 6 days of reaction in kinetic resolution. On the other hand and more interestingly, the following plants: *Arracacia xanthorrhiza* Bancroft, *Zingiber officinale* Roscoe and *Polymnia sonchifolia* Poepping & Endlicher showed the characteristic behaviour of a possible deracemization reaction. The formation of a single enantiomer in high yields (>90%) from the racemate could proceed through a two-step redox-sequence (one enantiomer is selectively oxidized to the corresponding ketone which is reduced in a second step by another redox enzyme displaying the opposite stereochemical preference).

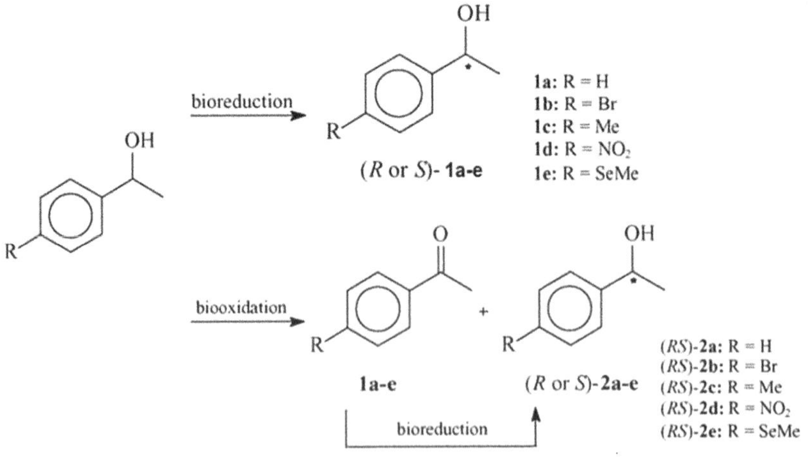

Figure 13.1   Bioreduction and biooxidation of alcohols (*RS*)-1a–e and 2a–e (adapted from Andrade *et al.*[117]).

In the study of biotransformation using whole plants, some important points can be considered. A detailed study of the physiology and ecology of plant growth is necessary. In most cases, enzyme activity is not found in the non-germinated seed state. However, this activity is present after germination.[150] An exception for this case was demonstrated by Lin *et al.*[151]; castor bean lipase was shown to be active in dormant seeds. It is present in the membrane of lipid bodies and is active under acidic conditions with an optimum pH of 4.1. The authors demonstrated that the lipase had a preference for the hydrolysis of ricinoleic acid which constitutes about 90% of castor bean oil, however the regioselectivity was not good.[151]

Another relevant point in studies of whole plants is their interaction with organic solvents. It is possible to observe a limitation of the reaction due to the toxic effect of the solvent on the enzyme activity. On the other hand, organic solvents may absorb inside the cell membrane, leading to a change in membrane fluidity and ease in substrate uptake, resulting in activity retention.[110]

References

1. E. Fumagali, R. A. C. Gonçalves, M. d. F. P. S. Machado, G. J. Vidoti and A. J. B. d. Oliveira, *Rev. Bras. Farmacogn.*, 2008, **18**, 627–641.
2. A. Giri, V. Dhingra, C. C. Giri, A. Singh, O. P. Ward and M. L. Narasu, *Biotechnol. Adv.*, 2001, **19**, 175–199.
3. H. Dörnenburg and D. Knorr, *Enzyme Microb. Technol.*, 1995, **17**, 674–684.
4. N. Pras, *J. Biotechnol.*, 1992, **26**, 29–62.
5. S. Banerjee, S. Singh and L. U. Rahman, *Biotechnol. Adv.*, 2012, **30**, 461–468.
6. J. Zhao, L. C. Davis and R. Verpoorte, *Biotechnol. Adv.*, 2005, **23**, 283–333.
7. S. Zafar, S. Yousuf, H. Kayani, S. Saifullah, S. Khan, A. Al-Majid and M. Choudhary, *Chem. Cent. J.*, 2012, **6**, 1–8.
8. S. Musharraf, S. Naz, A. Najccb, S. Khan and M. Choudhary, *Chem. Cent. J.*, 2012, **6**, 1–6.
9. K. Ishihara, H. Hamada, T. Hirata and N. Nakajima, *J. Mol. Catal. B: Enzym.*, 2003, **23**, 145–170.
10. I. Smetanska, in *Food Biotechnology*, ed. U. Stahl, U. B. Donalies and E. Nevoigt, Springer, Berlin, Heidelberg, 2008, vol. 111, pp. 187–228.
11. C. Simões, N. Albarello, T. Castro and E. Mansur, in *Biotechnological production of plant secondary metabolites*, ed. I. E. Orhan, Bentham Science Publishers, 2012, pp. 67–86.
12. M. Lucchesini and A. Mensuali-Sodi, in *Bio-Farms for Nutraceuticals*, ed. M. Giardi, G. Rea and B. Berra, Springer, US, 2010, vol. 698, pp. 185–202.
13. F. DiCosmo and M. Misawa, *Biotechnol. Adv.*, 1995, **13**, 425–453.
14. M. E. L. Costa and S. Raposo, in *Reactores Biológicos: Fundamentos e Aplicações*, ed. M. M. Fonseca and J. A. Teixeira, Lidel, Lisboa, 2007, pp. 441–454.

15. E. McCoy and S. O'Connor, in *Natural Compounds as Drugs Volume I*, ed. F. Petersen and R. Amstutz, Birkhäuser Basel, 2008, vol. 65, pp. 329–370.

16. P. J. Weathers, M. J. Towler and J. Xu, *Appl. Microbiol. Biotechnol.*, 2010, **85**, 1339–1351.

17. F. Sato and Y. Yamada, in *Advances in Plant Biochemistry and Molecular Biology*, ed. H. N. Hans, J. Bohnert and G. L. Norman, Pergamon, 2008, vol. 1, pp. 311–345.

18. J. Zhao, Q. Hu, Y. Q. Guo and W. H. Zhu, *Appl. Microbiol. Biotechnol.*, 2001, **55**, 693–698.

19. A. C. Torres, A. T. Ferreira, F. G. Sá, J. A. Buso, L. T. Caldas, A. S. Nascimento, M. M. Brígido and E. Romano, *Glossário de Biotecnologia Vegetal*, Editora Embrapa, Brasília, 2000, pp. 1–128.

20. P. Bhatia, N. Ashwath, T. Senaratna and D. Midmore, *Plant Cell, Tissue Organ Cult.*, 2004, **78**, 1–21.

21. G. Phillips, *In Vitro Cell. Dev. Biol.: Plant*, 2004, **40**, 342–345.

22. O. Kayser and W. J. Quax, in *Biotechnological Approaches for the Production of some Promising Plant-Based Chemotherapeutics*, ed. A. Baldi, V. S. Bisaria and A. K. Srivastava, Wiley Online Library, 2006, pp. 117–156.

23. A. Navaratne and G. A. Rechnitz, *Anal. Chim. Acta*, 1992, **257**, 59–66.

24. R. R. Termignoni, *ultura de Tecidos Vegetais*, Editora da UFRGS, Porto Alegre, 2005.

25. E. F. George, in *Plant propagation by tissue culture: Volume 1. The Background*, ed. E. F. George, M. A. Hall and G.-J. De Klerk, Springer, Netherlands, 2008, vol. 1, pp. 1–28.

26. M. E. Kolewe, V. Gaurav and S. C. Roberts, *Mol. Pharmaceutics*, 2008, **5**, 243–256.

27. K. Saito and H. Mizukami, in *Plant Biotechnology and Transgenic Plants*, ed. K.-M. Oksman-Caldentey and W. H. Barz, CRC Press, 2002, pp. 77–109.

28. M. Vanisree, C.-Y. Lee, S.-F. Lo, S. M. Nalawade, C. Y. Lin and H.-S. Tsay, *Bot. Bull. Acad. Sin.*, 2004, **45**, 1–22.

29. S. Ramachandra Rao and G. A. Ravishankar, *Biotechnol. Adv.*, 2002, **20**, 101–153.

30. M. M. Yeoman and C. L. Yeoman, *New Phytol.*, 1996, **134**, 553–569.

31. H. N. Chang and S. J. Sim, *Curr. Opin. Biotechnol.*, 1995, **6**, 209–212.

32. S. C. Roberts and M. L. Shuler, *Curr. Opin. Biotechnol.*, 1997, **8**, 154–159.

33. J. W. Choi, G. H. Cho, S. Y. Byun and D. I. Kim, *Adv. Biochem. Eng./Biotechnol.*, 2001, **72**, 63–102.

34. J.-J. Zhong, in *Plant Cells*, ed. J. J. Zhong, S. Y. Byun, G. H. Cho, J. W. Choi, J. R. Haigh, H. Honda, E. James, J. W. Kijne, D. I. Kim, T. Kobayashi, J. M. Lee, M. Kino-oka, J. C. Linden, C. Liu, J. Memelink, N. Mirjalili, H. Nagatome, M. Taya, M. Phisaphalong, R. van der Heijden and R. Verpoorte, Springer, Berlin, Heidelberg, 2001, vol. 72, pp. 1–26.

35. V. Mulabagal and H.-S. Tsay, *Int. J. Appl. Sci. Eng.*, 2004, **2**, 29–48.

36. A. Frydman, O. Weisshaus, D. V. Huhman, L. W. Sumner, M. Bar-Peled, E. Lewinsohn, R. Fluhr, J. Gressel and Y. Eyal, *J. Agric. Food Chem.*, 2005, **53**, 9708–9712.

37. L. G. Zhou and J. Y. Wu, *Nat. Prod. Rep.*, 2006, **23**, 789–810.
38. R. Eibl and D. Eibl, in *Cell and Tissue Reaction Engineering: Principles and Practice*, ed. R. Eibl, D. Eibl, R. Pörtner, G. Catapano and P. Czermak, Springer-Verlag, Berlin, Heidelberg, 2009, pp. 315–356.
39. B. Ruffoni, L. Pistelli, A. Bertoli and L. Pistelli, in *Bio-Farms for Nutraceuticals*, ed. M. Giardi, G. Rea and B. Berra, Springer, US, 2010, vol. 698, pp. 203–221.
40. T. Suga and T. Hirata, *Phytochemistry*, 1990, **29**, 2393–2406.
41. H. Hamada and T. Furuya, in *Plant Cell and Tissue Culture for the Production of Food Ingredients*, ed. T.-J. Fu, G. Singh and W. Curtis, Springer, US, 1999, pp. 113–120.
42. R. A. C. Gonçalves, A. C. Cunha, A. J. B. Oliveira, M. F. P. Machado and R. A. M. Santos, in *Biocatálise e Biotrasnformação : Fundamentos e Aplicações*, ed. A. J. Marsaioli and A. L. M. Porto, Editora Schoba, São Paulo, Brasil, 2010, vol. 1, pp. 102–157.
43. D. P. Piet, R. Schrijvers, M. C. R. Franssen and A. de Groot, *Tetrahedron*, 1995, **51**, 6303–6314.
44. E. Moyano, J. Palazon, M. Bonfill, L. Osuna, R. M. Cusido, K. M. Oksman-Caldentey and M. T. Pinol, *J. Plant Physiol.*, 2007, **164**, 521–524.
45. T. Furuya, K. Sakamoto, K. Iida, Y. Asada, T. Yoshikawa, S. Sakai and N. Aimi, *Phytochemistry*, 1992, **31**, 3065–3068.
46. J.-H. Zhu, R.-M. Yu, L. Yang, Y.-S. Hu, L.-Y. Song, Y.-J. Huang, W.-M. Li and S.-X. Guan, *Process Biochem.*, 2010, **45**, 1652–1656.
47. J. M. Cooney, G. S. Hotter and D. R. Lauren, *Phytochemistry*, 2000, **53**, 447–450.
48. M. Lindmark-Henriksson, D. Isaksson, T. Vanek, I. Valterova, H. E. Hogberg and K. Sjodin, *J. Biotechnol.*, 2004, **107**, 173–184.
49. T. Vanek, I. Valterova and T. Vaisar, *Phytochemistry*, 1999, **50**, 1347–1351.
50. T. Horiike, M. Ohshiro and M. Kuroyanagi, *Phytochemistry*, 1997, **44**, 627–632.
51. H. Hamada, H. Yasumune, Y. Fuchikami, T. Hirata, I. Sattler, H. J. Williams and A. Ian Scott, *Phytochemistry*, 1997, **44**, 615–621.
52. A. Nasib, S. G. Musharraf, S. Hussain, S. Khan, S. Anjum, S. Ali, A. U. Atta-Ur-Rahman and M. I. Choudhary, *J. Nat. Prod.*, 2006, **69**, 957–959.
53. J. Dai, M. Ye, H. Guo, W. Zhu, D. Zhang, Q. Hu, J. Zheng and D. Guo, *Tetrahedron*, 2002, **58**, 5659–5668.
54. M. Ye, L. Ning, J. Zhan, H. Guo and D. Guo, *J. Mol. Catal. B: Enzym.*, 2003, **22**, 89–95.
55. V. Molmoori, K. Srisailam and V. Ciddi, *Appl. Biochem. Biotechnol.*, 2008, **144**, 201–212.
56. M. Ščigelová, T. Macek, A. Minghetti, M. Macková, P. Sedmera, V. Přikrylová and V. Křen, *Biotechnol. Lett.*, 1995, **17**, 1213–1218.
57. F. Sabir, R. Sangwan, J. Singh, L. Misra, N. Pathak and N. Sangwan, *Plant Biotechnol. Rep.*, 2011, **5**, 127–134.
58. Y. Zhan, J.-H. Zou, X. Ma and J. Dai, *J. Mol. Catal. B: Enzym.*, 2005, **36**, 43–46.

59. X. Zhang, M. Ye, Y. H. Dong, H. B. Hu, S. J. Tao, J. Yin and D. A. Guo, *Phytochemistry*, 2011, **72**, 1779–1785.

60. C. Vega-Jarquin, L. Dendooven, I. Magaña-Plaza, F. Thalasso and A. Ramos-Valdivia, *Environ. Toxicol. Chem.*, 2001, **20**, 2670–2675.

61. D. Dagnino, J. Schripsema and R. Verpoorte, *Phytochemistry*, 1994, **35**, 671–676.

62. W. Zhu, G. Asghari and G. B. Lockwood, *Fitoterapia*, 2000, **71**, 501–506.

63. M. Dvorakova, I. Valterova, D. Saman and T. Vanek, *Molecules*, 2011, **16**, 10541–10555.

64. L. Yang, J. Dai, J. Sakai and M. Ando, *Biotechnol. Lett.*, 2005, **27**, 793–797.

65. V. Srivastava, A. S. Negi, P. V. Ajayakumar, S. A. Khan and S. Banerjee, *Appl. Biochem. Biotechnol.*, 2012, **166**, 1401–1408.

66. Y. Yin, R. Chen, D. Zhang, L. Qiao, J. Li, R. Wang, X. Liu, L. Yang, D. Xie, J. Zou, C. Wang and J. Dai, *J. Mol. Catal. B: Enzym.*, 2013, **89**, 28–34.

67. T. Vanek, M. Novotny, R. Podlipna, D. Saman and I. Valterova, *J. Nat. Prod.*, 2003, **66**, 1239–1241.

68. D. Caron, A. Coughlan, M. Simard, J. Bernier, Y. Piché and R. Chênevert, *Biotechnol. Lett.*, 2005, **27**, 713–716.

69. A. A. Orden, C. Magallanes-Noguera, E. Agostini and M. Kurina-Sanz, *J. Mol. Catal. B: Enzym.*, 2009, **61**, 216–220.

70. W. Chai, H. Hamada, J. Suhara and C. Akira Horiuchi, *Phytochemistry*, 2001, **57**, 669–673.

71. S. Patel, R. Gaur, P. Verma, R. S. Bhakuni and A. Mathur, *Biotechnol. Lett.*, 2010, **32**, 1167–1171.

72. A. A. Orden, F. R. Bisogno, D. A. Cifuente, O. S. Giordano and M. Kurina Sanz, *J. Mol. Catal. B: Enzym.*, 2006, **42**, 71–77.

73. M. Łuczkiewicz and A. Kokotkiewicz, *Plant Sci.*, 2005, **169**, 862–871.

74. S. Patel, R. Gaur, M. Upadhyaya, A. Mathur, A. K. Mathur and R. S. Bhakuni, *J. Nat. Med.*, 2011, **65**, 646–650.

75. B. Xue, L. Zhou, J. Liu and R. Yu, *Die Pharmazie*, 2012, **67**, 467–471.

76. I. S. Nunes, J. M. Faria, A. C. Figueiredo, L. G. Pedro, H. Trindade and J. G. Barroso, *Planta Med.*, 2009, **75**, 387–391.

77. H. Kanho, S. Yaoya, N. Kawahara, T. Nakane, Y. Takase, K. Masuda and M. Kuroyanagi, *Chem. Pharm. Bull.*, 2005, **53**, 361–365.

78. K. Shimoda, Y. Kondo, T. Nishida, H. Hamada, N. Nakajima and H. Hamada, *Phytochemistry*, 2006, **67**, 2256–2261.

79. K. Shimoda, N. Sato, T. Kobayashi, H. Hamada and H. Hamada, *Phytochemistry*, 2008, **69**, 2303–2306.

80. K. Shimoda, N. Kubota, K. Taniuchi, D. Sato, N. Nakajima, H. Hamada and H. Hamada, *Phytochemistry*, 2010, **71**, 201–205.

81. W. Li, K. Koike, Y. Asada, T. Yoshikawa and T. Nikaido, *J. Mol. Catal. B: Enzym.*, 2005, **35**, 117–121.

82. W. Li, K. Koike, Y. Asada, T. Yoshikawa and T. Nikaido, *Tetrahedron Lett.*, 2002, **43**, 5633–5635.

83. Z. György, A. Tolonen, P. Neubauer and A. Hohtola, *Plant Cell, Tissue Organ Cult.*, 2005, **83**, 129–135.

84. K. Shimoda, Y. Kondo, M. Akagi, K. Abe, H. Hamada and H. Hamada, *Phytochemistry*, 2007, **68**, 2678–2683.

85. C. X. Peng, J. S. Gong, X. F. Zhang, M. Zhang and S. Q. Zheng, *Afr. J. Biotechnol.*, 2010, **7**, 211–216.

86. W. Li, K. Koike, Y. Asada, T. Yoshikawa and T. Nikaido, *Carbohydr. Res.*, 2003, **338**, 729–731.

87. D. A. Casas, S. I. Pitta-Alvarez and A. M. Giulietti, *Appl. Biochem. Biotechnol.*, 1998, **69**, 127–136.

88. L. Yang, C. Yan, J. Zhu, L. Song and R. Yu, *World J. Microbiol. Biotechnol.*, 2010, **26**, 1201–1205.

89. K. Shimoda, S. Kwon, A. Utsuki, S. Ohiwa, H. Katsuragi, N. Yonemoto, H. Hamada and H. Hamada, *Phytochemistry*, 2007, **68**, 1391–1396.

90. X. Chen, J. Zhang, J.-H. Liu and B.-Y. Yu, *J. Mol. Catal. B: Enzym.*, 2008, **54**, 72–75.

91. C.-Y. Yan, R.-M. Yu, Z. Zhang and L.-Y. Kong, *J. Integr. Plant Biol.*, 2007, **49**, 207–212.

92. K. Kawaguchi, T. Watanabe, M. Hirotani and T. Furuya, *Phytochemistry*, 1996, **42**, 667–669.

93. V. Srivastava, R. Kaur, S. K. Chattopadhyay and S. Banerjee, *Ind. Crops Prod.*, 2013, **44**, 171–175.

94. H. Katsuragi, K. Shimoda, N. Kubota, N. Nakajima, H. Hamada and H. Hamada, *Biosci., Biotechnol., Biochem.*, 2010, **74**, 1920–1924.

95. S. T. Häkkinen, P. Lackman, H. Nygren, K. M. Oksman-Caldentey, H. Maaheimo and H. Rischer, *J. Biotechnol.*, 2012, **157**, 287–294.

96. R. M. de Pádua, N. Meitinger, J. D. de Souza Filho, R. Waibel, P. Gmeiner, F. C. Braga and W. Kreis, *Steroids*, 2012, **77**, 1373–1380.

97. A. C. Figueiredo, M. J. Almendra, J. Barroso and J. C. Scheffer, *Biotechnol. Lett.*, 1996, **18**, 863–868.

98. S. Mathieu, J. Wirth, F.-X. Sauvage, J.-P. Lepoutre, R. Baumes and Z. Gunata, *Plant Cell, Tissue Organ Cult.*, 2009, **97**, 203–213.

99. J. M. Faria, I. S. Nunes, A. C. Figueiredo, L. G. Pedro, H. Trindade and J. G. Barroso, *Biotechnol. Lett.*, 2009, **31**, 897–903.

100. K. Shimoda, T. Harada, H. Hamada, N. Nakajima and H. Hamada, *Phytochemistry*, 2007, **68**, 487–492.

101. J.-W. Choi, Y.-K. Kim, H.-K. Park, W. Lee and D.-I. Kim, *Biotechnol. Bioprocess Eng.*, 1999, **4**, 281–286.

102. C. Magallanes-Noguera, M. M. Ferrari, M. Kurina-Sanz and A. A. Orden, *J. Biotechnol.*, 2012, **160**, 189–194.

103. T. Vaněk, I. Valterová, R. Pospíšilová and T. Vaisar, *Biotechnol. Tech.*, 1994, **8**, 289–294.

104. S. R. Rao and G. A. Ravishankar, *Process Biochem.*, 1999, **35**, 341–348.

105. H. J. Woerdenbag, N. Pras, H. W. Frijlink, C. F. Lerk and T. M. Malingré, *Phytochemistry*, 1990, **29**, 1551–1554.

106. C. V. F. Baldwin, R. Wohlgemuth and J. M. Woodley, *Org. Process Res. Dev.*, 2008, **12**, 660–665.

107. A. Vandenberghe, I. E. Markó, F. Lucaccioni and S. Lutts, *Ind. Crops Prod.*, 2013, **42**, 380–385.

108. F. Baldassarre, G. Bertoni, C. Chiappe and F. Marioni, *J. Mol. Catal. B: Enzym.*, 2000, **11**, 55–58.

109. R. Bruni, G. Fantin, A. Medici, P. Pedrini and G. Sacchetti, *Tetrahedron Lett.*, 2002, **43**, 3377–3379.

110. E. Majewska and M. Kozłowska, *Tetrahedron Lett.*, 2013, **54**, 6331–6332.

111. A. M. Fonseca, F. J. Q. Monte, M. d. C. F. de Oliveira, M. C. de Mattos, G. A. Cordell, R. Braz-Filho and T. L. G. Lemos, *J. Mol. Catal. B: Enzym.*, 2009, **57**, 78–82.

112. G. A. Cordell, T. L. G. Lemos, F. J. Q. Monte and M. C. de Mattos, *J. Nat. Prod.*, 2007, **70**, 478–492.

113. B. Leroy, *Chim. Nouv.*, 2006, **92**, 84–90.

114. B. Baskar, S. Ganesh, T. S. Lokeswari and A. Chadha, *J. Mol. Catal. B: Enzym.*, 2004, **27**, 13–17.

115. W. K. Mączka and A. Mironowicz, *Tetrahedron: Asymmetry*, 2004, **15**, 1965–1967.

116. D. Scarpi, E. G. Occhiato and A. Guarna, *Tetrahedron: Asymmetry*, 2005, **16**, 1479–1483.

117. L. H. Andrade, R. S. Utsunomiya, A. T. Omori, A. L. M. Porto and J. V. Comasseto, *J. Mol. Catal. B: Enzym.*, 2006, **38**, 84–90.

118. J. S. Yadav, P. T. Reddy, S. Nanda and A. B. Rao, *Tetrahedron: Asymmetry*, 2002, **12**, 3381–3385.

119. J. S. Yadav, G. S. K. K. Reddy, G. Sabitha, A. D. Krishna, A. R. Prasad, U. R. R. Hafeez, K. Vishwaswar Rao and A. Bhaskar Rao, *Tetrahedron: Asymmetry*, 2007, **18**, 717–723.

120. B. Bohman, L. R. Cavonius and C. R. Unelius, *Green Chem.*, 2009, **11**, 1900–1905.

121. G. Zhai, H. J. Lehmler and J. L. Schnoor, *Environ. Sci. Technol.*, 2010, **44**, 3901–3907.

122. X. Z. Yu, J. D. Gu and S. Liu, *J. Hazard. Mater.*, 2007, **147**, 838–844.

123. E. P. Best, S. L. Sprecher, S. L. Larson, H. L. Fredrickson and D. F. Bader, *Chemosphere*, 1999, **39**, 2057–2072.

124. S. Liawrungrath, P. Purachat, W. Oungpipat and C. Dongduen, *Talanta*, 2008, **77**, 500–506.

125. D. C. Wijesuriya and G. A. Rechnitz, *Biosens. Bioelectron.*, 1993, **8**, 155–160.

126. A. M. C. Bizerra, G. d. Gonzalo, I. Lavandera, V. Gotor-Fernández, M. C. de Mattos, M. d. C. F. de Oliveira, T. L. G. Lemos and V. Gotor, *Tetrahedron: Asymmetry*, 2010, **21**, 566–570.

127. W. K. Maczka and A. Mironowicz, *Zeitschrift fur Naturforschung. C, J. Biosci.*, 2007, **62**, 397–402.

128. W. K. Mączka and A. Mironowicz, *Tetrahedron: Asymmetry*, 2002, **13**, 2299–2302.

129. T. Olejniczak, A. Mironowicz and C. Wawrzeńczyk, *Bioorg. Chem.*, 2003, **31**, 199–205.
130. T. Utsukihara, S. Watanabe, A. Tomiyama, W. Chai and C. A. Horiuchi, *J. Mol. Catal. B: Enzym.*, 2006, **41**, 103–109.
131. M. Nagaki, H. Imaruoka, J. Kawakami, K. Saga, H. Kitahara, H. Sagami, R. Oba, N. Ohya and T. Koyama, *J. Mol. Catal. B: Enzym.*, 2007, **47**, 33–36.
132. Z. H. Yang, R. Zeng, G. Yang, Y. Wang, L. Z. Li, Z. S. Lv, M. Yao and B. Lai, *J. Ind. Microbiol. Biotechnol.*, 2008, **35**, 1047–1051.
133. X. Liu, B. Zhang, Y. Xia and J. Xu, *Acta Chim. Sin.*, 2009, **67**, 1492–1496.
134. A. Kergomard, M. F. Renard, H. Veschambre, D. Courtois and V. Petiard, *Phytochemistry*, 1988, **27**, 407–409.
135. K. Nakamura, H. Miyoshi, T. Sugiyama and H. Hamada, *Phytochemistry*, 1995, **40**, 1419–1420.
136. R. Villa, F. Molinari, M. Levati and F. Aragozzini, *Biotechnol. Lett.*, 1998, **20**, 1105–1108.
137. M. Takemoto, Y. Yamamoto and K. Achiwa, *Chem. Pharm. Bull.*, 1998, **46**, 419–422.
138. Y. Naoshima and Y. Akakabe, *Phytochemistry*, 1991, **30**, 3595–3597.
139. Y. Naoshima, Y. Akakabe and F. Watanabe, *Agric. Biol. Chem.*, 1989, **53**, 545–547.
140. X. Chang, Z. Yang, R. Zeng, G. Yang and J. Yan, *Chin. J. Chem. Eng.*, 2010, **18**, 1029–1033.
141. L. L. Machado, J. S. N. Souza, M. C. de Mattos, S. K. Sakata, G. A. Cordell and T. L. G. Lemos, *Phytochemistry*, 2006, **67**, 1637–1643.
142. J. S. N. Souza, Master's thesis, Universidade Federal do Ceará, 2003.
143. B. Xie, J. Yang, Q. Yang and W. Yuan, *J. Mol. Catal. B: Enzym.*, 2009, **61**, 284–288.
144. A. Zilinskas and J. Sereikaite, *J. Mol. Catal. B: Enzym.*, 2013, **90**, 66–69.
145. M. Breuer, K. Ditrich, T. Habicher, B. Hauer, M. Kesseler, R. Sturmer and T. Zelinski, *Angew. Chem., Int. Ed. Engl.*, 2004, **43**, 788–824.
146. X. Liu, Y. Wang, H. Y. Gao and J. H. Xu, *Chin. Chem. Lett.*, 2012, **23**, 635–638.
147. T. Daußmann, T. C. Rosen and P. Dünkelmann, *Eng. Life Sci.*, 2006, **6**, 125–129.
148. A. J. Palazzo and D. C. Leggett, *J. Environ. Qual.*, 1986, **15**, 49–52.
149. A. Mironowicz, *Phytochemistry*, 1998, **47**, 1531–1534.
150. P. Villeneuve, *Eur. J. Lipid Sci. Technol.*, 2003, **105**, 308–317.
151. Y. H. Lin, C. Yu and A. H. Huang, *Arch. Biochem. Biophys.*, 1986, **244**, 346–356.

Development of Processes for the Production of Bulk Chemicals by Fermentation at Industrial Scale – An Integrated Approach

JØRGEN MAGNUS*[a]

[a]Bayer Technology Services GmbH, 51368 Leverkusen, Germany
*E-mail: jorgen.magnus@bayer.com

14.1 Introduction

14.1.1 The Potential of White Biotechnology for the Production of Bulk Chemicals

Microorganisms have a high potential for producing bulk chemicals due to the wide diversity of enzymes found in nature. These enzymes enable the synthesis of a range of interesting chemicals. Today, the vast majority of bulk chemicals are produced by chemical synthesis from fossil resources. It is possible to find biosynthetic pathways to many of these chemicals using enzymes known in the literature. When these pathways are implemented in a microorganism like *E. coli* or *Saccharomyces cerevisiae*, the chemical can be produced by fermentation using a renewable raw material, such as sugar. Considering

RSC Green Chemistry No. 45
White Biotechnology for Sustainable Chemistry
Edited by Maria Alice Z. Coelho and Bernardo D. Ribeiro
© The Royal Society of Chemistry 2016
Published by the Royal Society of Chemistry, www.rsc.org

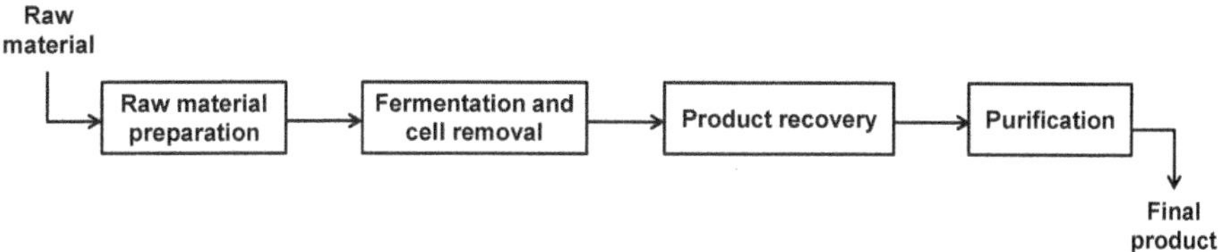

Figure 14.1 Generic block diagram of a bioprocess.

that only a fraction of the enzymes in nature are known, the possibilities for replacing chemical synthesis by biosynthesis are substantial. Producing bulk chemicals by fermentation becomes not so much a question of whether it is theoretically possible, but rather whether a technically and economically feasible process can be developed within a reasonable time and budget. To that end, efficient methods are needed in molecular biology and process engineering.

Some examples of already established large-scale production processes of bulk chemicals by fermentation include:

- 1,3-Propanediol by DuPont using a genetically modified *E. coli* strain.[1] Used for the production of polymers and other chemical products.
- Amino acids, in particular glutamate and lysine, produced mostly by *Corynebacterium glutamicum* and *E. coli*, developed either by the classical method of random mutagenesis and selection/screening, or by metabolic engineering.[2] Glutamate and lysine are used as food and feed additives.
- Lactic acid produced by lactic acid bacteria or genetically modified yeast.[3] Lactic acid is used to produce polylactic acid, a biodegradable plastic.
- Citric acid by *Aspergillus niger* developed by random mutagenesis.[4]
- Succinic acid by genetically modified *E. coli*, *Corynebacterium glutamicum* or yeast.[5]
- Vitamins, in particular vitamin C, produced by a combination of microbial and chemical steps.[6]
- Penicillin production by *Penicillin chrysogenum* and other penicillin molds developed mainly by the classical method of random mutation.

Refer to Sun and Alper[7] for a more complete list of production processes. Figure 14.1 shows the major process steps in a typical production process.

14.1.2 Setup of a Development Project

Developing a large scale bioprocess can roughly be divided into two tasks: developing a good production strain that converts a raw material, such as sugar, into the product molecule, and developing the process technology that allows large scale production. In the past, these two tasks have often been approached separately. One would typically start with the strain development and develop the strain until it had obtained a very good performance.

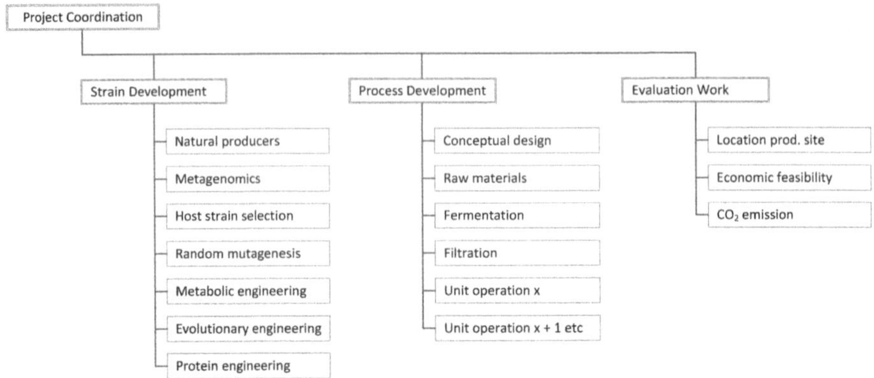

Figure 14.2 An example of the work breakdown structure of a development project.

After that, the process technology would be developed according to the characteristics of the strain. It will be argued here that integrating the strain development and the process technology development, and performing them simultaneously, is a more fruitful approach. The technical challenges can be attacked on two fronts, the possibility for optimizing the cost-effectiveness is greater and the total development time is shorter. This point will be discussed later in the chapter. The rapid development in molecular biology in recent years has allowed much faster strain development than was the case only ten or twenty years ago. The simultaneous development of the production strain and the process technology therefore becomes more and more feasible.

Figure 14.2 displays the work streams that would be included in a development project. Strain Development is divided into work packages corresponding to the techniques typically used in strain development. Process Development has Conceptual Design for the entire process as a general work package and the different unit operations as separate specific work packages. In addition to the two main work streams, Evaluation Work is a separate work stream. Here, the connection is made between the technical issues treated in the first two work streams and the business drivers of the project.

This chapter shows how the development of a process for bulk chemicals based on fermentation can be approached and gives an overview of the development work that needs to be done.

14.2 Steering the Direction of the Development Project

14.2.1 The Three Typical Main Business Drivers in Large Scale Bioproduction

From the business perspective of the producing company, there are typically three main drivers for establishing large scale production of a bulk chemical by fermentation.

14.2.1.1 Reduce Cost of Production

A bioprocess often has the potential of being cheaper than the corresponding chemical process. When the economy of scale is exploited, small molecules with a low or modest standard Gibbs free energy of formation can often be produced for less than 1.50 € kg^{-1} (considering raw material prices in 2012–2014), which in several cases is less than the cost of production by a chemical process. The cost of production for a bioprocess must be lower than the cost of production by classical chemical synthesis. Otherwise, it is difficult for a company to justify the increased risk associated with using a new technology such as biotechnology. A company can, in general, not expect that the customers will be willing to pay more for a biochemical than for a petrochemical (the "no bio-bonus" rule). Some exceptions to this rule exist, in particular for products with a high visibility for the end user.

14.2.1.2 Reduce CO_2 Emissions

Depending on how energy efficiently the production process can be run, the CO_2 emissions of a bioprocess have the potential to be lower than those for the equivalent chemical process since renewable raw materials are used. The CO_2 emissions for the production of 1,3-propanediol by fermentation are 20% less than those of the petrochemical process.[8]

14.2.1.3 Become Independent of Fossil Resources

In the long perspective, it is an advantage to switch from fossil resources, which are finite and at some point will become scarce, to renewable resources.

14.2.2 The Three Typical Main Parameters for Reducing the Cost of Production

The three typical main process parameters with the highest impact on the cost of production are product yield, productivity (space–time yield) and product concentration in the fermenter.

14.2.2.1 Product Yield

The product yield, $Y_{p/s}$, is the amount of product obtained per kilo of spent raw material. It is normally calculated with respect to the main raw material which is the carbon- and energy-source (*e.g.* sugar). The reciprocal of $Y_{p/s}$ indicates how many kilos of sugar are needed to produce one kilo of product. For large scale facilities, the cost of sugar will be the main contributor to the total cost of production, often making up more than 50% of the total production cost. The product yield is therefore the single most important parameter for the production cost. In the early stages of strain development, the yield is typically very poor, often less than 0.1 kg kg^{-1}. Improving the product yield is thus a key task in development.

The maximum theoretical product yield for a given biosynthetic pathway can be calculated based on a stoichiometric model of the reaction network in the cell. When biomass production is included in the model and a good estimate for the sugar consumption due to the maintenance metabolism can be made, the metabolic rates during a fermentation run can be simulated according to the envisioned fermentation procedure. In this way, a realistic estimate of the achievable product yield can be made at an early stage, which is imperative for a reliable cost estimate.

14.2.2.2 Productivity (Space–Time Yield)

The productivity of a fermenter is the amount of product formed per time and volume unit in the fermenter. It is given in kg product per hour and m^3. For a given plant capacity, the productivity decides the total fermentation volume needed. Fermenters are a significant cost item for investment costs. In addition, the energy associated with running fermenters can make a considerable contribution to both the production cost and the CO_2 emissions. Fermenters running aerobically need to be both stirred and aerated. The power input to large fermenters is normally in the range of 0.5 to 3 kW m^{-3} fermenter volume (in special cases, such as penicillin production with filamentous fungi, power inputs of up to 10 kW m^{-3} may be required). The energy needed for compressing the air depends on the aeration rate and the geometry of the fermenters, but is also a significant factor for some processes. Reducing the total size of the fermenters (*i.e.* optimizing the productivity) is therefore an important task.

14.2.2.3 Product Concentration in the Fermenter

One of the major challenges in downstream processing is the separation of the product from the water phase in the fermenters. The product concentration is always very dilute (10–15% w/w is considered a good product concentration). The separation task becomes more expensive the more dilute the product concentration is, and the development should therefore aim to optimize the product concentration.

14.2.3 Alignment with Business Drivers

The development work must be aligned with the business drivers of the producing company. The work should focus on optimizing the parameters that contribute most to the business drivers. As stated above, the reduction of production costs and CO_2 emissions will normally be central business drivers. It is therefore useful to have a calculation of the production costs and the CO_2 emissions which is updated continuously during development. The cost and CO_2 calculations are based on a model of the process. In this way,

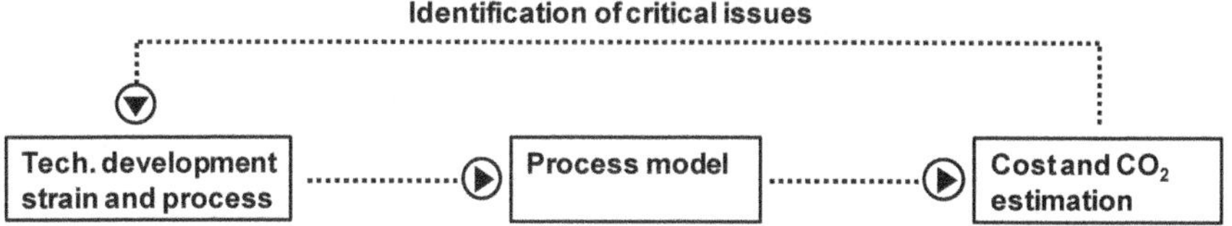

Figure 14.3 Aligning the technology development with the business drivers production cost and CO_2 emissions.

it will be visible what issues (process steps, strain features, *etc.*) are critical to costs and CO_2 emission. This information is then fed back into the strain and process development so that the developers can focus on these issues (see Figure 14.3).

The production costs and CO_2 emissions must be calculated so that they can be compared directly to the competing petrochemical process. Some standards for CO_2 calculations exist (*e.g.* the British standard PAS 2050:2011). A complete life cycle analysis as described in ISO standards 14040:2006 and 14044:2006 might be too extensive and premature for early stage development.

The production cost calculation can be performed by considering the following items:

- Raw materials (sugar source, ammonia, salts, *etc.*)
- Consumables (filter aid, chromatography resin, *etc.*)
- Utilities/energy (electricity, cooling water, steam, *etc.*)
- Personnel
- Maintenance
- Depreciation of the production plant.

A very rough estimation of the production cost can be carried out using the following three steps:

1. Estimate the yield that can be realistically achieved by strain development. Consider the biochemistry and the thermodynamics of the biosynthesis pathway. Consider also the maintenance metabolism and the production of biomass.
2. Estimate the current sugar price based on corn in the USA or sugar cane in Brazil. Look at publications from the bioethanol industry.
3. Look in the literature for reports on cost structures for similar bioproducts. If these cannot be found, assume that the sugar cost is 60% of the total cost (this should be a good first guess for C_4–C_6 molecules).

Example: with a yield coefficient of 0.4 kg kg^{-1}, a sugar price of 300 US$ per ton and the assumption that the sugar cost makes up 60% of the total, the production cost is estimated as 1250 US$ per ton of product. A more accurate estimate must be based on a model of the process.

14.3   Strain Development

The kinds of methods needed for strain development depends on the targeted product. In the following, some of the most common methods and considerations are described.

14.3.1   Search for Natural Producers

If the product is a natural part of the metabolism of microorganisms, one might be able to find natural producers in the environment and use these as a starting point for strain development. Samples may be taken from environments where the natural producer is expected to occur. These could include environments with a high or low pH, a high or low temperature, low or zero concentration of oxygen, contamination by organic waste *e.g.* from a refinery, *etc.* Natural producers are isolated from samples by a suitable screening method as described below in the section on screening.

A natural producer may be optimized further by random mutagenesis (or metabolic engineering if it can be modified genetically). Another option is to analyze it, for instance by measuring its -omics features, to understand why it is a good producer, and to use this knowledge for metabolic engineering of a different, more suitable host strain. In particular, sequencing the whole genome of the microorganism has become an option in recent years as the cost of sequencing has been reduced dramatically. When the appropriate tools in bioinformatics are applied, an essential biological understanding can be extracted and used in metabolic engineering.

14.3.2   Metagenomics

When a new enzyme not known in the literature is needed in order to construct a biosynthesis pathway, the metagenomic approach may be used.[9] DNA is extracted from samples from the environment. The DNA is fragmented, cloned and expressed, *e.g.* in a standard laboratory *E. coli* strain. The clones are screened for the desired activity and positive hits are sequenced. An alternative approach is to sequence the DNA fragments directly and identify potentially new enzymes by homology alignment.

14.3.3   Host Strain Selection

The strain hosting the biosynthetic pathway leading to the desired product, as well as the genetic modifications enabling highly efficient production, must be carefully selected. Common host strains in large scale bioproduction of bulk chemicals include:

- *Escherichia coli*—model bacterium in microbiology. Used for a range of products because of the well-established methods for genetic modification.

- *Corynebacterium glutamicum*—used mainly for amino acid production due to its strong amino acid metabolism.
- *Aspergillus niger*—natural producer of citric acid and other organic acids.
- *Saccharomyces cerevisiae*—established as an ethanol producer. Facultative anaerobe and tolerant to low pH.

Other strains that have been studied in depth and might come into question include:

- *Bacillus subtilis*—established as an enzyme producer.
- *Pseudomonas putida*—resistant to certain organic solvents.
- *Clostridium ljungdahlii*—strict anaerobe which can be used for synthesis gas fermentation.[10]
- *Pichia pastoris*—used for the production of recombinant proteins. Can grow on methanol.

When choosing the right host strain one should consider the following aspects:

- Natural ability to produce the product.
- Accessibility for genetic modification.
- Resistance towards the possible toxicity of the product.
- Possibility for production at extreme pH or temperature.
- Particular features of the strain's metabolism.

E. coli is the most widely used host strain and the organism that normally allows the fastest strain development due to the availability and range of methods for genetic modification. There is currently a trend to use *S. cerevisiae* because it allows fermentation at low pH which eases the sterility requirements for the fermentation and in some cases allows acids to be produced as pure acids rather than salts. Thermophilic strains also allow reduced sterility requirements. Kern *et al.*[11] give an extensive review of different host strains in industrial biotechnology.

14.3.4   Random Mutagenesis

Random mutagenesis is the classical method of strain development in which random mutations are introduced into the genome of the strain and the mutants with the desired phenotype are identified. The identified mutant is then subjected to mutagenesis again, and the cycle is repeated over and over. The success of the random mutagenesis approach depends on how efficient the selection method is. A very efficient selection method allows for the screening of billions of mutants.

Several large scale fermentation processes running today use strains developed by random mutagenesis. In processes where the remaining biomass is sold as feeding stuff, there might be regulatory requirements prohibiting the use of recombinant DNA technology, and random mutagenesis is then the only option for strain development. Random mutagenesis is also the logical option when an uncommon host strain is chosen for which there are no established tools for targeted genetic modification.

The method of mutation is typically either treatment with UV-C light or with a chemical such as *N*-methyl-*N'*-nitro-*N*-nitrosoguanidine (NTG) or ethyl methanesulfonate (EMS). Other methods include transposon mutagenesis or the use of mutator strains. In addition, plasmid DNA may be chemically mutagenized by treatment with hydroxylamine. The next step is to isolate the strains with the desired phenotype. This can be done by a selection method or a detection method (or a combination of both). In the optimization of amino acid producers, selection by an amino acid analogue has been used successfully.[12] The analogue inhibits growth, but may be outcompeted by the natural amino acid when this is present in higher concentrations. By that mechanism the best producers will also be the mutants that grow faster in a screening assay. Once the correlation between production rate and growth rate is established, the best producers can be selected by cultivating a mutation library in the presence of the amino acid analogue.

Connor *et al.*[13] demonstrated the development of an *E. coli* strain for the production of 3-methyl-1-butanol by random mutagenesis. They used NTG for the mutagenesis and used 4-aza-D,L-leucine for selection by the same strategy as is used for amino acid producers. Hughes *et al.*[14] used UV-C mutagenesis to develop yeast strains that can grow anaerobically on xylose. A famous example of strain development by random mutation is the development of high-producing penicillin strains. The productivity of the fungus *Penicillium chrysogenum* was increased more than 500 times from that of the original strain[15] and the titer by more than 4000 times.[16]

14.3.5   Screening

Screening of libraries is an essential step in the search for natural producers, in strain development by random mutagenesis, in metagenomics and in directed evolution. The efficiency of the screening depends on the assay. Particularly efficient are assays that select the strains that exhibit the desired characteristic, *i.e.* the assay is designed so that only strains with this characteristic can grow. When selection is used, large collections of microorganisms can be screened efficiently on agar plates or in liquid culture.

If assays that select cannot be developed, a detection assay can be used. The desired characteristic is typically the ability to produce a metabolite. In that case, the detection assay could be the direct measurement of the product molecule, *e.g.* by fluorescent or adsorption spectroscopy, HPLC, MS, *etc.* Measurement of a secondary effect can also be used. If, for instance, the product is an acid, a pH indicator can be used to identify the producing colonies on an agar plate. Alternatively, a reagent, or an enzyme that transforms the product molecule further to yield a color change, can be used (*e.g.* the formation of melanin from tyrosine as reported by Santos and Stephanopoulos[17]). Additionally, auxotrophic indicator strains may be used in an overlay assay, *e.g.* amino acid auxotrophic strains which can be used to detect amino acid producers.

Biosensors are microorganisms that give a fluorescent output in the presence of the product molecule. Either the host cell itself can be modified to function as a biosensor, or the biosensor strain can be incubated together with the mutants. The mechanism of the biosensor is typically based on ligand interaction with a transcriptional regulator, but can also be ligand interaction at the RNA level, or it can be based on an autofluorescent protein pair that makes use of Förster resonance energy transfer.[18] In any case, the presence of the product molecule results in the expression of a fluorescent protein such as GFP (green fluorescent protein). Two recently reported examples of the use of biosensors include the detection of flavonoids[19] and the detection of methionine and branched-chain amino acids.[20]

14.3.6 Metabolic Engineering

Metabolic engineering emerged as its own scientific discipline in the early 90s. It deals with the improvement of cellular activities by manipulation of enzymatic, transport and regulatory functions in a cell through the use of recombinant DNA technology.[21] In other words it is about rational improvement of a cell by targeted genetic manipulation, and differs from random mutagenesis where the genetic changes are not targeted. The cell is viewed as a chemical reactor which can be designed to produce a specific molecule. Following that logic, the metabolic engineering approach can roughly be divided into four possible actions:

1. Include new reactions in the cell (by introducing new genes into the genome)
2. Enhance or weaken existing reactions (by overexpressing or down-regulating enzymes)
3. Remove reactions (by deleting or disrupting genes on the genome)
4. Change metabolic regulation (by adding or deleting genes with a regulatory function, or by specifically changing the amino acid sequence of an enzyme, *e.g.* to change allosteric regulation)

As in other engineering disciplines, metabolic engineering goes through the two steps of analysis (what genetic modifications need to be carried out) and synthesis (implementing the genetic modifications in the cell).

14.3.6.1 *Analysis: Identifying the Targets for Genetic Modification*

In the early stages of a development project, the first targets for genetic modifications are often set based on experience and literature references. A typical example is 3-deoxy-D-arabino-heptulosonate-7-phosphate synthase

(DAHP synthase). This enzyme is often the first target when the flux through an aromatic amino acid pathway is optimized since it has long been known to have a large portion of the control of the flux through this pathway.[22–24] However, after the first evident targets have been identified, it soon becomes necessary to take a systematic approach. Thus, the further identification of targets should be based on a holistic understanding of the metabolism of the cell according to the way of thinking in systems biology.[25]

The complexity of metabolism, and in particular of the intracellular reaction network in a cell, can best be captured by a mathematical model. Only through modeling and simulation can the systemic properties of the reaction network be analyzed and understood.[26]

Large scale stoichiometric models have been developed for several common microorganisms.[27] A stoichiometric model can be used to analyze the different possible pathways to the desired product molecule. Elementary flux mode analysis is a useful tool in this respect.[28] The most energy efficient pathway to the product molecule can be identified. In metabolic flux analysis (MFA), a stoichiometric model is used to calculate the intracellular fluxes. Knowledge of the fluxes in the cell gives insight into the physiology of the cell, which can be used to identify targets for metabolic engineering. For instance, the flux distribution between the glycolysis and the pentose phosphate pathway or the fluxes through different anaplerotic pathways will indicate which pathways are particularly active and relevant as targets. In addition, the demand for co-metabolites, such as NAD(P)H and ATP, is calculated which gives further hints about where the limiting steps are.

An MFA can be based on measurements of the observable rates (*e.g.* glucose uptake, product and by-product excretion rates, CO_2 excretion, biomass production (growth), *etc.*). This approach does not allow all the intracellular rates to be calculated. The use of ^{13}C labelled glucose gives a more detailed and accurate determination of the fluxes, but requires a more extensive experimental data basis and more specialized software for calculations. In the case of a stationary ^{13}C flux analysis, experimental data may be obtained from the labelling pattern of the amino acids in the protein of the cells.[29] In the case of an instationary ^{13}C flux analysis, the labeling pattern of intra- and extracellular metabolites is used.[30] Mass spectroscopy is commonly used for sample analysis.

The next step is the development of a predictive model. The model may include the reaction network of the entire cell, or be limited to certain pathways. By defining rate equations for each reaction in the stoichiometric model, a dynamic model can be developed.[31] Once the kinetic parameters of the model have been determined, the model can be used for predicting the phenotype of the strain (*e.g.* the production rate) after one or several genetic modifications have been implemented. In this way, possible genetic targets can be evaluated *in silico* without the need for cloning experiments. The dynamic model can further be used for metabolic control analysis. The theory of metabolic control analysis defines coefficients that quantify the degree of control that a system parameter (such as an enzyme level) has on a variable (such as a flux). The concentration control coefficient, C^C, is the relative change of a metabolite concentration x_i, with the relative change of a parameter p_j and is defined in eqn (14.1). The flux control coefficient, C^F, is the relative change of a reaction rate r_i, with the relative change of a parameter p_j and is defined in eqn (14.2) (see Visser and Heijnen[32] for a review of metabolic control analysis). The flux control coefficients (eqn (14.2)) are particularly useful when they are calculated as the sensitivity of the production rate with regard to the different enzyme levels. In that case they will give direct indications of what enzyme levels should be increased by overexpression. Concentration control coefficients (eqn (14.1)) can be useful to evaluate if the availability of a certain metabolite is a limiting factor. Finally, the dynamic model may be used to calculate the optimal distribution of enzyme activities.[33]

$$C^C_{i,j} = \frac{p_j}{x_i}\frac{\mathrm{d}x_i}{\mathrm{d}p_j} \tag{14.1}$$

$$C^F_{i,j} = \frac{p_j}{r_i}\frac{\mathrm{d}r_i}{\mathrm{d}p_j} \tag{14.2}$$

A dynamic model is a very strong tool as it allows the analysis described above. However, its predictions must be interpreted with care. Setting up and defining a model that gives accurate predictions can be very difficult. The effort needed for the development of such a model is extensive and may not always be justified in a development project.

Omics technologies can also give insight into the physiology of a cell which is useful for the determination of the genetic targets. Genomics has become increasingly applicable as the price for sequencing has fallen dramatically. Today, sequencing of complete genomes is no longer a significant cost factor in a strain development project. The NCBI database currently has more than 30 000 partially or completely sequenced prokaryotic genomes (around 3800 are completely sequenced, see www.ncbi.nlm.nih.gov). Genomic data becomes useful when the genes of the production strain are annotated and their functions can be predicted. Transcriptomics is the measurement of the mRNA levels to analyze the expression of genes. Proteomics quantifies the amounts of protein expressed in a cell and gives a further indication of what genes are expressed. Metabolomics is the quantification of the metabolites in a cell. A large intracellular pool of a metabolite may signify a bottleneck in the reaction pathway. Transcriptomics, proteomics and metabolomics can be used to analyze the difference in expression levels or metabolite levels between a reference strain and a strain with one or more genetic modifications. Also, the difference between two different physiological states is useful to investigate, *e.g.* to understand the effect of limiting a nutrient in the fermentation broth. A metabolic model may also be used to get a better interpretation of omics data. The detailed biological understanding gained in this way forms the basis for the rational identification of further targets for modification.

14.3.6.2   Synthesis: Implementing the Targets

The four actions in metabolic engineering mentioned above are carried out by using the tools of molecular biology.

In the early phase of a development project, it is convenient to express new genes, or overexpressing existing ones on a plasmid to get a rapid evaluation of the effect *in vivo*. Also, mutated enzymes with altered allosteric regulation can be evaluated in this way. As one would typically investigate many different targets (and combinations thereof) and a certain degree of trial and error is necessary, it is important to be able to clone genes on plasmids rapidly. The recent decrease in cost for synthetic genes has increased the speed at which clones can be generated, and allows a more extensive and faster testing of possible targets than what was the case ten years ago.

In a later phase, the most effective genetic changes should be incorporated in the chromosome to yield a stable strain which does not have any selection markers such as antibiotic resistances. The production strain should not be dependent on antibiotics. The use of high copy number plasmids can also result in too high expression of an enzyme.

Recombineering[34] can be used to delete, replace or insert a gene in the chromosome. In *E. coli* the most frequently used method is based on λ-red recombinase,[35] which is commonly used for gene deletion (knock-out) or for efficient integration of new genes in the chromosome.[36]

The ability to adjust the expression levels of enzymes to a specific level can become important after the first targets have been implemented. This can be achieved by promoter fine-tuning using a synthetic promoter library. Alper *et al.*[37] used mutagenesis of a constitutive promoter to create a synthetic promoter library and used it to identify the optimal levels of phosphoenolpyruvate carboxylase and deoxy-xylulose-P synthase in *E. coli*. Braatsch *et al.*[38] used a similar approach to create a synthetic promoter library and showed how this could be used to fine-tune the expression level of phosphoglucose isomerase in *E. coli*.

These and other advanced molecular tools in metabolic engineering have been reviewed by Tyo *et al.*[39] and Galanie *et al.*[40]

14.3.6.3   An Example of Metabolic Engineering: The Biosynthesis of Phenol

A biosynthesis pathway converting chorismate to phenol was implemented in *E. coli* to allow the production of phenol from sugar by fermentation.[41] The metabolic engineering followed the four steps listed below (refer to Figures 14.4 and 14.5):

1. New reactions: The conversion of 4-hydroxybenzoate (4-HB) to phenol by 4-HB decarboxylase was included in the reaction network by expressing the gene *hbdBCD*. This gene was taken from another strain and is only found naturally in a handful of strains.[42] In addition, the gene *ubiC* encoding the chorismate lyase that converts chorismate to 4-HB was

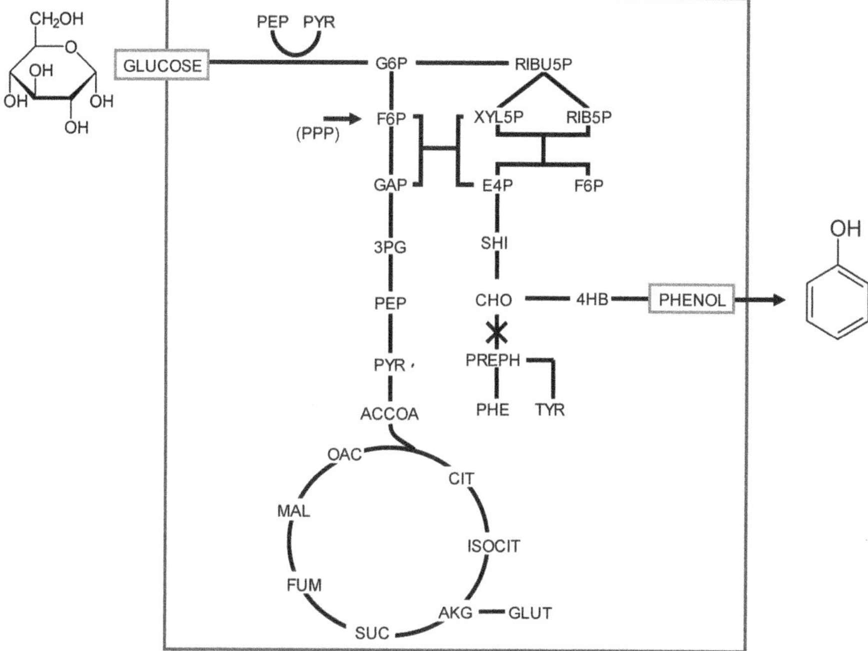

Figure 14.4 The reaction network in an engineered cell, showing the whole cell conversion of glucose to phenol.

overexpressed. *ubiC* is part of the genome of *E. coli*, but normally has a very low expression level since the cell only needs very small amounts of 4-HB. By including these two reactions, the complete pathway from glucose to phenol was established.

2. Enhanced reactions: The genes *aroG*, *aroB* and *aroL* were overexpressed on a plasmid or by chromosomal integration to increase the flux through the aromatic amino acid pathway towards phenol.

3. Removed reactions: The genes *pheA* and *tyrA* were deleted from the genome to remove the chorismate mutase reaction that converts chorismate to prephenate. By doing this, more chorismate is available for phenol production.

4. Changed regulation: A mutated *aroG* gene coding for a feedback resistant form of the AroG enzyme was used. Unlike the natural form, this enzyme is not inhibited by phenylalanine. In addition, the *tyrR* gene regulating the expression of the genes in the aromatic amino acid pathway was deleted.

14.3.7 Evolutionary Engineering

Evolutionary engineering is a technique used to develop strains with particular desired properties by "...continuous evolution procedures that rely on the use of an appropriate selection pressure towards a desired phenotype in continuously growing cultures" as defined by Cakar *et al.*[43] Different

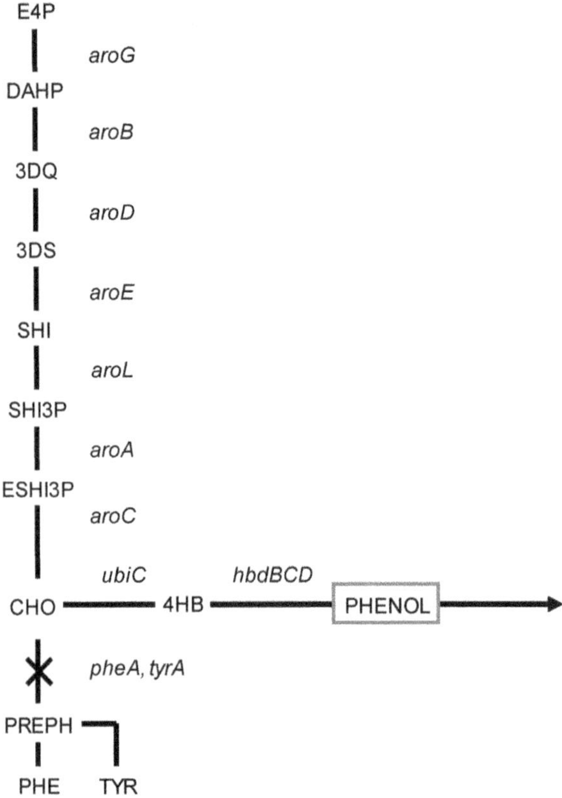

Figure 14.5 The aromatic amino acid pathway with bifurcation towards phenol. Genes are written in italics. Metabolites in capital letters as abbreviations: E4P = erythrose-4-phosphate, DAHP = 3-deoxy-D-arabino-heptulosonate-7-phosphate, 3DQ = 3-dehydroquinate, 3DS = 3-dehydroshikimate, SHI = shikimate, SHI3P = shikimate-3-phosphate, ESHI3P = 5-enolpyruvyl-shikimate 3-phosphate, CHO = chorismate, 4HB = 4-hydroxybenzoate, PREPH = prephenate, PHE = phenylalanine, TYR = tyrosine.

types of mutagenesis methods and selection techniques can be employed, as reviewed by Sauer.[44] Evolutionary engineering is particularly suited to deal with the challenge of product inhibition which is encountered in most strain development projects. It should be employed early in development to create a host strain that is tolerant towards the product and which can subsequently be optimized towards production by metabolic engineering. Another option is to characterize the tolerant strain to understand the mechanism of its tolerance and transfer this mechanism to another host strain. For instance, if the strain has developed a particularly efficient pump to keep the *in vivo* concentration of the product low, it can normally be identified by genome sequencing and transferred to another organism. In any case, it is sensible to investigate the molecular mechanism behind the increased tolerance as this could be unfavorable for production (*e.g.* further catabolism of the product molecule). Evolutionary engineering might also be used to develop resistances to other types of stress that can be beneficial in a production scheme or to improve the uptake of nutrients, as demonstrated by Smith *et al.*[45]

14.3.8 Protein Engineering

Protein engineering can be used to develop enzymes with improved or new properties. A typical application of protein engineering within strain development is to remove an allosteric regulation mechanism, such as feedback inhibition, when this is counterproductive to the production pathway. Protein engineering may also be used to increase the enzyme activity to overcome rate limiting steps. A further application is to expand or narrow the substrate range in order to convert other substrate variants or to exclude unwanted reactions for enzymes that catalyze more than one reaction. Finally, developing an enzyme that catalyzes a fully new reaction, for which there are no known enzymes in nature, is also possible. However, only a few examples of successful attempts to create such new enzymes are known. Two examples are given by Savile *et al.*[46] and Westphal *et al.*[47]

Either rational design by site-directed mutagenesis or directed evolution by random mutation can be used to modify enzymes. When the structure of the enzyme is known, specific amino acids can be exchanged by applying point mutations (see Figure 14.6). Often, the enzyme structure is not known

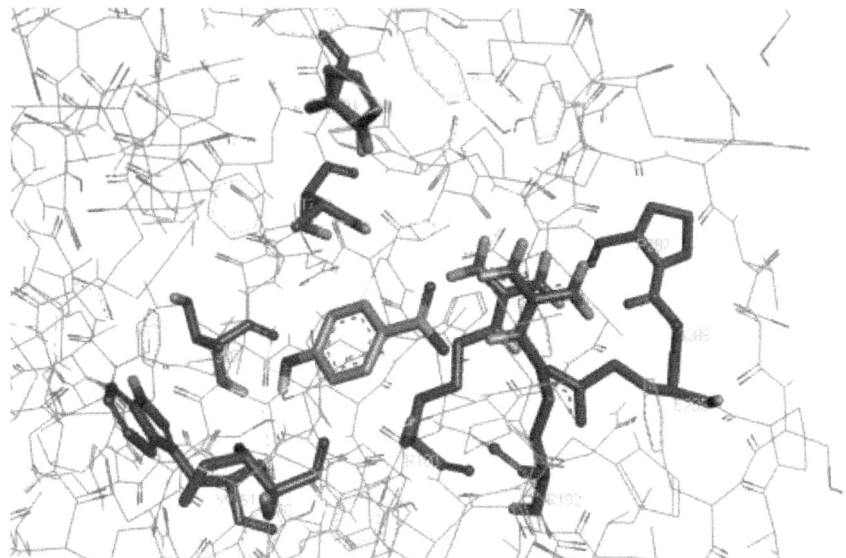

Figure 14.6 Protein structure of the 4-hydroxybenzoate decarboxylase mentioned in the example. The active site and the substrate and some of the amino acids in the active site that come into question for site-directed mutagenesis are shown. However, it is often the case that amino acids far outside the active site must be changed to yield the desired modification of the enzyme property.

or the predicted changes in enzyme properties are not achieved by site-directed mutagenesis. In this case, random mutagenesis, *e.g.* by error prone PCR, followed by selection or high throughput screening is used.

Savile *et al.*[46] used a combination of rational design and directed evolution to develop a transaminase that allows conversion of ketones to chiral amines. Westphal *et al.*[47] used rational design to develop an enzyme that catalyzes the carboligation of aromatic aldehydes to (*S*)-benzoins.

Reviews on protein engineering are given by Arnold,[48] by Behrens[49] and by Lane and Seelig.[50]

14.4   Process Technology Development

The kind of process technology needed in a production facility will depend on the product and its specifications. Figure 14.1 shows a generalized block diagram of the processing steps in a bioprocess. In the following, the approach on how to develop the process technology for a bioprocess is described.

14.4.1   Conceptual Design

Conceptual Design is defined as its own work package and generates the overall production concept. The work package Conceptual Design gives an overview of the entire process and has a coordinating role for the technical development of the process.

Modelling and simulation is used to design and optimize the process and to set up energy and material balances. The different processing options are evaluated, regarding technical feasibility, production cost, energy efficiency and CO_2 emissions. More detailed design of the unit operations is carried out by subject experts in separate work packages, as depicted in Figure 14.2.

14.4.2   Raw Materials

From a technical point of view it makes sense to choose the raw materials early in the development, as both the strain and the process technology must be adapted to the raw material. However, the choice of raw material depends on the location of the production site. A decision regarding the production site is normally taken at a later stage when technical and economic feasibility has been proven. So, the developers must often consider several possible raw materials during development.

The main raw material functions as a carbon and energy source for the production strain. The following raw materials come into question for this purpose:

1. Sucrose- and oligosaccharide-containing plants such as sugar cane and sugar beet
2. Starch-containing plants such as corn, wheat and rye
3. Lignocellulose from wood, grass, straw or bagasse
4. Glycerol
5. Synthesis gas

Many large scale facilities use starch refined by a first generation biorefinery as raw materials. Molasses and sugar cane juice are used in particular for the production of bioethanol in Brazil. Corn and molasses or sugar cane juice are currently the most economic raw materials. Molasses or sugar cane juice can be used for the production of bulk chemicals in just the same way that they are used for the production of bioethanol. The advantage of lignocellulose as a raw material is that it does not compete with the production of food or animal feeding stuff. Several technologies have been developed to convert lignocellulose into fermentable sugars, but the cost and energy consumption associated with that process are still a challenge. Glycerol has the potential of being an inexpensive alternative to sugar as it is a by-product of biodiesel production. Synthesis gas (a mixture of H_2, CO and CO_2) can be used in fermentations with anaerobic bacteria such as clostridia. This opens up the possibility of using pyrolysed biomass or steel mill waste gas as raw material.

The method for converting the raw material of choice into fermentable sugar can often be licensed from a company with expertise in that area. However, this process step has an impact on the other unit operations and must therefore be considered when developing the process.

When sugar cane is the raw material, the molasses (a by-product of sugar production) can be used directly to feed the fermenter. Alternatively, all the sugar in the sugar cane can be used as feed. In that case, the sugar cane goes through a milling step and the sugar is extracted into water before the resulting sugar cane juice is concentrated by evaporation in a process similar to the first steps of sugar production.

Starch containing plants are treated in a first generation biorefinery by milling (wet or dry), liquefaction (addition of α-amylase to break up the starch polymer) and saccharification (addition of glucoamylase to convert the oligosaccharides into glucose). Johnson[51] gives a review of this process.

Lignocellulose is converted to fermentable sugars in a second generation biorefinery by a physical, chemical or enzymatic pretreatment step, followed by enzymatic hydrolysis by the addition of cellulases (see Alvira *et al.*[52] for a review). In addition the pentoses found in hemicellulose can be used for fermentation to utilize a larger portion of the lignocellulose.[53]

Glycerol and synthesis gas may need purification before being fed into the fermenter. Steel mill waste gas often contains sulphuric compounds that may be toxic to microorganisms.

In addition to a carbon and energy source, a nitrogen source is needed, especially if the product molecule has one or more nitrogen atoms. Ammonia in gas form can be fed directly to the fermenters. Another option is to use a complex medium such as corn steep liquor.

14.4.3   Fermentation

The fermentation process should be developed so that the three main parameters mentioned above, product yield, space–time yield and product concentration, are optimized. The fermentation process and the characteristics of the production strain should be adapted to each other.

14.4.3.1 Operation Mode

The fermenter can be operated as a batch, fed-batch, repeated fed batch or continuous fermentation. Krahe[54] reviews the different operation modes. This subject is also treated in standard textbooks on biochemical engineering (*e.g.* Bailey and Ollis[55]). The fed-batch mode has obvious advantages over pure batch operation as it will normally prolong the production period in the fermenter and increase the final titer. The increased production period reduces the total downtime calculated over a whole year which results in an increased space–time yield. This effect might be even more prominent in repeated fed-batch fermentations as this mode of operation reduces the downtime further. In cases where toxic substances accumulate in the fermenter, or where the production strain is very sensitive towards the product molecule, a simple batch fermentation may still be the best choice. Continuous fermentation may be run with or without cell retention. A continuous fermentation with cell retention has the biggest potential to achieve high product and space–time yields. There are higher sterility requirements for continuous fermentations than for batch or fed-batch fermentations.

14.4.3.2 Growth Limitation

Running the fermenter at growth limiting conditions increases the product yield as less carbon is used for the generation of biomass and more for the generation of product. A prerequisite is that the strain can maintain a high production rate under growth limiting conditions, *i.e.* that it has been developed to have a production rate which is decoupled from the growth rate. Growth limitation is easy to apply when minimal medium is used. In that case, a growth factor other than the carbon and energy source, *e.g.* phosphate, can be supplied in limiting amounts. In large scale fermenters, where transfer rates of nutrients and oxygen are often limited for technical reasons, growth limitation is sometimes inevitable.

14.4.3.3 Aerobic/Anaerobic Fermentation

The stirring of the bioreactor and the compression of air for aeration are both significant contributors to the total energy usage in a bioprocess. Anaerobic fermentation may therefore be more energy efficient than aerobic fermentation, provided that the strain can produce the product efficiently under anaerobic conditions. Anaerobic conditions may also be used to limit growth. Performing the growth phase under aerobic conditions to generate biomass quickly, and then shifting to anaerobic conditions for the production phase can be a favorable method if the production strain allows it. The succinic acid process uses this strategy.[56,57]

14.4.3.4   Scale Up

Fermentation is a unit operation which is challenging to scale up. Process development is done in small scale fermenters (*e.g.* 1 L) to allow a large number of development runs. Production fermenters may be as large as 1000 m^3. Scale up can be based on different coefficients that are kept constant when going from small to large scale. The two most common scale up coefficients are oxygen mass transfer ($k_L a$) and power input per volume (P/V). Schmidt[58] provides a list of scale up coefficients and discusses the different approaches. However, no generally applicable strategy has been established.

A particular challenge when scaling up is the existence of concentration gradients in large scale reactors. The larger the volume, the poorer the mixing quality will be. When the air is sparged only at the bottom of the vessel and the feed is added at the top of the vessel, there will be a concentration gradient over the whole reactor (the sugar concentration is high at the top and near zero at the bottom and *vice versa* for the oxygen concentration). In addition, there will be radial gradients and smaller local regions with variations in not only nutrients and dissolved gasses, but also in pH, shear stress and temperature. As a result the cells are experiencing continuously changing conditions which puts them in a physiological state which could be very different to the constant and well-defined state established in a laboratory fermenter. This could result in different production characteristics.[59] Takors[60] discusses how to quantify and analyze the biological impact on the cells by the heterogeneities in large scale fermenters.

The design and operation of large scale bioreactors are much more limited than for small scale reactors (for instance with respect to the volumetric power input). Therefore, it might be useful to think the other way around and to establish a scale down model of a production fermenter (see Lara *et al.*[59] for a discussion of this approach). In this way, the suboptimal conditions in a production fermenter can be simulated in lab scale. Computational fluid dynamics is also a very useful tool for scale up or scale down studies as discussed by Formenti *et al.*[61]

14.4.3.5   Cell Removal

Cells need to be removed from the fermentation broth before it can be processed further in the downstream section. Cross flow filtration in microporous tubular filter modules is often used.[62] Alternatively, centrifugation, *e.g.* in a stacked disc centrifuge, can be used. Stacked disc centrifuges, which are also used in waste water treatment plants, have a very high capacity and can process large volumes in relatively small and inexpensive apparatus. However, the cell separation is not 100%, and additional filtration, such as a dead end filter, might be necessary.

14.4.4   Product Recovery

Product recovery is normally the first step after the fermentation. Sometimes, depending on the separation method used and the purity requirements of the final product, it is necessary to remove protein and some of the other impurities prior to the product recovery step.

Separating the product from the water phase, *i.e.* recovering the product, is a crucial step in the downstream part of a bioprocess. The product will seldom have a concentration higher than 100–150 kg m^{-3} and it is important to find a cost and energy efficient way of getting rid of the 85–90% water in the fermentation broth. A number of different technologies may be used depending on the product molecule:

- Evaporation
- Extraction
- Reactive extraction
- Precipitation
- Acidification and crystallization
- Chromatography
- Electrodialysis

14.4.4.1   *Evaporation*

Evaporation is an option for removing most of the water when the product molecule has a low vapor pressure in the water mixture, and product losses during evaporation are insignificant. The 1,3-propanediol process by DuPont uses flash evaporation to concentrate the product solution from about 90% moisture to about 20% moisture.[63] Heat integration is important in order to limit energy consumption which can be substantial considering that the amount of water that needs to be evaporated is several times as much as the amount of product.

14.4.4.2   *Extraction and Reactive Extraction*

Extraction into an organic phase is an option when an organic liquid can be found for which the product molecule has a good distribution coefficient. The distribution coefficient, $k_{(op/wp)}$, is defined as the ratio of the product concentrations in the organic phase and the water phase at thermodynamic equilibrium. With a high distribution coefficient, an efficient extraction procedure can be designed using *e.g.* a mixer–settler or an extraction column. Other physical properties of the organic phase such as density, viscosity and solubility in water must also be considered for the design of the extraction equipment. Also, the properties relating to the subsequent separation of the product from the organic phase must be considered. Normally, this separation is a distillation, in which case the boiling point and possible azeotropes

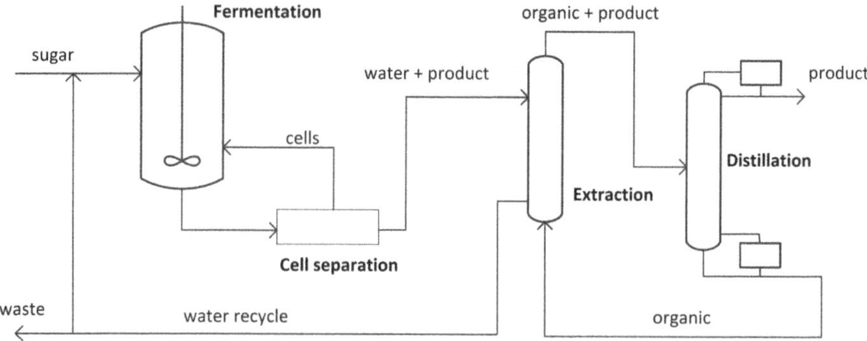

Figure 14.7 An example of an integrated process with continuous fermentation and continuous product removal. The product is a substance that can be extracted efficiently into an organic solvent. The organic solvent has a lower density and a higher boiling point than the product and has a low solubility in water. Further distillation steps might be required to purify the product or to purify the organic before it re-enters the extraction.

with the product molecule must be evaluated. The organic solvent should have a low solubility in water in order to limit the consumption of organic solvent. The price of the organic solvent should also be low. If the water phase leaving the extraction step is recycled into the fermenter, the toxicity of the organic solvent for the production strain becomes important (see Figure 14.7). The usefulness of an extraction step depends on whether or not a good organic solvent can be found that meets all the criteria mentioned above.

When the product molecule is bound to a carrier molecule that facilitates the transition from the aqueous phase into the organic phase, it is referred to as reactive extraction. The product molecule is extracted back into a second aqueous phase in a second extraction unit. Rüffer *et al.*[64] describe a reactive extraction process for phenylalanine. Kurzrock and Weuster-Botz[65] describe a reactive extraction process for succinic acid.

14.4.4.3 *Precipitation*

Many of the most relevant fermentation products are acids, either organic acids or amino acids. These are present in the fermentation broth in their ionized form as salts in solution. By adding a salt with a cation that forms an insoluble salt with the product ion, the product can be precipitated and thereby separated from the water phase. A recovery method based on precipitation was developed for succinic acid.[66] Calcium dihydroxide is added to the fermentation broth to form calcium succinate which precipitates. Precipitation with calcium hydroxide is also the recovery method in the standard lactic acid process[67] and in the citric acid process.[68]

14.4.4.4   *Acidification and Crystallization*

Several organic acids have a low solubility in water as pure uncharged acid, but high solubility when they are present as a salt in solution. In these cases, acidification and subsequent crystallization can be used to separate the acid from the water. The pH is lowered below the pK_a of the acid, and the acid will change from its dissociated form to its undissociated form and precipitate as crystals due to the low solubility of the pure acid. The crystals can then be removed from the slurry by centrifugation or filtration, and washed and dried to give a product with high purity. When acidification is used, it is a good idea to use NaOH to adjust the pH in the fermenter and HCl to lower the pH during acidification. After crystallization, NaCl will remain in the mother liquor. As NaCl does not contain nitrogen or carbon, the cost for waste water treatment will be low (no waste water treatment is necessary for NaCl dissolved in water). If ammonia is used in the fermentation and an acid containing carbon is used for acidification, the resulting cost for waste water treatment could be substantial, given the high amounts of salt formed (on a molar basis the amount of salt will be greater than the amount of product).

14.4.4.5   *Chromatography*

Charged molecules that cannot be separated from the water phase by acidification and crystallization, for example because the undissociated form also has high solubility, can be removed by ion exchange chromatography. Amino acids are often recovered in this way. An example is the production of lysine.[69,70]

14.4.4.6   *Electrodialysis*

Electrodialysis can be used for charged molecules. It has been suggested as a separation method in the production of lactic acid.[71,72]

14.4.5   **Purification**

Depending on the purity reached in the recovery step, it is sometimes necessary to purify the product further to reach the purity requirements demanded *e.g.* in the production of polymers. Distillation or other classical unit operations in chemical engineering can be used. These are covered in textbooks on chemical engineering. Refer to Figure 14.7 for a general example of a process with product recovery and purification.

14.5   **The Integrated Approach: Developing Microbiology and Process Technology in Parallel**

Following the reasoning that the cell is a chemical reactor that can be designed, the strain characteristics and the process operation mode should be adapted to each other. The optimal combination of strain and production technology has the highest potential for yielding an economically favorable process. Some aspects regarding the integration of strain and process development are discussed in the following sections. An example of an integrated process is given in Figure 14.7.

14.5.1 Product Inhibition

Product inhibition is a classical challenge in bioprocesses. As the concentration of the product in the fermenter increases, it will at some point become inhibiting to the production rate and/or to the growth of the strain. This problem can be approached either by developing a strain (*e.g.* by evolutionary engineering) which is tolerant to the product, or by using *in situ* product removal (ISPR) to keep the product concentration in the fermenter low during production. A combination of both is also a possibility.

Both approaches could have disadvantages. The increased tolerance to the product could mean that the cell spends more energy *e.g.* to maintain a concentration difference between its intra- and extra-cellular concentration of the product. This will in turn decrease the product yield. The ISPR could lead to the accumulation of toxic substances in the fermenter (see Figure 14.7) and by that add a new requirement to the strain development, namely the tolerance towards the new substance. It would also increase the complexity of the process and could lead to an increase in production cost and capital expenditure.

The best solution is found by calculating the effect on the production cost of the different approaches. For this, a process model is needed in which the interaction between the process technology and the strain is taken into account.

14.5.2 Fermentation Operating Mode

As stated above, a continuous fermentation with cell retention has the biggest potential to achieve high product and space–time yields. However, this operation mode sets certain requirements on the strain which must be considered by the team developing the strain. First of all, the product excretion rate must be decoupled from the growth rate. Otherwise, cell retention has no benefits. Secondly, the continuous fermentation should be run at a high product concentration in order to deliver a concentrated product solution to the downstream section of the plant. The cell specific product excretion rate should therefore not be inhibited (too much) by the product. Thirdly, the strain must be stable and should not lose its production capabilities over time. With the advanced tools in molecular biology, and the biological understanding available today, it is possible to develop strains with these characteristics. However, in the traditional approach of developing the strain first, and later adapting the process, these characteristics are not prioritized because the operation mode of the production fermenters is not considered during strain development. When both fermentation procedure and strain are developed simultaneously and in agreement, such opportunities can be realized.

14.5.3 Unit Operations in the Downstream Part of the Plant

Some of the problems encountered in the product recovery and purification part of the plant can be solved during strain development. For instance, if the typical by-products of a fermentation such as lactic acid or acetic acid are difficult to separate from the product, the strain may be genetically modified to avoid these by-products. The same argument goes for the fermentation medium. If certain medium components cause problems in the downstream part, the strain can be adapted to use a different component.

On the other hand, the downstream processes might solve problems encountered in strain development too, *e.g.* by including a chemical conversion step. If the desired fermentation product is toxic towards the cells, or the last reaction step in the biosynthesis pathway is difficult to achieve in the cell, the strain could be designed to produce a precursor, and the final reaction step can be performed as a classical chemical conversion step.

14.5.4 Holistic Understanding of Biology and Process Technology

A bioreactor can be analyzed on two levels. On the micro level, the cells have an intracellular reaction network in which hundreds of reactions take place simultaneously, eventually resulting in the formation and excretion of the product molecule. On the macro level, the bioreactor is a 3-phase stirred tank reactor where the cells act as catalysts to convert a reactant to a product. On the micro level, the variables are the intracellular metabolite concentrations and reaction rates, while the parameters typically considered are the enzyme properties such as activities and kinetic constants. On the macro level the process parameters are the aeration rate, feeding rate, stirring rate, pH, temperature, *etc.* The variables are the consumption rates of O_2 and nutrients, the excretion rates of products and by-products and the formation of biomass. Establishing a link between the micro and the macro level by defining a model that connects the variables and parameters on both levels with each other can be a powerful tool when developing a fermentation process. A holistic understanding of the bioreactor allows a more scientific approach to scale up and optimization of the fermentation procedure.

A holistic understanding can be obtained by using computational fluid dynamics combined with a model of the intracellular reaction network as demonstrated by Schmalzriedt *et al.*[73] and by Lapin *et al.*[74] The incorporation of signal transduction pathways and the use of online measurements of the omics features are further possibilities in this respect.

14.6 Conclusions

A large number of bulk chemicals can theoretically be produced by microorganisms. However, in most cases, it is challenging to develop a bioprocess which is cost competitive with the established petrochemical process. The petrochemical process often has several decades of process optimization behind it. Keeping the development time short and the cost of the development project low are further challenges relating to the development project itself. A structured project organization with integrated strain- and process development is the best way to meet these challenges. A strong holistic understanding of both biology and process technology, and a clear focus on the main cost drivers throughout the development is essential. Some of the leading companies with extensive experience and resources in white biotechnology are already following the integrated approach. This can be crucial in enabling more bioprocesses to reach industrial production scale.

Acknowledgements

The author is grateful to Prof. Dr.-Ing. Ralf Takors, Institute of Biochemical Engineering, University of Stuttgart, Prof. Dr. Georg Sprenger, Institute of Microbiology, University of Stuttgart and Dr.-Ing. Helmut Brod, Fermentation Technologies, Bayer Technology Services for reviewing the manuscript.

References

1. C. E. Nakamura and G. M. Whited, *Curr. Opin. Biotechnol.*, 2003, **14**, 454–459.
2. H. Kumagai, *Amino Acids Production*, Springer-Verlag, Berlin, Heidelberg, 2013.
3. M. Ilmen, K. Koivuranta, L. Ruohonen, V. Rajgarhia, P. Suominen and M. Penttila, *Microb. Cell Fact.*, 2013, **12**, 53.
4. L. P. S. S. Vandenberghe, C. Rodrigues, A. Pandey and J. M. Lebeault, *Braz. Arch. Biol. Technol.*, 1999, **42**, 263–276.
5. S. Okino, R. Noburyu, M. Suda, T. Jojima, M. Inui and H. Yukawa, *Braz. Arch. Biol. Technol.*, 2008, **81**, 459–464.
6. G. H. H. Pappenberger and H. P. Hohmann, *Industrial Production of L-Ascorbic Acid (Vitamin C) and D-Isoascorbic Acid*, Springer-Verlag, Berlin, Heidelberg, 2014.
7. J. Sun and H. S. Alper, *J. Ind. Microbiol. Biotechnol.*, 2015, **42**, 423–436.
8. A. Chilton, B. Ellison, S. Milne, W. Soutter, K. Walter and J. White, in *AZO Materials*, 2007-6-12.
9. W. R. Streit and R. Daniel, *Metagenomics: Methods and Protocols*, Springer Protocols, Humana Press, New York, 2010.

10. H. N. Younesi, G. Najafpour and A. R. Mohamed, *Biochem. Eng. J.*, 2005, 110–119.

11. A. Kern, E. Tilley, I. S. Hunter, M. Legisa and A. Glieder, *J. Biotechnol.*, 2007, **129**, 6–29.

12. M. Ikeda, *Adv. Biochem. Eng./Biotechnol.*, 2003, **79**, 1–35.

13. M. R. Connor, A. F. Cann and J. C. Liao, *Appl. Microbiol. Biotechnol.*, 2010, **86**, 1155–1164.

14. S. R. Hughes, W. R. Gibbons, S. S. Bang, R. Pinkelman, K. M. Bischoff, P. J. Slininger, N. Qureshi, C. P. Kurtzman, S. Liu, B. C. Saha, J. S. Jackson, M. A. Cotta, J. O. Rich and J. E. Javers, *J. Ind. Microbiol. Biotechnol.*, 2012, **39**, 163–173.

15. J. Nielsen, *Biotechnol. Bioeng.*, 1998, **58**, 125–132.

16. S. Parekh, V. A. Vinci and R. J. Strobel, *Appl. Microbiol. Biotechnol.*, 2000, **54**, 287–301.

17. C. N. Santos and G. Stephanopoulos, *Appl. Environ. Microbiol.*, 2008, **74**, 1190–1197.

18. M. Schallmey, J. Frunzke, L. Eggeling and J. Marienhagen, *Curr. Opin. Biotechnol.*, 2014, **26**, 148–154.

19. S. Siedler, S. G. Stahlhut, S. Malla, J. Maury and A. R. Neves, *Metab. Eng.*, 2014, **21**, 2–8.

20. N. Mustafi, A. Grunberger, D. Kohlheyer, M. Bott and J. Frunzke, *Metab. Eng.*, 2012, **14**, 449–457.

21. J. E. Bailey, *Science*, 1991, **252**, 1668–1675.

22. T. Ogino, C. Garner, J. L. Markley and K. M. Herrmann, *Proc. Natl. Acad. Sci. U. S. A.*, 1982, **79**, 5828–5832.

23. K. Backman, M. J. O'Connor, A. Maruya, E. Rudd, D. McKay, R. Balakrishnan, M. Radjai, V. DiPasquantonio, D. Shoda, R. Hatch, *et al.*, *Ann. N. Y. Acad. Sci.*, 1990, **589**, 16–24.

24. G. A. Sprenger, *Appl. Microbiol. Biotechnol.*, 2007, **75**, 739–749.

25. M. Papini, M. Salazar and J. Nielsen, *Adv. Biochem. Eng./Biotechnol.*, 2010, **120**, 51–99.

26. M. Cvijovic, J. Almquist, J. Hagmar, S. Hohmann, H. M. Kaltenbach, E. Klipp, M. Krantz, P. Mendes, S. Nelander, J. Nielsen, A. Pagnani, N. Przulj, A. Raue, J. Stelling, S. Stoma, F. Tobin, J. A. Wodke, R. Zecchina and M. Jirstrand, *Mol. Genet. Genomics*, 2014, **289**, 727–734.

27. A. M. Feist, M. J. Herrgard, I. Thiele, J. L. Reed and B. O. Palsson, *Nat. Rev. Microbiol.*, 2009, **7**, 129–143.

28. C. T. Trinh, A. Wlaschin and F. Srienc, *Appl. Microbiol. Biotechnol.*, 2009, **81**, 813–826.

29. A. Marx, A. A. de Graaf, W. Wiechert, L. Eggeling and H. Sahm, *Biotechnol. Bioeng.*, 1996, **49**, 111–129.

30. K. Noh, K. Gronke, B. Luo, R. Takors, M. Oldiges and W. Wiechert, *J. Biotechnol.*, 2007, **129**, 249–267.

31. J. B. Magnus, D. Hollwedel, M. Oldiges and R. Takors, *Biotechnol. Prog.*, 2006, **22**, 1071–1083.

32. D. Visser and J. J. Heijnen, *Metab. Eng.*, 2002, **4**, 114–123.

33. J. B. Magnus, M. Oldiges and R. Takors, *Biotechnol. Prog.*, 2009, **25**, 754–762.

34. G. Pines, E. F. Freed, J. D. Winkler and R. T. Gill, *ACS Synth. Biol.*, 2015, DOI: 10.1021/acssynbio.1025b00009.

35. K. A. Datsenko and B. L. Wanner, *Proc. Natl. Acad. Sci. U. S. A.*, 2000, **97**, 6640–6645.

36. C. Albermann, N. Trachtmann and G. A. Sprenger, *Biotechnol. J.*, 2010, **5**, 32–38.

37. H. Alper, C. Fischer, E. Nevoigt and G. Stephanopoulos, *Proc. Natl. Acad. Sci. U. S. A.*, 2005, **102**, 12678–12683.

38. S. Braatsch, S. Helmark, H. Kranz, B. Koebmann and P. R. Jensen, *BioTechniques*, 2008, **45**, 335–337.

39. K. E. Tyo, H. S. Alper and G. N. Stephanopoulos, *Trends Biotechnol.*, 2007, **25**, 132–137.

40. S. Galanie, M. S. Siddiqui and C. D. Smolke, *Curr. Opin. Biotechnol.*, 2013, **24**, 1000–1009.

41. J. Magnus, Method for producing phenol from renewable resources by fermentation, *World Pat.*, WO2014076113A1, 2012.

42. B. Lupa, D. Lyon, M. D. Gibbs, R. A. Reeves and J. Wiegel, *Genomics*, 2005, **86**, 342–351.

43. Z. P. Cakar, U. O. Seker, C. Tamerler, M. Sonderegger and U. Sauer, *FEMS Yeast Res.*, 2005, **5**, 569–578.

44. U. Sauer, *Adv. Biochem. Eng./Biotechnol.*, 2001, **73**, 129–169.

45. J. Smith, E. van Rensburg and J. F. Gorgens, *BMC Biotechnol.*, 2014, **14**, 41.

46. C. K. Savile, J. M. Janey, E. C. Mundorff, J. C. Moore, S. Tam, W. R. Jarvis, J. C. Colbeck, A. Krebber, F. J. Fleitz, J. Brands, P. N. Devine, G. W. Huisman and G. J. Hughes, *Science*, 2010, **329**, 305–309.

47. R. Westphal, C. Vogel, C. Schmitz, J. Pleiss, M. Muller, M. Pohl and D. Rother, *Angew. Chem.*, 2014, **53**, 9376–9379.

48. F. H. Arnold, *Acc. Chem. Res.*, 1998, **31**, 125–131.

49. G. A. H. Behrens, A. Hummel, S. K. Padhi, S. Schatzle and U. T. Bornscheuer, *Adv. Synth. Catal.*, 2011, 2191–2215.

50. M. D. Lane and B. Seelig, *Curr. Opin. Chem. Biol.*, 2014, **22**, 129–136.

51. D. L. Johnson, *The Corn Wet Milling and Corn Dry Milling Industry – A Base for Biorefinery Technology Developments*, Wiley-VCH Verlag GmbH & Co, Weinheim, 2006.

52. P. Alvira, E. Tomas-Pejo, M. Ballesteros and M. J. Negro, *Bioresour. Technol.*, 2010, **101**, 4851–4861.

53. J. Zaldivar, J. Nielsen and L. Olsson, *Appl. Microbiol. Biotechnol.*, 2001, **56**, 17–34.

54. M. Krahe, in *Ullmann's encyclopedia of Industrial Chemistry*, Weinheim, 2002.

55. J. E. Bailey and D. F. Ollis, *Biochemical Engineering Fundamentals*, McGraw-Hill, N.Y., 1986.

56. R. Datta, D. A. Glassner, M. K. Jain and J. R. Vick Roy, Fermentation and purification process for succinic acid, *US Pat.*, 5 168 055, 1992.

57. H. L. Song and S. Y. Lee, *Enzym. Microb. Technol.*, 2006, **39**, 352–361.
58. F. R. Schmidt, *Appl. Microbiol. Biotechnol.*, 2005, **68**, 425–435.
59. A. R. Lara, E. Galindo, O. T. Ramirez and L. A. Palomares, *Mol. Biotechnol.*, 2006, **34**, 355–381.
60. R. Takors, *J. Biotechnol.*, 2012, **160**, 3–9.
61. L. R. Formenti, A. Norregaard, A. Bolic, D. Q. Hernandez, T. Hagemann, A. L. Heins, H. Larsson, L. Mears, M. Mauricio-Iglesias, U. Kruhne and K. V. Gernaey, *Biotechnol. J.*, 2014, **9**, 727–738.
62. C. Charcosset, *Biotechnol. Adv.*, 2006, **24**, 482–492.
63. D. Adkesson, A. Alsop, T. Ames, L. Chu, J. Disney, B. Dravis, P. Fitzgibbon, J. Gaddy, F. Gallagher, W. Lehnhardt, J. Lievense, M. Luyben, M. Seapan, R. Trotter, G. Wenndt and E. Yu, Recovering 1,3-propanediol from fermentation broth comprising: subjecting broth to filtration, ion exchange, and distillation, *US Pat.*, 20050069997 A1, 2005.
64. N. Rüffer, U. Heidersdorf, I. Kretzers, G. A. Sprenger, L. Raeven and R. Takors, *Bioprocess Biosyst. Eng.*, 2004, **26**, 239–248.
65. T. Kurzrock and D. Weuster-Botz, *Bioprocess Biosyst. Eng.*, 2011, **34**, 779–787.
66. D. A. Glassner and R. Datta, Process for the production and purification of succinic acid, *US Pat.*, US5143834 A, 1992.
67. K. Buchta, *Lactic acid*, Verlag Chemie, Weinheim, 1983.
68. B. A. Matityaju and A. M. Eyal, Concurrent production of citric acid and alkali citrate, *Eur. Pat.*, EP 0432610 B1s, 1989.
69. I. Lee, K. Lee, K. Namgoong and Y. S. Lee, *Enzyme Microb. Technol.*, 2002, 798–803.
70. T. Hermann, *J. Biotechnol.*, 2003, **104**, 155–172.
71. V. M. Hábová, K. Melzoch, M. Rychtera and B. Sekavová, *Desalination*, 2004, 361–372.
72. M. Moresi and F. Sappino, *J. Food Eng.*, 1998, **35**, 75–90.
73. S. Schmalzriedt, M. Jenne, K. Mauch and M. Reuss, *Adv. Biochem. Eng./Biotechnol.*, 2003, **80**, 19–68.
74. A. Lapin, J. Schmid and M. Reuss, *Chem. Eng. Sci.*, 2006, 4783–4797.

Trends and Perspectives in Green Chemistry and White Biotechnology

BERNARDO DIAS RIBEIRO[a] AND MARIA ALICE ZARUR COELHO*[a]

[a]Biochemical Engineering Department, School of Chemistry, Federal University of Rio de Janeiro, Brazil
*E-mail: alice@eq.ufrj.br

15.1 Ultrasound

Ultrasound is defined as sound of a frequency beyond that to which the human ear can respond and generally is considered to lie between 20 kHz and 500 MHz. Power ultrasound (high intensity, low frequency, 20 kHz–1 MHz) enhances the chemical reactivity in a liquid medium through the generation and destruction of cavitation bubbles. Like any sound wave, ultrasound is propagated *via* a series of compression and rarefaction waves induced in the molecules of the medium through which it passes. At sufficiently high power, the rarefaction cycle may exceed the attractive forces of the molecules of the liquid, and cavitation bubbles will form (Figure 15.1). It is the fate of these cavities when they collapse in succeeding compression cycles that generates the energy for chemical and mechanical effects. The collapse is thought to generate very high local temperatures (around 5000 °C) and pressures

RSC Green Chemistry No. 45
White Biotechnology for Sustainable Chemistry
Edited by Maria Alice Z. Coelho and Bernardo D. Ribeiro

Published by the Royal Society of Chemistry, www.rsc.org

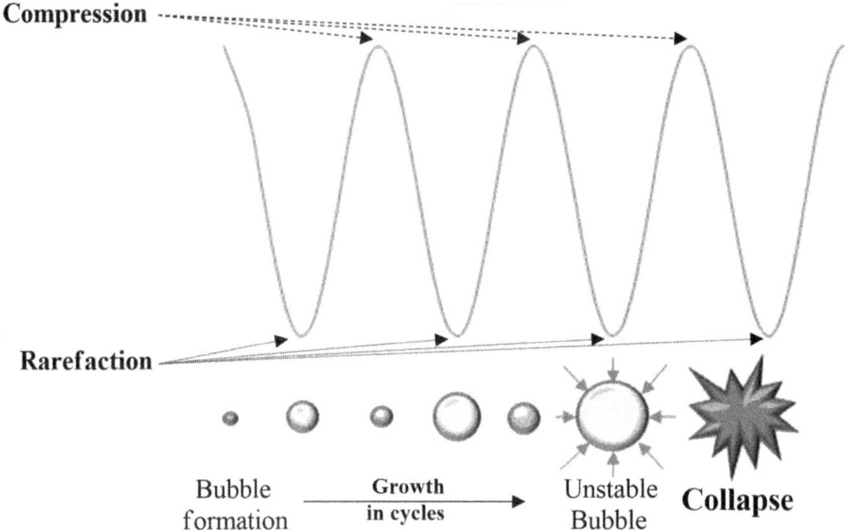

Figure 15.1 Formation and collapse of a cavitation bubble (inspired by Mason *et al.*[2]).

(in excess of 1000 atmospheres). The main applications of power ultrasound are in emulsification, surface cleaning, extraction, plastic welding, therapeutic medicine, degassing, food processing and catalysis.[1,2]

Ultrasound has a range of effects on chemical reactions in treated materials, such as the formation of free radicals (hydroxyl radicals from water), the increase of reaction rates (high shear created by cavity collapse), improvement of catalyst efficiency, alteration of the reaction pathway, and production of sonoluminescence, and can even affect chemical processes at surfaces, causing mechanical damage to the solid material, with shock waves and microjets, such as pitting of solid surfaces, fragmentation of brittle materials, and deaggregation of groups of particles. In heterogeneous reactions, the application of ultrasonic waves has the same physical effect as a high-speed agitator or a homogeniser in which fluids do not cavitate.[1,3]

For actual application of ultrasound waves, there is a need to use instruments known as transducers (convert mechanical or electrical energy into high-frequency sound) and suitable equipment, including whistle reactors, ultrasonic baths, and probe systems. In a whistle reactor, a stream of liquid flows, and passes a metal blade, in order to produce vibrations; whistle reactors are normally applied for homogenization, emulsification, and dispersion. Ultrasonic baths are cheap, simple, and versatile, and comprise a metal bath with one or more transducers attached to the walls of the tank. Items to be treated can be directly immersed in the bath and subjected to ultrasound wave propagation. Probe systems consist of a metal horn coupled to an ultrasonic transducer, used to amplify the vibration produced by an electrostrictive material, which is generally a piezoelectric material, in the transducer.[3]

For most commercial probe systems, the frequency is 20 kHz, and for baths it is around 40 kHz.[1,2]

Ultrasound can affect positively or negatively the activity and stability of many enzymes (Table 15.1). Typically, enzymes can be readily denatured by slight changes in environmental conditions, including temperature, pressure, shear stress, pH and ionic strength. Under these extreme conditions, sonication could cause the breakdown of hydrogen bonding and van der Waals interactions in polypeptide chains, leading to the modification of the secondary and tertiary structure of the protein; besides, acoustic micro-streaming and free radicals formed may react with some amino acid residues that participate in enzyme stability, substrate binding or in the catalytic function with a consequent change in biological activity. However, free radicals generated during cavitation do not affect the structure of all enzymes, with the effect being dependent on the chemical structure of the protein. In some systems, it is more appropriate to focus on substrate pre-treatment rather than using ultrasound to increase the activity of enzymes, in order to make the substrate more available to enzyme reaction processes, increasing mass transfer. Enzymes immobilized at surfaces are typically surrounded by boundary layers, and at the outer boundary layer phenomena such as cavitation-induced micro-jets and micro-streaming probably accelerate the transport of substrates to the enzyme and product removal from the enzyme, and are much less likely to be attacked directly by hydroxyl radicals produced by cavitation events.[4,5]

15.2 Fluorous Solvents

Fluorous solvents, such as perfluoro-substituted alkanes, dialkyl ethers and trialkylamines, are a class of nonpolar, hydrophobic, chemically inert, easily recyclable, and nontoxic solvents, with a higher density than the corresponding nonfluorinated solvents. The unique property of fluorous solvents to be miscible or immiscible with organic solvents as a function of temperature is a key concept for developing multiphase synthetic processes. The operational strategy begins when a catalyst is dissolved in a fluorous solvent and combined with substrates dissolved in the organic solvent to form a biphasic system. By warming the reaction system, the two phases become miscible and form one phase. The catalytic reaction can then occur in the homogeneous system and, finally, products can easily be recovered by simple recooling of the reaction mixture, because the two phases separate and the catalyst remains in the fluorous phase ready to be reused in another cycle.[30,31]

In a system with perfluorohexane and hexane, Beier and O'Hagan[30] investigated enantiomeric partitioning in transesterification reactions with vinyl 2-methylpentanoate and esterification reactions with 2-methylpentanoic acid and 2-methylhexanoic acid using *Candida rugosa* lipase and fluorinated decanol. The two liquid phases (hexane–PFH) became homogeneous at 30 °C and the reactions were carried out at 40 °C. After completion, all reactions were

Table 15.1 Enzymatic reactions assisted by ultrasound.

Enzyme	Source	Reaction	Equipment	References
α-Amylase, amyloglucosidase, and invertase	*Bacillus* sp., *Rhizopus* sp. and *Saccharomyces cerevisiae*	Hydrolysis of soluble potato starch, glycogen and sucrose	Ultrasonic bath (38 kHz, 60 W)	6
α-Amylase and amyloglucosidase	*Bacillus licheniformis* and *A. niger*	Hydrolysis of potato soluble starch	Ultrasonic bath (40 kHz, 130 W)	7
α-Amylase	*Trichoderma reesei*	Cold hydrolysis of starch	Ultrasonic bath (40 kHz, 130 W)	8
Cellulase	*Trichoderma reesei*	Hydrolysis of bleached cotton fabric	Probe (40 kHz, 90–100 W)	9
Cellulase	Novozymes (China)	Hydrolysis of carboxymethyl cellulose	Probe (24 kHz, 15–60 W)	10
Cellulase	*Trichoderma reesei*	Hydrolysis of filter paper in the presence of compressed liquefied petroleum gas	Ultrasonic bath (40 kHz, 150 W)	11
Cellulases	Advanced Biochemicals (India)	Biodegradation of distillery wastewater	Ultrasonic bath (22.5 kHz)	12
Dextranase	*Chaetomium erraticum*	Degradation of dextran	Probe (25 kHz, 40 W)	13
Laccase	*Trametes villosa*	Bleaching of cotton fabrics	Probe (20 kHz, 7 W)	14
Lipase	*Rhizomucor miehei*	Production of diacylglycerol by hydrolysis of palm oil	Probe (23 kHz, 200 W)	15
Lipase	*Pseudomonas* sp.	*N*-Alkylation of primary aromatic amines	Probe (22 kHz, 32 W)	16
Lipase B	*Candida antarctica*	Glycerolysis of olive oil	Ultrasonic bath (37 kHz, 130 W)	17
Lipase B	*Candida antarctica*	Esterification of rutin and naringin with unsaturated fatty acids	Probe (25 kHz, 150–200 W)	18

Lipase B	*Candida antarctica*	Acetylation of resveratrol with vinyl acetate	Ultrasonic bath (40 kHz, 150 W)	19
Lipase B	*Candida antarctica*	Esterification between acetic acid and butanol	Ultrasonic bath (40 kHz)	20
Lipase B	*Candida antarctica*	Transesterification of glycerol and methyl benzoate	Ultrasonic bath (37 kHz, 50 W)	21
Papain, pectinase, cellulase and α-amylase	Sinopharm (China)	Extraction of polysaccharides from *Epimidium* leaves	Ultrasonic bath (40 kHz, 300 W)	22
Pancreatin	Porcine pancreas	Extraction of iodinated amino acids from edible seaweeds	Ultrasonic bath (45 kHz)	23
Peroxidase	*Armoracia rusticana*	Degradation of textile dyes acid red and malachite green	Ultrasonic bath (35 kHz, 35 W)	24
Phospholipase A1	Genetically modified *Aspergillus oryzae*	Degumming of rapeseed oil	Ultrasonic bath (40 kHz, 210 W)	25
Protease	*Bacillus lentus*	Extraction of tamarind seed xyloglucan	Probe (20 kHz, 125–250 W)	26
Proteases	*Bacillus subtilis* and *B. lentus*	Degradation of sericin	Ultrasonic bath (70 W)	27
Trypsin and α-chymotrypsin	Bovine pancreas	Hydrolysis of peanut protein allergens	Ultrasonic bath (50 kHz)	28
Tyrosinase	*Agaricus bisporus*	Formation of dopachrome	Ultrasonic bath (40 kHz, 100 W)	29

filtered to remove the enzyme and the homogeneous medium was cooled (0 °C) and the phases left to partition (30 min). In general, the enantiomeric excesses of the product esters were high. Swaleh *et al.*[32] also reported the use of fluorinated substrates, racemic esters, in enzymatic chiral resolution with lyophilized *Candida antarctica* lipase B in a system with *n*-butanol and acetonitrile (or *t*-butyl methyl ether). Luo *et al.*[33] innovated and used *Candida antarctica* lipase B in a kinetic resolution of fluorous ester of *rac*-1-(2-naphthyl)ethanol with enantioselective deacylation in a triphasic system: methanol/chloroform, FC-72 (perfluorohexanes) and methanol/sodium methoxide.

Panza *et al.*[34] reported the use of fluorinated nicotinamide adenine dinucleotide (NAD) as a soluble coenzyme of horse liver alcohol dehydrogenase in oxidation/reduction reactions in a system with methoxynonafluorobutane (HFE) and supercritical CO_2. Maruyama *et al.*[35] evaluated the enzymatic activity of complexes of lipases in poly(ethyleneglycol) 20 000 in the direct presence of perfluorohexane (FC-72) and perfluorooctane (FC-77 and FC-3255) in the reaction between vinyl cinnamate and benzyl alcohol.

Teo *et al.*[36] described the dynamic resolution of secondary alcohols (phenylethanol) through enzymatic stereoselective transesterification and heterogeneously catalyzed racemization of the alcohol over several zirconia-containing catalysts, using a fluorous phase-switching technique (2'2'2'-trifluoroethanol 1*H*,1*H*,2*H*,2*H*-perfluorundecanoate) coupled with continuous extraction in a membrane contactor, allowing recovery of the fluorous tagged species in a scalable operation. Shipovskov[37] reported the formation of noncovalent complexes between lipase from *Burkholderia cepacia*, and fluorinated ionic surfactant KDP 4606 was shown to promote solubilization of the enzyme in the fluorinated solvent perfluoro(methylcyclohexane) (PFMC) and its operation as a catalyst in the fluorous PFMC/hexane biphasic system. In the reaction of esterification of 1-phenylethanol and vinyl acetate, the solubilized lipase showed high stereospecificity (*ca.* 99%).

15.3   Aphrons

Colloidal gas aphrons (CGAs) are microfoam or microbubble dispersions created by intense stirring (5000–10 000 rpm) of a surfactant solution to obtain bubbles with diameter of 10–100 μm. The intense stirring is carried out by a horizontal disc capable of rotating at very high speeds, which causes gas entrainment and microbubble formation. The proposed structure of CGAs consists of a gaseous inner core surrounded by a thin aqueous surfactant film or shell composed of two surfactant layers and, in addition, a third surfactant layer that stabilizes this structure (Figure 15.2).[38] The commonly used methods for their preparation include mechanical agitation, sonication, and pressurized gas–liquid mixing systems, which usually result in the formation of microbubbles with wide size distributions. Microfluidic technologies are currently the primary methods for preparing monodisperse microbubbles.[39,40]

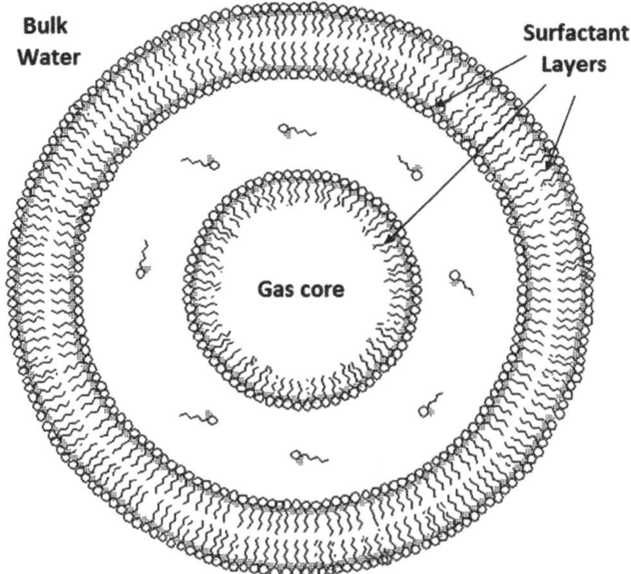

Figure 15.2   General structure of a colloidal gas aphron (inspired by Jauregi and Varley[38]).

Compared to conventional bubbles in the millimeter range, CGAs or microbubbles offer novel and unique properties:[38,40]

> › Large interfacial area per unit volume for the adsorption of molecules as a result of their small size and high gas hold-up [*i.e.* a volumetric gas ratio in the gas–liquid dispersion typically around 50%];
> → Slower rising velocity in the liquid phase, meaning a relatively high stability;
> → Flow properties similar to those of water (*e.g.* CGAs can be pumped easily, without collapse, from one location to another);
> → Easy separation from the bulk liquid phase because of their buoyancy; typically, the liquid–dispersion interface will start to rise in less than a minute;
> → Higher internal pressure, meaning that they shrink when their size is below a critical value.

As a consequence of these properties, researchers have considered various applications for CGAs, mainly in downstream processing such as recovery and concentration of proteins and enzymes, potential delivery systems for drugs and genes, separation of cells from fermentation broths, bioremediation, flotation for the removal of biological and nonbiological products, and enhancement of gas-to-liquid mass transfer, such as oxygen and ozone for aerobic fermentation, aquaculture, hydroponic cultivation, and aerobic treatment of sewage.[38,40–42]

Jauregi and Varley[43] studied the recovery of lysozyme from egg white using sodium bis-(2-ethyl hexyl) sulfosuccinate (AOT) aphrons. High recoveries (95%) and enrichments (19×) were generally obtained at protein concentrations ≤ 0.41 mg mL^{-1}, and surfactant concentrations >0.11 mg mL^{-1}, caused probably by electrostatic and/or hydrophobic interactions between the enzyme and surfactant. Zidehsaraei *et al.*[44] reported the use of CGAs for the extraction and recovery of glucoamylase produced by *Aspergillus niger* in solid state fermentation of 50% wet bran. Aphrons were generated using SDS and tetradecyl trimethyl ammonium bromide at the critical micellar concentration, adding aluminum sulfate (15 ppm) at pH 3.5 to obtain higher enzyme recovery and activity.

Lamb and Stuckey[45] used aphrons for enzyme immobilization. Polyaphrons were prepared by dropwise addition of *n*-decane containing a non-ionic alcohol ethoxylate surfactant (1% w/v Softanol 30) into a foaming aqueous solution containing an ionic surfactant (1% w/v sodium dodecyl sulphate, SDS, in deionised water) and 20 mg mL^{-1} of enzyme (*Bacillus* species α-amylase, *Aspergillus oryzae* β-galactosidase, lysozyme from chicken egg white, *Candida rugosa* lipase, bovine pancreas ribonuclease-A, and porcine pancreas trypsin). The initial volume of the aqueous phase was typically 2 mL, which was stirred at ~8000 rpm, and the organic phase was added at an average flow rate of 0.5 mL min^{-1} until a phase–volume ratio (PVR = V_{org}/V_{aq}) of ~4 was reached. Ribonuclease-A was found to retain partial activity when immobilized, whilst α-amylase and β-galactosidase exhibited large increases in activity of 15- and 6-fold, respectively, due to the beneficial effects of SDS on the protein structure. In 2000, the same authors continued studying β-galactosidase immobilization in aphron systems, evaluating some parameters such as enzyme concentration, the pH and ionic strength of the bulk aqueous phase, types of solvents and surfactants, and thermal stability.

Weber and Agblevor[46] improved oxygen transfer to aerobic microorganisms, such as *Trichoderma reesei*, when CGAs were connected to a conventional air sparger, showing 5× higher $k_L a$ values and nearly 2× higher cell mass productivities, whereas cellulase activities were similar in relation to conventional sparged fermenters.

15.4 Glycols

15.4.1 Glymes

Glymes (*i.e.* glycol diethers) are saturated polyethers containing no other functional groups. As compared to glycols (such as polyethylene glycols), glymes do not carry free hydroxyl groups and thus are aprotic, polar and chemically inert compounds. Most glymes are completely miscible with both water and organic solvents (such as ethanol, acetone, benzene, and octane), and tend to solvate cations, acting like crown ethers. In addition, glymes possess many other favorable properties, such as a wide range of temperature in which they remain liquid (typically >200 °C except monoglyme), low viscosities, high chemical and thermal stability, relatively low vapor pressure and low toxicity. In addition, these solvents are capable of dissolving triglycerides, including soybean oil, increasing the oil–methanol mutual solubility.

Due to these excellent solvent properties, glymes find many laboratory and commercial applications in areas such as reaction media (*i.e.* for organometallic reactions, polymerization, and reactions involving alkali metals, oxidations and reductions), extraction solvents (for metals and organics), gas purification, absorption refrigeration, the formulation of adhesives and coatings, textiles, solvents for the electronics industry, pharmaceutical formation, batteries, and cleaning solutions.[47,48]

The use of glymes as solvents for biocatalysis is very recent. Poulhès *et al.*[49] reported the synthesis of oligomeric polyetheramides from amino ethylesters or diethyl esters and diamines based on ethylene- or diethyleneglycol moieties to ensure better solubility of polymeric materials in organic solvents compatible with lipase B from *Candida antarctica*, such as glyme (1,2-dimethoxyethane) and diphenyl ether at 80 °C under 10 mbars for 240 h. In 2013, the same authors[50] studied lipase Novozym 435 for enzymatic polymerization in the presence of glyme, however, with ω-amino-α-alkoxyacetate as monomer, decreasing the reaction time to 30 min.

Tang *et al.*[47] presented glymes as alternative benign solvents for lipase-catalyzed transesterification, using as reagents ethyl sorbate and 1-propanol, and soybean oil and methanol. Many lipases were tested, with immobilized *Candida antarctica* lipase B (Novozym 435) leading to higher enzyme activities and stabilities than *t*-butanol and ionic liquids, 77% higher in the presence of diethylene glycol dibutyl ether (G2-Bu). Furthermore, glymes aided lipase to tolerate high methanol concentrations (up to 60–70% v/v), and nearly stoichiometric triglyceride conversions could be obtained under mild reaction conditions. Other glymes such as triethylene glycol dimethyl ether (G3), tetraethylene glycol dimethyl ether (G4), ethylene glycol diethyl ether (G1-Et), diethylene glycol dimethyl ether (G2-Et), higlyme (MW > 400; methyl ether of high molecular weight highly ethoxylated alcohol), and dipropylene glycol dimethyl ether (P2) also activated lipase, giving higher yields in transesterification (95.5%) using P2 as solvent.

15.4.2 Liquid Polymers

Poly(ethyleneglycol) (PEG) has been used in downstream processes, in aqueous biphasic systems, or in enzyme modification. However, low molar mass PEGs can also be alternative nonvolatile solvents, known as liquid polymers, including poly(propylene glycol) (PPG) and poly(tetrahydrofuran) (PTHF), that could be used in biocatalytic processes. Substituting ionic liquids, the combination of PEG and supercritical CO_2 constitutes a cheap and benign biphasic solvent system. However, because liquid polymers are less polar than ILs, the polymers should be seen as complementary to, rather than as replacements of ILs. LPs are tunable over a wide range of polarities by modification of the repeating unit and by expansion of the polymer using dissolved CO_2. Low molecular weight liquid PEGs can be regarded as protic solvents with aprotic sites of binding constituted by EO units.[31,51]

PEG has a reasonable claim to the label "green solvent" because it is easily soluble in water and many organic solvents (toluene, dichloromethane, alcohol, and acetone), nonvolatile, nonflammable, nontoxic to humans, animals and aquatic life, and biodegradable by bacteria in soil and sewage. Besides, PEG has been found to be stable towards acid, base, high temperature, O_2, H_2O_2 high oxidation systems, and $NaBH_4$ reduction systems, and can be recovered from aqueous solution by extraction with a suitable solvent or by direct distillation of water or the solvent. The similar small molecular weight liquid polypropylene glycols (PPGs) 425 and 1000 have also found application in solvent substitution. PPG is a viscous liquid with a negligible vapor pressure, stable, and easily recovered. Low molecular weight PPG-250 and -425 are in fact water soluble, but PPG shows an inverse temperature–solubility relationship, along with a rapid decrease in water solubility as the molecular weight increases.[51] Carvalho *et al.*[52] investigated PPG as a solvent for the solubilization of sitosterol and bioconversion media with *Mycobacterium* sp. NRRL B-3805 cells for sitosterol side-chain cleavage to 4-androstene-3,17-dione. Reetz and Wiesenhöfer[53] tested the enzymatic acylation of 1-phenylethanol with vinyl acetate in a biphasic solvent system composed of PEG 1500 and supercritical carbon dioxide (50 °C/150 bar) using lipase B from *Candida antarctica*.

15.5 Alkyl Carbonates

Alkyl carbonates are derivatives of neutral CO_2 produced by reaction with alcohols, for shifting an unfavorable equilibrium or separating a product, such as for example dimethyl carbonate, which has attracted some interest as an environmentally benign substitute for highly toxic phosgene and dimethyl sulfate in carbonylation and methylation reactions, as in polyurethane synthesis; as a green replacement for halogenated solvents due to its similarity to these types of liquids, as well as ethers and esters; as a solvent for lithium batteries, supercapacitors and dye-sensitized solar cells, due to its good ability to dissolve salts, and particularly lithium ions; as a monomer for several types of polymers, and as an intermediate in the synthesis of pharmaceutical and agricultural chemicals. Nowadays, three preparation routes have been commercialized to produce DMC: a phosgene process, oxidative carbonylation of methanol, and a two-step transesterification process (Figure 15.3), in which CO_2 is inserted into epoxides followed by transalkylation of the cyclic carbonate using methanol.[54,55]

Enzymes are involved with alkyl carbonates, not only as solvents and acyl acceptors, but also in their synthesis. Liu *et al.*[55] used lipase from *Penicillium expansum* immobilized on a film based on a blend of carboxymethyl cellulose

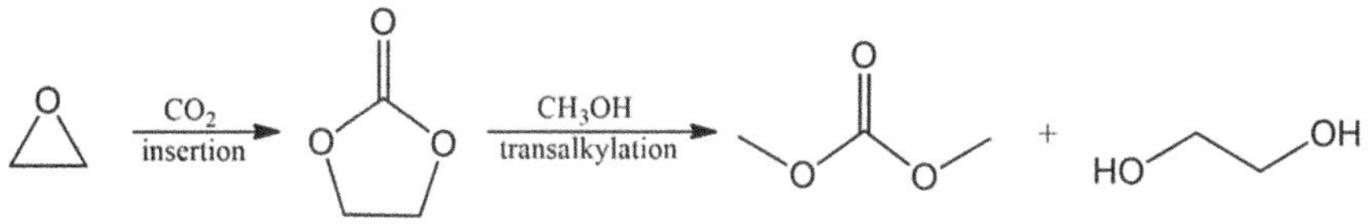

Figure 15.3 Two-step transesterification process for dimethyl carbonate production.

and polyvinyl alcohol for the synthesis of dimethyl carbonate using ethylene carbonate and methanol as substrates.

The main application of alkyl carbonates in enzymatic reactions is related to the preparation of biodiesel (fatty acid methyl esters) by transesterification. Furthermore, in these reactions, glycerol carbonates and their derivatives are also generated. Su *et al.*[56] tested different vegetable oils (soybean, rapeseed, corn, castor, sunflower, cottonseed, olive, peanut and sesame) using as solvent *n*-heptane and petroleum ether, comparing lipases from different sources (*Candida antarctica*, *Aspergillus niger*, *Mucor miehei*, porcine pancreas and *Candida* sp.) and acyl acceptors (dimethyl carbonate, methanol and methyl acetate), obtaining two times more FAME yield when DMCs were used. Zhang *et al.*[57] investigated a solvent-free system for enzymatic transesterification between dimethyl carbonate and palm oil, obtaining a yield of 90.5% using as conditions: molar ratio DMC:oil = 10:1, 20% Novozym 435 (based on the oil weight), 55 °C and 24 h. The same group, Sun *et al.*[58] reported the same reaction, evaluating the kinetic model (ordered bi–bi mechanism without substrate inhibition) and the influence of agitation speed. Similar work was performed by Go *et al.*[59] using soybean oil, obtaining yields of 96.4% FAME and 92.5% glycerol carbonate, differing in conditions by the addition of 0.7% water at 60 °C and 48 h of reaction. Gharat and Rathod[60] used waste cooking oil with DMC, obtaining 77.9% yield of fatty acid methyl esters in a solvent-free system containing 10% Novozym 435, molar ratio DMC:oil 6:1, reacted for 24 h at 60 °C. These results were improved in their subsequent work,[61] when they used ultrasound irradiation (25 kHz, 200 W), achieving 86.6% yield of FAMEs in 4 h of reaction. Not only DMC can be used, but also diethyl carbonate in biodiesel production with fatty acid ethyl esters using camellia oil soapstocks and Novozym 435 (5%).[62]

Other groups aimed only at the production of glycerol carbonate, reacting glycerol with dimethyl carbonate, since it is investigated as an additive for coatings, paints and detergents, as a component of gas separation membranes, as a biolubricant and as a precursor for the polymer industry. Jung *et al.*[63] reported the enzymatic production of glycerol carbonate using Novozym 435 (75 g L^{-1}), reaching 96.25% yield, under these conditions: dimethyl carbonate: glycerol molar ratio 2:1, 60 °C, 48 h, addition of acetonitrile and Tween 80 (10% v/v). Lee *et al.*[64] tested a solvent-free system with silica-coated glycerol using Novozym 435 (5 and 20% w/w) at 70 °C over 48 h, at a DMC:glycerol molar ratio of 10:1, obtaining a yield of 93%. Tudorache *et al.*[65] also tested a solvent-free system, however they immobilized a lipase from *Aspergillus niger* onto a functionalized magnetic micro-/nanoparticle surface for this reaction. In 2013,[66] the same authors used lipase from *A. niger* immobilized onto a cross-linked enzyme aggregate on magnetic particles, achieving a yield of 55% glycerol carbonate. Tudorache *et al.*[67] substituted pure glycerol for waste glycerol collected from different feedstock patterns (*e.g.* soybean, sunflower, rape, corn, olive, palm, and residual oil) and the same group[68] used pure glycerol and glycerol from residual sunflower oil with different alkyl carbonates (diethyl carbonate and dibenzyl carbonate), maintaining the same immobilized enzyme. Other types of organic carbonate can be produced enzymatically with dimethyl carbonate, such as diphenyl and vinyl carbonates.[69]

15.6   Other Applications

15.6.1   Tunable Solvents

Organic aqueous tunable solvent (OATS) systems are engineered to couple a reaction and separation, using water, CO_2 and solvents like acetone, acetonitrile, tetrahydrofuran and dioxane, as seen in Figure 15.4. These OATS mixtures allow homogeneous reactions between hydrophobic and hydrophilic components, using homogeneous catalysts which are generally more active and more selective than heterogeneous catalysts; furthermore, they are especially useful for asymmetric synthesis as they almost always result in superior enantiomeric excesses (ee). Homogeneous catalysts avoid limitations such as active-site heterogeneity or mass transport limitations of the reactants and products to and from the active site. However, a major barrier is

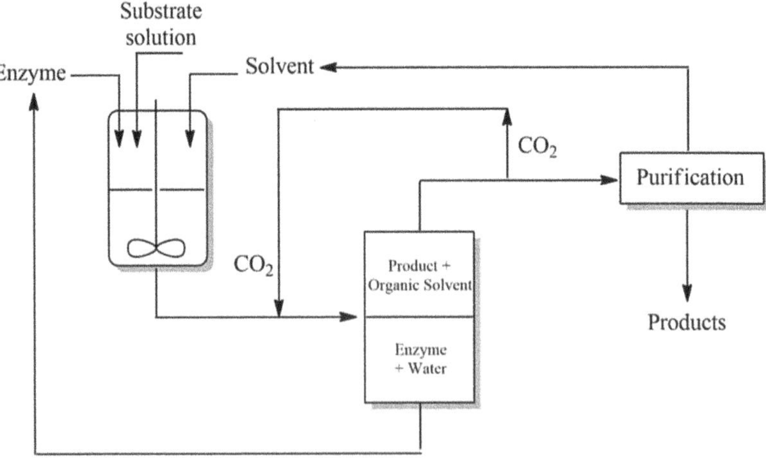

Figure 15.4   OATS system (inspired by Hill *et al.*[70]).

often their expense and/or toxicity; for either reason, they must be separated from the products and, if possible, recycled for reuse. OATS can reduce the ecological footprint and cost associated with asymmetric transformations of a hydrophobic substrate by:[70,71]

1. Allowing recycling of both catalyst and solvents;
2. Using highly selective enzymatic catalysts to reduce the production of byproducts requiring downstream purification;
3. Replacing a portion of the volatile organic solvent used in reactions with environmentally and chemically benign water and CO_2, which are both virtually non-toxic, non-flammable, and relatively inert;
4. Applying modest CO_2 pressure as a phase switch for product recovery, in contrast to the massive amounts of organics consumed in conventional liquid–liquid extractions.

Broering *et al.*[70] and Hill *et al.*[71] used *Candida antarctica* lipase B in an OATS system with dioxane (40% and 30% vol, respectively) and phosphate buffer (pH 7.1) for hydrolysis of 2-phenylethyl acetate to 2-phenylethanol and acetate, with further addition and pressurization of CO_2 to 50 bar for phase separation. Broering *et al.*[70] tested the reactivity, pH-stability, and recyclability of the biocatalyst, and the partitioning behavior of the substrate and product in an OATS system, whereas Hill *et al.*[71] verified the chiral resolution of 2-phenylethyl acetate with high enantiomeric excess.

15.6.2　Biodesalination

Groundwater is the primary source of drinking water in many countries, and often has high levels of hardness, which is caused by a variety of dissolved multivalent metallic ions, predominantly calcium and magnesium. Therefore, prior to residential distribution and consumption, the hardness concentration should be reduced and hard water should be softened. Two major methods are typically used to remove hardness: lime soda softening and ion exchange softening. Lime soda softening is used mostly for municipal purposes; it employs chemical precipitation, in which lime is added to hard water to precipitate calcium ions as calcium carbonate and magnesium ions as magnesium hydroxide. The ion exchange process is primarily employed for residential water softening. The softening system consists of salt-saturated (*e.g.*, sodium chloride) polymer resin beads and a brine tank to regenerate the resin bed; in this process, each divalent hardness ion (Ca^{2+} or Mg^{2+}) in the water is replaced by two sodium ions, thereby softening the water. Other residential water softening methods include distillation, nanofiltration, electrodialysis, carbon nanotubes, capacitive deionization, and reverse osmosis, which consume a large amount of energy, and operation and maintenance of these systems can be expensive.

To avoid introducing additional salts such as sodium into drinking water during the softening process, saltless water softening technologies must be developed as an alternative to the ion exchange process. Microbial desalination cells (MDCs) are derived from microbial fuel cells (MFCs), which are bioelectrochemical devices that use microorganisms as biocatalysts to convert chemical energy into electrical energy. The principle behind the operation of an MFC is the generation of electrons from the catabolic action of microorganisms, a mixture of aerobic and anaerobic sludge, which can be transferred through the cell membrane to the anode electrode. The anode is connected to the cathode *via* an external circuit through which electrons flow based upon the difference in redox potential that exists between their dissimilar liquid solutions. By installing an additional chamber between the anode and the cathode, an MFC is converted to an MDC with the function of desalination. MDC technology has been previously employed to soften hard water; however, the presence of microorganisms in an MDC is inappropriate for residential applications. Another method for hardness removal through electricity generation is to use immobilized enzymes on the anode, an enzymatic oxidation of glucose with encapsulated NAD-dependent glucose dehydrogenase enzymes with the cofactor NAD^+.[72,73]

15.6.3 Nanotechnology

There have been many approaches to improve enzyme stability: medium and protein engineering, and more commonly, enzyme immobilization and modification. Enzyme modification is defined by the covalent reactions of the enzyme molecule, adding functional groups or polymers on the surface of enzyme molecules to change their surface properties, resulting in an improvement of enzyme stability. Enzyme immobilization represents the attachment or incorporation of enzyme molecules onto or into large structures *via* simple adsorption, covalent attachment, or encapsulation. Recent developments with cross-linked enzyme crystals (CLECs) and cross-linked enzyme aggregates (CLEAs) are based on multipoint attachment between enzyme crystals (or molecules).[74] Nanotechnology is used precisely to introduce new supports for enzyme immobilization.

Nanotechnology can be defined as "Technology development at the atomic, molecular, or macromolecular range of approximately 1–100 nanometers to create and use structures, devices, and systems that have novel properties".[75] Its progress was followed by the rapid growth of nanobiotechnology. In the initial stages of nanobiocatalysis, enzymes were immobilized on various nanostructured materials using conventional approaches, such as simple adsorption and covalent attachment. This approach gathered attention by immobilizing enzymes onto high surface area nanostructured materials, such as carbon nanotubes, nanoporous materials, electrospun nanofibers, mesoporous silica and magnetic γ-Fe_2O_3 nanoparticles. This allowed improved enzyme loading, minimum diffusional limitation, resistance to proteolytic digestion, and the control over size at the nanometer-scale, such as the pore size in nanopores, the thickness of nanofibers or nanotubes and the particle size of nanoparticles. Other approaches to immobilize enzymes in nanoparticles are nanoentrapment using water-in-oil micro-emulsion systems (reverse micelles), which leads to discrete nanoparticles through polymerization in the water phase or at the water–oil interface, and single enzyme nanoparticles (SENs), where an organic–inorganic hybrid polymer network of a thickness of less than a few nanometers is built up from the surface of the enzyme. Recently, nanobiocatalytic approaches have evolved beyond simple enzyme immobilization strategies to include enzyme stabilization, wired enzymes, the use of enzymes in sensitive biomolecular detection, artificial enzymes, nanofabrication and nanopatterning, and new applications such as trypsin digestion in proteomic analysis, antifouling and biofuel cells.[76-79]

References

1. T. J. Mason and P. Cintas, in Sonochemistry, *Handbook of Green Chemistry and Technology*, ed. J. Clark and D. MacQuarrie, Blackwell Science, Oxford, 2002, ch. 16.
2. T. J. Mason, L. Paniwnyk, F. Chemat and M. A. Vian, Ultrasonic Food Processing, in *Alternatives to Conventional Food Processing*, ed. A. Proctor, RSC Publishing, Cambridge, 2011, ch. 10.
3. E. Ortega-Rivas, *Non-thermal Food Engineering Operations*, Springer, New York, 2012.
4. R. Mawson, M. Gamage, N. S. Terfee and K. Knoerzer, in *Ultrasound Technologies for Food and Bioprocessing*, ed. H. Feng, G. V. Barbosa-Cánovas and J. Weiss, Springer, New York, 2011, ch. 14.
5. B. Kwiatkowska, J. Bennett, J. Akunna, G. M. Walker and D. H. Bremner, *Biotechnol. Adv.*, 2011, **29**, 768–780.
6. S. Barton, C. Bullock and D. Weir, *Enzyme Microb. Technol.*, 1996, **18**, 190–194.
7. E. X. Leaes, D. Lima, L. Miklasevicius, A. P. Ramon, V. Dal Prá, M. M. Bassaco, L. M. Terra and M. A. Mazutti, *Biocatal. Agric. Biotechnol.*, 2013, **2**, 21–25.
8. M. Souza, E. T. Mezadri, E. Zimmerman, E. X. Leaes, M. M. Bassaco, V. Dal Prá, E. Foletto, A. Cancellier, L. M. Terra, S. L. Jahn and M. A. Mazutti, *Ultrason. Sonochem.*, 2013, **20**, 89–94.
9. O. E. Sazbó and E. Csiszár, *Carbohydr. Polym.*, 2013, **98**, 1483–1489.
10. Z. Wang, X. Lin, P. Li, J. Zhang, S. Wang and H. Ma, *Bioresour. Technol.*, 2012, **117**, 222–227.
11. J. R. F. Silva, K. C. Cantelli, M. V. Tres, C. D. Rosa, M. A. A. Meirelles, M. B. A. Soares, D. Oliveira, J. V. Oliveira, H. Treichel and M. A. Mazutti, *Biocatal. Agric. Biotechnol.*, 2013, **2**, 102–107.
12. P. C. Sangave and A. B. Pandit, *J. Environ. Manage.*, 2006, **80**, 36–46.
13. M. Bashari, A. Eibaid, J. Wang, Y. Tian, X. Xu and Z. Jin, *Ultrason. Sonochem.*, 2013, **20**, 155–161.
14. C. Basto, T. Tzanov and A. Cavaco-Paulo, *Ultrason. Sonochem.*, 2007, **14**, 350–354.

15. J. A. Awadallak, F. Voll, M. C. Ribas, C. Silva, L. Cardozo Filho and E. A. Silva, *Ultrason. Sonochem.*, 2013, **20**, 1002–1007.

16. H. R. Lobo, B. S. Singh, D. V. Pinjari, A. B. Pandit and G. S. Shankarling, *Biochem. Eng. J.*, 2013, **70**, 29–34.

17. K. G. Fiametti, M. K. Ustra, D. Oliveira, M. L. Corazza, A. Furigo Jr. and J. V. Oliveira, *Ultrason. Sonochem.*, 2012, **19**, 440–451.

18. M. M. Zheng, L. Wang, F. H. Huang, P. M. Guo, F. Wei, Q. C. Deng, C. Zheng and C. Y. Wan, *J. Mol. Catal. B: Enzym.*, 2013, **95**, 82–88.

19. C. H. Kuo, F. W. Hsiao, J. H. Chen, C. W. Hsieh, Y. C. Liu and C. J. Shieh, *Ultrason. Sonochem.*, 2013, **20**, 546–552.

20. A. B. Martins, M. F. Schein, J. L. R. Friedrich, R. Fernandez-Lafuente, M. A. Z. Ayub and R. C. Rodrigues, *Ultrason. Sonochem.*, 2013, **20**, 1155–1160.

21. G. Ceni, P. C. Silva, L. Lerin, J. V. Oliveira, G. Toniazzo, H. Treichel, E. G. Oestreicher and D. Oliveira, *Enzyme Microb. Technol.*, 2011, **48**, 169–174.

22. R. Chen, S. Li, C. Liu, S. Yang and X. Li, *Process Biochem.*, 2012, **47**, 2040–2050.

23. V. Romaris-Hortas, P. Bermejo-Barrera and Moreda-Piñeiro, *J. Chromatogr. A*, 2013, **1309**, 33–40.

24. R. Patidar, S. Khanna and V. S. Moholkar, *Ultrason. Sonochem.*, 2012, **19**, 104–118.

25. X. Jiang, M. Chang, X. Wang, Q. Jin and X. Wang, *Ultrason. Sonochem.*, 2014, **21**, 142–148.

26. S. Poommarinvarakul, J. Tattiyakul and C. Muangnapoh, *J. Food Sci.*, 2010, **75**(5), E253–E260.

27. N. M. Mahmoodi, M. Arami, F. Mazaheri and S. Rahimi, *J. Cleaner Prod.*, 2010, **18**, 146–151.

28. H. Li, J. Yu, M. Ahmedna and I. Goktepe, *Food Chem.*, 2013, **141**, 762–768.

29. Z. L. Yu, W. C. Zeng and X. L. Lu, *Ultrason. Sonochem.*, 2013, **20**, 805–809.

30. P. Beier and D. O'Hagan, *Chem. Commun.*, 2002, 1680–1681.

31. P. Lozano, *Green Chem.*, 2010, **12**, 555–569.

32. S. M. Swaleh, B. Hungerhoff, H. Sonnenschein and F. Theil, *Tetrahedron*, 2002, **58**, 4085–4089.

33. Z. Luo, S. M. Swaleh, F. Theil and D. P. Curran, *Org. Lett.*, 2002, **4**(15), 2585–2587.

34. J. L. Panza, A. J. Russell and E. J. Beckman, *Tetrahedron*, 2002, **58**, 4091–4104.

35. T. Maruyama, T. Kotani, H. Yamamura, N. Kamiya and M. Goto, *Org. Biomol. Chem.*, 2004, **2**, 524–527.

36. E. L. Teo, G. K. Chuah, A. R. J. Huguet, S. Jaenicke, G. Pande and Y. Zhu, *Catal. Today*, 2004, **97**, 263–270.

37. S. Shipovskov, *Biotechnol. Prog.*, 2008, **24**, 1262–1266.

38. P. Jauregi and J. Varley, *Trends Biotechnol.*, 1999, **17**, 389–395.

39. M. Vignes-Adler and D. Weaire, *Curr. Opin. Colloid Interface Sci.*, 2008, **13**, 141–149.

40. Q. Xu, M. Nakajima, Z. Liu and T. Shiina, *Int. J. Mol. Sci.*, 2011, **12**, 462–475.

41. M. A. Hashim, B. S. Gupta and S. V. Kumar, *Biotechnol. Tech.*, 1995, **9**(6), 403–408.

42. S. B. Lamb and D. C. Stuckey, *Enzyme Microb. Technol.*, 2000, **26**, 574–581.

43. P. Jauregi and J. Varley, *Biotechnol. Bioeng.*, 1998, **59**(4), 471–481.

44. A. Z. Zidehsaraei, M. Moshkelani and M. C. Amiri, *Sep. Purif. Technol.*, 2009, **67**, 8–13.

45. S. B. Lamb and D. C. Stuckey, *Enzyme Microb. Technol.*, 1999, **24**, 541–548.

46. J. Weber and F. A. Agblevor, *Process Biochem.*, 2005, **40**, 669–676.

47. S. Tang, C. L. Jones and H. Zhao, *Bioresour. Technol.*, 2013a, **129**, 667–671.

48. S. Tang, H. Zhao, Z. Song and O. Olubajo, *Bioresour. Technol.*, 2013b, **139**, 107–112.

49. F. Poulhès, D. Mouysset, G. Gil, M. P. Bertrand and S. Gastaldi, *Polymer*, 2012, **53**, 1172–1179.

50. F. Poulhès, D. Mouysset, G. Gil, M. P. Bertrand and S. Gastaldi, *Polymer*, 2013, **54**, 3467–3471.

51. J. Chen, S. K. Spear, J. G. Huddleston and R. D. Rogers, *Green Chem.*, 2005, **7**, 64–82.

52. F. Carvalho, M. P. C. Marques, C. C. C. R. Carvalho, J. M. S. Cabral and P. Fernandes, *Bioresour. Technol.*, 2009, **100**, 4050–4053.

53. M. T. Reetz and W. Wiesenhöfer, *Chem. Commun.*, 2004, 2750–2751.

54. B. Ferrer, M. Alvaro and H. Garcia, in *Green Solvents I: Properties and Applications in Chemistry*, ed. A. Mohammad, Inamuddin, Springer, Dordrecht, 2012.

55. J. Liu, H. Guo, Q. Zhou, J. Wang, B. Lin, H. Zhang, Z. Gao, C. Xia and X. Zhou, *J. Mol. Catal. B: Enzym.*, 2013, **96**, 96–102.

56. E. Z. Su, M. J. Zhang, J. G. Zhang, J. F. Gao and D. Z. Wei, *Biochem. Eng. J.*, 2007, **36**, 167–173.

57. L. Zhang, S. Sun, Z. Xin, B. Sheng and Q. Liu, *Fuel*, 2010, **89**, 3960–3965.

58. S. Sun, L. Zhang, X. Meng and Z. J. Xin, *Renewable Sustainable Energy*, 2013, **5**, 033127.

59. A. R. Go, Y. Lee, Y. K. Kim, S. Park, J. Choi, J. Lee, S. O. Han, S. W. Kim and C. Park, *Enzyme Microb. Technol.*, 2013, **53**, 154–158.

60. N. Gharat and V. K. Rathod, *J. Mol. Catal. B: Enzym.*, 2013a, **88**, 36–40.

61. N. Gharat and V. K. Rathod, *Ultrason. Sonochem.*, 2013b, **20**, 900–905.

62. P. Wang, *Curr. Opin. Biotechnol.*, 2006, **17**, 574–579.

63. H. Jung, Y. Lee, D. Kim, S. O. Han, S. W. Kim, J. Lee, Y. H. Kim and C. Park, *Enzyme Microb. Technol.*, 2012, **51**, 143–147.

64. K. H. Lee, C. H. Park and E. Y. Lee, *Bioprocess Biosyst. Eng.*, 2010, **33**, 1059–1065.

65. M. Tudorache, L. Protesescu, A. Negoi and V. I. Parvulescu, *Appl. Catal., A*, 2012, **437– 438**, 90–95.

66. M. Tudorache, A. Nae, S. Coman and V. I. Parvulescu, *RSC Adv.*, 2013, **3**, 4052–4058.

67. M. Tudorache, A. Negoi, B. Tudora and V. I. Parvulescu, *Appl. Catal., B*, 2014a, **146**, 274–278.

68. M. Tudorache, A. Negoi, L. Protesescu and V. I. Parvulescu, *Appl. Catal., B*, 2014b, **145**, 120–125.
69. R. L. Rodney, J. L. Stagno, E. J. Beckman and A. J. Russell, *Biotechnol. Bioeng.*, 1999, **62**(3), 259–266.
70. J. M. Broering, E. M. Hill, J. P. Hallett, C. L. Liotta, C. A. Eckert and A. S. Bommarius, *Angew. Chem., Int. Ed.*, 2006, **45**, 4670–4673.
71. E. M. Hill, J. M. Broering, J. P. Hallett, A. S. Bommarius, C. L. Liotta and C. A. Eckert, *Green Chem.*, 2007, **9**, 888–893.
72. M. A. Arugula, K. S. Brastad, S. D. Minteer and Z. He, *Enzyme and Microbial Technology*, 2012, **51**, 396–401.
73. K. S. Brastad and Z. He, *Desalination*, 2013, **309**, 32–37.
74. J. Kim, J. W. Grate and P. Wang, *Chem. Eng. Sci.*, 2006a, **61**, 1017–1026.
75. L. Pray and A. Yaktine, *Nanotechnology in Food Products*, National Academies Press, Washington, 2009.
76. Y. Wang and X. Cao, *Bioresour. Technol.*, 2011, **102**, 10173–10179.
77. J. Kim, H. Jia and P. Wang, *Biotechnol. Adv.*, 2006b, **24**, 296–308.
78. J. Ge, C. Yang, J. Zhu, D. Lu and Z. Liu, *Top. Catal.*, 2012, **55**, 1070–1080.
79. J. Kim, J. W. Grate and P. Wang, *Trends Biotechnol.*, 2008, **26**(11), 639–646.

Subject Index